T0143148

The Science of Energy

A complete account of our knowledge of energy and its transformations would require an exhaustive treatise on every branch of physical science, for natural philosophy is simply the science of energy.

William Garnett, *Encyclopaedia Britannica* (9th edition), 1879

The Science of Energy

A Cultural History of Energy Physics in Victorian Britain

CROSBIE SMITH

THE UNIVERSITY OF CHICAGO PRESS

CROSBIE SMITH is reader in history of science,
University of Kent at Canterbury.

The University of Chicago Press, Chicago 60637
The Athlone Press, London NW11 7SG
© Crosbie Smith 1998
All rights reserved. Published 1998
Printed in Great Britain

07 06 05 04 03 02 01 00 99 98   1 2 3 4 5 6

ISBN: 0-226-76420-6 (cloth)
ISBN: 0-226-76421-4 (paperback)

Library of Congress Cataloging-in-Publication Data

Smith, Crosbie.
The science of energy: a cultural history of energy physics in
Victorian Britain / Crosbie Smith.
        p.   cm.
Includes bibliographical references and index.
ISBN 0-226-76420-6 (alk. paper). – ISBN 0-226-76421-4 (pbk.: alk. paper)
1. Force and energy – History – 19th century.   2. Power resources –
History – 19th century.   3. Power (Mechanics) – History –
19th century.   I. Title.
QC72.S58   1998
621.042 – dc21   98-24960
                        CIP

This book is printed on acid-free paper.

For my mother and late father

Contents

List of Figures

Preface

A principal claim of this book is that a history of energy in the nineteenth century makes little sense if it is treated only as an internal story of physics, without regard for the cultural and economic concerns of the time. The contextual approach to the history of science advocated here has been used in recent years with stunning effect to cast new light on scientific knowledge and practice in a whole variety of fields from the 'Scientific Revolution' in the sixteenth and seventeenth centuries to evolutionary debates in the nineteenth century and beyond. One of the central features of a contextual history is to present scientific work not as the product of isolated individuals but as crucially contingent upon the cultural resources of the age in which it was produced. For a history of energy these resources involve such seemingly diverse ingredients as industrial machines, social and institutional networks, and religious and political ideologies.

With this central concern in mind, the contextual historian may begin to set up his or her topic in terms of a group of scientific practitioners working within a localized context. Seeking credibility for their scientific activity, these practitioners will often address local audiences which range from the critical expertise of a learned scientific society to the popular demands of a lay constituency. They will have shaped their public performances by a combination of various cultural resources, as well as by a degree of stage direction and an awareness of the needs of the audience. I have chosen to develop a similar pattern for my history of energy. I argue that the 'science of energy' did not arise inevitably and spontaneously in the mid-nineteenth century. Instead, I suggest that it was contingent upon an identifiable, though informal, network of scientific practitioners whose early local audiences included industrialists and engineers. From a mainly Scottish location, the key players not only added to their number, but, through careful stage management, attracted national and international recognition. As a result of their strategy, they enhanced their scientific credibility to such a degree that even in their own lifetimes they became international stars of science assured of a place in every hall of fame of physics: most notably, James Joule, William Thomson, Macquorn

Rankine and James Clerk Maxwell. Alongside them went a cast of earnest co-stars, less well known to posterity but no less significant in the historical process: P.G. Tait, James Thomson and Fleeming Jenkin in particular.

During the preparation of this book I have been highly conscious of the very many scholarly papers on the history of energy published in specialist history of science journals over the past three decades. While I have endeavoured to render many of the insights offered by these scholarly papers accessible to a wider audience, postgraduate students in the field and other specialist readers may also wish to study such material for themselves. With this in mind, I have drawn attention in the notes to scholarship of particular worth. At the same time, I have been conscious that recent trends in professional history of science has demanded considerable reappraisal of the historical presentations in the scholarly literature such that a straightforward integration has not been possible or desirable. What began a few years ago as a project designed to synthesize and present scholarly material for a non-specialist readership has therefore demanded a careful appraisal of much of the core history of energy.

To Norton Wise I am especially grateful for several of the interpretations presented in this book, especially his insights into the central role of 'work' in nineteenth-century British physics and its relation to the science of political economy. I also draw extensively on material developed at length in our joint biographical study of William Thomson (later Lord Kelvin), *Energy and Empire*. Specific debts are acknowledged in the footnotes. I am also conscious of my dependence upon the recent scholarly work of Boyd Hilton and Anne Scott (ch.2), Ivor Grattan-Guinness (ch.3), Iwan Morus (ch.4), Timothy Lenoir (ch.6), Ben Marsden (chs 5–8), Simon Schaffer (ch.13) and Bruce Hunt (ch.14).

I am indebted above all to Ben Marsden for his critical reading of the entire typescript and for his wealth of perceptive insights. In addition, his work on Macquorn Rankine has been a great stimulus to the writing of this book. The extremely thorough reading by the University of Chicago Press's referee of the draft typescript, together with the resulting insights and suggestions, has been invaluable for the shaping and preparation of the final version. I am also grateful to other colleagues in the former History of Science Unit and in its successor the Centre for History & Cultural Studies of Science at Kent: to Alex Dolby for invaluable discussions of various historiographical issues including 'simultaneous discovery'; to Graeme Gooday for all his constructive suggestions over six years; and not least to Maurice Crosland for his general encouragement and for his toleration, even approval, of my unorthodox methods over a long period.

Other colleagues across Britain, notably Geoffrey Cantor, Simon Schaffer, Jim and Anne Secord, Andrew Warwick, Robert Fox, Jon Hodge and Bob Olby, have been most supportive. I am also indebted to Peter Harman and David Wilson for their early encouragement of my work on energy physics. For their positive roles and valuable comments, as well as for practical assistance at critical points in the execution of this project, I thank Trish Hatton, Ian Higginson, Gerald Schmidt, Hugh Cunningham, Alan Armstrong, David Birmingham, Richard Eales, Maurice

Raraty, Roger Clark, Ralf Müller, Mathias Dörries, Otto Sibum, Jon Agar, Michael Neal, Michael Griffiths, Corinna See and Adrian Adair.

Funding was provided by the Royal Society History of Science Grants Committee to enable extensive periodical research. I am as always grateful to the respective staff of the Cambridge University Library Manuscripts (especially Godfrey Waller) and Rare Books Rooms, Glasgow University Library Special Collections Department, and the Templeman Library at the University of Kent for their assistance. For permission to reproduce Figures 5.1, 5.2 and 6.1, I thank Cambridge University Press.

Crosbie Smith
Centre for History & Cultural Studies of Science
School of History
University of Kent at Canterbury

Abbreviations

Ann. Electricity	*Annals of Electricity*
Ann. Phys. Chem.	*Annalen der Physik und Chemie*
Ann. Sci.	*Annals of Science*
Arch. Hist. Ex. Sci.	*Archive for History of Exact Sciences*
Arch. int. d'hist. des sci.	*Archives internationales d'histoire des sciences*
BAAS Rep.	*Report of the British Association for the Advancement of Science*
BJHS	*British Journal for the History of Science*
BJPS	*British Journal for the Philosophy of Science*
CDMJ	*Cambridge and Dublin Mathematical Journal*
CMJ	*Cambridge Mathematical Journal*
DNB	*Dictionary of National Biography*
DSB	*Dictionary of Scientific Biography*, ed. C.C. Gillispie (16 vols., New York, 1970–80)
Edinburgh New Phil. J.	*Edinburgh New Philosophical Journal*
Enc. Brit.	*Encyclopaedia Britannica*
Glas. Phil. Soc. Proc.	*Proceedings of the Glasgow Philosophical Society*
HSPBS	*Historical Studies in the Physical and Biological Sciences*
HSPS	*Historical Studies in the Physical Sciences*
Hist. Sci.	*History of Science*
Hist. Tech.	*History of Technology*
Mem. Man. Lit. & Phil.	*Memoirs of the Manchester Literary and Philosophical Society*
Notes and Records RSL	*Notes and Records of the Royal Society of London*
Phil. Mag.	*Philosophical Magazine*
Phil. Trans.	*Philosophical Transactions of the Royal Society*
Proc. Camb. Phil. Soc.	*Proceedings of the Cambridge Philosophical Society*
Proc. RI	*Proceedings of the Royal Institution*
Proc. Roy. Soc.	*Proceedings of the Royal Society*
Proc. RSE	*Proceedings of the Royal Society of Edinburgh*
SHPS	*Studies in History and Philosophy of Science*
Tech. Cult.	*Technology and Culture*
Trans. Camb. Phil. Soc.	*Transactions of the Cambridge Philosophical Society*
Trans. RSE	*Transactions of the Royal Society of Edinburgh*
ULC	University Library Cambridge
ULG	University Library Glasgow

CHAPTER 1

Introduction: a History of Energy

Not Copernicus and Galilei [sic], when they abolished the Ptolemaic system; not Newton, when he annihilated the Cartesian vortices; not Young and Fresnel, when they exploded the Corpuscular Theory; not Faraday and Clerk-Maxwell, in their splendid victory over *Actio in distans* – more thoroughly shattered a malignant and dangerous heresy, than did Joule when he overthrew the baleful giant FORCE, and firmly established, by lawful means, the beneficent rule of the rightful monarch, ENERGY! Then, and not till then, were the marvellous achievements of Sadi Carnot rendered fully available; and Science silently underwent a revolution more swift and more tremendous than ever befel a nation. But this . . . must be a theme for the Poet of the Future!

Anonymous review of Joule's Scientific Papers in the Phil. Mag. (1884)[1]

From the early 1850s the Glasgow professor of natural philosophy, William Thomson (1824–1907), and his ally in engineering science, Macquorn Rankine (1820–72), began replacing an older language of mechanics with terms such as 'actual' and 'potential energy'. In the same period, Rankine constructed a new 'science of thermodynamics' by which engineers could evaluate the imperfections of heat engines of all conceivable varieties. Within a very few years, Thomson and Rankine had been joined by like-minded scientific reformers, most notably the Scottish natural philosophers James Clerk Maxwell (1831–79) and Peter Guthrie Tait (1831–1901) and the engineer Fleeming Jenkin (1833–85). As individuals, as partners in various scientific or commercial projects, and as an informal group with strong links to the British Association for the Advancement of Science (BAAS), these 'North British' physicists and engineers were primarily responsible for the construction of the 'science of energy'. As such, they form the core of this historical study.

'During the fifty years' life of the British Association, the Advancement of Science, for which it has lived and worked so well, has not been more marked in any department than in one which belongs very decidedly to the Mathematical and Physical Section – the science of Energy'. With these carefully chosen words Sir William Thomson opened his address as president of Section A, Mathematics and Physics, of the BAAS at its 50th annual meeting in 1881. From that vantage point, the phrase 'science of energy' was of comparatively recent pedigree. Rankine's paper, 'Outlines of the Science of Energetics', had been presented to the Glasgow Philosophical Society in 1855. Although largely confined to a local audience of Clydeside industrialists, engineers and academics, Rankine aimed to formulate 'a science whose subjects are material bodies and physical phenomena in general'. About two years later, Maxwell began a private draft manuscript on mechanical principles with the words: 'sketch of an Introduction to the Higher Parts of Mechanics, being the Science of Energy as it relates to the Motion and Rest of a Single Particle, of Two Particles or a system of Particles, this system being either of invariable form, elastic or fluid'.[2]

Only from the 1860s, however, was the phrase given a high public profile, particularly after the publication of Tait's *Sketch of Thermodynamics* (1868) in which the second chapter, 'Historical Sketch of the Science of Energy', occupied a substantial forty pages. Even more significantly, Maxwell's Cambridge friend (and later one of his biographers) William Garnett (1850–1932) penned the article 'Energy' for the scholarly ninth edition of the *Encyclopaedia Britannica* which began rolling off the presses in 1875. Garnett provided one of the strongest statements of 'the science of energy':

> A complete account of our knowledge of energy and its transformations would require an exhaustive treatise on every branch of physical science, for natural philosophy [physics] is simply the science of energy. There are, however, certain general principles to which energy conforms in all the varied transformations which it is capable of undergoing, and of these principles we propose to give a brief sketch.[3]

Construction of 'the science of energy' involved far more than linguistic reform. The new language of energy was symptomatic of a series of profound conceptual shifts which resulted in a whole new scientific vision, with accompanying changes in scientific practice, quite unlike anything that had preceded it. Although fundamentally mechanical in nature, the universe would now be understood neither in terms of action-at-a-distance forces nor in terms of discrete particles moving through void space, but as a universe of continuous matter possessed of kinetic energy. Governed by basic laws of matter and energy ordained by God, such a cosmos contrasted with the deterministic universe of the French astronomer Pierre Simon de Laplace (1749–1827) and the German physicist–physiologist Hermann Helmholtz (1821–94) by ensuring a role for human free will in directing energy during its transformation from states of intensity to states of diffusion.

Fundamental to my interpretation of the history of energy in the nineteenth century is the argument that energy physics was not the inevitable consequence of the 'discovery' of a principle of energy conservation in mid-century, but the product of a 'North British' group concerned with the radical reform of physical science and with the rapid enhancement of its scientific credibility. As a result of careful dissemination of the energy principles through well-chosen forums such as the BAAS, the energy proponents succeeded in redrawing the disciplinary map of physics and in carrying forward a reform programme for the whole range of physical and even life sciences. 'Energy' therefore became the basic intellectual property of these elite men of science, a construct rooted in industrial culture but now transcending that relatively-local culture to form the core of a science with universal character and universal marketability.

Throughout this study I pursue the claim that construction of the science of energy should be understood in intimate relation to its audiences. I also advance the argument that the capitalistic contexts of Victorian Scotland in particular justify the deployment of the language of the market place. More specifically, however, my analysis requires recognition of the importance of, first, local contexts, and second, national, imperial and international contexts. With respect to local audiences, I distinguish three 'North British' sites for the construction and dissemination of the new natural philosophy: Scottish universities, marine engineering works, and scientific societies. With respect to national (and imperial/international) audiences, I focus on the role of the British Association. But in order to capture more systematically the historical character of these complex interactions among the protagonists and their contexts, recent sociological analysis will inform my account.

In their *Laboratory Life* Bruno Latour and Steve Woolgar argue that scientists are engaged in a quest for 'credibility'. Rather than suppose that scientists are motivated merely by a search for reward ('credit' as 'recognition of merit'), they develop an economic model of scientific activity in terms of 'credit as credibility'. The model crucially integrates three meanings of credibility as believability, personal power based on the confidence of others, and business trust. 'Credit as credibility' here functions as an exchangeable commodity: credit can be shared, stolen, accumulated or wasted. When, for example, a scientist embarks on a project, (s)he becomes the investor of credibility. If the knowledge produced proves valuable to another scientist, the credibility of the producer is raised. Credit as credibility is thus to be understood not simply as monetary capital but (following Pierre Bourdieu) as symbolic capital. A cycle of credibility is created by the investment of symbolic capital in a project. This in turn attracts investment in the form of grants to fund equipment and it is this equipment which in turn produces data, arguments and publications. If marketable, these results raise the value of the original stock of capital and so the cycle can begin anew.[4]

The original model, however, requires adaptation to the present case. Latour and Woolgar's examples relate primarily to rather closed communities of twentieth-century scientists, whereas in Victorian Britain the boundaries between science and its publics tend to be highly permeable. Elsewhere I have attempted to modify the 'cycle of credibility' as a 'spiral of credibility' in relation to William Thomson's classroom and laboratory activities. In that case study, I argued that Thomson returned to Glasgow as new natural philosophy professor in 1846 with an already high credibility: second wrangler and first Smith's prizeman in the Cambridge mathematical tripos (1845), author of a series of state-of-the-art original articles on electricity in the *Cambridge Mathematical Journal*, and as someone who had recently acquired experimental expertise in exact measurement in Victor Regnault's physical laboratory in Paris. Investing that credibility in the classroom and its supporting spaces enabled the new professor over the next decade to build up a multi-functional and territorially expanding laboratory which brought him increasing returns not only in symbolic but also in material capital, especially through the patenting of instruments for the burgeoning ocean telegraphic industry.[5]

In the present study I suggest that the 'credibility cycle' may similarly be adapted in order to illuminate the social processes by which an individual, and more especially a group such as the scientists of energy, rises in status and authority both within the scientific community and in relation to wider public audiences. The case of James Prescott Joule (1818–89) in the 1840s provides a good example of such an individual rise in status (ch.4). More broadly, however, this study of the rise in authority of the North British energy group utilizes the 'spiral of credibility' in a number of phases which I now sketch in summary form.

First, James Thomson (1822–92) embarked on a quest for engineering credibility in the early 1840s. His goal was to acquire both practical engineering skills and a knowledge of scientific principles which would provide him with the credibility to pursue patent inventions for economical forms of motive power, most notably water-wheels. Without such credibility, he would have no more status or authority than that of the ubiquitous 'ingenious inventors' who frequented the Literary and Philosophical Societies of most provincial towns. The quest took him from Clydeside to Thames iron and steam shipbuilding and thence back to Glasgow via William Fairbairn's Manchester engine-building works. In the course of that trajectory through several different local contexts, James encountered on the one hand the latest demands for economy and efficiency to make long-distance ocean steam navigation possible and profitable and on the other hand the theory of the motive power of heat put forward 20 years earlier by the little-known Sadi Carnot (1796–1832) (ch.3). The most immediate consequence, however, was not career advancement or invention but a growing enthusiasm, shared with his Cambridge undergraduate brother William, for the new theory. As a result, James, for long lacking the confidence of his younger brother, received a boost in personal authority.

Second, as the Thomson brothers recognized the relationship of the independent work of Carnot and Emile Clapeyron (1799–1864), Joule, Rankine and others to their own interests over the next six years, they conferred credibility upon these practitioners. For Joule in particular, that credibility was especially welcome after a long struggle to persuade sceptical members of the scientific communities in Manchester and London (ch.4). Third, William Thomson, working closely with his brother, conducted his own laboratory trials and communicated the results both to his class and to his newly established correspondence network. A process of *private, internal* credibility enhancement was set in train: Thomson, for example, commended Rankine for his treatment of Joule's results; Joule learnt more of

Rankine's work, and so on. The class too was impressed, or so the professor believed. Mutual credibility was on the increase (ch.5).

Fourth, an internal market of this kind was too limited, too like a mutual admiration society. The strategy now shifted to public spaces such as the meetings of the Glasgow Philosophical Society but most notably to those of the Royal Society of Edinburgh where Professor J.D. Forbes (1809–68), a British Association grandee and occupant of the Edinburgh natural philosophy chair, conferred public credibility by invitations to his former pupil Rankine and to his Glasgow opposite number William Thomson to address the Society and publish in its *Transactions* on subjects related to the Carnot–Clapeyron theory (ch.5).

Public credibility was also obtained in a rather different setting, that of Clyde-side marine engineering. Ever since the *Comet* (1812) had initiated steam navigation on the River Clyde, Glasgow engineers had been quick to promote the deployment of steam power on estuarial, cross-channel and ocean trades. But in 1850 the well-named iron screw steamer *City of Glasgow*, built locally by Tod & Macgregor to their own account, entered trans-Atlantic passenger service. Unlike previous trans-Atlantic steamers, the *City of Glasgow* operated without mail subsidy and by the end of the year had been bought by William Inman and partners of Liverpool, the first vessel to serve the new Liverpool & Philadelphia Steamship Company (later known as the Inman Line). For Clydeside engineers and ship-builders, this was the moment of great expectations for ocean steam navigation. Among the scientists of energy, it was Rankine who simultaneously promoted the new science of thermodynamics as a primary tool in the hands of marine engine-builders for the production of the most efficient and compact heat engines for long-distance navigation (ch.7). Thus did the heat engine derive credibility from thermodynamics and thermodynamics credibility from the heat engine.

For Thomson (and later Tait), however, the key local context was the university. Most Scottish Arts undergraduates aspired to the ministry of the Scottish Kirk and attendance at natural philosophy classes traditionally formed part of the broad philosophical education which preceded a more specialized preparation in theology. Scottish university professors (and most of the 'scientists of energy' held Scottish chairs at one time or another) were heavily dependent for their income upon student fees paid directly to the professor. But the launching of a new natural philosophy was not simply a matter of increasing student numbers and hence raising the level of professorial income. Rather, the period from 1840 to 1860 was one of exceptional instability in Scottish academic culture. The so-called 'Disruption' in 1843, led by the distinguished preacher and theologian Thomas Chalmers (1780–1847), split the Scottish church into Established and Free branches. The turmoil occasioned by this division threatened to deprive the Scottish universities of the proportion of their traditional clients now destined to become ministers of the Free Kirk (ch.2).

After Chalmers's death in 1847, factions within the Free Kirk became more assertive in their biblical literalism, especially concerning the *Genesis* account of the Creation, in the face of external forces of materialism recently exemplified in the anonymous *Vestiges of the Natural History of Creation*. At the same time, university reformers desirous of abolishing all religious tests for academic chairs threatened to alienate even the Established Kirk. As Scottish presbyterian culture manifested every sign of disintegration, the promoters of the science of energy began to represent the new natural philosophy as a counter to the seductions of enthusiast biblical revivals on the one hand and evolutionary materialism on the other. Unlike the largely defensive roles assigned to natural philosophy in the new Free Church Colleges, however, the science of energy was not deployed merely to shore up existing theological doctrines. Instead, its promoters were actively involved with the building of a new 'moderate' presbyterianism associated with the leadership of the Glasgow minister and Moderator of the Scottish Church, Norman Macleod (1812–72) (ch.2).

In pursuit of national credibility, the scientists of energy soon mobilized the British Association. At first, they successfully won support from the old 'lions' such as Col. Edward Sabine (1788–1883), Astronomer Royal G.B. Airy (1801–92) and William Hopkins (1793–1866), all of whom had been active in the first two decades of the Association. As a result of such powerful support, resistance was minimal and sporadic. In 1865, for example, the astronomer Sir John Herschel (1792–1871) complained that the principle of conservation of energy, 'now so universally recognized as a dominant one in physics', was a mere verbal truism. Its tautological character derived from the introduction of the phrase 'potential energy' which accounted for apparent losses in kinetic energy or *vis viva*:

> This principle, so far as it rests upon any scientific basis as a legitimate conclusion from dynamical laws, is no other than the well known theorem of the conservation of *vis viva* . . . *supplemented, to save the truth of its verbal enunciation*, by the introduction of what is called 'potential energy', a phrase which I cannot help regarding as unfortunate, inasmuch as it goes to substitute a truism for the announcement of a great dynamical fact.[6]

Rankine, who had coined the phrase 'potential energy', chose to respond. He acknowledged that Herschel's point had all the authority of 'an author who is not less eminent as a philosopher than as a man of science, and whose skill in the art of expressing scientific truth in clear language is almost unparalleled'. He also admitted 'that the use of the term "potential energy" tends to make the statement of the law of conservation of energy wear, to a certain extent, the appearance of a truism'. But, he argued, 'such must always be the effect . . . of drawing up precise and complete definitions of physical terms'. In the present case:

> 'actual energy' and 'potential energy' are defined in such a way as to make the proposition: That what a body or system of bodies gains in one form of energy through mutual actions, it loses in another form – in other words, that the sum of actual and potential energies is 'conserved' – follow from the definitions, so as to sound like a truism; but when it is proved by experiment and observation that there are relations amongst real bodies agreeing with

the definitions of 'actual energy' and 'potential energy', that which otherwise would be a truism becomes a fact.[7]

Again, the celebrated Michael Faraday (1791–1867) remained implacably opposed to the new energy doctrines. His objections rested largely on his dislike of any mechanical philosophy, such as that of Joule, Rankine or Thomson which interpreted nature in terms of matter and motion or matter and mechanical force. Portraying God as a God of *power* rather than a divine engineer or architect, Faraday's vision of nature reflected the biblical perspective inherent in his particular version of Christianity known as Sandemanianism and as such contrasted sharply with that of the science of energy. For good reason, then, did Forbes remark in 1856 to William Whewell (1794–1866), Master of Trinity College Cambridge, that 'Thomson (admirer as he is of Faraday) would certainly not allow to see [Faraday] superior to Joule, about whom he is, to my conception, scarcely rational'.[8]

In recent studies of Charles Darwin (1809–82) and his circle, the battle lines have been drawn between the Cambridge clerical dons, represented by Whewell and Adam Sedgwick (1785–1873), and the rising generation of scientists, represented by T.H. Huxley (1825–95) and fellow members of the so-called X-Club in London who aspired to a professionalised science free from the perceived shackles of Anglican theological doctrines. In taking the science of energy story into the 1860s, I argue that these battle-lines need to be redrawn to include as a third group our North British scientists of energy, a group not assimilable to Cambridge Anglicanism even though alliances between Scotland and Cambridge were commonplace. I claim instead that the science of energy offered a powerful rival reform programme, in direct competition with scientific naturalism (chs 9 and 10). Once again, the peripatetic BAAS (rather than the Royal Society) provided a national forum for the debates, just as it also offered the national, imperial and international forum for the massive campaign to embody the science of energy in an absolute system of physical units of measurement (ch.13).

Probably the most prestigious, and certainly the most enduring, kind of national and international credibility was the embodiment of the science of energy, the new natural philosophy, in a definitive treatise or series of treatises, most notably Thomson and Tait's *Treatise on Natural Philosophy* (1867) and Clerk Maxwell's *Treatise on Electricity and Magnetism* (1873). But such textual forms proved extraordinarily malleable. In Maxwell's case especially, the death of the author literally unleashed a flood of successors who, as mathematical physicists and later as historians of physics, would compete voraciously to claim the Maxwellian legacy. Such reinventions were symptomatic that the scientists of energy, overtaken by old age and death, were no longer on an upward spiral of credibility. As new approaches competed for the mastery of nature, the fate of the science of energy was sealed. Far from being a timeless enterprise, its heyday was almost certainly past by the 1880s (ch.14).

Yet it did not seem thus at the 1881 meeting of the British Association in York. The year and the place of Sir William Thomson's presidential address to Section

A of the BAAS were highly symbolic. Fifty years earlier the Association had held its first annual meeting in the same ancient English city at the geographical centre of gravity of the United Kingdom of Great Britain and Ireland and thus a common meeting place for scientific people without regard for provincial academic, local commercial or metropolitan factions.[9]

The BAAS had been the forum at which the young Professor Thomson had encountered the views of Joule during the Oxford meeting of 1847 (ch.5), had presided over Rankine's presentation of the new energy doctrines at the Belfast meeting of 1852, had raised questions about the finite age of the sun and solar system at the Liverpool meeting of 1854 and the Manchester meeting of 1861 (ch.7), and had promoted through the Association's Committee on Electrical Standards a new system of energy-based electrical units (ch.13). Now in 1881 Thomson reminded his specialist audience that 'The very name energy, though first used in its present sense by Dr Thomas Young about the beginning of this century, has only come into use practically after the doctrine which defines it had . . . been raised from a mere formula of mathematical dynamics to the position it now holds of a principle pervading all nature and guiding the investigator in every field of science'.[10]

Although Thomson here effaced any claim to a leading role in the establishment and promotion of the science of energy, he had for some three decades done much to propagate and sustain an 'official' version of its history, a version which, though challenged in its own time, remained more or less the standard account well into the twentieth century. Like all history told by the winning side, it was a story of progress and a story with heroes. For Thomson, it was also a story with predominantly British characters. In his paper 'On the Dynamical Theory of Heat' (1851), Thomson offered the following account which in later years acquired almost biblical authority in textbooks concerned with energy:

> SIR HUMPHRY DAVY by his experiment of melting two pieces of ice by rubbing them together, established the following proposition:– '. . . caloric [heat treated as a weightless substance] does not exist'. And he concludes that heat consists of a motion excited among the particles of bodies. . . . The dynamical theory of heat, thus established by Sir Humphry Davy, is extended to radiant heat by the discovery of phenomena, especially those of the polarization of radiant heat, which render it excessively probable that heat propagated through 'vacant space' . . . consists of waves of transverse vibrations in an all-pervading medium [ether]. The recent discoveries made by Mayer and Joule, of the generation of heat through the friction of fluids in motion, and by the magneto-electric excitation of galvanic currents [as in a dynamo], would either of them be sufficient to demonstrate the immateriality of heat; and would so afford, if required, a perfect confirmation of Sir Humphry Davy's views.[11]

Thomson here made no reference to his own recent doubts regarding the validity of Davy's claims (ch.6). Nor did he explain why Davy's results had apparently lain unappreciated for half a century. Similarly, he avoided any reference to the fact that both the German physician Julius Robert Mayer (1814–78) and the English natural philosopher Joule had, in different ways, been largely ignored and even scorned by their respective German and British scientific peers (ch.4) and that Thomson himself had taken almost four years to accept Joule's arguments (ch.6). In due course Count von Rumford (1753–1814) also took a posthumous place alongside Davy as one of the most active pioneers of the new doctrine of heat, but only by way of reinforcing the original account (chs 4 and 7).

Scientists of energy had particular reasons for representing the historical story in this way. Their reforming predecessors in Section A of the BAAS had long admired the pre-eminence of French mathematical physics exemplified in Laplace's *Mécanique céleste*. Equally, however, they had become increasingly dissatisfied with the basis of the Laplacian doctrines, which assumed action between point atoms over empty space as the explanatory framework for *all* natural phenomena from light to electricity and from astronomy to cohesion. For most of the 1840s, Thomson and his Cambridge friends had looked to alternative French doctrines, such as those of Joseph Fourier (1768–1830), for a formulation which appealed directly to observation and experiment.[12] If, by contrast with the hypothetical basis of the Laplacian scheme, the science of energy could be shown to rest on a strong and uncontroversial experimental basis, then its truth would be rendered visible for all scientific practitioners to see for themselves.

No member of the group, however, could take for granted automatic acceptance of the new energy doctrines. Joule, for instance, had spent several years attempting to win credibility from the elite Royal Society for his views on the convertibility of heat and work (ch.4). It was not therefore surprising that scientists of energy should wish to present a history of the dynamical theory of heat as an uncontroversial progression to the 'true' theory. This rhetorical strategy conformed also to a widespread belief in progress through knowledge. Thomson especially espoused in politics *and* science those values known as 'whig' or 'liberal'. Whig values placed a premium on the inexorable march of human progress from darkness to light, from error to truth. Nineteenth-century whigs used history in general, and history of science in particular, to legitimate their firm beliefs in the *necessity* of progress.[13] That they succeeded with respect to the history of energy is illustrated by remarks on the subject made by John Theodore Merz (1840–1922) in his four volume *A History of European Thought in the Nineteenth Century* (1904–12):

> none of the three great generalisations which we have so far reviewed have been the creations of the philosophers of the nineteenth century . . . The unification of scientific thought which was gained by any of these three views, the astronomical, the atomic, and the mechanical, was thus only partial. A more general term had to be found . . . which would give a still higher generalisation, a more complete unification of knowledge. One of the principal performances of the second half of the nineteenth century has been to find this more general term, and to trace its all-pervading existence on a cosmical, a molar, and a molecular scale. It will be the object of this chapter to complete the survey of those sciences which deal with lifeless nature by

tracing the growth and development of this greatest of all exact generalisa-
tions – the conception of energy.[14]

Merz, then, represented the history of energy as the route to the crowning
achievement of the exact sciences in the nineteenth century.

By the mid-twentieth century 'whig history of science' was becoming less and
less acceptable to most (though by no means all) professional historians. In 1959
Thomas Kuhn's 'Energy Conservation as an Example of Simultaneous Discovery'
offered a fresh perspective on the history of energy. Kuhn's approach no longer
sought to defend the priority claims of national heroes, but aimed to view the
history of energy in a European-wide intellectual and social context. He therefore
named twelve European men of science and engineering 'who, within a short
period of time, grasped for themselves essential parts of the concept of energy and
its conservation'.[15]

Joule, Helmholtz, Mayer and the Danish engineer L.A. Colding (1815–88)
combined generality of formulation with concrete quantitative applications. Sadi
Carnot, Marc Séguin (1786–1875), Karl Holtzmann (1811–?) and G.A. Hirn
(1815–1890), all of whom had strong engineering interests, lacked that generality
but 'recorded their independent convictions that heat and work are quantitatively
interchangeable, and all computed a value for the conversion coefficient or an
equivalent'. A third group of individuals consisted of the British natural philosophers William Robert Grove (1811–96) and Faraday, the German chemist Justus
von Liebig (1803–73) and the German natural philosopher C.F. Mohr (1806–?79).
This group lacked quantitative evaluation but 'all described the world of phenomena as manifesting but a single "force", one which could appear in electrical,
thermal, dynamical, and many other forms, but which could never, in all its transformations, be created or destroyed. That so-called force is the one known to later
scientists as energy'.[16]

In Kuhn's account, then, history of science offered 'no more striking instance of
the phenomenon known as simultaneous discovery' than the history of energy
conservation. Looking for elements in the climate of European scientific thought
during the period 1830–50 'able to guide receptive scientists to a significant new
view of nature', Kuhn isolated three such factors which he termed the 'availability
of conversion processes', the 'concern with engines', and the 'philosophy of
nature'. 'Availability of conversion processes' meant the many experimental discoveries after about 1800 which appeared to involve the conversion of one physical
phenomenon into another: the forces of chemical affinity into an electric current as
in the battery, for example. 'The concern with engines' referred to the increasing
role of waterwheels and heat engines during the Industrial Revolution, and to the
demand for an accurate measure of their performance. The 'philosophy of nature'
or *Naturphilosophie* concerned a deeply held, German-centred cultural belief in a
single unifying principle for all natural phenomena.[17]

The skill and novelty of Kuhn's approach, like that of his better known
Structure of Scientific Revolutions (1962), has generated a variety of critical
responses. Kuhn himself anticipated unease with the notion of 'simultaneous dis-

covery': in 'the ideal case of simultaneous discovery two or more men would announce the same thing at the same time and in complete ignorance of each other's work, but nothing remotely like that happened during the development of energy conservation . . . no two of our men even said the same thing'. Thus 'what we see in their works is not really the simultaneous discovery of energy conservation' but rather 'it is the rapid and often disorderly emergence of the experimental and conceptual elements from which that theory was shortly to be compounded. It is these elements that concern us'.[18]

Critics have seized upon Kuhn's apparent hostage to fortune. Yehuda Elkana, for example, claimed that 'the principle of energy conservation is not a case of mutiple discovery, but that in the span of 1840–60 different problems were bothering different groups of people in different places, and they came up with different answers. The answers turned out to be related . . . [and thus] the model suggested for simultaneous discoveries is not the correct one'.[19] Other historians of science have presented their criticisms of Kuhn's original account through studies of individual members of the twelve 'simultaneous discoverers', including Faraday, Joule, Grove, Colding, Mayer and Helmholtz.[20]

Opponents of Kuhn's 'simultaneous discovery', however, have generally ignored the point that he was employing the notion as a kind of shorthand device aimed at highlighting some peculiar features of the story of energy conservation, that is, that over a period of two decades so many relatively independent individuals should have apparently 'grasped for themselves essential parts of the energy concept and its conservation'. The fundamental point of the paper was not to defend 'simultaneous discovery' as an abstract sociological or philosophical term, but to attempt an explanation of the historical phenomenon in terms of the three factors (conversion processes, concern with engines, and philosophy of nature) which were characterized by frequency of occurrence, specificity to the period, and a decisive effect upon individual research.

In an explanatory note added to his paper, Kuhn distinguished between what he now termed *prerequisites* to the discovery (that is, developments within the sciences such as experimental calorimetry needed before the 'discovery' could be made), and *trigger-factors* responsible for simultaneous discovery (that is, the immediate stimuli which triggered the work of the twelve 'pioneers'). He thus emphasized that his paper was not concerned with shared prerequisites but with the three trigger-factors which distinguished pioneers from predecessors.[21]

Throughout all of Kuhn's analysis, however, runs the implicit thread that energy conservation was a matter of 'discovery' like that of a precious gem hidden *in nature*. Thus he asserted that we know why the experimental and conceptual elements 'were there: Energy *is* conserved; nature behaves that way'. The real question for Kuhn was then 'why these elements suddenly became accessible and recognizable'.[22] Indeed, Kuhn's choice of the label 'pioneers' for his group of twelve carries the assumption that each person was, like the proverbial sleepwalker, moving towards the full energy principle. This assumption is unfortunate in that it interprets the contemporary scientific practice of each of the twelve *with the benefit of hindsight*. Furthermore, philosophers of science, followed by sociologists of science, have long been concerned with the problematic nature of all discovery. Gems are neither 'discovered' in nature nor do they possess intrinsic value. They have to be crafted by human agency and taken to the market place of a commodity culture.[23] Thus the 'discovery' of energy conservation becomes instead a construction which has to be formulated, articulated and promoted and which might be said to constitute a simulation or representation of the external world of 'nature' rather than some kind of essential truth about 'reality'.

A more persuasive historical account is now possible. First, we should abandon terms such as 'pioneers' and 'discoverers'. We should forget presuppositions about facts and principles literally discovered *in* nature. We should not misrepresent the activities of those scientific practitioners who were unaware that later generations would see their work as contributing to quite a different set of truths from the one they themselves sought to reveal.

Second, we must accept that later nineteenth-century scientific communities *did* portray individuals, such as Mayer, Colding and Joule, as independent discoverers and pioneers. Indeed, the individuals themselves often wanted their own work to be construed retrospectively in this manner. Kuhn in fact used such judgments as a criterion for selecting most of his twelve 'pioneers': 'I have tried to include all the men who were thought by their contemporaries or immediate successors to have reached independently some significant part of energy conservation'.[24] Such contemporary judgments, however, do not mean that the chosen candidates themselves, at the time, were discoverers of energy conservation. Rather, a persuasive historical account must try to explain why these 'immediate successors' should have been so concerned to invent pioneers of energy conservation (chs 9 and 10).

In this account of the construction of energy physics in the nineteenth century I have adopted a historicist technique which minimizes the role of an omnipresent narrator. The significance of this approach means that the reader will gain a stronger appreciation of the historical processes by which the science of energy was brought into being and made credible. By avoiding a strictly sequential narrative, I aim to represent the historical events as they were perceived by the protagonists of energy physics themselves. This approach therefore contrasts strikingly with the omnipresent narrator required both for simple chronological narratives taking us forward from one 'important landmark' to the next over a long period and for 'simultaneous discovery' accounts treating of several 'independent' discoverers all sleepwalking towards the same physical truth. A few brief examples, developed in subsequent chapters, will illustrate the technique.

First, an omnipresent narrator would take us linearly from Sadi Carnot's *Reflexions* (1824) through Clapeyron's analytical reformulation (1834) to William Thomson's 'Account of Carnot's Theory' (1849). In this study, with the focus on the North British natural philosophers and engineers, the account is non-linear: what matters first is the Thomson brothers' recovery of Carnot's theory through their reading of Clapeyron. Only later do they obtain the original text and proceed to reinvent Carnot in a manner suitable to the wants of the new thermodynamics (chs 3 and 5). Second, an omnipresent 'simultaneous discovery' narrator largely ignores the *interactions* between Joule and Mayer whereas in this account it is precisely the controversies over priorities fought initially in the pages of the *Comptes rendus* around 1848 that in practice enhance the credibility of the hitherto largely ignored protagonists (ch.4).

In view of long-running debates over whether or not Hermann Helmholtz should be regarded as the 'true' discoverer of energy conservation, the third example is perhaps the most striking illustration. With the benefit of hindsight, an omnipresent narrator would place Helmholtz's '*Über die Erhaltung der Kraft*' (1847) chronologically 'in advance' of the North British natural philosophers and engineers. But Helmholtz's essay had received little attention in local German-speaking contexts. The essay had to be 're-read' from a different perspective. Thus very early in the 'career' of the science of energy, Thomson enlisted the support of Helmholtz's essay which, though based in part on Laplacian doctrines, was soon made to serve the new cause of energy (ch.7). While Thomson willingly promoted Helmholtz's views on the mechanical origin of the sun's heat in the 1860s and beyond, Helmholtz in turn became the most ardent Continental promoter of the new doctrines. Maxwell, for example, expressed the opinion in 1862 that Helmholtz was 'one of the first, and is one of the most active, preachers of the doctrine that since all kinds of energy are convertible, the first aim of science at this time should be to ascertain in what way particular forms of energy can be converted into each other, and what are the equivalent quantities of the two forms of energy'.[25]

As close personal friends, Thomson and Helmholtz, together with Thomson's North British colleagues Rankine, Tait, Maxwell, Jenkin and Balfour Stewart (1828–87), dominated the British and German markets for the new science of energy. Textbooks at a variety of levels, popular expositions and lectures, and embodiment in practical problems of heat engines, telegraphy and electrical engineering all served to secure the territory of energy physics. But the outward spiral of credibility was not a self-expanding process. It was a campaign which was to be hotly contested.

Perceiving and exploiting differences between the North British physicists and their German advocate, the Royal Institution's John Tyndall (1820–93) adopted a far more radical version of energy physics by seizing upon Helmholtz's commitment to scientific determinism. Such a position was anathema to the scientists of energy who retaliated with several vigorous attacks on Tyndall and his promotion of determinism as a necessary consequence of conservation of energy. Thus Thomson, Tait, Jenkin, and Maxwell all argued that human free will was fully compatible with the energy doctrines (ch.12). Again, Tyndall forcefully promoted the cause of Mayer in the discovery of energy conservation disputes, underpinning his attempt to distance himself from the dominant northern group (ch.9).

From this introductory discussion, it should be clear that the history of energy is no simple story of scientific genius and experimental discovery. There was no discovery in that sense. Discoverers were found and pioneers invented during the promotion of the new doctrines at meetings of the BAAS and through the *Philosophical Magazine* from 1852. The new word 'energy' was William Thomson's, but neither he nor Rankine made claims to discovery.[26] In practice, this form of 'self-effacement' had the effect of dramatically increasing the credibility of the science of energy. By avoiding any personal claims, the scientists of energy transferred authority to a much wider and seemingly 'objective' scientific community which contained weighty names from the past as well as the present and which had been largely invented for the purpose. Thus the science of energy was presented as a new programme for scientific practice, universal in its claims, but with an impeccable British pedigree.

Recognizing the potency of the new science, critics tried too late to halt the spread of the energy doctrines. Rivals attempted instead to annex the new science for their own ends. But not until 'Energetics' in Germany at the end of the century did the science of energy lose much of its original shape (ch.14). At the same time, the science of energy came to be presented by historians such as Merz as the most far-reaching achievement of nineteenth-century physical science. Or, in the words of Sir Joseph Larmor (1857–1942), heir to the Lucasian Chair of Mathematics at Cambridge once occupied by Newton: 'This doctrine [of energy] has not only furnished a standard of industrial values which has enabled mechanical power . . . to be measured with scientific precision as a commercial asset; it has also, in its other aspect of the continual dissipation of mechanical energy, created the doctrine of inorganic evolution and changed our conceptions of the material universe'.[27] This book, then, is a history of the science of energy in the nineteenth century.

CHAPTER 2

From Design to Dissolution: Scotland's Presbyterian Cultures

> nature contains within itself the rudiments of decay . . . unless renewed by the hand of the Almighty, the earth on which we are now treading must disappear in the mighty roll of ages and of centuries . . . we may be prepared to believe that the principles of destruction are also at work in other provinces of the visible creation – and that though of old God laid the foundation of the earth, and the heavens are the work of his hands, yet they shall perish; yea, all of them shall wax old like a garment, and as a vesture shall He change them, and they shall be changed.
>
> *Rev Thomas Chalmers preaches (c.1830s) on the transitoriness of visible things*[1]

The cultural transformation of early nineteenth-century Scotland was rapid and profound. In the West, the increasing industrialization of the Clyde valley intensified social and political conflict within the new urban spaces. But even in the genteel Georgian grandeur of Edinburgh's New Town the illusory stability of the Scottish Enlightenment began to fragment with the rise of powerful new cultural and religious movements. In England during the same period a safe cultural distance separated Anglican academic establishment, with its Oxford and Cambridge axis, and the radical dissenting entrepreneurs of the industrializing North. No similar separation existed in Scotland. Within both Edinburgh and Glasgow, academic and religious debates raged amid visible poverty, spectacular economic growth and dramatic political upheaval as Chartists mobilized against the perceived injustices of aristocratic establishment.[2]

South of the border, the Archdeacon of Carlisle, William Paley, had embodied many of the values of liberal Anglicanism in his *Natural Theology* (1802). All manner of plants and animals testified to the power, wisdom and goodness of the Creator. Just as a watch could not have come about by chance, ran the standard argument from design, so particular organs such as eyes or ears exhibited the character of an artefact, created for the purposes of seeing or hearing. And just as a watch required a watchmaker, so plants and animals had been created by a

Designer whose wisdom and power was evidenced by the ever-accumulating data of natural history. Furthermore, the goodness of that Designer was demonstrated in the sum total of pleasure and happiness manifest in nature, far outweighing any corresponding quantity of pain and misery. Paley's perspective tended above all to support a fixed order of nature conducive to the preservation of a hierarchical society in which change and improvement was neither necessary nor desirable.[3]

By the 1820s and 1830s many of Paley's assumptions came under intense scrutiny. The fossil findings of the new science of geology brought to light species which no longer roamed the earth. New speculations were rife concerning the origin of the earth as a molten mass and its subsequent cooling over millions of years, allowing for the progressive introduction of different species suited to different climatic epochs. Astronomers too provided evidence that the solar system might not be as eternally stable and enduring as Enlightenment models had once suggested. Change and decay, rather than unchanging stability, had to be accommodated within any revisionist natural theology. Published during the early 1830s, the *Bridgewater Treatises* were ostensibly an up-dating of Paley for a new generation, but in many respects represented a shift away from a static cosmology to one of temporal change.[4]

Since the Reformation, presbyterianism had lain at the heart of Scottish culture. Unlike Anglicanism, the congregation of each parish church enjoyed considerable autonomy in the election of its ruling elders and in the choice of its minister. Neither congregation nor minister was subject to the hierarchical authority of a bishop. Central authority on earth rested instead with the General Assembly which met annually as a church parliament to debate and decide policy and to elect for the forthcoming twelve months a new Moderator from among its clergy. Again in contrast to England where clergymen were frequently the second or third sons of landed gentry, a low-cost Scottish university education allowed the most able sons of small farmers to rise to study for the ministry of the Kirk.[5]

In the second half of the eighteenth century Scotland enjoyed one of the most tranquil periods in a turbulent history. While France, Ireland and the United States endured Revolution, Scottish academic and commercial culture flourished as the bitterness of old religious and political disputes diminished. 'Moderate' presbyterianism, allied to elite intellectual culture in the cities and universities, predominated. But an alliance between the 'moderates' and aristocratic culture ran counter to the traditions of democratic presbyterianism. These tensions surfaced in the 1830s over the issue of patronage. Although in principle the choice of a new minister rested with the people of the parish, aristocratic patrons exercised a powerful voice in the process, provoking secession of some parishes and ultimately the 1843 'Disruption' when Thomas Chalmers led his supporters out of the established church and founded the Free Church of Scotland.[6]

The 'Disruption' was symptomatic of the growing tensions between the ageing 'Moderates' and the new 'evangelicals' within the Church. Associated with rationalist, Enlightenment values and with natural theology, 'Moderates' were well represented in the five Scottish university colleges. They therefore played a key role in the education and training of future ministers of the Kirk. But by the 1830s the Scottish universities seemed to many critics to have their best years behind them, leaving the echoes of Adam Smith, Joseph Black, John Robison, James Watt and other distinguished individuals of the late eighteenth century to grow ever fainter. The 'Moderate' theologians in control of Glasgow College in particular appeared more interested in preserving their own privileges and power than in promoting the spiritual well-being of their students. Intellectual rituals, whether in logic, mathematics, theology, moral philosophy or natural philosophy, seemed increasingly irrelevant in a climate of emotive issues rangeing from poor law reform to the origin of man.[7]

The making of an evangelical culture

After attending the University of St Andrews in the 1790s, Thomas Chalmers entered Edinburgh University at the turn of the century. In 1805 he was unsuccessful against John Leslie as a candidate for the Edinburgh chair of mathematics, left vacant by John Playfair's move to natural philosophy as successor to the eminent John Robison, close friend of Watt and Black. Chalmers subsequently became minister of a country parish before being called to Glasgow in 1814. There he made his name as a powerful and eloquent preacher in evangelical mould, with emphasis on the fallen, sinful state of man and the need for individual redemption through Christ.[8] Although convinced that no intellectual or rational path to God was adequate and that nature did not provide a clear insight into 'the character of Him who willed such an enigmatical world as ours into existence', Chalmers nevertheless held that natural theology could serve 'as a harbinger to the higher lessons of the gospel'. One such harbinger, *Astronomical Discourses*, went through nine editions in one year.[9]

Chalmers returned to the academic fold as professor of moral philosophy at St Andrews in 1823. Four years later he was offered the moral philosophy chair at the new University of London, but took instead the chair of divinity at Edinburgh, a post which he held until the Disruption. Most of his contributions to political economy and natural theology, including the first *Bridgewater Treatise*, were published in the 1830s.[10]

In his sermons and writings Chalmers drew upon a Scottish distinction, first published in John Robison's *Encyclopaedia Britannica* articles on 'Physics' and 'Philosophy', between natural history and natural philosophy. Natural history treated of pure descriptions and arrangements as illustrated by the categories of living things: kingdoms, classes, genera and species, for example. As the science of *contemporaneous* nature, natural history served to classify the resemblances existing among objects, both animate and inanimate. It was therefore concerned with the structure, collocation and disposition of objects, whether of planets, molecules or animals: 'Natural History takes cognizance of all those characters that exist together at the instant, and which may be described without reference to time – as smell, and colour, and size, and weight, and form, and relation of parts, whether of the simple inorganic or more complex organic structures'.[11]

Natural philosophy, on the other hand, was the science of *successive* nature, treating of the incessant but regular and law-like changes observed in nature. This science was further divided into natural philosophy, 'strictly and indeed usually so called' which dealt with the laws of sensible motions, and chemistry which concerned the visible effects of insensible motions.[12] Chalmers's central claim was that the laws of natural philosophy by themselves were inadequate to account for the *arrangements* of our existing natural history. Furthermore, were these arrangements to be destroyed, the powers of natural philosophy could not restore them. This assertion was central for 'the cause of natural theology'. Although natural philosophy may 'account for the evolution of things or substances collocated in a certain way . . . they did not originate the collocations':

> The laws of nature may keep up the working of the machinery – but they did not and could not set up the machine. . . . For the continuance of the system and of all its operations, we might imagine a sufficiency in the laws of nature; but it is the first construction of the system which so palpably calls for the intervention of an artificer, or demonstrates so powerfully the fiat and finger of a God.[13]

Natural history, rather than natural philosophy, thus provided the basis for a limited natural theology. Our observation of the adaptation of parts to *an* end led us from the order of nature to a designing power. God's 'natural' attributes of omniscience, power, omnipresence, unity, eternity, and self-existence all followed from natural theology as well as being known from revelation through the scriptures. Unlike Paley, however, Chalmers acknowledged that the 'moral' attributes of God flowed less easily from nature. A God with attributes of justice, truth, righteousness, holiness and goodness did not follow from natural proofs. Adaptation by itself failed to prove benevolence, for an apparatus of torture could be well-designed for its purpose. Not for Chalmers, then, the utilitarian weighing of pleasure and pain:

> Amid the vast capacities for enjoyment. . . . We need not speak of that countenance of menace and boding disaster which is put on by inanimate things – or, for the smile and the verdure and the gracefulness of nature in her happier moods, tell also of her angry tempests, of her wasteful volcanoes, of her sweeping hurricanes and floods. . . . It is enough faithfully to record the moral perversities wherewith the social state of man is vexed and agitated[14]

Nature was thus transitory and finite, subject to decay, disease and death. Her beautifully designed dispositions were not eternal: 'There is we think an utter derangement into which nature has been thrown – so that all her elements are impregnated with disease'. Reacting to Enlightenment stability and perfection in Nature, the evangelical Chalmers indeed constantly stressed the 'principles of destruction . . . at work' in the visible creation. Although doctrines of election and predestination (whereby God has pre-ordained who shall be among the elect and who shall be damned) do not have the same prominence in Chalmers's preaching as in earlier and far more extreme presbyterian sermons, Chalmers's sermons appear to retain a strong Calvinist sense of an universe cursed by depravity, imperfect in its parts and destined for inevitable decay.[15]

Chalmers wanted to leave no room for materialism or deism in his philosophy of nature. On the one hand, Revelation taught us that the laws of natural philosophy were wholly contingent upon God: 'We admit that His creative energy originated all, and that His sustaining providence upholds all'.[16] On the other hand, the evidence of natural history pointed firmly to both the beginnings and the endings of the present order of things. Any materialist or deist creed, that the laws of nature were self-sufficient and eternal, was undercut by the claim that these laws could not originate the economies of nature which were themselves subject to processes of decay and derangement.

With respect to beginnings, Chalmers both looked at the historical evidence, provided by ancient texts, for the comparative recency of the present order of things and considered the observational evidence in the form of 'the natural and geological proofs for a commencement of our present terrestrial economy'. Although he favoured the claim that the first chapter of Genesis described literally a week of miracles – 'the period of a great creative interposition, during which by so many successive evolutions, the present economy was raised out of the wreck and materials of the one which had gone before it' –, he did not regard potential conflict between scripture and geology on the antiquity of the earth and of the human species as an ultimate threat to the argument for creative interposition. Chalmers's preferred interpretation was that there may have been an interval between the primal creation and the six days work as set out in Genesis, a perspective explored as early as 1814 in a critical review of Cuvier's theory of the earth. Such a view of the earth's origin allowed both for a long geological time-scale and for creative interposition.[17]

Chalmers treated the concomitant problem of endings in a sermon on a well-known text from second Corinthians 4:18: 'the things which are seen are temporal; but the things which are not seen are eternal'. Stressing the transitory nature of all visible things, he pronounced his belief that the words held true in an absolute and universal sense: visible things 'had a beginning and they will have an end'. Only the 'uncreated God' was absolutely eternal and enduring. Moreover, he interpreted the text as referring not simply to the passing away of visible things but to the possible transfiguration of matter without its actual destruction. His claims reflected contemporary speculations within the British scientific establishment. In particular, William Whewell and other Cambridge gentlemen of science were of the opinion that the observed behaviour of Encke's comet appeared to show that bodies (whether comets or planets) orbiting the sun were subject to the retardation effects of an etherial medium and that therefore the solar system was not an eternally stable perpetual motion machine. Not content with such far-distant cosmic dissolution alone, however, Chalmers pronounced the decay of all of nature's economies.[18]

In a complementary sermon Chalmers considered the parallel question of the renewal of a decayed or disordered nature through divine action. His text was second Peter 3:13: 'Nevertheless, we, according to his promise, look for new heavens and a new earth, wherein dwelleth righteousness'. The convulsions of the last day, he claimed, may break the world down from its present arrangements – already in decay within themselves – and the whole of the existing framework may be utterly dissolved in a fervent heat:

> And thus may the earth become without form, and void, but without one particle of its substance going into annihilation. Out of the ruins of this second chaos, may another heaven and another earth be made to arise; and a new materialism with other aspects of magnificence and beauty emerge from the wreck of this mighty transformation.[19]

Whether by divine intervention on the last day or whether by the slow process of decay, the dissolution of the present order was certain. Only by divine command would a new order, a new economy of nature, arise as an abode not of rebellion, sin, and evil, but of righteousness. All things would become new just as individuals could become new through spiritual rebirth rather than by rational or natural means.[20]

As Boyd Hilton has argued, Chalmers's evangelical theology also had major ramifications for his diagnosis of the ills of society, imperilled as it was in these years by gloomy prognostications of overpopulation, famine, and cholera. Hilton locates Chalmers in opposition to extreme evangelicals who, committed to an interventionist God 'constantly directing earthly affairs by special warnings and judgments', held that human governments should also take an interventionist approach to social and economic problems. In contrast Chalmers took a moderate evangelical position characterized by a desire to make society operate, like nature, in accordance with divinely chosen laws. Chalmers's society would therefore function best without arbitrary interventions by governments. Free trade, *laissez-faire* and self-help doctrines were desirable not because they led naturally (as utilitarians and secular political economists generally held) to an increase in the material stock of individual and social happiness but because they would facilitate individual morality.[21] Thus an interventionist Poor Law and misconceived charity would 'enlist the human will on the side, if not of poverty, at least of the dissipation and indolence which lead to poverty'.[22]

Hilton has further shown that Chalmers's evangelical model of Free Trade was essentially static and cyclical, powerfully regulated by economic conscience. The temptations, trials and sufferings characteristic of the free market would constantly nurture that conscience, productive of abstinence and moral restraint as the counters to greed and indulgence. Guided by his mentor Thomas Robert Malthus (1766–1834), Chalmers envisioned a world of finite resources in which the goal had to be the restoration of equilibrium, whether of population or production. The contrast with the Free Trade model of the professional economists could scarcely have been greater: theirs was 'expansionist, industrialist, competitive, and cosmopolitan', a model of optimistic economic growth through capital accumulation and international division of labour leading to more wealth and happiness for all.[23] Chalmers's model was of course a blueprint for the moral regeneration of society, an ideal dependent for its fragile stability not on anything inherent in human nature but ultimately in the acceptance by an individual human being of the gift of divine grace that created each person anew, reborn from sin and death.

Such a concept of divine 'gift', whether of grace or nature, was central to presbyterian theology. In order to elucidate the significance of the concept, I draw on the work of anthropologists such as Marcel Mauss and of historians of science (notably Mario Biagioli) on the theme of gift-exchange. Much of the literature focuses on the reciprocal nature of gift-exchange. For example, Biagioli quotes Galileo's friend Ciampoli: 'There is something that holds true everywhere in the world: you need to give gifts to those in power . . . Blessed are those who can accelerate their success by giving gifts'.[24] As Biagioli shows, seventeenth-century court culture required players to engage in a complex game of gifts and countergifts in which patrons and their clients *appear* to be giving and returning gifts voluntarily while in fact acting in accordance with powerful cultural conventions.

Galileo, however, soon moved into a sphere of power above that of the normal patron and involving a code of behaviour above that of knightly honour. Biagioli develops the issue in relation to the culture of absolutism within which he places Galileo. 'The absolute prince resembles a god', he explains. 'More precisely, a Protestant god; one who is infinitely powerful and great so that his terrestrial clients cannot expect him to react to what they are doing to please him'.[25] Thus the absolute ruler bestows a gift upon a favoured courtier in the same way that a Christian receives the gift of Divine grace: the prospective recipient has done nothing to earn the gift (as reward for good works) but can only accept or reject it. In the words of the early nineteenth-century hymn writer Reginald Heber (1783–1826), enjoyed by presbyterian congregations: 'Vainly with gifts would His favour secure'.[26]

We find the presbyterian representation of 'giving' in Chalmers's sermons, preached under the text from Acts 20:35: 'It is more blessed to give than to receive'. Seeking to be remade in the image of God, Chalmers argued, Christians possess nothing that they did not receive from God. In contrast to man, God

> is the unfailing fountain out of which every thing flows. All originates in Him. A mighty tide of communication [transmission . . .] from God to His creatures, has been kept up incessantly from the first hour of creation. It flows without intermission. It spreads over the whole universe He has formed. It carries light, and sustenance, and enjoyment, through the wide dominions of Nature and of Providence.[27]

The protestant, presbyterian God thus bestows His gifts, freely given, to His universe. Human beings can only draw upon those divinely filled storehouses of nature. Man was in no position to enter into a gift-exchange with the Almighty, still less to enter into some form of trading relationship akin to that of the sale of indulgences. As Chalmers put the point: 'Heaven, my brethren, is not so purchased [by good works such as charity]. You are made meet for heaven by the Spirit working in your soul a conformity to the image of the Saviour'. But man had nevertheless an obligation not to *waste* those gifts of God, whether physical or spiritual. The great storehouses of nature, the gifts of God to mankind, were to be used, like Bacon's 'knowledge', 'for the glory of the Creator, and the relief of man's estate'.[28]

According to Chalmers, all men who had *received* from God were in some position to act as givers to others. He first argued against the notion that a Christian should crudely renounce his worldly wealth, for example by burning it. 'It is not enough . . . that you prove your indifference to this world's wealth by parting with it', he told his congregation. '[Y]ou must have an object in parting with it'. Second, he claimed that 'the feeding of the poor is only one of the many objects, for which you are entrusted with the gifts of Providence'. Implicitly invoking the gospel of Malthus, Chalmers proclaimed that if we were to direct all our disposable wealth to feed the poor, 'Idleness and profligacy would lay hold of the great mass of our peasantry . . . and the people of the land, thrown loose from every call to the exertions of regular industry, would spread disorder over the whole face of the country'.[29]

The best utilization of wealth, then, demanded a 'practicable scheme for the interests of humanity, and for the alleviation of all its sufferings'. This exhortation meant that some of an individual's wealth would go towards the improvement of commerce and industry, some towards the support of the family, some towards the Church and missionary work, and some towards feeding the poor. Among the poor themselves were those worthy souls 'who, by the struggles of a painful and honourable industry, have just kept themselves above the necessity of receiving'. In so doing they too became 'givers' of the sum which would otherwise have been used to support them. Even among those unable to labour, there may be some who succeed in minimizing their claims. In such cases God 'will put down to their account all that they have saved to the givers; and He will say, that, by the whole amount of what is thus saved, they have fed the stream of that benevolence which is directed to other objects'.[30]

In this presbyterian economy, there were two distinct processes: first, that of 'gift-giving' rather than 'gift-exchange'; and second, that of the trading or exchange economy. The first has to be understood in terms of a downwards-directed structure. God has provided mankind with both natural and spiritual gifts. But individual men, possessed of free will, have the choice of refusing or accepting such gifts. To refuse divine gifts is to remain a lost and fallen soul, condemned to sin, idleness, and poverty. To accept is not to seek to return the gift to God in exchange for salvation but rather on the one hand to engage in the exchange economy with other men of similar wealth and power, and on the other hand to pass down to others some of the wealth gained from turning the gift to beneficial and profitable effect: essentially a version of the parable of the Talents in Matthew 25: 14–30. In such a way will all the strata of society benefit as each lower layer receives its gifts from above and seeks to turn those gifts to useful effect, for the creation of wealth and the benefit of those less fortunate, all the while seeking to minimize those who, unable to give, are able only to receive. It was thus an economy which sought to maximize the virtues of useful work and minimize the vices of idleness and waste (ch.3).

Universities in crisis

Chalmers had been addressing early nineteenth-century tensions in Scottish culture brought about by the increasing industrialization of the lowlands. His prognostications appealed to diverse audiences, troubled by the disturbing signs of the times in and beyond Scotland.[31] In contrast, the old 'Moderate' theology, coldly intellectual and elitist, was ever more out of step with the demands for social and political change. Nowhere was this trend more evident than in the venerable Glasgow College on the High Street, surrounded by a city whose population in the 1840s was probably well in excess of 300,000. The redoubtable Principal Duncan Macfarlan (1771–1857), himself a Glasgow student in the 1780s when the town's population barely exceeded 40,000 and Moderator of the Church during the year of the Disruption, led the Moderate professors in a hard-fought campaign to preserve their privileges and stem the tide of reform in the College. For their part, Whig and Tory Governments shared the view that all was not well in the Scottish universities. Yet despite their best efforts to appoint a succession of reform-minded academics to 'regius' (crown-funded) chairs in a variety of scientific and medical fields, the College professors managed to retain control over all important decisions including the substantial annual income from land.[32]

Glasgow's academic aristocracy boasted Tory and Moderate principles. But the University's students were for the most part sons of merchant or commercial fathers in the West of Scotland aspiring to the status of ministers within the established Church of Scotland or one of its breakaway denominations. By the early 1840s Glasgow College's 'Moderates' had reason to feel threatened. Since around half a College professor's annual income derived from student fees, a decline in student numbers during the 1830s was having a direct effect. Moreover, old age had finally begun to take its toll among the College professors. Worse still, not all the new professors turned out to be College loyalists. Two radical reformers, James Thomson (mathematics) and John Pringle Nichol (astronomy), arrived in the 1830s. While the 'Moderates' successfully rejected the candidature of the distinguished Thomas Chalmers in favour of a relatively obscure Moderate in 1840, another radical, William Thomson (medicine) was in a College chair soon after. By the middle of the new decade the reformers began to command at least equal numbers of votes in the College. Central to their agenda was the abolition of the religious tests by which new professors had to sign the Westminster Confession of Faith, professing their commitment to presbyterianism in its most traditional form.[33]

Chalmers's Free Kirk, however, posed by far the greatest threat to the future of all the Scottish Universities. Eventually almost every parish church throughout the land would have its rival Free Kirk. The Glasgow College reformers perceived that if the College were to remain the preserve of the Moderates the result would be a sectarian institution, forced to compete for student business with a potential proliferation of free and secessionist colleges. The rapid establishment of an independent Free Church College in Edinburgh provided a powerful argument for the reformers in their campaign to remove all traces of sectarianism from the

University and to rebuild its reputation as a national institution suited to the needs of an industrial age.[34]

Natural philosophy classes formed an integral part of the traditional core Arts curriculum that provided a broad philosophical education to Bachelor of Arts level after which aspiring ministers generally attended theology classes to Master of Arts level. Natural philosophy therefore complemented courses in such subjects as mathematics, logic, Greek, humanity and moral philosophy. In contrast, chemistry was allied to the medical curriculum and played no role in the education of future ministers of the Kirk. All Arts subjects at the established universities were therefore highly vulnerable to the Disruption. As Anna Thomson, daughter of the mathematics professor, wrote to her brother in 1843:

> Papa's [mathematical] classes threaten at present to be rather smaller than usual and indeed this is to be expected as many of the Free Church students will go to Edinburgh [Free Church College], and the Assembly agreed to allow them to study divinity and nat[ural] phil[osophy] in the same year . . . so as to allow them to get on quickly to the church and this arrangement takes some of them to Edinburgh sooner than they would otherwise have gone.[35]

Moreover, the Free Kirk was not deficient in scientific talent. Chalmers himself placed a high value on the role of the sciences within an evangelical culture, while scientific men moving to the Free Kirk included the venerable Scottish experimental natural philosopher Sir David Brewster and the distinguished geologist Hugh Miller. Guided by the example of Chalmers, both men engaged with questions of the consistency of science and scripture. Far from by-passing such controversial issues or dismissing the claims of geology in favour of a simple biblical literalism, these evangelicals seemingly offered radical and challenging options suited to the needs of a new generation of students no longer prepared to buy the dry and staid natural theology of the university Moderates.[36]

Glasgow natural philosophy was already in a weakened state with the incapacity since 1839–40 of Professor William Meikleham (appointed 1803). The College reformers prepared to replace him at the earliest opportunity with someone of distinction such as the Episcopalian James David Forbes, Edinburgh professor since 1831, a man with an extensive correspondence network, active in the early years of the British Association, and an experimentalist of European reputation. By means of this 'Forbes model' the reformers were declaring both their primary goal of raising the credibility of the subject from a very low level and their secondary aim of avoiding the appointment of a presbyterian merely concerned to preserve the old Moderate image. Natural philosophy, they saw, had a key role to play in Glasgow. On the one hand, it needed to be reformed to provide, post-Disruption, a non-sectarian foundation to the Arts curriculum and hence to future ministers of the various Kirks. On the other hand, with its connections to industry and engineering, it was seen by the whig reformers as having a leading role to play in the progress of the City of James Watt.[37]

Three years after the Disruption, the reformers found their candidate not in an established man of science but in William Thomson, a 22-year-old recent graduate of Cambridge and son of the mathematics professor. As Chalmers, for many years a close friend of the family, wrote to Thomson's father during his son's first session (1846–7) in the natural philosophy chair:

> I was much interested by the comments which I received of your son the Professor. I have ever had a pre-eminent respect for high mathematical talent; and I am aware of the transcendental and lofty walk on which he has entered . . . may your son be guided by the wisdom from above to the clear and influential discernment of this precious truth – that a sound faith and a sound [natural] philosophy are one.[38]

Expectations of the new professor were high. Responding to those expectations, Professor Thomson's inaugural Latin dissertation combined his expertise in Fourier's theory of heat conduction with the question of the earth's progressive cooling, a theme closely related to the earth's origins and one which was to become prominent in the science of energy (chs 5 and 6).[39]

With Chalmers's death in 1847, the Free Kirk lost for a time the leadership which had given it intellectual as well as spiritual force. Miller, for example, found himself increasingly at odds with the new regime. Factions within the Kirk moved to reaffirm fundamental doctrines based on a literal reading of the scriptures and enshrined in the Westminster Confession and *The Larger Catechism* (1648):

> Q.15. *What is the work of creation?*
> A. The work of creation is that wherein God did in the beginning, by the word of his power, make [out] of nothing the world, and all things therein, for himself, within the space of six days, and all very good.

Various well-publicized 'heresy' trials soon followed over the next couple of decades as some Free Church College scholars offended against the strictures of the Kirk's doctrines.[40]

Scotland, moreover, was not without its radical evolutionary theorists. Within Glasgow College, the radical Nichol, friend of John Stuart Mill (1806–73), had already denounced Chalmers's claim that the laws of nature were powerless to set up the original dispositions or collocations. Nichol instead advocated a new kind of natural law, embodying temporality, which would account for the past development and future progress of the universe, including society and man. Nichol's promotion of a nebular hypothesis in which the solar system had developed progressively from an originally diffused gaseous state provided the anonymous and highly controversial *Vestiges of the Natural History of Creation* with the basis for its notorious law of development, embracing everything from molecules to man. Published the year after the Disruption, numerous editions of *Vestiges* followed. Although not removing deity entirely from the system of nature, *Vestiges* was often read as having made materialistic and naturalistic claims by conflating God to the Law of Development.[41]

The challenge facing the university reformers was inseparable from the construction of a new 'Moderatism' that would simultaneously combat the threats from both biblical literalism on the one extreme and aggressive materialism or naturalism on the other. In short, the goal was to restore and enhance the reputation and credibility of the universities for the Victorian age by reasserting a strong theology and a strong natural philosophy, both in harmony with the scriptures *liberally* interpreted. Glasgow found its champions of the new order in the theology of the Reverend Norman Macleod and in the energy physics of Professor William Thomson and a growing circle of close associates.

Recreating a 'Moderate' culture

Not all the younger generation of clergy followed Chalmers into the Free Kirk. Recognizing the shortcomings of the old Moderates, there were those who wanted to see the regeneration of a united and national Church of Scotland which combined sound scholarly foundations with a strong spirituality in opposition to revivalist enthusiasm and biblical literalism on the one hand and cold doctrinal intellectualism on the other. One such minister was Norman Macleod, later Moderator of the Church of Scotland, editor of the widely read presbyterian magazine *Good Words*, and close friend of the Glasgow College reformers. No one embodied this new presbyterian style better than Macleod, of whom the Broad Church Anglican Dean of Westminster, Arthur Stanley, wrote c.1870:

> He was the chief ecclesiastic of the Scottish Church. No other man during the last thirty years in all spiritual ministrations so nearly filled the place of Chalmers. . . . Macleod represented Scottish Protestantism more than any other single man. Under and around him men would gather who would gather round no one else. When he spoke it was felt to be the voice, the best voice of Scotland.[42]

Born into the Manse in Campbeltown in 1812, Norman Macleod entered Glasgow University around 1827. After four years of Arts, he moved to Edinburgh where he studied divinity under Chalmers. Macleod's biographer noted that 'Chalmers had a peculiar power over him, for professor and student had many similar natural characteristics. The large-heartedness of the teacher, his missionary zeal, and the continual play of human tenderness pervaded by the holy light of divine love, roused the sympathies of the scholar'. During the next six years Macleod studied and travelled widely. His reading included Charles Lyell's *Principles* (1830–3) soon after its publication and, upon returning to Glasgow College to resume his theological studies, he submitted to Professor William Fleming (moral philosophy) an 80-page answer, 'crammed with geology', to a set essay title on 'the Mosaic account of creation'.[43]

Macleod's father moved to a Glasgow parish in 1835, just prior to being elected Moderator of the Church of Scotland. With family roots firmly in Highland (Gaelic) culture, Norman Macleod nevertheless would spend much of his career within the vibrant 'Second City of the Empire'. When in 1851 he became minister of Glasgow's Barony parish church, close to the ancient cathedral and not far from the College, he participated fully in the life of the industrial city. According to his brother, he would sit in his study in the quiet of a winter morning and know 'that six o'clock had struck by hearing, far down below him in the Valley of the Clyde, the *thud* of a great steam hammer, to which a thousand hammers, ringing on a thousand anvils, at once replied, telling that the city had awakened to another day of labour' while on Saturday evenings 'he would wander with delight among the ships and sailors, criticising hulls and rigging, and looking with boyish wonder at the strange cargoes that were being discharged from the foreign traders'.[44]

Although rarely involved in party politics during his subsequent professional career, in the late 1830s Macleod became 'leader of the Peel party' in Glasgow College which identified him with the values of the new reforming conservatism against the old die-hard toryism characteristic of the Moderate faction in the University. As his brother put it, Macleod was 'not a Moderate of the accepted type'. He was rather a 'moderate' in the original sense: 'The extreme views of truly good and spiritual men in the Church, and those of truly bad and material men in the State, will bring on a gale which will capsize her', he wrote in 1840. At the time of the Disruption, he lent his support to the policies of 'The Forty', a group of ministers who sought moderate reform without fracturing the Kirk.[45]

Macleod's Christianity also shaped his response to Chartists. In his first parish, Loudoun in Ayrshire, skilled weavers, their trade 'being gradually extinguished by the great factories', had turned to radical politics and away from traditional religion. Like Chalmers he saw Christianity as the redeeming power. But unlike Chalmers he had little faith in the universal moral power of *laissez faire*. Responsibility lay with the whole nation and not least with the Church whose role was less one of preaching divine justice than of teaching the love of God:

> We demand from them patience while starving – do we meet their demands for bread? We demand from them obedience to law – do we teach them what they are to obey? We demand from them love of man – have we taught them the love of God? What is the nation to do for these men, who made the nation anxious, and the Exchange of the world oscillate. . . . Hast thou [addressing an imaginary 'Special' constable] ever troubled thyself about healing his broken heart as thou hast about giving him a broken head?[46]

Witnessing at first hand a serious chartist riot in Glasgow in 1848, he saw the leading thoroughfares 'swept by a torrent of men and women of a type utterly different from the ordinary poor. Haggard, abandoned, ferocious, they issued from the neglected haunts of misery and crime, drove the police into their headquarters, and, for a while, took possession of the streets'. Such a spectacle reinforced his conviction that the Established Church had so far failed in its responsibilities towards the urban poor.[47]

Macleod thus attempted to manoeuvre the Church back to the middle ground and away from its cultural and political elitism. An ardent supporter of the broad 'Evangelical Alliance' from the 1840s until his disillusionment with what he saw as its increasing exclusiveness in the early 1860s, he was consistent in his opposition to religious elitism and authoritarianism. 'You know my latitudinarian principles in regard to Church government', he wrote to a close friend at the time of the Disruption. 'I value each form in proportion as it gains the end of making man more meet for Heaven'.[48] Some forms, however, were more conducive to those ends than others:

> The tendency of ultra-Calvinism . . . is to fill the mind with dark views of the Divine character; to represent Him as grudging to make men happy; as exacting from Christ stripe for stripe that the sinner deserved. Hence a Calvinistic fanatic has the same scowling, dark, unloving soul as a Franciscan or Dominican fanatic who whips himself daily to please Deity . . . they more easily and readily entertain doctrines which go to prove how many may be damned than how many may be saved; because all this seems to suit their views of God's character[49]

On the other hand, he confessed with respect to Roman Catholicism that 'If I could tomorrow believe on the authority merely of the Church, and that private judgment were not my duty, I would turn Papist'. Here indeed was the essence of presbyterianism. 'Private judgment' had powerful implications for the interpretation of scipture and nature: 'Wherein', he asked, 'lies the difference between assenting to the Principia of Newton, because written by a great mathematician and not because I see them to be true, and my assenting to the Bible, because written by inspired men and not because I see how truly they spoke?'[50]

By the mid-1850s, the Scottish University Tests had been repealed, freeing the University of Glasgow from the constraints which had deterred, for example, the Anglican George Gabriel Stokes (1819–1903), close Cambridge friend of William Thomson, from becoming a candidate for the mathematics chair following James Thomson's death from cholera in 1849.[51] Within the Established Church, however, the 'old guard' of Moderates even proposed to form a new university for Scotland which would retain the ancient links with the Church. Distancing himself from such calls, Macleod addressed the General Assembly of the Church of Scotland in 1854:

> He objected to the national Church throwing herself loose from the national Universities, and sinking down to the position of a mere sect, and handing over the Universities to other parties. He warned them that if there issued from this House [the General Assembly] opinions which obtained no more sympathy in the country, instead of gaining a hold on the affections of the people, they would come to have no more influence on the nation than the weather-cock on the top of the steeple affected people passing in the street. . . . He was one of those, moreover, who believed that the Legislature [the British Parliament] had a perfect right to modify such institutions as the Universities to meet the wants of the age . . . if it had been proposed to place a Jesuit in a Medical Chair, and . . . if his limb were to be operated on, he should prefer a skilful Jesuit to an unskilful Protestant.[52]

Such views accurately capture the anti-sectarian, inclusive rather than exclusive, 'popular' presbyterianism which increasingly characterized the Church of

Scotland in Macleod's remaining years. Correspondingly, the new ethos in Glasgow University itself closely matched this style which mirrored 'the wants of the age' and in particular the wants of industrial society. A forceful instance of the new harmony being cultivated between Church and University from the mid-1850s can be found in the career of Rev. John Caird (1820–98), principal of the University of Glasgow from 1873.

In a sermon entitled 'The Christian's Heritage', one of a series collected for publication in 1858, Caird wrote:

> the history of the Church but too often exhibits the strange anomaly of a religion of love producing the keenest haters, and a gospel of peace engendering strifes and animosities more bitter than the disputes and rivalries of the profane. . . . The Christians at Corinth had quarrelled with each other on the merits of their respective teachers – each party boasting of the pre-eminent wisdom or eloquence of its own head, and contemning the gifts of his supposed rivals . . . In the pursuit of wealth it may be natural, however culpable, to begrudge another his gains, or to be elated at our own; for wealth is a limited good . . . But with respect to spiritual good . . . These belong to that class of blessings which possess the qualities of universality and inexhaustibleness. The light of the sun is not the less bright to me that it beams at the same moment on millions of my fellow-men.[53]

Here Caird enunciated the values of the new version of presbyterianism against the background of strife, fragmentation and exclusiveness within Scottish presbyterianism. It was a decisive message, proclaiming the universality and inexhaustibleness of God's goodness rather than its limitation to the chosen few.

Caird, a son of the Greenock family of engineers and shipbuilders with considerable practical engineering experience himself, had been a class-mate of William Thomson at Glasgow College in the late 1830s. Graduating Master of Arts at Glasgow in 1845, Caird entered the ministry of the Established Church just two years after the Disruption. Appealing to intellectual rather than popular audiences, he returned to Glasgow as professor of theology in 1862, succeeding to the principalship in 1873. His theme was to show 'that Christianity and Christian ideas are not contrary to reason, but rather in deepest accordance with both the intellectual and moral needs of men'. In the same period, his brother Edward (later Jowett's successor as Master of Balliol College, Oxford) became professor of moral philosophy while in 1869 George Macleod (1828–92), a brother of Norman, succeeded Sir Joseph Lister as regius professor of surgery.[54] The replacement of the old guard of 'Tory Moderates', led by Principal Macfarlan and Professor William Fleming, by this new generation of reformers could scarcely have been more striking.

Within this context of 'liberal' presbyterianism, extending as it did from the universities through the theological students to the wider middle classes, particularly in the prosperous industrial West of Scotland, lay the popular magazine *Good Words*, edited by Norman Macleod from 1860. The professed aim of the work was to publish (in the words of Dr Thomas Arnold) 'on common subjects, written with a decidedly Christian [incarnationalist] tone'. Questions of physical and social science as well as poetry and ficion were included. In an earlier experiment with the *Edinburgh Christian Magazine* (circulation about 4,000), Macleod had attempted to emancipate 'cheap religious literature from the narrowness and weakness to which it had come'. The experiment had apparently not been a financial success. But with *Good Words*, the story was different: circulation was 110,000 per month.[55]

On the other hand, Macleod soon encountered the wrath of what he called the 'Evangelical party' (as distinct from the 'Evangelical Alliance' which he had long supported) who objected not only to the blend of the secular and scriptural but also to the theological 'liberalism' which *Good Words* promoted. The notorious *Record* newspaper in particular attacked the contributions of Caird and Macleod. Macleod noted that he 'was threatened in London that unless I gave up [Dean] Stanley and [Charles] Kingsley I should be "crushed". . . . It might make the devil laugh and the angels weep'.[56] Macleod perceived the opposition thus:

> The so-called 'Evangelical party' – for, thank God, they are but a small clique – are becoming the worshippers of mere Shibboleths – phrases . . . though a man believes, as I do, with his whole soul the doctrines of the Scriptures, yet woe to him unless he believes the precise philosophy, or the systematic form of those doctrines held by the clique! . . . They will tell you that you deny the atonement unless you believe that Christ on the cross endured the punishment which was due to each sinner of the elect for whom he died; which, thank God, I don't believe, as I know He died for the whole world. And so as regards every other doctrine . . . you must believe in literal fire and brimstone: a lake of fire, into which infants even may be cast, or you are not 'Evangelical!'[57]

Recreating a Moderate culture, then, required the avoidance on the one hand of 'the exclusively narrow religious ground' and on the other of 'whatever was antagonistic to the truths and spirit of Christianity'. It was nothing less than the remaking of Scottish presbyterianism 'suitable to the wants of the age'. Those wants, as Macleod saw them in the early 1850s, were inseparable from 'the waking up of the great city, the thundering of hammers from the boilers of great Pacific and Atlantic steamers – a music of humanity, of the giant march of civilisation; far grander to hear at morn than even the singing of larks'.[58]

CHAPTER 3

Recovering the Motive Power of Heat

The city of Glasgow has somehow or other got such a very bad name for its weather and its morality, that one would suppose, from the statements made in some quarters, we sat soaking in water all the day, and soaking in whisky all the night, and on Sabbath-day sat sulky and gloomy in the house.

Rev. Norman Macleod reflects on the condition of Glasgow for the benefit of the General Assembly of the Church of Scotland, 1859[1]

During the late 1830s and early 1840s, demand for motive power soared. Railway engineers aimed to construct more powerful, more compact locomotives to supply the seemingly insatiable market for the transport of freight and passengers throughout Britain, continental Europe and the New World. Natural philosophers and 'ingenious gentlemen' alike were captivated by the promise of more economical alternatives to steam in the form of electromagnetic and air-engines. But above all, steamships, Clydeside's special gift to the nation, had begun to cross oceans and to challenge the supremacy of sail for the privilege and profit of carrying the trade of empires.[2] Within these new contexts the brothers James and William Thomson eventually came together with Macquorn Rankine to forge the new sciences of energy and thermodynamics.

Glasgow work

James Thomson and his younger brother William were born into a family which espoused Scottish Enlightenment beliefs of religious toleration, political reform, social progress and commercial prosperity. Their father, James Thomson (1786–1849), had risen from farm labourer in the tranquil Ulster countryside of the eighteenth century, to Glasgow University student, Belfast mathematics teacher, and ultimately to mathematics professor in Glasgow University from 1832.

This spectacular advance in social status had been made possible by the democratic values of Scottish education whereby the sons of crofters and artisans could attain, through a relatively cheap and accessible university system, professional status as medical practitioners, clergymen, lawyers and (less easily) university professors. In Scotland, education was often seen as the route to individual improvement as well as the means by which society itself would progress in wealth and happiness. Correspondingly, Margaret, mother of the Thomson brothers, hailed from a Glasgow commercial family and it was soon after her death in 1830 that the family of Professor Thomson and his surviving six children moved to the town which had become known as 'the Second City of the Empire'.[3]

Glasgow had long been a prosperous commercial centre and site of one of Scotland's five ancient university colleges with an academic tradition extending back to the fifteenth century. By the beginning of the nineteenth century, civil engineers had remoulded the shallow River Clyde into something like a ship canal without locks, such that deep-sea vessels could for the first time reach the heart of the city. Glasgow had been transformed from Scottish market town to international seaport.

Cotton, chemical, coal and iron industries marked the rise of Glasgow and the Clyde Valley as a major industrial region. The population of the city increased from around 84,000 in 1801 to over 270,000 by 1840 as work-hungry Highlanders and Irish immigrants crowded aboard the new passenger steamships to satisfy the ever-increasing demand for labour. From the 1840s it was heavy industry – iron manufacture, railways, steam engine building, and above all iron shipbuilding – which gave Glasgow its reputation as a city second only to London in prestige and second to none in industrial progress.[4]

From the age of fourteen James had displayed a passionate interest in practical engineering. The very earliest extant letter written by William informed their sister that 'We have not begun the steam-engine, for papa was not wanting us to do it'. Although James's earliest inventions never reached the patenting stage, they all embodied his concern to minimize waste of useful work and to maximize economy of operation. Thus observation of the way in which the paddle wheels of Clyde steamers 'wasted' power in lifting water prompted him in 1836 to devise a self-adjusting paddle blade which would shed the water quickly and so minimize losses of useful work. In another invention (1840) he aimed to replace the labour required to pole barges upstream with mechanical legs powered by the downward flow of the current.[5]

For its tenth annual meeting (1840) the British Association chose Glasgow. The local Philosophical Society had been instrumental in persuading the Association's managers to come to Clydeside. James Thomson, who had just graduated Master of Arts with honours in mathematics and natural philosophy from Glasgow College, became secretary of the models and manufactures committee for the forthcoming meeting. An exhibition gave pride of place both to the Newcomen steam-engine model which had inspired Watt during his association with the University many years before and to the marine steam engine of Henry Bell's *Comet* which had come to symbolize the beginning of practical steam navigation and the dawn of the steamship age. The Thomson brothers had been well and truly initiated into the rituals of Britain's most powerful scientific organization.[6]

The 'Address of the General Secretaries' by two of the BAAS's leading savants, the geologist Roderick Murchison (1792–1871) and Major Edward Sabine (1788–1883), united the city's industrial prowess with its debt to 'science' and above all to James Watt:

> raised through the industry and genius of her sons, to a pinnacle of commercial grandeur, well can this city estimate her obligations to science! . . . . For she it is, you all know, who nurtured the man whose genius has changed the tide of human interests, by calling into *active energy* a power which (as wielded by him), in abridging time and space, has doubled the value of human life, and has established for his memory a lasting claim on the gratitude of the civilised world. The names of Watt and Glasgow are united in imperishable records![7]

Born in the old Clydeside seaport and shipbuilding town of Greenock, James Watt had begun life in a maritime culture that combined useful mathematics (especially in relation to navigation and surveying) with an austere Scottish presbyterianism which frowned upon 'wasteful' and idle pursuits. Indeed, the young Watt carried with him a 'Waste Book' in order to note down all expenditure, and sources of waste, which he had incurred and which in future might be reduced. In later years his concern with moral and business economy was everywhere apparent. Considering the wisdom of employing a certain John Southern (1758–1815) as assistant, for example, Watt wrote to his partner Matthew Boulton (1728–1809) in the early 1780s: 'If you have a notion that young Southern would be sufficiently sedate, would come to us for a reasonable sum annually, and would engage for a sufficient time, I should be very glad to engage him . . . provided he gives bond to give up music, otherwise he will do no good, it being the source of idleness'.[8]

In the university session 1763–64 the Glasgow natural philosophy professor had apparently asked Watt to repair a model Newcomen engine. This 'atmospheric' engine required the condensation of steam *within* the cylinder in order to create a partial vacuum such that atmospheric pressure would act to push the piston into the cylinder. Watt then began a series of experimental investigations guided by questions of economy and power which led to his introduction of a *separate* condenser. In Watt's arrangement, greater economy was possible because heat was not 'wasted' in reheating the cylinder of the engine for each stroke.

Watt later denied that his invention derived directly from the experimental researches of the Glasgow University chemist Joseph Black (1728–99) on 'latent heats' whereby the transformation of ice into water at the freezing point, or water into steam at the boiling point, required a large quantity of heat which did not manifest itself in a rise in temperature. But Watt was so closely connected to the tightly knit scientific community at Glasgow College that historical debates over his debt to Black's 'physics' are somewhat misplaced. Far from being an isolated mechanical genius, Watt's interests in economy, steam power and heat were also the interests of academic peers such as Black and Robison. Because of Watt's proximity to this community, Black could remind him in 1780: 'You haveing [sic] been early acquainted with and directed by it [the doctrine of latent heat] in your pursuit of improvements upon the Steam Engine and incited to make Experiments with regard to it'. Black furthermore provided Watt with funding for his early experiments and remained a life-long correspondent.[9]

Formed in 1775, the Birmingham partnership of Boulton and Watt coincided with the extension of Watt's original patent (1769) for a further 25 years. During this period, Watt introduced engines capable of providing rotative motion for driving mill machinery rather than the simple reciprocating action employed in pumping engines and he constructed double-acting engines in which the steam entered the cylinder on alternative sides of the piston. The testing and measurement of engine performance also featured prominently in the Watt regime. Rotative engines were rated in 'horses', where each horse was assumed to be capable of raising 33,000 pounds one foot high per minute. By the 1790s, John Southern, now manager of Boulton and Watt's Birmingham works, modified a pressure-indicator device in such a way that a pencil, moving with the piston, traced a diagram of pressure against displacement. This 'indicator diagram', the area of which represented the power developed by the engine, allowed for adjustment of the engine in order to maximize useful work.[10] The indicator diagram, more widely known from the 1820s, would subsequently become, in Rankine's terminology, the 'diagram of energy' (ch.8).

As Murchison and Sabine's address suggested, however, Watt's migration to Birmingham was no obstacle to the making of Watt into a Glasgow legend. The iconography of Watt made engineering and invention into a noble pursuit for aspiring Scottish sons, combining as it did the virtues of hard work with the values of economic progress in a presbyterian culture which frowned upon sin and idleness. James Thomson's encounters at the 1840 BAAS meeting, moreover, would reinforce his conviction that Glasgow engineers, rather than Cambridge mathematicians, stood as the real heroes of scientific and social progress.[11]

According to James's recollection of the meeting, he attempted to outline to Philip Kelland (1808–79), Cambridge-educated professor of mathematics at Edinburgh, some of his ideas on 'the possibility of having a tidal mill which could perpetually grind corn or overcome friction, and that this must involve an expenditure of the power stored in the motions of the earth or moon or of one of them'. Kelland, however, 'insisted that I [James] was wrong and that the water could effect no deduction from the power or work stored in the motions of the earth and moon jointly'.[12]

In contrast to this brusque encounter with one of the BAAS's 'lions', 'Mechanical Science' (section G) provided James with far greater inspiration. Heading the list of papers were two by the naval architect and shipbuilder John Scott Russell (1808–82). The emphasis in both papers was on quantitative estimates to enable comparisons of marine engine performance with a view to minimize 'a loss of fuel and power' and to maximize economy. As an example, Russell estimated the coal consumption of two fictitious trans-Atlantic steamers (Samuel Cunard's first trans-Atlantic steamships, built and engined on the Clyde, actually entered service that year). The first, of lesser power, consumed less coal per day but took longer in all weathers. The second ship, of greater power, demonstrated its economic advantage in foul weather by completing the voyage in 25 per cent less time than the first ship.[13] These issues, of the optimum proportion of horse-power to hull tonnage and of the economy of marine steam engines for long deep-sea voyages, would form the principal context for the Thomson brothers' interest in the Carnot–Clapeyron theory from 1844 when James was serving his engineering apprenticeship at the Thames shipbuilding works of William Fairbairn.

Section G also included a paper by the Glasgow professor of civil engineering and mechanics, Lewis Gordon (1815–76), who had been appointed as recently as August 1840 to the first such chair in a British university. As Ben Marsden has shown, Gordon brought to Glasgow first-hand knowledge of several continental engineering schools. His BAAS paper, 'On the Turbine Water-wheel', presented an account of the *Turbine-Fourneyron* based on the principle 'by which the maximum of useful effect is obtained from a given fall of water'. To achieve this aim, the arrangement had to be such that 'the water enters the wheel without shock, and quits it again without velocity'. Although Gordon did not cite any source, this criterion for maximum economy derived from the hydraulic investigations of the French engineer Lazare Carnot (1753–1823). Gordon instead considered the practical means by which this criterion could be met:

> A notion of its construction may readily be formed, by supposing an ordinary water-wheel laid on its side, the water being made to enter from the interior of the wheel by the inner circumference of the crown, flowing along the buckets, and escaping at the outer circumference. Then centrifugal force becomes a substitute for the force of gravity.[14]

A characteristic feature of the turbines of Fourneyron was the curvature of the buckets 'on to which . . . the water entered at every point of the inner circumference, and flowing along the buckets, escaped at every point of the outer circumference', producing maximum effect. In practice, turbines of about 60 horse-power were found to yield efficiencies of 70–80 per cent of the theoretical effect. Gordon referred especially to turbines at St Blasier in the Black Forest where a large fall of 345 feet 'serves a [textile] factory in which are 8000 water spindles, 34 fine and 36 coarse carding-engines, 2 cleansers, and other accessories'.

Gordon's continental encounters during the 1830s revealed a whole brave new world of engineering science in France. It was a world that the Thomson brothers met with increasing enthusiasm. Especially after the 1789 Revolution, the French state had created a strong demand for engineering skills which would be deployed to serve a variety of military, civil, bureaucratic and educational functions. Engineering mechanics, embodied in new institutions and new treatises, flourished in this post-Revolutionary era. In particular, Lazare Carnot, one of the leading figures behind the new Ecole Polytechnique instituted for the training of military engineers in the Napoleonic regime, maintained close links with engineering and emphasized the mechanics of machines and geometrical analysis.[15]

Carnot's *Essai sur les machines en général* (1783) and his *Principes fondamentaux de*

l'équilibre et du mouvement (1803) offered a general theory of machines. In the latter
treatise, *moment d'activité* (or 'latent *vis viva*') served as a force times distance term.
Carnot then claimed that 'the moment of activity which you would consume will
be half of the forces [*vires vivae*] that you would arouse' in the case of machines at
rest. However, the rather abstract nature of Carnot's treatises probably did little to
bridge the gap between theory and practice or between rational mechanics and
practical engineering. As one of Carnot's contemporaries noted: 'These principles
[of equilibrium] are indispensable for the knowledge and theory of machines; but
yet they are far from furnishing us with practical notions on *working machines*.
They [machines] are *never* in equilibrium. The machine must *move* in order to
work'.[16]

Carnot's reference to 'half' *vis viva* suggested that this venerable term was losing
some of its conceptual primacy. Introduced by G.W. Leibniz (1646–1716) as an
integral part of his complex metaphysical system of a perfect creation, *vis viva*
(measured as mass times velocity squared) was conserved throughout the universe
and not merely in idealized mechanical systems where no collision occurred. At
the same time, however, Leibniz had made no claim for conversion of *vis viva* into
heat. Founded upon a God who acted in accordance with 'rational' principles such
as 'sufficient reason' and 'plenitude', Leibniz's system possessed little appeal for
the anti-metaphysical, empirically minded mathematicians of the Enlightenment.
After the passing of Leibniz and Newton, disputes over the proper measure of
force (momentum or *vis viva*) smouldered through the eighteenth century. Increasingly, however, conservation of *vis viva* came to be regarded not as a fundamental
law but as a useful, derivative principle applicable to mechanical systems.[17] There
was thus no simple line of descent from Leibniz's principle of conservation of *vis
viva* to nineteenth century energy conservation.

A later generation of French academic engineers transformed engineering
mechanics into a new science of work. In the period 1819–39 G.G. de Coriolis
(1792–1834) and J.V. Poncelet (1788–1867), both former pupils at the *Ecole Polytechnique*, introduced work terms as *quantité de travail* and *travail mécanique*
respectively. *Vis viva* was redefined as $1/2mv^2$ such that 'work' took priority over
the old mv^2. The equation between work done and half *vis viva* was explicitly
formulated. Unlike Coriolis, Poncelet was based not in Paris but in Metz, located
in the western industrial province of Lorraine. From 1824 he taught applied
mechanics at the *Ecole du Génie* (one of the so-called 'applied schools') in Metz.
His student textbook, *Cours de mécanique industrielle*, first appeared in 1829. Industrial mechanics, he there explained, 'consists principally of studying the diverse
transformations ... which the work of engines can undergo by the means of
machines or tools, [and aims] to compare between them the quantities of work, to
evaluate them in money or in production of such and such a kind, etc'. Application
of the principle of transmission of work to industrial mechanics constituted, in
Poncelet's explicit phrase, the 'science of the work of forces'.[18]

Poncelet did not confine his activities at Metz to teaching. He built up an
impressive group of engineers, all educated at the *Ecole Polytechnique*, concerned
with engine performance. Well known for his own various state-of-the-art water-

wheel designs, Poncelet's group aimed to develop more efficient waterwheels and water turbines.[19] Engine performance was no longer measured as in the eighteenth century by means of a weight raised to a height, but by an instrument known as a dynamometer. The principle involved application of a brake to a revolving shaft which enabled the force exerted by the shaft to be measured. The essential feature of waterwheels and water turbines, as well as windmills, is that they yield direct rotary motion, whereas steam engines produce linear motion which may be converted into rotary motion via a crank. Thus it is easy to consider the weight of water lifted directly by a steam engine to a given height in a given time, but much less easy to measure the direct rotary power supplied by a turbine. In the context of French hydraulic engineering, members of the Metz group devised some of the most widely acclaimed dynamometers of the period.[20] In his BAAS paper, Gordon thus drew special attention to the quantitative investigations of the French engineer A.J. Morin (a member of Poncelet's Metz group concerned with water power) using his 'brake dynamometer, or friction strap'.[21]

Gordon's presentation stoked James Thomson's growing zeal for engineering. James had been on a walking tour of the Black Forest with his brother a few months earlier, although there is no record of them having seen the St Blasier fall. Shortly after his return from Germany, James began work under John Macneil (1793–1880) who was directing the construction of a major railway in Ireland (north from Dublin) on behalf of the government.[22] Forced to abandon this first experience of practical engineering on account of a knee injury sustained during the Black Forest tour, James instead identified himself with Gordon's interests by attending the Glasgow University engineering class in 1841–2 and composing an 'Essay on Overshot Water Wheels' in May 1842 at the end of the session.

While not naming Lazare Carnot, James there explained that 'The water should be projected on the wheel with a velocity very slightly greater than that of the wheel itself'. He furthermore argued that 'in overshot wheels, the vis viva of the water entering the wheel may be regarded as very nearly a dead loss [altogether lost], part of it passing away in the velocity of the water leaving the wheel and most of the remainder being destroyed by the shock at entering'. Well designed overshot wheels, he believed, could yield very high useful effects of some 84 per cent. These reflections formed the prelude to a long period of designing water turbines, the first patent for his 'vortex turbine' being taken out in 1850.[23]

Again following in his mentor's footsteps, James soon began to involve himself in the proceedings of the Glasgow Philosophical Society. Established in 1802 'for the improvement of science', the Glasgow Philosophical Society was simultaneously practical and gentlemanly. Membership of the Society consisted of three classes: resident, honorary and corresponding. Thus 'Gentlemen dwelling in or near Glasgow, who have rendered the Society some essential service, made improvements or discoveries in the [manufacturing] arts or sciences; or by whose admission the Society is likely to derive considerable benefit, may be admitted as resident members'. Likewise, 'No Gentleman can be admitted as an Honorary Member, unless he has, in an eminent degree, promoted the advancement of the arts and sciences, or of this Society, and, in other respects, may be considered as

worthy'. Finally, corresponding members 'Must be respectable Scientific Gentlemen, residing at a distance, either in this or any other country; from whose communications, early and beneficial intelligence on philosophical subjects may from
time to time be expected'. Only resident members paid a three-guinea fee as entry
money plus a half-guinea annual membership fee. These fees, prohibitive for the
lower social orders, nonetheless approximated to the annual fees (low by Oxford
and Cambridge standards) paid by Glasgow students to attend university lecture
courses.[24]

Although the Society had gained a membership list of about 130 at the end of its
first decade, its future was in real doubt by the early 1830s. In 1833 it only
managed to stage one meeting instead of the weekly meetings originally envisaged.
Out of less than 50 registered members in 1832, half were directly engaged in
manufactures of some kind, reflecting the interests of Glasgow, but academics and
other 'professional men' were poorly represented. As we have seen, many of the
Glasgow College professors were Moderates of the old school, hostile to the
reforming values of the new Glasgow (ch.2). The fortunes of the Society were only
revived following the appointment of Thomas Thomson, reforming regius professor of chemistry in the University, as president for life in late 1834.[25]

Other professors also joined: the chemist Thomas Graham (1834), the astronomer J.P. Nichol (1836), the mathematician James Thomson (1839), Lewis
Gordon (1840) and the humanist William Ramsay (1841). Suddenly the same
professors active in the reform of the ancient University of Glasgow became leading members of the rapidly reviving Philosophical Society. By the early 1840s, the
Society began publishing a *Proceedings* at a time when the market for all kinds of
literature was rapidly expanding.[26]

Elected in 1841, the young James Thomson thus entered a particularly dynamic
local scientific society of academics, medical practitioners, manufacturers and
other 'respectable' gentlemen of Glasgow whose shared interest was 'the
improvement of science' coupled to the 'improvement of society'. By the early
1870s, indeed, these commitments had become formalized in the objects of the
Society: 'to aid the advancement of the mathematical, physical, and natural Sciences, with their applications, and to promote the diffusion of scientific
knowledge'.[27]

A study of the 1841–4 *Proceedings* has shown that 'improvement' consisted of
complementary *economic* and *moral* components. Improvement or progress
entailed the maximization of 'useful work' or labour and the minimization of
'waste'. On the land, 'improvement' demanded the use of fertilizers to render
otherwise waste lands productive of food. Among human populations,
unproductive groups, such as the blind, could be 'improved' by means of special
alphabets of raised letters, especially with a view to their spiritual 'improvement'
through a reading of the Bible. The huge labouring populations could be morally
improved through the 'rational amusement' of models and manufactures exhi-
bitions rather than wasteful amusements of 'injurious description' that might
encourage social disorder through drunkenness, debt and crime.[28]

Social and environmental conditions could also be improved to reduce the incidence of diseases wasteful of human labour. Above all, the questions of economy in prime movers – the steam engines and water wheels which drove the mills and created the wealth of Clydeside – had to be addressed. The all important theme was that of maximizing the useful work obtainable for a given 'fall' of water or for a given weight of coal and minimizing the waste from whatever cause.[29]

James himself reported to his younger brother, then a Cambridge undergraduate far removed from the world of iron and coal, of just such a paper read to the Society at one of its now fortnightly meetings. The paper 'was on a curious method of raising water to the tops of mills from the cisterns on the tops of engine houses. . . . There is evidently a great loss of power . . . Mr Gordon remarked that a similar apparatus had been used for raising water for irrigation'.[30] Concern with 'losses of power' in engines and machinery, as well as with problems of waste in general, was the hall-mark of James's activity throughout the 1840s.

Recovering the Carnot–Clapeyron theory

James Thomson left Glasgow in 1842 to spend time in the West Midlands town of Walsall with J.R. McClean who had been a class-mate in Professor Thomson's mathematics class and who by 1844 was resident engineer in charge of the construction of a railway near Fleetwood in Lancashire. James explained to William that the 'works which Mr McClean has the charge of are some canals – I go out often, between breakfast and dinner to one of the locks which was begun about the time I came here, and I take measurements of the different parts as they are built and draw them when I come home'.[31]

According to James's later account, the brothers had a special interest in the question of 'loss of power' during this period:

> Even you and I at Walsall (in 1842 I think) when watching the consumption of power in the flow of water into a canal lock were speculating on what became of the power as we could not suppose the water *worn* and therefore altered like as solids might be supposed to be when power is consumed in their friction.[32]

The immense surge of water involved in filling a lock (with or without lifting a barge from one contour to another) could in principle have been utilized to turn a waterwheel and thus produce useful work. But in practice it seemed to the brothers that the power was being wasted in eddies and spray. As yet they had no means of accounting for this apparent loss of power.

The reference to the wearing of solids from friction suggests they may have had in mind William Whewell's Cambridge text, *The Mechanics of Engineering* (1841), in which Whewell had written with respect to solids: 'work consists in shaping or moving certain portions of matter. Thus, we have to grind bodies, to polish them, to divide them into parts. . . . In these cases we have to overcome the cohesion of matter, inertia, elasticity, weight'.[33] The Thomsons, however, could not see how such alterations could be used to account for *fluid* friction.

By early 1843, James entered the world of heavy engineering. Lodging at the Dickensian-sounding Mrs Grim's in Church Lane, he joined the Horseley Iron Works at Tipton, Staffordshire, a firm which claimed to have built in 1821 the first iron steamer to put to sea. Very soon he reported to his brother on what he saw as the firm's rough and wasteful methods of estimating the costs of an important bridge contract, worth twenty or thirty thousand pounds. The firm apparently lost the contract, ran short of orders, and found itself in the hands of its creditors.[34] Already imbued with the imperative to maximize useful work, James saw the fall of the Horseley Iron Works as a powerful lesson in the evils of inaccuracy and waste in practical engineering.

Towards the late summer of 1843, James's father paid £100 under the 'premium apprenticeship' system (by which young men paid for the privilege of being apprenticed to a prestigious firm) for his son to serve his time within the shipbuild-ing subsidiary, at Millwall on the Thames, of William Fairbairn's Manchester machinery-, locomotive-, and boiler-making enterprise. An expatriate Scot, Fairbairn was well known in Manchester Literary and Philosophical Society circles as well as the BAAS. His investment of upwards of £50,000 in the Millwall iron shipbuilding and marine engineering 'factory' from the mid-1830s, however, turned out to be a costly venture. With a labour force of around 2,000, modern plant, and plentiful orders in its early years, the enterprise might have proved highly profitable at a time when British, especially Clyde, shipbuilders were show-ing the practicability of iron warships and ocean-going passenger steamers. But the capital-intensive business was difficult to sustain against local competition. After only a few months in Millwall, James told his brother that the Works were to be sold, that Fairbairn was to hand over the Manchester business to his sons, and that James himself would probably transfer there. In fact, several months were to lapse before James left Millwall.[35] During the interval, the Thomson brothers initiated a dialogue on the Carnot–Clapeyron theory.

To begin with, James attempted to put into practice some of the lessons learned from Lewis Gordon's classes, especially with regard to the measurement of mechanical effect. Thus he informed William: 'One day last week I showed Mr Fairbairn [junior] a drawing of my modifications of Morin's dynamometer, and proposed that, if they retain the boiler yard here and get it driven by the present engine which will be let to other people, this dynamometer might be useful for measuring the exact quantity of mechanical effect they used'. Here James was suggesting a means by which the excess work produced by the yard's steam engine could be prevented not merely from going to waste, but could actually be sold in measurable quantity to neighbours in need of motive power. He was also adopting Gordon's vocabulary through the term 'mechanical effect', which the engineering professor had taken from the German '*mechanische Wirkung*'.[36]

James then invited his brother to help solve a question 'which will require a good deal of applied mathematics and one which is of great importance at present'. The problem concerned the design of a vibrating or oscillating engine in which the whole cylinder oscillated about a fixed axis in such a way that the piston rod drove the crank directly rather than via connecting rods (Figure 3.1). Whatever the prac-

Figure 3.1 James Thomson's hand-drawn sketch of a 'vibrating' or 'oscillating engine' in which C is the cylinder, P the bearing for the cylinder, B the valve box, and AA the steam passages. The steam enters by the hole shown black in one gudgeon and goes out by a similar hole in the other (James to William Thomson, 19 June 1844, T401, Kelvin Collection, ULC).

tical disadvantages of the vibrating engine (such as an increased risk of the piston rod breaking if a heavy sea brought the paddle wheels to a dead stop), the brothers appear to have been greatly interested in the possibilities offered by the design for minimizing friction and hence one cause of 'waste' in marine steam engines.[37]

Now thoroughly involved in the business of marine steam engines for long-distance ocean navigation, James's immediate concerns were taken up with copying drawings for Robert Murray, one of Fairbairn's draughtsmen, directed towards a prize competition offered by the Peninsular Steam Navigation Company (later P&O) 'for the best plans for a steam boat of 1200 tons'. With the same breath, he asked his brother:

I would like if you would tell me who it [is] that has proved that there is a definite quantity of mechanical effect given out during the passage of heat from one body to another. I am writing an article which I think I shall give in to the Artizan about its proposal as to the possibility of working steam-engines (theoretically) without fuel by using over again the heat which is thrown out in the hot water from the condenser, and I shall have to enter on the subject of the paper you mentioned to me.[38]

Evidently William had mentioned to James the translation of Emile Clapeyron's 'Memoir on the Motive Power of Heat' in Taylor's *Scientific Memoirs*.

James explained the motive power of heat thus: 'That, *during the passage of heat from a given state of intensity to a given state of diffusion a certain quantity of mec[hanical] eff[ect] is given out whatever gaseous substances are acted on, and that no more can be given out when it acts on solids or liquids*'. This claim, James continued, was all that he could prove because he did not know whether in solids or liquids the fall of a certain quantity of heat would produce a certain quantity of mechanical effect, and that the same mechanical effect '*will give back as much heat*'. That is, 'I don't know that the heat and mec eff are interchangeable in solids and liquids, though we know that they are so in gases'.[39] In contrast to Joule (ch.4), we must note that James was not claiming here an *interconversion* of heat and work, but rather a relation between the fall in intensity of heat and the mechanical effect produced in an ideal engine.

Implicitly recalling lessons imbided from Gordon, he called William's attention to a powerful analogy for heat engines: 'The whole subject you will see bears a remarkable resemblance to the action of a fall of water'. Whether we let water 'fall from one level to another' or whether we 'let heat fall from one degree of intensity to another', in both cases a definite quantity of mechanical effect 'is given out but we may get more or less according to the nature of the machines we use to receive it'. A water mill, for example, 'wastes part by letting the water spill from the buckets before it has arrived at the lowest level' while a steam engine 'wastes part by throwing out the water before it has come to be of the same temperature as the sea'. Echoing his 1842 essay on overshot waterwheels for Gordon, he also emphasised that in a waterwheel 'much depends on our not allowing the water to fall through the air before it commences acting on the wheel', while in a steam engine 'the greatest loss of all is that we do allow the heat to fall perhaps from 1000° to 220°, or so, before it commences doing any work'. Practical barriers to the utilization of mechanical effect from such high levels existed: 'we have not the materials by means of which we are able to catch the heat at a high level' and in any case 'if we did generate the steam at 1000° a great part of the heat would pass unused up the chimney'. Likewise, a waterwheel near the source of a stream would 'waste all the tributary streams which run in at a lower level'.[40]

By late October 1844, James had moved to the fitting shop of Fairbairn's Manchester business. He reported to his brother just before Christmas that there were a 'great many engines' under construction at the works, including a four horse-power one 'for Lord Rosse for grinding his speculums'. Thus even the mighty telescopes of Lord Rosse, designed to resolve the celestial nebulae, were dependent on the pervasive steam-powered industrial culture which had become so integral a part of the Thomsons' scientific lives in these years. James also reported, however, that 'The engine which drives the works has got a disease in her boilers, on account of which the works will have to be stopped for ten or twelve days'. As a result James returned to Glasgow where it was discovered that he was suffering himself from 'a quickness of the pulse' such that he had to abandon his apprenticeship at Fairbairn's. Indeed his Calvinist doctor apparently recommended that he should henceforth cease to think of this life and 'prepare himself for the other world'.[41]

Though incapacitated physically, James quickly discovered ways of averting the evils of time wasted. While in Glasgow, he took up a subject of vital importance to the economic viability of ocean-going steamships, that of the distillation of fresh from salt water. Unlike land engines or railway locomotives, steamships had no readily available supply of fresh water to replenish their boilers which salt water would rapidly render useless. If the valuable cargo space were not to be entirely taken up with coal and water, some efficient means of distilling fresh from salt water had to be found. His old friend at Millwall, Robert Murray, wrote to him about the matter in October 1845. James's response shows just how deep were his concerns with economy, and how close were the links between human labour and mechanical effect.

Both methods which he discussed involved 'the same principle of making the heat which is given out in the condensation of the steam serve to generate new steam'. The first method separated the fresh from the salt water 'by the expenditure of a certain quantity of heat'. In the other, preferred, method, human labour would be employed to pump steam from the salt water boiler to a vessel placed within it. The effect would be to raise the pressure and hence the temperature of the steam. As heat conducted from the inner vessel to the boiler, condensation of steam in the former would produce fresh water, while the heat transferred would serve to generate more steam in the latter. The crucial difference in the two methods rested on the former's dependence upon coal as the source of heat (the commodity most precious on a long ocean voyage) and the latter's dependence upon human labour.

Of critical advantage, 'The labour of the sailors I suppose is not of very much consequence, as they have but little to do except in stormy weather, and when going into and out of ports'. James, the good economist, had perceived that the labour that sailors had once expended aboard sailing ships in making sail, trimming the yards, reefing, and so on, to take full advantage of the motive power of the wind, need not be wasted aboard steamships, but could be usefully exploited in ways which would conserve coal bunkers at no extra cost to the owners! He even attempted to make calculations as to the relative quantities of fresh water obtainable by the two methods, but the vagueness of the data prevented him from reaching a definitive conclusion.[42]

As a result of continuing poor health, James's engineering career was now redirected towards invention – notably towards development of a new kind of waterwheel which became known as the vortex turbine. Once again this line of investigation connected with the earlier interests of Lewis Gordon. By April 1847 James told McClean that he had been testing the performance of a large model of his wheel constructed at the works of Walter Crum, Vice-president of the Glasgow Philosophical Society and owner of a large calico-printing and cotton works in the Lanarkshire village of Thornliebank to the south of Glasgow. Measurement had been made by a friction dynamometer and the model, working to a tenth of a horse-power, yielded about 70 per cent of 'the total work due to the water expended' compared to about 75 per cent for the best overshot waterwheels.[43]

James gave his brother the task of exploring the possibility of a French patent during William's visit to Paris in the summer of 1847. William took the opportunity of discussing in detail the latest French innovations with the ageing Poncelet. It became evident that James's design of horizontal waterwheel differed from Fourneyron's in that the water moved inwards from the circumference rather than outwards from the axis, and hence became known as the vortex turbine on account of its similarity to a natural vortex.[44] The engineer thus aspired to follow nature's economy as closely as possible in the quest for maximum work and minimum waste.

Meanwhile, the brothers' parallel interest in the Carnot–Clapeyron theory had been continuing. Very soon after completion of the Cambridge Senate House Examination in mathematics, William had travelled to Paris early in 1845 with a view to improving his knowledge of experimental physics, and thus his prospects of succeeding the ailing Glasgow professor of natural philosophy in a chair which required its incumbent to present relatively popular lectures to students with little preparation in mathematics.[45]

It was while in Paris that he became acquainted with the physical laboratory of Victor Regnault (1810–78) where large-scale experiments were being carried out on the properties of gases and vapours, including steam. Greatly impressed by Regnault's experimental method, that of the very accurate measurement of physical properties, William would later institute his own physical laboratory in Glasgow employing a very similar style. In such an environment, William apparently read Clapeyron's paper in its original French version, and, as he later recalled, searched in vain for Carnot's text:

> I went to every book-shop I could think of, asking for the *Puissance Motrice du Feu*, by Carnot. 'Caino? Je ne connais pas cet auteur'. . . . . 'Ah! Ca-rrr-not! Oui, voici son ouvrage', producing a volume on some social question by Hippolyte Carnot [Sadi's brother]; but the *Puissance motrice du feu* was quite unknown.[46]

James meanwhile located the original Clapeyron paper for himself, remarking to his brother that the 'preliminary part, of w^h you told me the substance at Knock [near Largs on the Firth of Clyde], is I think a very beautiful piece of reasoning and of course is perfectly satisfactory'.[47]

Clapeyron, a mining engineer, had published his analytical version of Carnot's theory in the *Journal de l'Ecole Polytechnique* in 1834. He introduced Carnot's work with enthusiasm: 'The idea taken for the basis of his [Carnot's] researches appears to me fertile and incontestible: his demonstrations are founded on the absurdity which arises from admitting the possibility of producing absolutely either the motive force or the heat'. But he also warned that Carnot, 'dispensing with mathematical analysis', had arrived at his results 'by a series of delicate reasonings difficult to apprehend'.[48] He therefore set about remedying these perceived deficiencies in the original formulation.

Linking work ('quantity of action' or 'effect') to *vis viva*, Clapeyron argued that 'there is a loss of *vis viva*, of mechanical force, or of quantity of action, whenever immediate contact takes place between two bodies of different temperatures and heat passes from one into the other without traversing an intermediate body'. This insight, analogous to that of Lazare Carnot for waterwheels, was drawn from Sadi Carnot's seemingly unobtainable memoir. As a result, Clapeyron continued, 'the *maximum* of the effect produced cannot be obtained but by means of a machine in which only bodies of equal temperature are brought into contact'.[49]

Clapeyron reformulated Carnot's original cycle of operations for the production of motive power from the fall of caloric in terms of a potent graphical representation. Although closely resembling Watt's 'indicator diagram' of pressure and volume, he made no reference to his sources.[50] The representation (Figure 3.2) consisted of four stages. In the first stage a gas enclosed in 'an extensible vessel' (that is, a cylinder with piston) at temperature T is brought into contact with a heat source A at the same temperature. The gas expands along the path CE, the pressure falling according to the law of Mariotte (Boyle's law) but the temperature being kept constant by the supply of heat from A. During this expansion, the quantity of mechanical action, $\int P.dV$, will be the area BCED under the curve CE. In the second stage the source of heat A is removed and the expansion of the gas continues without loss of heat. This time the temperature falls and again the pressure continues to fall but at a faster rate along EF 'according to an unknown law'. When the temperature has fallen to t, also that of a second source of heat B, the quantity of mechanical action will be the area DEFG.

In the third stage, B is brought into contact with the gas which is then compressed. The gas remains at temperature t but its pressure rises according to the law of Mariotte along FK. Clapeyron supposed that the compression continued 'until the heat disengaged [by compression] and absorbed by the body B is precisely equal to the heat communicated to the gas by the source A during its dilatation in contact with it in the first part of the process'. The mechanical action required for this stage will be the area GFKH.

Finally, we remove B and compress the gas without gain of heat to the original volume AB and pressure BC. During this fourth stage the temperature will return to T. At the end of the 'circle of operations' the gas, with respect to temperature, pressure and volume, will be 'precisely in the same state in which it was originally'. There will have been a net gain of mechanical action represented by the area CEFK while 'the entire quantity of heat furnished by the body A to the gas during its dilatation by contact with it passes into the body B during the condensation of the gas, which takes place by contact with it'. Significantly, then, 'we have mechanical force [work] developed by the passage of caloric from a hot to a cold body, and

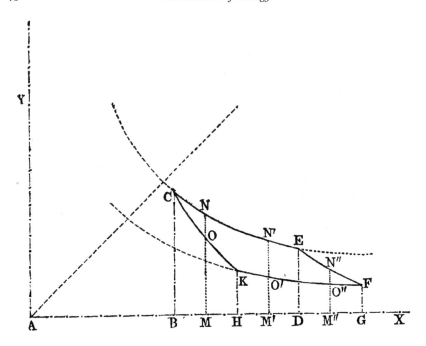

Figure 3.2 Clapeyron's own graphical representation of Carnot's cycle of operations. In the diagram, the 'X'-axis represents volume and the 'Y'-axis pressure
(Clapeyron 1837: 350).

this transfer is effected without the contact of bodies of different temperatures'.
Thus the ideal condition for maximum effect had been realized.[51]

Clapeyron was aiming to translate Carnot's results into analytical form and to
'deduce from them the expression of the *maximum* quantity of action produced by
the passage of a given quantity of heat from a body maintained at a determinate
temperature, to another body maintained at a lower temperature'. By considering
an infinitesimal temperature difference, he deduced expressions for the quantity of
action developed and for the quantity of heat needed to produce this 'effect'. The
ratio of the two expressions yielded 'the expression of the *maximum* effect which
can be developed by the passage of a quantity of heat equal to unity, from a body
maintained at the temperature t to a body maintained at the temperature $t - dt$'.
From Carnot's results this ratio was independent of the agent employed (the
working substance of the engine), but could not be assumed independent of
temperature: the expression for the heat required 'ought to be equal to an
unknown function of t, which [function] is the same for all the gases'.[52] He further
deduced that the ratio of effect produced to heat used was equal to dt/C:

> The function C . . . is, as we see, of great importance; it is independent of
> the nature of the gases, and is a function of the temperature alone; it is

essentially positive, and serves as a measure of the *maximum* quantity of action developed.[53]

In later sections of the paper he attempted to determine values of the function C for certain values of temperature. These results tended to confirm his (and Carnot's) view that C increased with temperature, that is, that less motive power would be produced at higher temperatures for a given quantity of heat.[54]

Clapeyron concluded with a review of the consequences of Carnot's theory for actual heat engines. First, he discussed the advantage of expansive working without which 'the sensible heat of the vapour . . . is entirely lost'. Thus 'To render useful all the motive force at our disposal, the detent [expansion] should be continued until the temperature of the vapour be reduced to that of the condenser', though 'practical considerations . . . [requiring a cylinder of enormous length] prevent the attainment of this limit'. And second, he emphasized the fact that the temperature of the furnace was 1,000°C to 2,000°C higher than that of steam boilers resulting in 'an enormous loss of *vis viva* in the passage of heat from the furnace into the boiler'. He therefore ended with a seductive appeal to innovative engineers: 'It is therefore only from the employment of caloric at high temperatures, and from the discovery of agents proper to realize its motive force, that important improvements may be expected in the art of utilizing the mechanical power of heat'.[55]

The English translation in Taylor's *Scientific memoirs* appeared alongside C.F. Gauss's mathematical studies of terrestrial magnetism. The intended audience was that of mathematical natural philosophy. Since mathematical natural philosophy was largely restricted to a Cambridge-trained elite more interested in astronomy than engineering, and since practical engineers usually lacked mathematical expertise, only natural philosophers and engineers interested in the principles, or the principles and practice, of engineering would take an interest.[56] Fitting such a comparatively unusual characterization, James and his brother, together with Lewis Gordon, were anxious to see how Clapeyron's theory might be embodied in scientific and engineering practice. Within a year of recovering the original Clapeyron, though still not in possession of Sadi Carnot's treatise, they had begun that practical embodiment in a physical artefact known as the Stirling air-engine.

Carnot embodied: the case of Stirling's air-engine

Following his election to the Glasgow chair of natural philosophy, William Thomson joined his elder brother as a member of the Glasgow Philosophical Society in December 1846. At some point in the recent past, Professor Gordon had offered his views on the Carnot–Clapeyron theory to the Society. It is probable that discussion focused on the possibility that a so-called 'air engine' could be designed which would produce more mechanical effect from a given quantity of fuel than a steam engine because of the greater 'fall' of heat from high to low temperature.[57]

One such engine, artefact of a presbyterian minister from the town of Galston in Ayrshire, had already become known in Scotland. Rev. Robert Stirling (1790–1878) took out his first patent in 1816 for an engine which used heated air rather than steam. The engine incorporated an 'economizer' (or 'regenerator') designed to recycle the caloric (heat) that would otherwise be wasted. Joint patents with his civil engineer brother James for 'improved' hot-air engines followed in 1827 and 1840. One actual air-engine of 45 horse-power was employed in James Stirling's Dundee Foundry.[58]

Early in 1847, Professor William Thomson discovered a model Stirling air-engine in the College (Figure 3.3). He explained the find to his Edinburgh colleague, J.D. Forbes: 'I have found a Stirling's air engine in our Augean stables, and got it taken to pieces, as it was clogged with dust & oil, and I expect to have it going as soon as I have time'. The classical reference to the 'Augean stables' was apt. Home to 3,000 oxen and uncleansed for 30 years, the stables of Augeas, fabulous king of Elis, had been purified in a single day by Hercules' diverting of the river Alpheus through it. Glasgow natural philosophy, largely unreformed for more than three decades, was being cleansed by the reforming professor who was determined to transform classroom and research into an entirely new enterprise for the Victorian age. Presented to the natural philosophy class sometime in the late 1820s by its maker, the Stirling engine might yet become a symbol of a new era of motive power.[59]

Simultaneously Thomson raised with Forbes an issue which was to remain unresolved for several years: 'I think this consideration will make it clear that there is really a loss of effect, in the conduction of heat through a solid'. If, he suggested, we considered a fire or other source of heat in the interior of a hollow conducting shell, that heat could be used to melt ice at 32° Fahrenheit without the expenditure or gain of mechanical effect. Such an arrangement appeared to involve an unexplained loss of mechanical power or effect which might otherwise have been produced from the heat source: 'It seems very mysterious how power can be lost in such a way [by the conduction of heat from hot to cold], but perhaps not more so than that power should be lost in the friction of fluids (a plumb line with the weight in water for instance) by which there does not seem to be any heat generated, nor any physical change effected'.[60]

Here William had transferred the issue of 'waste of power' from the engineering context, so familiar through James's concerns, to a philosophical (scientific) context of natural laws and principles. Engineers sought the minimization of waste; philosophers sought to account for the waste, real or apparent, in terms of natural laws and causes. As in his earlier discussion of fluid friction with James at Walsall, there was the continuing mystery over the seemingly absolute loss of power. A few months later William would receive from Joule the assurance that generation of heat was the explanation for the apparent loss of power in fluid friction (ch.4). But neither had William as yet elucidated the connection between such losses of power by heat conduction and the Stirling air-engine. He did so a few months later in his first presentation to the Glasgow Philosophical Society to which he had recently been elected.

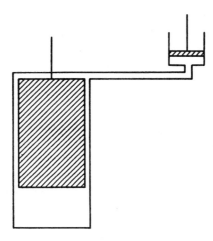

Figure 3.3 William Thomson gave the following account of Stirling's air-engine to his Glasgow students in January 1850: 'Here we have a large cylinder with a plunger in it. Suppose it to be at the top. There is a considerable quantity of air below. If we apply a spirit lamp below & heat that air it expands and rushes up along the sides of the plunger, along the tube and forces up the piston in the other, small cylinder. There is a wheel placed between the two cylinders (not represented). There is a crank attached to each end of the axle of the wheel. When the small piston rises it turns round the wheel which brings the plunger down and this drives out most of the heated air. The air in coming in contact with the cool metal at the top contracts and draws down the piston which raised the plunger and again the air is heated & so on. In order to condense the air better it is expedient to have a stream of water rushing over the upper part thus carrying away the heat' (William Smith, Lecture for 15 January 1850, in 'Notes of the Glasgow College natural philosophy class taken during the 1849–50 session', MS Gen. 142, Kelvin Collection, ULG).

William's professed aim was to offer 'some illustrations, afforded by the Air Engine, of general physical principles'. If the engine were turned *forwards*, he claimed, and if no heat were applied, the reservoir below the plunger would become colder than the surrounding atmosphere and the space above hotter. That is, the engine would turn forwards of itself until the temperatures of the upper and lower reservoirs were equalized. If we then continued to turn the engine forwards (by cranking it), the lower reservoir would cool further and the temperature relation would be reversed. Work would thus have to be done to turn the engine and establish the new temperature difference.

But, he argued, suppose the temperature in the upper reservoir were prevented from rising and the temperature of the lower from falling. Then 'the engine may be turned without the expenditure of any work (except what is necessary in an actual machine for overcoming friction &c.)'. He suggested letting a stream of water at

32° Fahrenheit run across the upper part of the engine, with the lower part held in a basin of water at the same temperature. With the engine turning forwards, heat would be taken from below the plunger and deposited in the space above such that all the water in the basin would be converted gradually into ice at 32° *without expenditure of work.*[61]

The physical principle, which the Stirling air engine illustrated, was that, consistent with the Carnot–Clapeyron theory, the making of ice involved no work, for it apparently required no temperature difference. On the other hand, as he had explained to Forbes, heat generated at high temperature, and then used to melt ice, yielded a 'loss' of the work which might have been produced if the temperature difference had been employed to drive an engine. James, however, soon pointed out that the conclusion that the making of ice required no work was not entirely adequate. Since water expanded on freezing, the freezing of water could actually be said to involve the *production* of work.[62]

Just prior to this conclusion, James had met Rev. Stirling in person and requested him 'particularly not to tell me anything that he did not regard as entirely public, because I had some ideas on the subject myself'. The Thomsons also took the precaution of having several witnesses: James's father, brother William and William's classroom assistant, Robert Mansell. All the evidence therefore points to the probability that James Thomson, stimulated by Clapeyron's analysis, was considering a new air-engine patent different from those of Stirling. For his part Stirling told James at this meeting of April 1848 that he was presently making some improvements on his air-engine. But James was unimpressed: 'I found that, as I had previously thought, he does not understand his own engine; not knowing at all the way in which the heat is expended in generating work'.[63]

In their discussion, as recorded by James, Stirling had apparently admitted that the changes of temperature produced by changes of pressure of the air had long perplexed him, and even alarmed him in regard to the perfection of the engine since 'it had appeared that the respirator would not even *theoretically* give back all the heat to the air; but that he is now inclined to think that "a sort of average is struck" or a compensation is made by which all the heat is really given back if the air passages be small enough, the metal perfectly absorbent and non-conducting &c'. In response, James had told him that 'some tranference of heat from the furnace to the water by means of the changes of temperature of the air is essential to the action of the engine; otherwise it would be theoretically a perpetual motion'. But Stirling had tried to argue 'that there are plenty of theoretical perpetual motions if we have friction, resistances &c out of consideration' to which James replied that 'there are these, but not perpetual sources of power'. Stirling therefore confessed that 'perhaps I [James] was correct and that he had never thought particularly on the difference between a perpetual motion and a perpetual source of power'.

Reflecting on the impossibility of a perpetual source of power in the light of his brother's consideration of Stirling's engine led James a month later to carry out an estimate of the lowering of the freezing point of water by the effect of pressure. William presented the formal paper on his behalf to the Royal Society of

Edinburgh in January 1849. There James explained that William's earlier conclusion 'at first appeared to me to involve an impossibility, because water expands on freezing; and therefore it seemed to follow, that if a quantity of it were merely enclosed in a vessel with a movable piston and frozen, the motion of the piston, consequent on the expansion, being resisted by pressure, mechanical work would be given out without any corresponding expenditure; or, in other words, a perpetual source of mechanical work, commonly called a perpetual motion, would be possible'. In order to avoid such a perpetual motion, long held to be impossible by most natural philosophers and engineers, James announced that 'it occurred to me that it is necessary farther to conclude, that the freezing point becomes lower as the pressure to which the water is subjected is increased'.[64]

A year later, William verified experimentally the approximate relation of the depression of freezing point and increased pressure which James had derived from an application of the 'Carnot cycle'. The result appeared to offer a striking confirmation of the relationship between work and temperature differences (ch.5).[65] But William had already been developing other practical embodiments of the Carnot–Clapeyron theory. Towards the close of 1847, William sent some notes on Clapeyron's paper to Lewis Gordon, adding in the accompanying letter:

> A good deal has been said in various treatises about fixing an absolute standard of temperature. The air thermometer is chosen merely for [convenience] of comparison. Now would it not be a good *absolute* definition of equal degrees, to say that they are such that the same *quantity* (determined in an absolute way by the melting of ice) descending a degree will always produce the same mech[l] effect?[66]

Following presentations to the Glasgow and Cambridge Philosophical Societies (1848), William explained in the published version that 'the scale which is at present employed for estimating temperature is that of the air-thermometer', which rested on the principle that 'equal absolute expansions of the mass of air or gas in the instrument, under constant pressure, shall indicate equal differences of the numbers on the scale; the length of a "degree" being determined by allowing a given number for the interval between the freezing- and the boiling-points'. However, William recognized that the scale was only 'an arbitrary series of numbered points of reference sufficiently close for the requirements of practical thermometry'. Is there, he then asked, 'any principle on which an absolute thermometric scale can be founded?'[67]

A positive answer was to be found in the Carnot–Clapeyron theory. The consequence of the theory which he propounded was summed up thus:

> The characteristic property of the scale which I now propose is, that all degrees have the same value; that is, that a unit of heat descending from a body A at the temperature $T°$ of this scale, to a body B at the temperature $(T-1)°$, would give out the same mechanical effect [motive power or work], whatever be the number T. This may justly be termed an absolute scale,

since its characteristic is quite independent of the physical properties of any specific substance.[68]

In other words, unlike the air-thermometer which depended on a particular gas, William was now employing the waterfall analogy to establish for the first time a scale of temperature independent of the working substance.

From mid-1847, however, the Thomson brothers' enthusiastic acceptance of the Carnot–Clapeyron theory became increasingly threatened by Joule's contrary claims for the mutual convertibility of heat and work without the 'fall' so central to the French doctrines. After two years of intensive debate with Joule, William at last obtained a copy of the original Carnot text and set about a thoroughgoing appraisal which he published as 'An Account of Carnot's Theory of the Motive Power of Heat' (1849). Inspired by this version, an ambitious young German physicist, Rudolf Clausius, announced the reconciliation of the apparently conflicting views of Joule and Carnot in 1850. At the same time, a dynamic Scottish railway engineer, Macquorn Rankine, began publishing on the subject and soon became one of the most avid builders of the new sciences of energy and thermodynamics.

CHAPTER 4

Mr Joule of Manchester

Science has thus, most unexpectedly, placed in our hands a new power of great but unknown energy. It does not wake the winds from their caverns; nor give wings to water by the urgency of heat; nor drive to exhaustion the muscular power of animals; . . . nor accumulate hydraulic force by damming the vexed torrents . . . but, by the simplest means – the mere contact of metallic surfaces of small extent, with feeble chemical agents, a power everywhere diffused through nature, but generally concealed from our senses, is mysteriously evolved, and by circulation in insulated wires, it is still more mysteriously augmented . . . until it breaks forth with incredible energy.

Professor Benjamin Silliman's reflections on the motive power of electricity (1837)[1]

As professor of chemistry and natural history at Yale University from 1802 and founder editor of the widely-read *American Journal of Science*, Benjamin Silliman (1779–1864) had become the leading promoter of the natural sciences in North America. His 1837 pronouncement, symptomatic of the increasing enthusiasm among experimental philosophers for electricity as a motive power, took the form of an editorial comment concerning a report on a certain 'Mr Davenport's electro-magnetic machine' published in his own *Journal*. In a tone of mounting excitement, Silliman therefore proclaimed in conclusion that 'Nothing since the discovery of gravitation, and of the structure of the celestial systems, is so wonderful as the power evolved by galvanism, whether we contemplate it in the muscular convulsions of animals, the chemical decompositions, the solar brightness of galvanic light, the dissipating consuming heat, and, more than all, in the magnetic energy'.[2]

Silliman's remarks were reprinted soon after in the second volume of a new British periodical with the grandiose title *Annals of Electricity, Magnetism, and Chemistry; and Guardian of Experimental Science*, edited by the popular electrical showman William Sturgeon (1783–1850). The same volume contained James Prescott Joule's first publication, presented in the form of a letter to the editor under the title 'Description of an Electro-magnetic Engine'.[3] Not yet 20 years old, Joule seemed possessed by the same kind of zeal for new forms of motive power that characterized Glasgow contemporaries. Unlike James Thomson's interests in the Carnot–Clapeyron theory of the motive power of heat, however, Joule had entered a veritable battlefield of competing theories and practices in which elite experimental philosophers with gentlemanly status fought to keep on the margins of respectability radical electricians whose livlihood depended upon the shocks and sparks of the new science.

After the construction around 1800 of the first voltaic cells, designed to produce 'galvanic electricity' from chemical sources rather than to generate static electricity by frictional means, the subject of electrical science had become one of central importance to natural philosophers of every variety across Europe and beyond. Interest had been further stimulated by the phenomenon of electromagnetism, beginning in 1820 with Oersted's reports of the magnetic effects of a wire carrying galvanic electricity.[4] Investigations into the causes, effects and uses of electricity often took competing forms in different scientific communities which vied with one another for the attention of public audiences. Sturgeon's *Annals* in Britain and the *Journal of the Franklin Institute* in North America catered for strong practical interests in the subject. The *Annals* placed great emphasis on 'the rise and progress of electro-magnetic engines for propelling machinery', making the contentious claim in 1839 that developments in the design of electromagnets offered 'new and inexhaustible sources of force which appeared easily and extensively available as a mechanical agent'.[5]

The *Journal of the Franklin Institute* similarly promised wondrous economic benefits from the new 'electro magnetic propelling machinery':

> If we hire a man by the day we must not allow him to be idle, as in that case we give our money for nothing. The current of his life flows on, and he must be fed and clothed or the stream will stop. But give us a machine which is not costly at first, and if it works but one hour in the twenty-four, will itself be a consumer in that proportion only; a machine which we can at any moment set to turn our lathes, our grindstones, our washing machines, our churns, our circular saws. . . . Our farmers, our mechanics, and our housekeepers generally, must also be supplied.[6]

In contrast, the steam engine could not be 'used to advantage where it has not the labour of several horses to perform' since it needed 'the constant attention of the engineer, or of the fireman, and is kept at work at an expense which is relatively increased as its power is diminished'. An electromagnetic engine, approximating to the power of a man, appeared to offer considerable economic advantage over an equivalent steam engine, provided that the cost of materials consumed 'in perform-ing the work of a day' was less than the daily cost of a man's labour. Critically, therefore, the economic success of the electromagnetic engine depended upon 'the store house of nature' enabling us 'to obtain zinc and sulphuric acid at a cheap rate'.[7]

Unlike Sturgeon's primary goals of improving display apparatus and electrical machines, elite London professors had built their credibility on a philosophical orientation towards experimental research: their concerns were primarily with the laws and causes of natural phenomena. Michael Faraday and Charles Wheatstone (1802–75), for example, had risen from humble origins by a process of self-improvement to positions which commanded considerable authority both in scientific circles and in fashionable society. Faraday in particular was generally acknowledged to be the leading British expert in the new science of electricity and magnetism.[8] Tensions between communities of popular electricians on the one hand and elite natural philosophers with secure institutional positions on the other poses a fundamental question for a historical account of James Prescott Joule's early scientific work: given his enthusiasm for electrical science, how did he relate to these different electrical communities?

Historians have typically represented Joule as a self-educated amateur who worked in a remote and leisurely world away from the competitive pressures of professional science. But Joule's world was far from being leisurely and remote. During the 1840s he worked to persuade his scientific peers, struggled to build alliances with his immediate acquaintances, and fought vigorously to defend hard-won scientific property against the threat of rival claimants to priority. In what follows, I argue that Joule gradually fashioned for himself a career, not as ingenious inventor of new forms of motive power, not as member of an elite professoriate, and not as a popular practitioner of electrical science, but as a 'gentlemanly special-ist', that is, a scientific practitioner of independent means (such as business, bank-ing or medicine) whose expertise had been gained within a fairly specialized branch of science and whose credibility had been acquired within the ethos of gentlemanly scientific societies and institutions.[9] Thus although Joule's scientific interests appeared to range widely, he spent much of this decade establishing considerable authority in those branches of natural philosophy (electromagnetism, galvanism, heat and chemistry) which promised new and efficient forms of motive power.

Manchester's 'Guardian of Experimental Science'

Joule's forebears had been attracted to Manchester in the late eighteenth century by the growing industrial city's apparently insatiable demands for food and drink. His grandfather had established a brewery across the River Irwell in adjacent Salford. An early indication of its prosperity and prestige was the installation of a steam engine in the 1790s. By the end of the 1820s, Joule's father had built up the largest brewery in Manchester and Salford.[10] Social aggrandizement began to accompany economic success. In the early 1820s the Joules moved from their home beside the brewery to a new and fashionable residence, 'Broom Hill' at Pendlebury, with six servants living in, well away from the industrial heart of Salford. In 1843, the family would move again, this time to 'Oakfield', a mile or so south of the city. 'Gentrification' was manifested in other ways too: a switch from religious non-conformity to the establishment Church of England in the same period, and the employment of private tutors for the education of the new generation of Joules in the 1830s.[11]

James Joule's private education included just over two years with John Dalton (1766–1844), whose international reputation and quaker non-conformity made him a perfect icon for the professed latitudinarianism of the Anglican-dominated BAAS. Dalton, who instructed private pupils on premisses of the Manchester Literary and Philosophical Society of which he was then president, taught Joule arithmetic and geometry over a two-year period. Then, just as they began chemistry, Dalton suffered a serious stroke in the spring of 1837. But Joule had already become acquainted with Dalton's characteristic style of chemistry and natural philosophy. His detailed quantitative experimentation was indeed more typical of French science in the early nineteenth century and contrasted with older, qualitative traditions of British experimental philosophy. Such practical skills (especially in quantitative chemistry) would be as important for the future brewery owner as for the aspiring man of science.[12]

From Dalton too, Joule learnt of his master's long-time support for caloric doctrines on the nature of heat, as the opening paragraph of his *New System of Chemical Philosophy* (1808) testified: 'The most probable opinion concerning the nature of caloric is that of its being an elastic fluid of great subtilty, the particles of which repel one another, but are attracted by all other bodies'.[13] Ironically, Dalton's death in 1844 would coincide with his former pupil's most ardent attacks on the old doctrines of heat.

When Joule began his own experimental researches in the late 1830s, Manchester had become a formidable power in Britain's industrial armoury. From a town of about 20,000 inhabitants in the mid-eighteenth century, it was now a manufacturing city of over 300,000 people. Its original wealth had been created around the booming cotton industry which embodied the technical changes involved in the transformation of a craft into a factory system. As the motive power of water yielded to the new power of steam, Manchester's industrial base expanded into steam engine construction and all manner of related trades from machine tool to boiler making. As early as 1764, the Bridgewater Canal had opened for the transport of coal from the Duke of Bridgewater's mines to satisfy the city's great appetite for fuel. A nation-wide zeal for canal construction ensued. By the 1830s Manchester was once again setting the pace in transport innovation with the Liverpool and Manchester Railway carrying passengers and freight between two of the largest industrial and commercial cities of the Victorian age.[14]

Scientific and educational markets of all sorts flourished within Manchester's thriving commodity culture. The famous Lit & Phil (founded 1781) could promote itself by means of the image of Dalton which symbolized the practicality, the utility, the down-to-earth simplicity and reliability of Manchester's philosophical genius. The associated Manchester Academy (founded 1792) catered for the education of boys for commerce and dissenting ministries. And the Manchester

Mechanics Institution (founded 1824) aimed at the 'improvement' of artisans who might otherwise pose a threat to social order.[15] In 1839 William Sturgeon arrived from London as superintendent of the new Royal Victoria Gallery for the Encouragement and Illustration of Practical Science, whose promoters included Manchester engineers such as William Fairbairn and Eaton Hodgkinson (1789–1861) and whose aims united progress, industry and experimental science through display apparatus of an electrical kind.[16] No stranger to electrical theatre, Sturgeon had lectured at the Royal Gallery of Practical Science, or Adelaide Gallery, off the Strand in London. The Gallery had embodied the interests of practical electricians, especially electrical instrument and apparatus makers, with its emphasis on displaying shocks and sparks to the public.[17]

In his *Annals*, Sturgeon exhibited major disagreements over electrical theory and practice with the eminent Faraday and with other Fellows of the Royal Society, including the Plymouth-based experimentalist William Snow Harris (1791–1867). Indeed, his move from London to Manchester might have been in part occasioned by his strained relations with Royal Society Fellows, as well as by the expectation of a more receptive audience in Manchester. But neither the journal nor the Gallery were to provide Sturgeon with the rich returns for which this investor of electrical credibility might have hoped in a city teeming with innovation. And so the one-time inventor of the electromagnet was to pass his final years in virtual scientific and commercial penury.[18]

Joule's early contributions to Sturgeon's *Annals* typically emphasized the potential advantages, especially the economic advantages for locomotives, of his electromagnet. 'A great saving of room is effected', he stressed in his first letter, 'and, consequently, the power relative to the weight of the engine is increased'. Furthermore, he claimed, the arrangements were such that 'wheels or paddles may be affixed so as to answer either locomotive or sailing [steamboat] purposes'. The letters, *visual* and *practical* throughout, concerned geometrical arrangements rather than theories and causes. Joule was addressing practical men, artisans and electricians, rather than gentlemanly natural philosophers. When, therefore, he claimed with respect to his electro-magnets that 'Their lifting power is very good', and when he promised to communicate upon completion of the engine 'a particular account of its duty', he spoke both the language of Sturgeon the inventor and of engineer–inventors in general.[19]

'Duty' had for long been a standard measure of engine performance, understood generally in terms of so many pounds in weight (the load) raised one foot in height by the burning of one bushel or pound of coal. In subsequent discussions Joule distinguished between 'duty' as 'pounds raised per second of time to one foot in height' and 'economical duty' as 'pounds raised to the height of one foot by the agency of one pound' of coal in the steam engine or of one pound of zinc consumed in the battery used to power the electromagnetic engine. It was this latter measure which would serve as his principal standard for comparing the two kinds of engine.[20]

In his 'Historical Sketch of the Rise and Progress of Electro-magnetic Engines' for the next volume of the *Annals*, Sturgeon himself quickly endorsed Joule's invention as an 'exceedingly ingenious arrangement of electro-magnets of soft iron', a 'beautiful instrument' in which the 'bars . . . are of a peculiar construction, and the transposition of their polarity effected by an exceedingly ingenious contrivance. Mr. Joule proposes to apply his engine both to locomotive carriages and to boats'.[21] Sturgeon's summary, presenting Joule as an ingenious inventor with the skills to devise clever arrangements for practical uses, thus emphasized the machinery rather than the philosophical principles and causes of its working.

Despite, or perhaps because of, Sturgeon's enthusiastic endorsement of his inventiveness, Joule seemed in his second letter not to welcome the epithet of 'ingenious inventor'. Far from presenting an optimistic report on the future development of his electro-magnetic engine, he reported that the performance had proved disappointing: 'the enormous friction of the engine was accurately measured and reckoned as the load; the velocity of the magnets was about 3½ feet per second'. He also explained that he had resolved before attempting to make another engine 'to satisfy myself by experiments, conducted on a smaller scale, how far it was possible to increase the velocity of rotation'. Of the many factors 'which prevent an infinite velocity', he noted, 'the resistance which iron opposes to the instantaneous induction of magnetism is of considerable importance'. His experiments were therefore aimed at comparing the performance of different arrangements of iron in the electro-magnet.[22]

Even in this early letter Joule had thus begun to represent himself not as an ingenious inventor but as an experimental philosopher, concerned less with commercial promotion and more with claims to scientific authority established through experiment. Here and in subsequent letters, Joule wrote repeatedly of 'my further investigations' on the subject, of 'research' in progress, of the 'sources of error' involved, and of the quantitative 'laws' enunciated. Above all, his original 'ingenious' arrangements for a new engine had to accept the verdict of experiment.[23]

The title of a series of three communications to Sturgeon's *Annals* in 1840, 'On Electro-magnetic Forces', expressed fully the shift away from descriptions of ingenious apparatus to experimental natural philosophy. Indeed, the whole mode of presentation had changed from that of personal letters to a series of consecutively numbered paragraphs. It was hardly coincidence that Joule's second communication followed immediately after a reprint of some of Faraday's 'Experimental Researches in Electricity' with its distinctive numbering of paragraphs. Joule's new and less personal style, appealing to the forces of nature rather than to human artefacts, suggests that Faraday, not Sturgeon, had now become his role model.[24] At the end of the first communication he openly admitted that his goals no longer coincided with the expectations of his original audience:

> I must apologize to the reader, that I have not relieved the tediousness of this paper, by a single brilliant illustration. I have neither propelled vessels, carriages, nor printing presses. My object has been, first to discover correct principles, and then to suggest their practical development'.[25]

Consistent with his refusal to wear the epithet of ingenious inventor, Joule's ambitions lay in a very different direction. As early as 1840 he attempted to present himself as a gentlemanly natural philosopher to the nation's premier scientific society. 'On the Production of Heat by Voltaic Electricity' (1840) appeared in summary form in the *Proceedings of the Royal Society*, but the Royal Society apparently refused to publish a longer version in its prestigious *Transactions*. According to the *Proceedings*:

> The conclusion he [Joule] draws from the results of his experiments is, that the calorific effects of equal quantities of transmitted electricity are proportional to the resistance opposed to its passage, whatever may be the length, thickness, shape, or kind of metal which closes the circuit; and also that . . . these effects are in the duplicate ratio of the quantities of transmitted electricity [that is, proportional to the square of the 'current']. . . . He also infers from his researches that the heat produced by the combustion of zinc in oxygen is likewise the consequence of resistance to electric conduction.[26]

Here, in its earliest version, was what subsequently became Joule's famous 'i^2R' law of the heating effects of an electric current (proportional to the square of the current and to the resistance). But the aloof reporting of the *Proceedings*, combined with the non-appearance of the paper in the *Phil. Trans.*, suggests that Mr Joule was very far from possessing the credibility and authority required to persuade the Royal Society of the truth and importance of his investigation.

Historians have explained Joule's failure to publish in the *Phil. Trans.* in terms of the Royal Society's conservatism. Yet the Royal Society at this time was by no means merely composed of non-scientific aristocrats and elderly physicians as Charles Babbage (1791–1871) had alleged almost ten years before. The Society included many of the most distinguished philosophers of the day, including elite Cambridge mathematicians such as Herschel, Whewell and Airy. As a 22-year-old, little-known provincial gentleman, whose alignments appeared to be with the 'practical' electrician Sturgeon, critic of Faraday and uncritical editor of a journal whose articles included reports of the resuscitation of executed criminal corpses, Joule was neither qualified by virtue of a Cambridge mathematical education, nor yet trusted by virtue of gentlemanly contributions to science, to receive elite philosophical status. Furthermore the theoretical speculation regarding the electrical nature of chemical combustion certainly transgressed the strict 'experimental' ethos of the Society's *Transactions*.[27]

Joule's self-fashioning had thus far been conspicuously unfruitful. He therefore pursued two strategies, the one local and the other national. Taking advantage of his continuing links with Sturgeon, he delivered his first public lecture at the Victoria Gallery early in 1841. The lecture opened with an explicit declaration of his philosophical authority: 'it is my intention to bring forward in this paper an electro-magnetic principle, in reference partly to the motion of machines'. His special concern was to investigate 'a novel form of electro-magnetic engine' suggested to him by 'an ingenious gentleman of this town'. Here was the voice of the aspiring philosopher seeking to investigate and pronounce upon the principles and practicality of that which the merely ingenious individual had suggested.[28]

Setting out his assessment of the prospects for ordinary electro-magnetic engines, Joule reminded his audience that 'as soon as the general principles of electro-magnetism were understood . . . a great number of ingenious individuals constructed various arrangements of machinery'. Recalling remarks of his own in Sturgeon's *Annals* as recently as 1839, he noted that 'At that period the expectations that electro-magnetism would ultimately supersede steam, as a motive force, were very sanguine'. There seemed very little *in principle* which could not be overcome in *practice* to prevent an enormous velocity of rotation 'and consequently an enormous power' except 'the resistance of the air which it was easy to remove, the resistance of iron to the induction of magnetism which I succeeeded in overcoming to a great extent by annealing the iron bars [of the electromagnet] very well, and the inertia of the electric fluid'.[29]

The St Petersburg-based engineering professor M.H. Jacobi (1801–74), however, had shown 'the principal obstacle to the perfection of the electro-magnetic engine', namely 'that the electric action produced by the motion of the bars operates against the battery current, and in this way reduces the magnetism of the bars, until, at a certain velocity, the forces of attraction become equal to the load on the axle, and the motion in consequence ceases to accelerate'. Attempting to reinforce his own philosophical credentials, Joule pointed out to his audience that Jacobi had neither 'given precise numerical details concerning the duty of his apparatus; nor had he then determined the laws of the engine'. In order to supply this deficiency in our knowledge, therefore, Joule had been 'induced to construct an engine adapted for experiment'.[30]

Joule's principal conclusions, expressed in terms of economical duty or the amount of useful work obtainable from a given quantity of fuel, highlighted the importance of battery *resistance* (alongside other forms of resistance in the electro-magnet or conductors). His philosophical investigations into electro-magnetic engines as an economical source of power had thus added a new dimension demanding an inquiry into the nature and causes of resistance within, as well as outside, the source of power. But practical consequences, expressed in the sober tones of an experimental philosopher, did not offer much comfort to ingenious inventors searching for an alternative to steam power. Joule's experimental 'apparatus', with magnets revolving at a steady eight feet per second, had yielded 'mechanical force' equal to the raising of a weight of 331,400 pounds to the height of one foot for every pound of zinc consumed in a Grove's battery. In contrast 'the duty of the best Cornish steam-engines is about 1,500,000 lbs. raised to the height of 1 foot by the combustion of each pound of coal, or nearly five times the extreme duty that I was able to obtain from the magnetic engine by the consumption of a pound of zinc'.[31]

With such a very unfavourable comparison, Joule confessed that he almost despaired 'of the success of electro-magnetic attractions as an economical source of power; for although my machine was by no means perfect, I do not see how the arrangement of its parts could be improved so far as to make the duty per pound of zinc much superior to the duty of the best steam-engines per lb. of coal'. In addition, the high cost of the zinc and battery fluids, compared to the price of coal, prevented the engine 'from being useful for any but very peculiar purposes'. He then offered an experimental diagnosis of the 'ingenious gentleman's' magnetic engine and authoritatively pronounced an even less favourable verdict: the force would be 'far too minute for the movement of machinery; and the duty per pound of zinc is vastly less than that of the common electromagnetic engine'.[32]

This public lecture also gave Joule an opportunity to enhance his local philosophical credibility by appraising the explanatory power of possible mechanical causes of the increased length of the magnetized iron bar. He did so by critical rejection of a theory of the celebrated French mathematical physicist A.-M. Ampère (1775–1836) who had supposed an atmosphere of electricity surrounding each atom of iron. The atmospheres moved in planes at right angles to the axis of the magnet (Figure 4.1a). To account for saturation, that is 'for the fact that iron, after receiving a certain quantity of magnetism, is incapacitated from receiving a further supply', the atmospheres were assumed to have a centrifugal tendency such that as their rotation increased a point would be reached when they began to interfere with their neighbours. The same assumptions, however, led to the conclusion that the bar would become *shorter* under magnetic influence. Joule also rejected Ampère's theory on the grounds that it demanded impossible physical conditions, contrary to 'the analogy of nature', by supposing 'motion, or at least an active force, to be continued against antagonist forces, for an indefinite length of time, without loss, in order to explain the phenomena exhibited by a hard steel magnet'.[33]

Joule adopted instead a version of the theory put forward by an eighteenth-century German natural philosopher, F.U.T. Aepinus (1724–1802), which proposed atoms of iron surrounded first by atmospheres of magnetism and, beyond these, still rarer atmospheres of electricity (Figures 4.1b and 4.1c). The space between each of these 'compound atoms' was filled with 'calorific ether in a state of vibration' or in some manner allowing vibration of the atoms themselves. Under a magnetic inductive influence, the atmospheres accumulate on one side of the atom in line with the axis of the bar. This theory explained the various phenomena of magnetism in simple mechanical terms of matter, space and motion. Saturation occurred when 'all the magnetism of each atom of iron is accumulated to one side'. The same explanation accounted for increased length.[34]

Furthermore, the hypothesis explained that 'at a certain degree of heat all the magnetic power of iron is destroyed'. The magnetic particles vibrated in the space between them: 'This vibration is called heat'. Increasing temperature caused these vibrations to increase in violence. However, the comparatively greater inertia of the iron atoms meant that the atmospheres of magnetism and electricity 'will be in a state of vibration, while the atoms of iron remain in a state of comparative quiescence'. Above a certain intensity of vibration, 'the inductive influence will not be able to arrange the magnetism in any definite direction with regard to the atoms of iron'.

In Joule's opinion, then, this kind of theory represented the most desirable type of explanation in terms of mechanical causes:

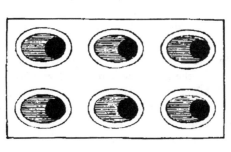

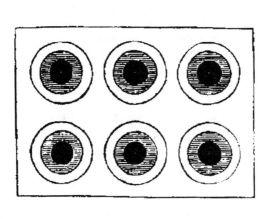

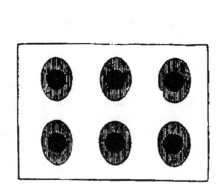

Figure 4.1a Joule's representation of Ampère's theory of magnetism in which an atmosphere of electricity surrounded each atom of iron (Joule 1842a).

Figure 4.1b In Joule's version of Aepinus's theory of magnetism the atoms of iron were surrounded by atmospheres of magnetism and, beyond these, rarer atmospheres of electricity (Joule 1842a).

Figure 4.1c Under magnetic influence, the atmospheres accumulate on one side of the atoms in line with the axis of the bar. A 'calorific ether in a state of vibration' filled the space between atoms (Joule 1842a).

Such a theory seems to me to afford a natural and complete expression of facts. It supposes nothing which we cannot readily comprehend, except the existence and elementary properties of matter, which are necessarily assumed by every theory, and which the Great Creator has placed utterly beyond the grasp of the human understanding.[35]

Joule had thus committed himself to a fundamentally mechanical nature whose basic building blocks of matter, together with its mechanical properties, had been created by God and thus admitted of no further human explanation or analysis. Although he shared with Faraday the belief that only God could create or destroy the basic entities of nature, he differed radically from Faraday in his assumptions concerning the mechanical character of those building blocks.[36] His mechanical commitments also differed strikingly from the assumptions of Julius Robert Mayer, subsequently his rival claimant to 'discovery' of the mechanical equivalent of heat. On the other hand, his perspective would appeal strongly to Scottish-centred natural philosophers within an engineering context, notably William Thomson and Macquorn Rankine (chs 5 and 6).

In the same year as his Victoria Gallery lecture, Joule published his investigations into the heating effects of electric currents in a well-established scientific periodical with national and international readerships. Unlike the *Phil. Trans.*, the *Philosophical Magazine* tended to welcome contributions of a more speculative character in which, for example, hypotheses were developed to account for physical and chemical effects. For the *Phil. Mag.* version Joule extended his investigations from solid conductors to electrolytes and arrived at an identical law for the heating effects. He quickly followed up this publication with three further studies, all of which shared a common investigation of electrical *resistances*, whether in metallic conductors, batteries, electrolytic cells or chemical combustion. The theme strongly suggests that in this period Joule's researches were motivated by a desire to enhance the economical duty of electro-magnetic engines through a philosophical understanding of the sources and nature of electrical resistances.[37]

Joule's advances to the British Association

By the end of 1841 Joule had enhanced his credibility on the Manchester stage. He had appeared in person at the Victoria Gallery to pass judgment upon the inventions of ingenious gentlemen. The Royal Society had taken notice of his investigations. The *Phil. Mag.* had published the extended memoir. Now, in November 1841, he read 'On the Electric Origin of the Heat of Combustion' before the Manchester Literary and Philosophical Society. Made up of gentlemen representing the city's most eminent and wealthy citizens – engineers such as Joseph Whitworth, William Fairbairn and Eaton Hodgkinson; chemical and cotton manufacturers, bankers, and of course John Dalton – the Lit. & Phil. had a membership of under 200, which, compared to the city population, was a reflection of its select character. On 25 January 1842 Joule was elected to that elite, if provincial body.[38]

His electrochemical researches found a receptive audience through the strong chemical interests of Lit. & Phil. members. Read to the Society in January 1843, 'On the Heat Evolved During the Electrolysis of Water' pursued his now-favourite theme of electrical resistances and concluded with a number of general observations on the subject. In the third such observation Joule inferred that whatever the arrangement of the apparatus 'the whole caloric of the circuit is exactly accounted for by the whole of the chemical changes', that is, he wanted to persuade himself and his readers that he had traced the heat produced or absorbed in every part of the circuit and had found that the gains and losses of heat were all balanced.[39]

Use of the terms 'caloric' and 'latent heat' here suggested that Joule's interpretation was still *compatible* with a material view of heat: that heat was simply transferred from one part of the circuit to another without net production or annihilation. A year later, however, after the presentation and publication of his first paper on the 'mechanical value of heat', he made clear that 'those words were only employed because they conveniently expressed the facts brought forward. I was then as strongly attached to the theory which regards heat as motion among the particles of matter as I am now'. Indeed, we have already discussed his 'vibrational' view of heat in relation to his enthusiasm for Aepinus' theory of magnetism. Furthermore, his recent series of papers had favoured an electrical theory of chemical heat (attributed to Davy and the Swedish electrochemist J.J. Berzelius (1779–1848)) by which 'Electricity may be regarded as a grand agent for carrying, arranging, and converting chemical heat'. Heat could thus be interpreted as vibrations within an electrical atmosphere.[40]

Having so far focused exclusively on the production and absorption of heat with respect to chemical changes, Joule shifted attention in his fourth general observation to the relation of heat and mechanical power. He first noted that both the mechanical and heating powers of a current were proportional to its intensity or electromotive force and therefore that the mechanical and heating powers were proportional to one another. The fifth general observation then introduced into the hitherto 'closed' electrical circuit an 'external' factor in the form of an electromagnetic engine:

> The magnetic electrical machine [generator] enables us to convert mechanical power into heat, by means of the electric currents which are induced by it. And I have little doubt that, by interposing an electro-magnetic engine in the circuit of a battery, a diminution of the heat evolved per equivalent of chemical change would be the consequence, and this in proportion to the mechanical power obtained.[41]

At this stage, Joule was clearly invoking his own authority and prestige as an electrical expert to legitimate a plausible assumption. On the one hand, then, a generator driven by mechanical power produced an electric current which in turn produced heating effects in the circuit; while on the other hand, an electromagnetic engine producing mechanical power would, in Joule's view, 'convert' a proportionate amount of heat into that mechanical power. In a footnote of February 1843 Joule indicated that he was 'preparing for experiments to test the accuracy of this proposition'.[42] The outcome was his first paper on 'the mechanical value of heat' presented to the Cork meeting of the BAAS six months later.

Joule's election to the Lit. & Phil. had coincided with an upsurge in its exclusive membership to a record all-time high of 191 members just ahead of the first Manchester meeting of the BAAS in 1842. Contrary to its professed aims of advancing science and its industrial uses, the BAAS had long delayed holding its annual meeting in Manchester. Eleven previous years had taken the Association on a circuit of capital cities (Edinburgh and Dublin), academic towns (Oxford and Cambridge), significant seaports and commercial, even industrial, centres (Bristol, Liverpool, Newcastle, Glasgow and Plymouth). With hostilities among its diverse scientific groups temporarily ceased, the turbulent industrial city could present a united welcome to the BAAS. Such a happy conjunction of local elite and national association had provided Joule with a golden opportunity. He had accordingly presented his first BAAS paper, 'On the Electric Origin of Chemical Heat', to a national audience of chemical philosophers in Section B.[43] In August 1843 he reappeared before Section B with a presentation 'On the Calorific Effects of Magneto-electricity, and on the Mechanical Value of Heat'. The full paper, dated July with a postscript of August 1843, appeared in three instalments in the *Phil. Mag.* during the same year.[44]

Joule there informed his readers that if 'we consider heat not as a *substance*, but as a *state of vibration*, there appears to be no reason why it should not be induced by an action of a simply mechanical character, such, for instance, as is presented in the revolution of a coil of wire before the poles of a permanent magnet'. His earlier investigations had referred 'to a particular part of the circuit only, leaving it a matter of doubt whether the heat observed was *generated*, or merely *tranferred from the coils* in which the magneto-electricity was induced, the coils themselves becoming cold'. Indeed, he confessed that his researches on electrolysis 'might seem to prove that *arrangement* only, not *generation* of heat, takes place in the voltaic apparatus, the simply conducting parts of the circuit evolving that which was previously latent in the battery'.[45]

Having implemented in practice his own suggestions made in the fifth observation of his 'Electrolysis of Water' paper, Joule could now communicate news of experiments relating to the revolution of a small electromagnet immersed in water between the poles of a powerful magnet. Using the electromagnet as a magneto-electric machine (generator), he carried out measurements both of the electricity generated (using an accurate galvanometer) and the calorific effect of the coil (measured by the change of temperature in the water surrounding it). He also introduced a battery into the magneto-electrical circuit. His main conclusion was that '*heat* is *generated* by the magneto-electrical machine, and that by means of the inductive power of magnetism we can *diminish* or *increase* at pleasure the *heat* due to chemical changes'. The heat was not merely *transferred* from one part of the arrangement to another as might be expected from a caloric theory of heat. This conclusion then prompted the further inquiry as to whether or not a constant ratio existed between the heat and 'the mechanical power gained or lost', that is 'a mechanical value of heat'.[46]

Adopting the mean result of thirteen experiments, Joule eventually presented his answer thus: 'The quantity of heat capable of increasing the temperature of a pound of water by one degree of Fahrenheit's scale is equal to, and may be converted into, a mechanical force capable of raising 838 lb. to the perpendicular height of one foot'. He admitted that there was a considerable difference between some of the results (which ranged from 587 to 1040) but the differences were not, he asserted, 'greater than may be referred with propriety to mere errors of experiment'. It is nevertheless easy to see why Joule's experimental results hardly spoke for themselves, requiring instead a trustworthy experimenter to assure his uneasy readers that the errors were indeed due to mere errors of experiment and not to some more fundamental cause.[47]

The chosen phrase 'mechanical value of heat' was also highly significant. If the meaning of 'value' was understood not simply in the *numerical* but also in the *economic* sense, then it is easy to see that Joule's investigations were being shaped by a continuing search for the causes of the failure of his electromagnetic engine to match the economy of heat engines. That search directed him to the 'mechanical value of heat', that is to the amount of work obtainable from a given quantity of heat which in turn derived from chemical or mechanical sources. Thus his primary concern was not with the conversion of work into heat as in frictional cases – the 'waste of useful work' which was of most interest to the Thomson brothers (chs 3 and 5) – but with maximizing the conversion of heat from fuel into useful work in various kinds of engines, that is, with 'economical duty'. Joule was therefore engaged in constructing a new theory of heat, not as an abstract and speculative set of doctrines, but as a means of understanding the principles which governed the operation and economy of electrical and heat engines of all kinds. Only in retrospect, furthermore, can Joule be represented as a 'discoverer' of the conservation of energy and a 'pioneer' of the science of energy.[48]

Two prominent practical conclusions added in the August postscript linked directly to Joule's earlier work on electromagnetic engines. First, he showed that the heat evolved by the combustion of a pound of Welsh steam coal was 'equivalent to the mechanical force capable of raising 9,584,206 lb. to the height of one foot, or about ten times the duty of the best Cornish engines'. Ninety per cent of the mechanical value of heat contained in the pound of coal was thus wasted in even the best steam engines. Second, Joule found that the mechanical forces derived from zinc batteries were equal to 1,106,160 lb (Daniell's) and 1,843,600 lb (Grove's) raised to the height of one foot, implying that electrical engines might be superior in principle to steam engines. But, he concluded, 'since it will be practically impossible to convert more than about one half of the heat of the voltaic circuit into useful mechanical power, it is evident that the electro-magnetic engine, worked by the voltaic batteries at present used, will never supersede steam in an œconomical point of view'.[49]

The August postscript also suggested that his readers would have to admit 'that Count Rumford was right in attributing the heat evolved by boring cannon to friction, and not . . . to any change in the capacity of the metal'. Since drafting his original paper, Joule had apparently 'discovered' the enigmatic Benjamin

Thompson, Count von Rumford (1753–1814) and was now appropriating him to Joule's own cause. Read before the Royal Society in 1798, Rumford's 'An Enquiry into the Source of Heat which is Excited by Friction' had as its central claims that 'the source of the heat generated by friction, in these experiments, appeared evidently to be *inexhaustible*' and that therefore heat could not be a material substance. He had further inferred that 'it appears to me to be extremely difficult, if not quite impossible, to form any distinct idea of any thing capable of being excited and communicated, in the manner the heat was excited and communicated in these experiments, except it be MOTION'. But Rumford, aligning himself with Newton's professed approach to gravitation, was reluctant to specify the particular kind of motion involved, preferring instead 'to investigate the laws of its operations'.[50]

At first sight it appeared that Joule had found the right company. With his paper published in full in the *Phil. Trans.*, with his seemingly aristocratic and trustworthy social status, with his emphasis on practical demonstration and economic utility, and with claims to be one of the founding fathers of the Royal Institution of Great Britain, Rumford seemed a perfect name for Joule to invoke in order to consolidate the credibility of the mechanical value of heat. He had furthermore established the Rumford Medal of the Royal Society to be awarded for researches in the fields of heat and light. But there had always been something faintly fraudulent about this American-born soldier of fortune who, though knighted by George III for his loyal service during the American Revolution, went on to head the Bavarian army and become a Count of the Holy Roman Empire in 1793. And as though having established his scientific prowess by an overthrow of Lavoisier's cherished caloric theory, Rumford later married the widow of the celebrated French chemist who had perished by the guillotine for his associations with the *ancien regime*.[51]

Rumford's experiments had been correspondingly very different in style from those of Joule. Never suffering from understatement, Rumford had described how he had formed a solid cylinder from a cast cannon in the Arsenal. Boring a hole about 10 cm in diameter and 18 cm in length in the cylinder, he had arranged a blunt steel borer, its axis turned by horses, to rub against the bottom of the hole. A mercury thermometer, located in a smaller hole beyond the end of the bore, measured a temperature rise from 60° to 130° Fahrenheit after 960 revolutions and 30 minutes. The resulting quantity of metal dust weighed 837 grains (about 54 grams). A more advanced version involved enclosing the cylinder in a wooden box filled with water. Rumford reported that from an initial 60° the water actually boiled after two and a half hours. Indeed, he claimed that it would have taken more than nine wax candles, 2 cm in diameter and burning for the same time, to have yielded as much heat.[52]

Given that the workshop location and large-scale character of Rumford's experiments had made them unsuited to demonstration before an audience of gentlemanly witnesses, it is not surprising to find that his claims failed to persuade his contemporaries. Adverse reactions are not difficult to find. Professor Thomas Thomson told his Glasgow University class of chemists around 1830 that a part at least, if not the whole, of the heat evolved in the experiments derived from an increase in the density of the solid metal cylinder and a consequent reduction of specific heat. Since Rumford had only examined the metal dust, his experiments did not 'furnish the demonstration of the immateriality of heat which these [natural] philosophers [Davy and Rumford] thought they did'.[53]

Joule's initial invocation of Rumford was thus a doubtful tactic. Only with the benefit of hindsight would Joule emerge well from a contrast in experimental styles between the two natural philosophers. In 1843 any alignment with the discredited Rumford was likely to prove less than helpful to Joule's cause. Perhaps recognizing the limited gains to be made here, Joule quickly cited his own attempt at (in Whewell's terminology) consilience. This second and very different (non-electrical) method measured the 'heat evolved by the passage of water through narrow tubes' and was therefore, like Rumford's approach, concerned with heating by frictional means. With a mechanical value of about 770 lbs., Joule claimed his result to be 'very strongly confirmatory' of his other results, and pledged to 'lose no time in repeating and extending these experiments, being satisfied that the grand agents of nature are, by the Creator's fiat, *indestructible*; and that wherever mechanical force is expended, an exact equivalent of heat is *always* obtained'.[54]

The Royal Society's proceedings

Guided by this powerful conviction that the exercise of divine will had established the unchangeable character of nature's fundamental agents, Joule had so far investigated the evolution of heat by electrical and by frictional means. He had now fashioned himself as a regular player on local and national stages, but was as yet a player who had not received any significant critical attention or acclaim. By 1844 he had embarked on a third method of construing the relationship between heat and mechanical force, a method which he would deploy for a second assault on the metropolitan stage. Once again, however, the Royal Society's *Phil. Trans.* rejected the full paper which, under the title 'On the Changes of Temperature Produced by the Condensation and Rarefaction of Air', would be taken instead by the *Phil. Mag.* But an abstract, prepared by P.M. Roget (1778–1869) of *Thesaurus* fame, appeared in the Royal Society's *Proceedings*.

Roget there explained that the author had 'contrived an apparatus where both the condensing-pump and the receiver were immersed in a large quantity of water, the changes in the temperature of which were ascertained by a thermometer of extreme sensitivity'. The apparatus provided a very different (non-electrical and non-frictional) method for determining the mechanical value of heat:

> By comparing the amount of force [measured as weight raised to a height] expended in condensing [compressing] air in the receiver with the quantity of heat evolved . . . it was found that a mechanical force capable of raising 823 pounds to the height of one foot must be applied in the condensation of air, in order to raise the temperature of 1 lb. of water 1° of Fahrenheit's scale.

In another experiment, when air condensed in one vessel was allowed to pass into another vessel from which the air had been exhausted, both vessels being immersed in a large receiver full of water, no change of temperature took place, no mechanical power having been developed.[55]

Roget concluded by stating that the author considered 'these results as strongly corroborating the dynamical theory of the nature of heat, in opposition to that which ascribes to it materiality'.

In the *Phil. Mag.* version, however, Joule introduced his new method very differently as one which involved 'the changes of temperature arising from the alteration of the density of gaseous bodies – an inquiry of great interest in a practical as well as theoretical point of view, owing to its bearing upon the theory of the steam engine'.[56] The presentation was in fact framed by concerns with the motive power of heat since the conclusion contained a vigorous attack on Clapeyron's theory (ch.3).

'It is the opinion of many philosophers', wrote Joule, 'that the mechanical power of the steam-engine arises simply from the passage of heat from a hot to a cold body, no heat being necessarily lost during the transfer'. In the course of its passage, the caloric developed *vis viva*. Joule, however, asserted that 'this theory, however ingenious, is opposed to the recognized principles of philosophy, because it leads to the conclusion that *vis viva* may be destroyed by an improper disposition of the apparatus'. Aiming his criticism at Clapeyron for a cleverly contrived theory, Joule explained that the French engineer had inferred that the fall of heat from the temperature of the fire to that of the boiler leads to an enormous loss of *vis viva*. Invoking a shared belief with two eminent Royal Society Fellows, Joule countered: 'Believing that the power to destroy belongs to the Creator alone, I entirely coincide with Roget and Faraday in the opinion that any theory which, when carried out, demands the annihilation of force, is necessarily erroneous'. His own theory, then, substituted the straightforward conversion into mechanical power of an equivalent portion of the heat contained in the steam expanding in the cylinder of a steam engine.[57]

As Roget had rather tersely implied, Joule's experimental method relied upon the accurate measurement of the heat produced by work done in compressing a gas. Conversely, the expansion of a gas against a piston would result in a loss of heat equivalent to the work done. On the other hand, the argument that no work was done by a gas expanding into a vacuum rested on the contentious claim – not yet a 'fact' – that no change in temperature had been or could be detected. Everything thus seemed to depend upon one's faith in the accuracy of the thermometers employed. As Otto Sibum has argued, Joule's own exacting thermometric skills can be located in the context of the family brewing business.[58] Such personal skills, however, evidently carried little weight with Joule's peers. Yet none of the potential difficulties with the method deterred Joule from concluding that the method yielded 'mechanical equivalents of heat' from 823 to 760, with an average of 798. This average , he asserted, was 'a result so near 838 lb., the equivalent which I deduced from my magnetical experiments, as to confirm, in a remarkable manner', this explanation of the heating effects of compression and rarefaction of air in terms of work done on or by the gas.[59]

Joule, however, took his inferences much further than simply claiming an equivalence value between heat and work. A year earlier he had argued that if heat were motion, the production of heat from mechanical force would follow. He now insisted that his results afforded 'a new and, to my mind, powerful argument in favour of the dynamical theory of heat which originated with Bacon, Newton, and Boyle, and has been at a later period so well supported by the experiments of Rumford, Davy, and Forbes'. Joule's appeal to historical precedent and pedigree of the most trustworthy kind was clearly not sufficient to counter his tendency to slide here into the kind of hypothetical reasoning for which experimental philosophers at the Royal Society frequently professed extreme distaste.[60]

Such hypothetical reasoning took the form of a specific theory of heat worked out during the summer of 1844 and incorporated both into his 'Rarefaction' paper and into an appendix to his 'Electrolysis' paper. Reading Faraday's finding that 'each atom is associated with the same absolute quantity of electricity' in a very literal manner, Joule assumed that 'these atmospheres of electricity revolve with enormous rapidity round their respective atoms; that the momentum of the atmospheres constitutes "caloric", while the velocity of their exterior circumference determines what we call temperature'.[61] The theory then accounted for the law of Boyle and Mariotte by supposing that in gases the attraction of atmospheres to their respective atoms and of atoms to one another was inappreciable. Thus 'the centrifugal force of the revolving atmospheres is the sole cause of expansion on the removal of pressure'. Joule further argued that it could easily be extended to radiation by supposing that the atmospheres possess, in a greater or lesser degree, 'the power of exciting isochronal undulations in the aether which is supposed to pervade space'.[62]

Relentlessly pursuing his conviction that the grand agents of nature were indestructible, Joule began work around 1845 on a fourth method for determining the mechanical value of heat. His second attempt to enter the sacred precincts of the Royal Society rebuffed, he returned to the BAAS's Section B. The 1845 meeting assembled in Cambridge, the first time in 12 years. 'I was sorry to hear of the poorness of the meeting. I fear the glory of the Association is gone', wrote William Thomson's younger brother John. Many of the founding gentlemen of science, most notably Whewell and Forbes, seemed less enthusiastic participants in the annual spectacle than they had been in the Association's earliest years.[63] A younger generation had not yet arisen to infuse new life into the scientific body. It was therefore not to be expected that Joule's latest efforts under the now-predictable title 'On the Mechanical Equivalent of Heat' would receive anything other than a polite but unremarkable reception.

This time the experiments involved a paddle-wheel placed in a can filled with water. Weights attached over pulleys working in opposite directions communicated motion to the paddle-wheel (Figure 4.2). Joule argued that 'the force spent in revolving the paddle-wheel produced a certain increment in the temperature of the water' and concluded that 'when the temperature of a pound of water is increased

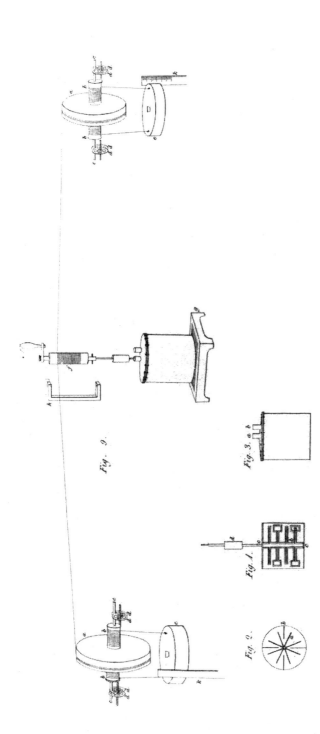

Figure 4.2 Joule's depiction of the paddle-wheel apparatus for his Royal Society paper (1850). In the principal diagram, the pulley-wheel arrangements for the descending weights are shown on the right and left, and the copper vessel containing the paddle-wheel is shown in the centre. The details of the paddle-wheel arrangement are displayed in the subsidiary diagrams (Joule 1850).

by one degree of Fahrenheit's scale, an amount of *vis viva* is communicated to it equal to that acquired by a weight of 890 pounds after falling from the altitude of one foot'.[64]

He had now firmly introduced the language of *vis viva* (measured as $1/2mv^2$) to describe the conversion of the motion of the falling weights into the heat of the water. He was thus identifying his analysis with a by now well-established point of view in engineering mechanics (ch.3), though with the new claim that the motion 'lost' was in fact converted into heat, understood as another form of *vis viva* (rather than momentum). It is likely that Joule's adoption of *vis viva* here was contingent upon the fact that his engineering friend, Eaton Hodgkinson, had read to the Lit. & Phil. as recently as April 1844 'Some Account of the Late Mr. Ewart's Paper on the Measure of Moving Force' which discussed the principle of *vis viva* in relation to the effects of machines.[65]

Undaunted by the seeming indifference to his claims, Joule seized an opportunity in the spring of 1847 to deliver a popular lecture entitled 'On Matter, Living Force [*vis viva*] and Heat' to a Manchester audience. The lecture gave clear expression of the full range of Joule's philosophy of nature, not least in respect of his cosmical perspective which differed considerably from that of the later energy cosmos (ch.7). Joule, a life-long Tory in politics and always concerned with the need for order and stability in a society frequently threatened by radical forces for social change, offered a conservative vision of the divinely ordained system of nature and allowed it to be published in a conservative newspaper, the *Manchester Courier*. His vision of the universe was of machinery working in the most ordered way imaginable, with no tendency to failure or decay. The principles of continual conversions or exchanges, established and maintained by God as the basis of nature's currency system, guaranteed this dynamic stability in nature's economy:

> Indeed the phenomena of nature, whether mechanical, chemical, or vital, consist almost entirely in a continual conversion of attraction through space, living force, and heat into one another. Thus it is that order is maintained in the universe – nothing is deranged, nothing ever lost, but the entire machinery, complicated as it is, works smoothly and harmoniously. And though . . . every thing may appear complicated and involved in the apparent confusion and intricacy of an almost endless variety of causes, effects, conversions and arrangements, yet is the most perfect regularity preserved – the whole being governed by the sovereign will of God.[66]

There was no suggestion here of any tendency for the economy of nature to decay. In this respect, Joule's perspective contrasted with Chalmers's evangelical vision (ch.2) and harmonized with John Playfair's system of nature earlier in the century (Joule had been collaborating with Playfair's nephew Lyon). Indeed, Joule took care to explain that his was a cyclical cosmos, a dynamic equilibrium, where 'the phenomena may be repeated in endless succession and variety'.[67]

By 1847, Joule had enhanced his reputation to the extent that he became a candidate for the natural philosophy chair at St Andrews University, one of Scotland's ancient universities of the period. Although he subsequently withdrew his application (possibly on account of imminent marriage and the demands of the brewery), he at last attracted the attention of elite philosophers of a new generation during the British Association meeting held in Oxford. The positive atmosphere contrasted with the sense of decline that had pervaded the Cambridge meeting two years earlier. *The Athenaeum* reported that the President of Section A (Mathematics and Physics), Professor Baden Powell, 'on taking the chair congratulated the Section on its vast increase in numbers since it last met in Oxford; at which period one small room was found adequate to accommodate them. The present room in its ample dimensions was filled to overflowing; and many of the most celebrated men of Europe congregated within its walls'. Eminent members present included Herschel, Whewell, Airy, William Rowan Hamilton (1805–65) and F.G.W. Struve (1793–1864).[68]

Although forced by pressure of business within the Section to make an abbreviated presentation, Joule nevertheless received welcome public attention from the media. 'Mr Joule', reported *The Athenaeum*, 'exhibited an instrument whereby the heat developed by fans moving in water, oil, and other liquids, could be referred to the distance through which weights descended while whirling the fans round'. *The Literary Gazette* identified Joule's communication as one 'worthy of notice'. It reported that he 'had detected this increase of heat in water to three-fourth's of a degree; the equivalent for which he found to be 775 4/10 lbs . . . The like equivalent for oil was 775 9/10 lbs. This almost exactness, and other tests, convinced him that he had expounded the mechanical equivalent of heat'.[69] It was not, however, the public reporting, nor indeed the presence of a distinguished cast of mathematicians and astronomers, but the notice of a young professor of natural philosophy which conferred upon Joule the credibility that he had long sought. Eager for the reform of British physical science, William Thomson and his associates would increasingly exploit Joule's insights as an integral part of an ambitious new science of energy. But the first appearance of a rival claimant to the mechanical equivalent doctrine threatened simultaneously to rob Joule of some at least of his hard-won credit and to serve his cause by providing it with much needed public attention.

Rivalry and recovery: the case of Mayer

In 1848 a German physician, Julius Robert Mayer, became acquainted with Joule's papers on the mechanical equivalent of heat. Seizing his opportunity to impress upon the scientific establishments the importance of his own contributions during the 1840s, he wrote to the French Academy of Sciences pointing out his claims to priority. Published in the *Comptes rendus* (the Academy's official reports), his letter drew a rapid defence from Joule. Joule's tactics, agreed in consultation with his new advocate William Thomson (ch.5), were to acknowledge Mayer's priority with respect to the idea of a mechanical equivalent, but to claim that he (Joule) had established it by experiment. As he expressed the point to Thomson in March 1851: '[I] shall be quite content to leave M. Mayer in the enjoyment of the credit of having predicted the law of the equivalency. But it would certainly be absurd to say that he has established it'.[70]

Although drawing upon resources common to European men of science (such as Gay-Lussac's experiment on the expansion and contraction of gases), Mayer's papers appeared unorthodox, unconvincing and even unoriginal to his critical readers. As a result, he was rejected in turn by most German and French scientific authorities, leaving him to fall back upon the last resort of private publication. The work of Mayer, outside the dominant schools of European mathematical and experimental science, nevertheless shared with that of his Prussian contemporary, Helmholtz, a straddling of the complementary fields of German physics and physiology.[71]

In very different local contexts, both Mayer and Helmholtz deployed physics to launch aggressive attacks on the notion that living matter depended on a special vital force, *Lebenskraft*. From about the mid-1820s, German physiologists had been reacting strongly against the 'speculative' and 'unscientific' doctrines of *Naturphilosophie* with its account of unity and organisation in Nature in terms of an immanent mind or *Geist*. Rather than advocate a purely reductive and mechanistic physiology, however, physiologists like Johannes Müller (1801–58) gave to living, organised beings a distinctive mode of organisation inexplicable in mechanical terms.[72]

Such reformulations of 'vital force' were problematic. *Lebenskraft* might be presented as nothing more than an expression of organization among related parts which carried no implication of a separate vital force located in the organism. On the other hand, *Lebenskraft* could be interpreted as a causal agency, responsible for generating, directing and sustaining the arrangement of living beings in contrast to purely physical and chemical processes. Tending towards the former, Justus von Liebig's *Lebenskraft* had no power of its own. It owed its effects to relationship with external causes. By analogy with the voltaic cell, the arrangement of the component materials represented a form of stored up activity which could be released by closure of the external circuit. Yet Liebig emphasised that it was in some sense 'a peculiar force, because it exhibits manifestations which are found in no other known force'. In their respective physiologies, however, neither Mayer nor Helmholtz left space for *Lebenskraft*.[73]

Although this new generation attempted to ensure the death of *Lebenskraft* and to distance themselves totally from *Naturphilosophie*, Mayer and Helmholtz formulated their physical doctrines within a German philosophical culture. Above all they shared a distinctively German interpretation of *Kraft*, or force, understood as requiring the separation of bodies in space.[74] For this reason, we should on no account substitute the term 'energy' for *Kraft* or force in the primary literature. Only as the priority disputes developed in the late 1840s, and especially in the 1860s, did the writings of these German protagonists begin to be re-read as 'pioneering contributions' towards the doctrines of energy physics (ch.9).

The case of Mayer dramatically illustrates this historical problem. Following university training as a physician and a year as ship's doctor on a voyage to the Dutch East Indies, he returned to his native Heilbronn early in 1841 to practise as a physician. Soon after, he submitted 'On the Quantitative and Qualitative Determination of Forces' to J.C. Poggendorff's *Annalen der Physik und Chemie*. The paper was rejected and not even returned to its author. Mayer had failed to meet several exacting criteria expected of an aspiring contributor to such a prestigious European scientific journal: he had not referred to any other writers on the subject; he had made no mention of original experimental research; and his terminology would have suggested to contemporary critics that his grasp of mechanical doctrines was slender.[75]

Mayer then undertook a thorough revision of strategy. 'On the Forces of Inorganic Nature' appeared in Liebig's *Annalen der Chemie und Pharmazie* for 1842. It was to be his only paper submitted directly to, and accepted for publication by, a national and international journal. As a professional physician, Mayer had chosen a journal whose audiences included strong medical interests. Once again, however, reference to previous writers was a notable omission from the text, making it difficult for contemporary readers to feel comfortable with its claims. Nor did it offer any physiological insights which might have attracted the attention of his own profession. This time, however, Mayer did introduce the concept of *vis viva* and its conservation, thereby to some extent locating his arguments within recognizable traditions of rational mechanics. But in a German scientific culture increasingly committed to empiricism, Mayer's rational metaphysics would have generated little sympathy.[76]

At first sight, indeed, Mayer's paper had little or nothing in common with Joule's post-1842 papers. For example, Mayer avoided any 'dynamical theory' of the nature of heat. In his view, heat was a force in its own right and not reducible to motion. As a force, therefore, heat was interpreted in relational terms, as 'separation in space'. Within this perspective, Mayer concluded with a brief outline of an experiment concerning the expansion and compression of a gas. 'The sinking of a mercury column by which a gas is compressed, he argued 'is equivalent to the quantity of heat set free by the compression'. Using as a basis the ratio of the capacity for heat of air under constant pressure to that under constant volume, he estimated that 'the warming of an equal weight of water from 0°C to 1°C corresponds to the fall of an equal weight from the height of about 365 metres'. Details of the calculation, however, were not published until 1845, and then only privately.[77]

Mayer's method of calculating the mechanical equivalent became associated with, in Thomson's terminology, 'Mayer's Hypothesis'. The 'hypothesis' was primarily concerned with the assumption of whether the effect of intermolecular forces could be ignored in the expansion and compression of gases. The label simultaneously detracted from Mayer's competence as an experimentalist. In a series of famous experiments on the physical properties of gases (particularly when expanding through small orifices), Joule and Thomson sought to test the validity of 'Mayer's Hypothesis' in the early 1850s.[78]

Mayer's talk of 'equivalents', transferred from chemical discourse, nevertheless provided the flashpoint between two very different approaches to cases of the relationship between heat and work, whether in heat engines or in living creatures.

The initial priority dispute thus served to bring to the attention of the international scientific communities the work of both Mayer and Joule, particularly at a time when Thomson, Rankine, Helmholtz, Clausius and others were reformulating theories of heat and motive power.

CHAPTER 5

Constructing a Perfect Thermo-dynamic Engine

There is an immensity of work in this subject which is enough to occupy a dozen scientific men at least.

James Joule strives to impress G.G. Stokes and the rising generation of Cambridge-educated 'scientific men' (1847)[1]

'When "thermal agency" is thus spent in conducting heat through a solid, what becomes of the mechanical effect which it might produce? Nothing can be lost in the operations of nature – no energy can be destroyed'.[2] First appearing in a footnote to William Thomson's 'Account of Carnot's Theory' (1849), these remarks not only introduced the term 'energy' into the debates over the motive power of heat, but summed up the principal conceptual dilemma which was to remain with Thomson until 1851: how to reconcile the apparent 'losses' of work through the processes of conduction and friction with a conviction that no work could be absolutely lost in the operations of nature.

Following his first meeting with Joule at the 1847 Oxford meeting of the British Association, Thomson was confronted with the conflicting claims of Joule and Clapeyron. Did the work produced by a heat engine depend upon the fall of a quantity of heat from a high to low temperature, that is, from a state of intensity to a state of diffusion? Or were heat and work mutually convertible terms, no fall of temperature being required for the production of mechanical effect? Finding in Joule's conversion of work into heat a persuasive solution to the old problem of fluid friction, Thomson nevertheless remained sceptical of the evidence which Joule had offered for the converse process of recovering work from heat. Applying the Carnot–Clapeyron theory to the problem of an absolute scale of temperature and finding experimental confirmation of the prediction that increase of pressure depressed the freezing point of water (ch.3), Thomson could not assent to Joule's blunt rejection of the Clapeyron theory. Only in 1850–1 did he resolve the dilemma, prompted by the competing investigations of Clausius and Rankine (ch.6).

All of these debates, however, took place within or between specific local sites. Thus Joule's Manchester settings, Thomson's Glasgow classroom and Forbes's Edinburgh location, for example, were brought into communication by means of a *private* correspondence network centred on Glasgow College. At this level of communication among the historical actors, conflicts were identified and engaged. William and James Thomson in particular seized upon one issue after another from Joule, who readily supplied relevant resources for discussion and demonstration. Within each site, laboratory work ranged from dramatic new apparatus designed to persuade sceptics of the mechanical equivalent to meticulous measurements with delicate thermometers to assess the validity of the Clapeyron theory embodied in particular physical processes such as the effects of pressure on the freezing point of ice. William's early Glasgow laboratory would be a central site for much of this work. From there results might be communicated initially to the natural philosophy class or to Forbes, Joule or Rankine. In this way mutual credibility was rapidly built up. Only as confidence increased did Thomson and his allies go public, usually at the Royal Society of Edinburgh where Forbes, in his capacity as Secretary, played a vital co-ordinating role as master of ceremonies and where the presentations would confer enhanced credibility upon all concerned.

Constructing knowledge in private: the Joule–Thomson debates (1847)

By the spring of 1847, Thomson had just completed his first session as professor of natural philosophy at Glasgow University. Having been elected to the chair in 1846, he had returned with a high stock of mathematical credibility acquired during his Cambridge years when he had not only graduated as Second Wrangler and First Smith's Prizeman, but had succeeded to the editorship of the *Cambridge Mathematical Journal* to which he had been contributing many original articles on heat and electricity. A brief period working in Victor Regnault's Paris physical laboratory had also made him well fitted for an overhaul of the experimental work of the Scottish chair. Shortly after arrival, therefore, he was allocated funding by the College to replace the obsolete and decayed stock of apparatus for lecture demonstrations. Having decided to invest heavily in the best electromagnetic apparatus available from the London instrument makers, Watkins and Hill, he was indeed well prepared for an appraisal of Joule's electrical researches.[3]

Towards the end of the session, he was engaged in an intensive mathematical correspondence with his Cambridge friend Stokes whose major current enthusiasm, closely related to his strong interest in the wave theory of light and the concomitant luminiferous ether, was hydrodynamics. 'I have been for a long time thinking on subjects such as those you write about [the mathematics of fluid flow or hydrodynamical theory], and helping myself to understand them by illustrations from the theories of heat, electricity, magnetism, and especially galvanism [current electricity]; sometimes also water', Thomson told Stokes in April 1847. 'I can strongly recommend [Fourier's] heat for clearing the head on *all* such considerations, but I suppose you prefer cold water'.[4] When his teaching duties ended,

Thomson left Glasgow for Cambridge where he spent much time discussing hydrodynamical problems with Stokes.[5] But in June the BAAS intervened and Thomson travelled across to Oxford.

Describing his experiences among the great and the good of the BAAS to his father at the end of the week, he laid considerable stress on the researches of Joule as being of special interest to his brother James. In William's judgment, Joule was 'wrong in many of his ideas, but he seems to have discovered some facts of extreme importance, as for instance that heat is developed by the fric^n of fluids in motion'.[6] This 'fact' connected on the one hand to the brothers' concern with problems of 'losses' of useful work in fluid friction (ch.3) and on the other hand to recent discussions with Stokes on hydrodynamical analogies for heat, magnetism and other physical phenomena.[7]

Shortly after their first meeting, Joule left Thomson copies of his 1843 and 1844 papers on the mechanical value of heat with the porter of the Oxford College where Thomson was in residence and added in the accompanying note: 'I have felt very much gratified in meeting with two at least, Mr Stokes and yourself, who enter into my views of this subject and hope to be able to cultivate an acquaintance I find so delightful'.[8] The succeeding years were to see just such a cultivation which would quickly enhance the mutual credibility of everyone who entered into the construction of the new science of energy.

So great was the interest aroused in Thomson's existing circle (ch.3) that, only a couple of days after Thomson's departure from Oxford, Lewis Gordon expressed the opinion that Joule's 'denial of Carnot's beautiful idea will not necessarily overthrow so fertile a theory. – I shall be glad to be re-enlightened on this subject by you'. Thomson then despatched Joule's papers to his brother James for authoritative appraisal. Remarking that they would astonish him, William believed 'at present some great flaws must be found' and urged James to 'look especially to the rarefaction and condensation of air, where something is decidedly neglected, in estimating the total change effected, in some of the cases'.[9]

In his private response James speculated that Joule might say that the transfer of a 'degree of heat' from a hot to a cold body would result in the cold body acquiring a greater 'absolute amount of heat than that lost by the hot one; the increase being due to the mechanical effect which might have been produced during the fall of heat from the high temperature to the low one'. In other words James attempted here a reconciliation of Joule and Clapeyron by discussing a case of the 'fall' of heat (conduction, for example) in which no mechanical effect was actually produced. In order to account for the non-production of mechanical effect, therefore, James distorted Joule's view such that the effect appeared as additional heat in the cold body, supplementing that which was merely being transferred by conduction from the hot body (Figure 5.1).[10]

For his part Joule lost no time in communicating further with Stokes who had also attended the BAAS meeting. He explained on 10 July 1847 that he was 'at present engaged in getting up an apparatus made of wrought iron in order to repeat the experiment on friction of fluids with mercury' and wanted to know if in fact it had been Stokes who had suggested this form of the

Joule's view:

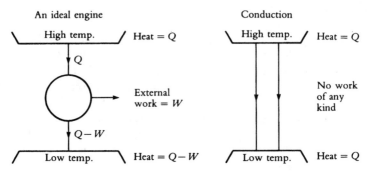

James Thomson's view:

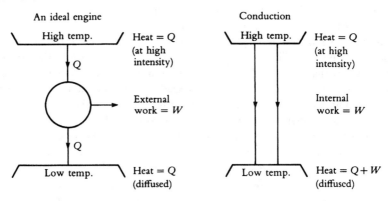

Figure 5.1 The conceptual gulf between Joule and James Thomson are here represented schematically in terms of differing relations between work and heat during the operation of an ideal engine and during conduction (Smith and Wise 1989: 309).

experiment at Oxford. He also stated his intention 'to repeat the expt with friction of metal against metal as suggested by Faraday'. Having seized a good opportunity to promote his views to the elite mathematical and experimental philosophers of the BAAS, Joule recognized that his best hopes for credibility lay with the new generation. Two weeks later, Joule wrote again to confirm that Stokes's intervening prediction 'respecting the success of the mercurial apparatus is fully verified by the only experiment hitherto made with it'. The specially constructed apparatus yielded an equivalent of 782, 'wonderfully close to 775', the equivalent already deduced.[11]

As his second teaching session got under way in late 1847, Thomson began to undertake his own experimental investigation into the mechanical value of heat by means of a manually driven paddle wheel apparatus situated in his Glasgow College apparatus room. He informed Forbes that by the end of the week he 'may have succeeded in boiling water by friction'. The significance to Thomson of this kind of experiment was its visible and dramatic demonstration of the conversion of mechanical effect into heat. Furthermore the prospect of actually boiling water in this way suggested that the cycle might in principle be completed if the steam generated were then employed in the production of useful work.

In practice, however, Thomson ran into difficulties. At first, results were promising as the temperature rose from about 47°C to 57°C at a rate of about one degree centigrade every five minutes. The second experiment set the initial conditions more favourably with the water at about 98°C. This time the temperature climbed 'at a much slower rate to about 99°'. At that moment 'part of the machine gave way before the experiment could be considered as quite decisive'. Undaunted, the professor called for a redoubling of efforts on the part of his labour force: '[My] assistant was preparing to make an experiment yesterday, to commence with water at 80° or 90°, & to go on grinding (along with another, for relief) for about 4 hours, but I have not heard yet whether it went on'.[12]

One year later he offered Joule an account of a development of his manually driven paddle wheel apparatus. A revolving 8-inch diameter disc of tin-plate with radial vanes on each side immersed in water contained in a circular box resulted in 'an immense fluid fric"'. The principal aim, however, was not a quantitative result, but a dramatic demonstration to convince a sceptical audience. Perhaps by now aware of the limitations of human labour, he exploited the cotton mill of his future father-in-law Walter Crum (1796–1867): 'I have had the apparatus adapted to being driven by machinery in a cotton mill, & I shall not be surprised if, by keeping the paddle going 10 times per second, & preventing as much as possible the loss of heat by radiation, we may be able to boil the water by friction alone'.[13]

Joule saw in this prospective demonstration the strategy to win over at last the sceptics in Manchester and London:

> I had found considerable difficulty in persuading our scientific folks here that the heat derived in my experiments was not derived from the friction of the bearings under water, but your experiments were not so easily cavilled at. I have also always found a difficulty in making people believe that fractions of a degree could be measured with any great certainty, but your experiments showing a rise of temperature of 30° or 40° would prove the truth of the fact by the warmth as felt by the hand[14]

From these remarks, it is evident that Joule had hitherto enjoyed limited success in persuading even his Manchester audience. Now he seized upon Thomson's strategy, dependent upon a visible change of state rather than upon a delicate thermometric instrument. He added, however, that he was 'working with a view to get the equivalent more exactly'. This work resulted in his definitive Royal Society paper 'On the Mechanical Equivalent of Heat', published in the *Philosophical Transactions* (1850), and key to his election as FRS in the same year.[15]

Back in 1847–8, however, Thomson was in no way convinced that the converse operation had been demonstrated. In a footnote to his 1848 paper on the absolute temperature scale (ch.3) he drew attention to Joule's 'very remarkable discoveries which he has made with reference to the *generation* of heat by the friction of fluids in motion, and some experiments with magneto-electric machines, seeming to indicate an actual conversion of mechanical effect into caloric'. In Thomson's view, 'No experiment however is adduced in which the converse operation is exhibited; but it must be confessed that as yet much is involved in mystery with reference to these fundamental questions of natural philosophy'. From the context it should also be noted that Thomson's use of the term 'caloric' did not presuppose a commitment to a material theory of heat: caloric here was used by him simply as a synonym for heat.[16]

Evidence that Thomson believed firmly in the principle that mechanical effect could not be absolutely lost appeared in a terse paper which Stokes presented on his behalf to the BAAS Swansea meeting (1848). Here he referred to the 'very beautiful theorem' of the German physicist Franz Neumann (1798–1895) expressing completely 'the circumstances which determine the intensity of the induced current' in a closed wire by a magnet in relative motion. The theorem expressed the induced current as rate of change of a potential function for the force between magnet and current. Thomson explained that 'a very simple *a priori* demonstration of this theorem may be founded on the axiom that the amount of work expended in producing the relative motion on which the electro-magnetic induction depends must be equivalent to the mechanical effect lost by the current induced in the wire'.[17]

He therefore assumed that the work done in producing the current had to be equal to the 'losses' which occurred during conduction of the current through the wire: 'In the first place, it may be proved that the amount of the mechanical effect continually *lost* or spent in some physical agency (according to Joule the generation of heat) during the existence of a galvanic current in a given closed wire is, for a given time, proportional to the square of the intensity of the current'. Thomson's paper extended earlier investigations of electric and magnetic forces as gradients of total potentials, that is, of the work done to assemble the systems. Now his demonstration was expressed in terms of rate of working.[18] But the paper offered no comment on the possibility of 'recovering' the mechanical effect 'lost' or 'spent' as heat. That question prevented total agreement between Thomson and Joule.

Responding quickly to Thomson's footnote of October 1848, Joule observed Thomson's continuing adherence to the Carnot–Clapeyron theory and stated his opinion that Thomson's views would 'lose none of their interest or value even if Carnot's theory be ultimately found incorrect'.[19] He was above all puzzled that the Glasgow professor had not been persuaded of the converse relation between heat and mechanical effect.

Joule for his part believed that he had experimentally demonstrated this claim with his electrical apparatus (1843). He had employed a battery first to generate heat in the coil of an electromagnetic engine held fixed and had then used the current to generate heat *and* work by allowing the coil to rotate. He had found the heating effect of the current reduced by the equivalent of the work done (ch.4). Now he tried to convince Thomson that the chemical forces between the atoms in the battery which in ordinary cases would be converted into heat were in this case turned into mechanical effect. Thomson simply responded that he would have to defer saying anything further about these experiments, for 'indeed I have not yet sufficiently considered the subject to see it in its bearings to our views on the Heat question'.[20]

The Manchester gentleman of science also reminded Thomson of the rarefaction and condensation of air experiments (1844) in which 'I thought I had proved the convertibility of heat into power; for I found that on letting the compressed air escape into the atmosphere, a degree of cold was produced *equivalent* to the mechanical effect estimated by the column of atmosphere displaced'. He even attempted a new argument to persuade Thomson of the necessity of admitting the heat–into–work relation:

> It appears to me that a theory of the steam engine which does not admit of the conversion of heat into power leads to an absurd conclusion. For instance, suppose that a quantity of fuel A will raise 1000 lbs. of water 1°. Then according to a theory which does not admit the convertibility of heat into power the same quantity A of fuel working a steam engine will produce a certain mechanical effect, and besides that will be found to have raised 1000 lbs. of water 1°. But the mechanical effect of the engine might have been employed in agitating water and thereby raising [say] 100 lbs. of water 1°, which added to the other makes 1100 lbs. of water heated 1° in the case of the engine. But in the other case, namely without the engine the same amount of fuel only heats 1000 lbs. of water. The conclusion from this would be that a steam engine is a *manufacturer* of heat, which seems to me contrary to all analogy and reason.[21]

On 27 October 1848 Thomson replied at length to Joule. The importance of this letter to Thomson can be gauged from the fact that he took the unusual trouble to preserve his own copy by pressing absorbent paper on to the original text before the ink had fully dried. In the opening paragraph he declared that he despaired of stating everything in one letter 'especially as I must think and work upon the subject a good deal longer before I can collect my ideas . . . and I now merely write a few remarks which will I hope lead towards an ultimate reconciliation of our views'.[22]

Having provided Joule with the strongest experimental endorsement yet of his claim for the generation of heat from mechanical effect (above), Thomson also admitted Joule's objection to the Carnot-Clapeyron theory 'because it leads to the conclusion that *vis viva* may be destroyed by an improper disposition of the apparatus'. In his 1844 paper, Joule had justified his objection by his *belief* 'that the power to destroy belongs to the Creator alone' and had claimed to coincide with Roget and Faraday, 'in the opinion that any theory which . . . demands the annihilation of force, is necessarily erroneous' (ch.4). Thomson assented to Joule's objection 'in its full force, agreeing as I do with you when you say you coincide with Faraday and Roget'.[23]

In this statement Thomson was affirming with Joule a commitment to a long theological tradition which proclaimed an omnipotent God as not only Creator but also Governor of the universe. The laws of nature were instruments of divine providence with no existence independent of the divine will which alone could alter or destroy them. Furthermore, the basic entities of the material world ('matter' and 'force') were subject to conservation laws such that they could neither be created nor destroyed by human beings. Here, then, was a reaffirmation of the belief in conservation.[24]

From Thomson's point of view, however, Joule had not offered a persuasive resolution of what happens when heat is conducted from high to low temperature rather than used to produce mechanical effect. He confessed that he had never seen any way of explaining the difficulty: 'although I have tried to do so since I read Clapeyron's paper . . . I do not see any modification of the general hypothesis which Carnot adopted in common with many others, which will clear up the difficulty'. This 'general hypothesis' did not necessarily mean a commitment to a material theory of heat but entailed the adoption of an assumption that any gain or loss of heat depended only on the initial and final states and not on the path between them (later called a 'state function'). To convince Joule that 'there really is a difficulty in nature to be explained with reference to this point (just as there is with reference to the loss of mechanical effect in fluid friction)', Thomson discussed a thought experiment treating two distinct processes which both produced the same change of state.[25]

The first process yielded work by a reversible compression and expansion (Figure 5.2). A mass of air filled a perfectly insulated cylinder of volume V_0. Thomson supposed the initial temperature of the mass of air to be lower than that of the sea. Then, he continued, 'let the piston be pushed down till the temperature of the air becomes that of the sea' and let the whole apparatus be plunged into the sea. A compression in volume without heat gained or lost by the apparatus accompanied a rise in pressure and temperature (path A).

Thomson next supposed that the 'bottom & sides of the cylinder become perfectly permeable to heat'. The expansion in volume of the mass of air in the cylinder accompanied a fall in pressure at constant temperature, the piston doing work against external pressure (path B). When the piston arrives at its original position, the mass of air will have returned to its original volume, V_0, but at a higher temperature. Because 'the work spent in compressing the air when its temperature was being raised is clearly less than the work obtained by allowing it to expand retaining the higher temperature', Thomson argued, 'there is an amount of work gained' in this process.

The second process, however, was different. Suppose 'The original mass of air, with the piston held fixed, . . . [had] at once been plunged into the sea & been allowed to have its temperature inc[reased] by conduction gradually' (path C).

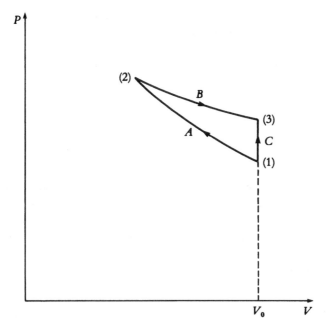

Figure 5.2   This pressure–volume diagram illustrates Thomson's analysis in his letter to Joule on 27 October 1848. The change of state of a cylinder of air from (1) to (3) can either take place along *A* (a path of adiabatic compression) or along *C* (a path of constant volume). Because he regarded heat as a state function, Thomson held that the heat absorbed in the two processes would have been the same (Smith and Wise, 1989: 313).

Thus the same 'effect might have been produced without getting any work'. Thomson therefore asked the fundamental question: 'What has become of the work that might have been gained in arriving at the same result after having gone through the process first described?' He saw as yet no solution to the dilemma, but insisted that there must be an answer: 'I do not see how it can be explained by saying that a greater quantity of heat is taken from the sea in the first place than in the second; and although perhaps an experimental test of the truth is necessary, I believe it would be universally admitted that the quantity of heat absorbed in the two cases would [be] precisely the same'. Thus he continued his allegiance to a view of heat as a 'state function' in such cases and found implausible Joule's claim for the actual conversion of heat (as *vis viva*, for example) into mechanical effect.[26]

Joule responded with the predictable comment that he would certainly say that more heat will in the first process 'be abstracted from the sea than if the temp of the air had been raised by it without any motion of the piston developing force'. Regarding a practical experiment as particularly worthwhile if the result could in any way yield proof against Carnot's theory, he suggested a bath of mercury to represent the sea. But Thomson was in no mood to abandon the theory in whole or in part. By 1848, indeed, his commitment to Carnot's theory had been strengthened when he received from Gordon a copy of Carnot's original text. The outcome was a reading of 'An Account of Carnot's Theory' to the RSE on 2 January 1849.[27]

Reading Sadi Carnot's Reflexions: *Thomson's 'Account' (1849)*

Sadi Carnot's *Reflexions on the Motive Power of Fire, and on Machines Adapted to Develop that Power* (1824) had been conceived, not as treatise in physical or mathematical science, but as a contribution to the engineering problem of fuel economy in heat engines. After 1815 French engineers had displayed a conspicuous enthusiasm for the compound steam engine of Arthur Woolf (1766–1837) on account of its greater economy. This high-pressure design, consisting of a high and a low pressure cylinder to permit the expansion of steam in two successive stages, was employed to pump water from the Cornish tin mines. In both Cornwall and France, where coal was not abundant locally, the superior economy of the Woolf engine was recognized as significant. As Robert Fox has shown, recognition by French engineers of the economic qualities of the Woolf engine provided one vital context for the *Reflexions*.[28]

Carnot's career had owed much to the influence of his father. Lazare's celebrated political and military exploits during the Revolutionary and Napoleonic periods (1789–1815) had included membership of the powerful Directory which governed France in the late 1790s, brief service as Napoleon's Minister of War, and defence of Antwerp against allied forces in 1814. Exiled from 1815 until his death eight years later, Lazare's revolutionary republicanism had found no favour in post-Napoleonic France. Educated largely by his father and then at the *Ecole Polytechnique* (1812–14), Sadi had also pursued a military career. But it was Lazare's concern with conditions for obtaining the maximum effect from hydraulic engines that provided one of the most important resources for Sadi's *Reflexions*.[29]

From 1819 Sadi had become associated with radical professors at the *Conservatoire des Arts et Métiers* (notably Charles Dupin and Nicolas Clément). These strong links, reinforced by his own republican sympathies, emphasized both Sadi's distance from the establishment *Académie des sciences* and his proximity to industrial economy. Lecturing on heat engines at the *Conservatoire* to non-academic audiences, Clément had avoided elite mathematics. Carnot's *Reflexions* pursued a similar strategy. He wrote simultaneously as a political economist and 'academic' engineer who believed that the wealth of a nation depended on production and distribution of marketable commodities, whether raw materials, manufactured articles, or agricultural produce. In each case that process was measured in terms of productive labour or useful work, whether by men, animals, or engines. And he claimed that Britain's economic prowess above all rested on her introduction of heat engines and especially on their development in the hands of practical engineers such as Newcomen, Watt and Woolf.[30]

Unlike Clapeyron's technically precise reworking of Carnot for engineers (ch.3), the original *Reflexions* had projected a powerful Enlightenment vision of the fruits of knowledge. Heat engines were not merely convenient and powerful artefacts which could replace water power in manufacturing industry. The heat engine also 'rapidly stimulates improvements in techniques and may even give birth to other techniques that are completely new'. Thus the use of the steam engine to pump coal mines rendered a service to an industry threatened by problems of flooding, and promoted a ten-fold increase in output. Similarly, the steamship was 'an entirely new advance that we owe to the heat engine':

> Already, as a result, we have been able to establish prompt, regular communications across the seas and on the great rivers of the Old and New Worlds. We have been able to traverse wild regions that were previously almost impenetrable, and to bear the fruits of civilization to parts of the globe where they would otherwise have remained unknown for many years to come. There is a sense in which steam navigation brings the most widely separated nations closer together. It tends to unite the people of the world and make them dwellers, as it were, in one country.[31]

Such remarks embody the 'Enlightenment' values often expressed in the rhetoric of American and French revolutionaries of the last quarter of the eighteenth century. Given the close associations of his father with the politics of revolutionary France, it was hardly surprising that Sadi should have reflected here on the new steamships as bearers of the fruits of 'scientific' civilization from France to all the world and as a means of advancing the cause of universal brotherhood.

Clément and Carnot also shared the aim of considering the maximum effect which could be derived from a quantity of fuel under ideal conditions. Above all Clément had been focusing on the free expansion of the working substance (steam, for example) after the supply had been cut off. Originally patented but not much utilised by Watt in 1782, this 'expansive principle' involved admitting steam to the cylinder for only part of the piston stroke. For the remainder of the stroke the steam was allowed to expand freely. The result was a considerable economy of fuel. Rejecting the traditional assumption that this expansion occurred at constant temperature, Clément claimed that the expansion took place without heat entering or leaving the cylinder (that is, 'adiabatically') and hence that the temperature of the working substance fell. These components of the stroke would provide a key resource for Carnot's cycle of operations.[32]

The engineer P.-S. Girard had formally presented Carnot's treatise to the *Académie* in June 1824 and soon after gave a report in the presence of elite members including Dulong, Laplace, Fourier and Arago. The reception accorded *Reflexions* was neither the result of Carnot's cultural isolation nor personal reticence. Rather it had simply failed to impress and persuade the scientific establishment. Perceived as long in its argument, poorly expressed in its formulation, and wrong in significant detail, its lack of success is not difficult to understand. On the other hand, engineers keen to patent air-engines – and thereby to replace steam engines with more economical forms of motive power – were probably far more positive in their reception of the new theory.[33]

'It is generally known that heat can be the cause of motion, and that it possesses great motive power'. Thus had Carnot begun his *Reflexions*. If the steam engines in 'widespread use' bore witness to that great motive power, so too was heat the cause of 'the great movements that we observe all about us on the Earth', movements that ranged from meteorological phenomena, such as rainfall, to earthquakes and volcanic eruptions. Echoing these cosmical claims for thermal agency, Thomson opened his 'Account' with the assertion that the 'presence of heat may be recognised in every natural object; and there is scarcely an operation in nature which is not more or less affected by its all-pervading influence'.[34]

Constructing an even stronger claim, Carnot had conceived of nature, and especially nature's sources of heat, as a 'vast reservoir' from which 'we can draw the moving force that is indispensable to us'. Nature, 'providing us everywhere with fuel', had 'given us the means to produce at any time and in any place both heat and the motive power to which heat gives rise'. The purpose of heat engines was therefore 'to develop this power and to harness it for our use'.[35] Consistent with Enlightenment ideology, Carnot's Nature, rather than the Christian God, was the great gift-giver and provider to mankind. Unlike Thomson a couple of decades later, however, Carnot had made no attempt to elucidate whether or not this reservoir was inexhaustible. Indeed, the analogy with water power throughout the *Reflexions* suggested a great natural cycle in which water not only ran down hill but was probably restored to its reservoir by the seemingly unlimited agencies of Nature. Significantly, both Lazare and Sadi chose to focus their analysis only on a limited portion of that cycle of nature, namely the 'fall' which could be harnessed by man for the production of motive power.

With characteristic public caution, Thomson's 'Account' avoided any mention, still less a discussion, of nature's 'vast reservoir'. The 'Account' instead acclaimed the virtues of Sadi's simplifying techniques which set aside the complexities of thermal phenomena in general. Because the various thermal effects were often produced together in 'the actual phenomena of nature', Thomson reasoned, it was desirable 'in laying the foundation of a physical theory of any of the effects of heat, to discover or to imagine phenomena free from all such complication, and depending on a definite thermal agency; in which the relation between the cause and effect, traced through the medium of certain simple operations, may be clearly appreciated'.[36]

In the case of the motive power of heat, therefore, the 'sole effect to be contemplated . . . is *resistance overcome*, or, as it is frequently called, "work performed", or "mechanical effect"'. Here lay the great appeal of Carnot's investigation of the motive power of heat in 'reducing' the complex phenomena of nature to the elegant practical simplicity of heat engines. 'Nothing in the whole range of Natural Philosophy', Thomson told the RSE in April 1849, 'is more remarkable then the establishment of general laws by such a process of reasoning'. From his reading of Carnot, then, Thomson distilled two questions, the resolution of which would provide the basis of a complete theory of the motive power of heat. First, 'What is the precise nature of the thermal agency by means of which *mechanical effect* is to be produced, without effects of other kinds?' And second, 'How may the amount of this thermal agency necessary for performing a given quantity of work be estimated?'[37]

Carnot's own reasoning had begun with seemingly very different questions: 'Can we set a limit to the improvement of the heat engine, a limit which, by the very nature of things, cannot in any way be surpassed? Or, conversely, is it possible for the process of improvement to go on indefinitely?' He had noted in particular that there had for many years been attempts to find working substances better than steam: 'Might air, for example, have great advantages in this respect?' He had therefore sought an answer of completely general form, independent of any specific mechanism or working substance such as steam: 'Arguments have to be established that apply not only to steam engines but also to any conceivable heat engine, whatever working substance is used and whatever operations this working substance is made to perform'.[38]

Carnot had then formulated his theory of the motive power of heat in the following way: 'The production of motion in the steam engine always occurs . . . when the equilibrium of caloric is restored, or . . . when caloric passes from a body at one temperature to another body at a lower temperature'. The term 'equilibrium' implied a balance – as in a lever or in a chemical reaction – such that a 'disturbance' of the balance led to a restoration of equilibrium. In this case the equilibrium of the heat or caloric 'has somehow been disturbed, for example by a chemical reaction such as combustion, or by some other means'. This concern with the 'balance of nature', with equilibrium systems capable of self-restoration, was entirely typical of the science of the period.[39] Carnot, however, admitted that 'In fact, the equilibrium of caloric between the bodies is only partially restored; the process [of restoration] is not complete' due to heat leaving through the chimney and in the condensing water. But only after Carnot did the balance model yield to a quite different model, that of 'progression', with the associated issue of 'dissipation' (ch.6).

In order to demonstrate his claim concerning the production of motive power from the passage of heat from a higher to a lower temperature, Carnot had reasoned that the 'caloric' produced in the furnace by combustion 'passes through the walls of the boiler and creates steam, becoming in a sense part of it'. The steam then transports the caloric first into the cylinder, where it drives the piston, and finally into the condenser where the steam is liquefied by contact with cold water. He had therefore concluded that 'the production of motive power in a steam engine is due not to an actual consumption of caloric but to its passage from a hot body to a cold one. It is due, in other words, to a restoration of the equilibrium of caloric'. The motive power of heat thus depended not only on the production of heat, but also on the availability of cold without which 'the heat is useless'. Furthermore, the claim that motive power is produced by the re-establishment of the equilibrium of caloric held good for any heat engine, that is, any engine driven by caloric, independent of the working substance.[40]

The language employed here also suggests that Carnot had been operating within a caloric (heat as a substance), rather than dynamical (heat as motion), theory. Two points, however, need to be noted. First, Carnot was using the term 'caloric' in a sense which did not imply anything about the ultimate nature of heat. He was representing heat in terms of a 'function of state', such that the quantity of heat at the beginning and end of a cycle of operations remained the same. And second, he added in a footnote to the *Reflexions* that 'the main principles on which the theory of heat is based need to be subjected to the most careful scrutiny'. His investigations later led to a private abandonment of a caloric theory in favour of a dynamical theory, but the manuscript notes were to remain unpublished until long after the establishment of the energy doctrines in the early 1850s.[41]

Thomson read this part of the *Reflexions* in the knowledge that there were now two distinct ways in which heat could produce mechanical effect: by means of the alterations of volume as in heat engines; and through the medium of electric agency. The latter method derived from Seebeck's report of thermoelectric currents (1821) which enabled us 'to conceive of an electro-magnetic engine supplied from a thermal origin being used as a motive power'. Because Carnot had not treated of electric agency, Thomson limited his analysis to the former method. He explained that the quantity of mechanical effect developed in such cases depended not only on the thermal agency but also on the alteration in the physical condition of the body. Thus in order to estimate the mechanical effect due solely to thermal agency, 'after allowing the volume and temperature of the body to change, we must restore it to its original temperature and volume'. Following 'almost universally-acknowledged principles', Carnot had assumed that at the end of any such cycle of operations the net quantity of heat remained the same as at the beginning of the cycle. For Thomson, this assumption became Carnot's 'fundamental principle' or 'axiom'.[42]

Since the *Reflexions*, however, Thomson argued that 'a most careful examination of the entire experimental basis of the theory of heat has become more and more urgent', especially with regard to 'all those assumptions depending on the idea that heat is a *substance*, invariable in quantity; not convertible into any other element, and incapable of being *generated* by any physical agency'. That Thomson had not committed himself to this material theory of heat is evidenced by his cautious acceptance of Joule's claim for the conversion of mechanical effect into heat: 'The extremely important discoveries recently made by Mr Joule of Manchester, that heat is evolved in every part of a closed electric conductor, moving in the neighbourhood of a magnet, and that heat is *generated* by the friction of fluids in motion, seem to overturn the opinion commonly held that heat cannot be *generated*'.[43]

Further evidence of Thomson's lack of commitment to a material theory of heat is provided by his view of electricity. In this period, electricity was for him not a substance but a state of a body. Net flux of electrical force into or out of a body measured its state of electrification. Similarly net flow of heat into or out of a body measured its thermal state. Indeed, since 1841 he had been developing views on electricity by analogy with heat conduction.[44] Moreover, he told his natural philosophy class in 1849 that matter was defined as that which 'may be perceived either directly or indirectly by means of the muscular sense of touch or the sensation of resistance'. By this definition 'electricity and heat are excluded and it is probable that neither of them are matter'. A few days afterwards he explained that 'Temperature may be defined as the state of a body as to heat or cold. We cannot explain it any more than that it is a state, a state which we are enabled to judge of by the [non-muscular] sense of touch'. For Thomson, then, quantity of heat entering or leaving a body depended only on its change of state.[45]

While expressing doubts about the material theory in his 'Account', Thomson made clear that as yet 'no operation is known by which heat can be absorbed into a body without either elevating its temperature, or becoming latent, and producing some alteration in its physical condition', that is, there was no persuasive evidence of the conversion of heat into mechanical effect. Carnot's principle thus 'may be considered as still the most probable basis for an investigation of the motive power of heat'.[46]

In order to answer his first question, Thomson now invited his readers to 'consider that machine for obtaining motive power from heat with which we are most familiar – the steam-engine. Following Carnot, Thomson then traced the cycle of heat from furnace to boiler to condenser and back to the boiler. Carnot's principle meant that the quantity of heat at the beginning and end of the cycle was constant. But, Thomson concluded, 'we perceive that a certain quantity of heat is *let down* from a hot body, the metal of the boiler, to another body at a lower temperature, the metal of the condenser; and that there results from this transference of heat a certain development of mechanical effect'. Given that all other cases of the production of mechanical effect from thermal agency showed a similar transference from high to low temperature, Thomson stated the answer to his first question: '*The thermal agency by which mechanical effect may be obtained, is the transference of heat from one body to another at lower temperature*'.[47]

Thomson's second question, concerning the amount of thermal agency required for a certain quantity of work, was shaped by his reading of the next section of *Reflexions*. There Carnot had posed the deceptively prosaic question: 'is the motive power of heat a fixed quantity, or does it vary with the working substance that is used?'. Obviously the answer had to be in terms of a particular amount of caloric and a specified difference of temperature. The answer assumed that with each different working substance, the engine was producing its 'theoretical maximum' of motive power, that is, without any 'practical losses' of power through friction or conduction. Carnot's conclusion had been that 'the maximum amount of motive power gained by the use of steam is also the maximum that can be obtained by any means whatsoever'.[48]

Carnot's 'proof' had derived from consideration of a simple reversible cycle of operations. Heat transferred from body A (furnace or boiler) to body B (condenser) could be used to produce a certain amount of motive power (W) which could then be used to restore the temperature difference in the amount of caloric used. If, however, we had a more efficient means of transferring the caloric to produce motive power ($W + w$), we could not only restore the temperature difference but would be left with surplus motive power (w). Continuing to run the cycle would produce motive power indefinitely, that is, we would have a perpetual

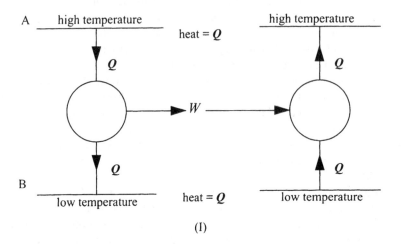

(I)

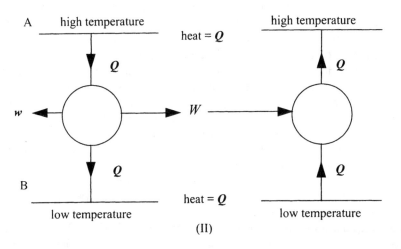

(II)

Figure 5.3 Although Carnot did not use visual representations, this figure helps
to illustrate the form of his textual argument. In (I), a perfectly reversible engine,
heat 'falling' from the boiler A to the condenser B produces work W which, if the
engine is reversed, may be used to raise the heat from B to A. In (II), an engine
more efficient than (I), the fall of heat produces work $W + w$, but the raising of heat
still requires only work W. If such an arrangement as (II) were possible (and
Carnot held that it was not), we would have a perpetual source of power, operating
without the expenditure of fuel other than that initially needed to produce W.

motion machine in the form of a perpetual source of power (Figure 5.3). This would be contrary to 'prevailing ideas, the laws of mechanics, and sound physics; it is inadmissible'.[49]

Appropriating Carnot's reasoning for his own purposes, Thomson represented the issue in terms of the *measurement* of thermal agency 'considered with reference to its equivalent of mechanical effect'. He then constructed what he termed Carnot's conception of 'a *perfect* thermo-dynamic engine'. Here Thomson used the term 'thermo-dynamic' for the first time. He explained that a perfect thermo-dynamic engine of any kind 'is a machine by means of which the greatest possible amount of mechanical effect can be obtained from a given thermal agency'. To answer the second question, we needed to 'construct or imagine a perfect engine which may be applied for the transference of a given quantity of heat from a body at any given temperature to another body, at a lower given temperature' and to 'evaluate the mechanical effect thus obtained'.[50]

Thomson pointed out, however, that such transference 'may be wholly, or partially, effected by conduction through a solid, without the development of mechanical effect'. Consequently, 'engines may be constructed in which the whole, or any portion of the thermal agency is wasted'. Hence Thomson expressed the 'criterion of a perfect engine' in his own formulation: '*A perfect thermo-dynamic engine is such that, whatever amount of mechanical effect it can derive from a certain thermal agency; if an equal amount be spent in working it backwards, an equal reverse thermal effect will be produced*'.[51] This proposition therefore defined a perfect thermo-dynamic engine against which real, less-than-perfect engines could be measured. The second question had apparently been satisfactorily answered.

Echoing recent remarks by Joule on the 'principal obstacle to the perfection of the electro-magnetic engine' (ch.4) and by James on 'the perfection of the [Stirling] engine' (ch.3), William Thomson's concept of a 'perfect thermo-dynamic engine' clearly owed much to the goal of maximum economy shared by his practical engineering contemporaries. But, drawing on his reading of Carnot's *Reflexions*, his concept, I suggest, had cosmological and spiritual, as well as practical, foundations. Recognizing the waste of useful work, and hence imperfections, of all humanly constructed engines, Thomson was drawing attention to a standard of perfection to which humans could only strive and aspire. Concomitantly he was claiming that the imperfections lay with man rather than in the creation. The parallel was with post-evangelical Christian imperatives, rooted in scriptural texts such as Matthew 5: 48: 'Be ye therefore perfect, even as your father in heaven is perfect'. Man's imperfections stood in contrast to the perfections of God, embodied in the perfect humanity of Christ, and His creation.[52]

This interpretation is fundamental to our understanding of a dramatic footnote to Thomson's definition of a 'perfect thermo-dynamic engine'. Having endorsed a few months previously Joule's view that mechanical effect was indestructible save by divine agency, Thomson now found himself with an apparently irreconcilable conflict between Carnot and Joule. By constructing a 'perfect thermo-dynamic engine' against which to evaluate all real thermo-dynamic engines, he had raised an as-yet unanswerable question:

When 'thermal agency' is thus spent in conducting heat through a solid, what becomes of the mechanical effect which it might produce? Nothing can be lost in the operations of nature – no energy can be destroyed. What effect then is produced in place of the mechanical effect which is lost? A perfect theory of heat imperatively demands an answer to this question; yet no answer can be given in the present state of science. A few years ago, a similar confession must have been made with reference to the mechanical effect lost in a fluid set in motion. . . . but in this case, the foundation of a solution of the difficulty has actually been found, in Mr Joule's discovery of the generation of heat. . . . Encouraged by this example, we may hope that the very perplexing question in the theory of heat, by which we are at present arrested, will, before long, be cleared up.[53]

Committed like Joule to the indestructibility of *vis viva* and mechanical effect, Thomson here drew upon Kelland's recent edition of Thomas Young's *A Course of Lectures on Natural Philosophy and the Mechanical Arts* which he had been reading in 1846 as preparation for the natural philosophy chair. Although the word 'energy' had occurred frequently in recent scientific discourses – as in W.R. Grove's 'galvanic energies' of zinc or Roderick Murchison's 'active energy' released from coal by the steam engine – it was Young who had applied the term to *vis viva*.[54] Thomson's choice of the term, therefore, brought together the long pedigree of *vis viva*, the legitimation afforded by a celebrated British natural philosopher, and a widespread qualitative usage in reference to the 'powers' stored up in chemical substances or fuels such as zinc and coal. When promoted vigorously by Thomson and Rankine after the resolution of the conflict between Joule and Carnot, 'energy' was to prove its worth as a term well adapted to wide audience appeal (ch.7).

In contrast Carnot had based his conclusion regarding the maximum of motive power derivable from a given quantity of heat falling through a given temperature upon the verdict that perpetual motion was 'improbable'. His judgment rested on empirical grounds alone: 'do we not know from experience that every attempt to produce perpetual motion has been futile, and that no one has ever managed to generate motion that is truly perpetual'.[55] Joule, Thomson and other British natural philosophers on the other hand often deployed theological arguments to deny not only any possibility of a creation of power from nothing by human agency but its annihilation also. The impossibility of such annihilation had indeed formed the foundation of Joule's rejection of Clapeyron (ch.4).

In 1849, however, reconciliation between Carnot and Joule was very much in doubt. In the same footnote Thomson again rejected Joule's wish to abandon Carnot's 'fundamental axiom' (heat as a 'state function'). On Carnot's theory, a perfect thermo-dynamic engine produced the maximum work for a given temperature 'fall' and a given quantity of heat. Less-than-perfect thermo-dynamic engines might still produce mechanical effect, but there was clearly a 'waste' or 'loss', representing the difference between the work actually produced and that which might have been produced by a perfect version of the thermo-dynamic engine.

Given that conservation dictated no absolute loss, what happened to the waste? Most fundamentally, could the heat conducted from high to low temperature be recovered and made to do useful work? Not until Thomson had formulated his theory of the dissipation of energy (ch.6) could he feel satisfied that these questions had been at least partially solved.

Thomson concluded the general introduction to his 'Account' by noting that Carnot had not concerned himself with the 'details of an actually practicable engine' of either steam or air and had instead 'confined himself to the ideal construction . . . of an engine in which the economy is perfect'. By thus determining 'the degree of perfectibility which cannot be surpassed' and by 'describing a conceivable method of attaining to this perfection by an air-engine or a steam-engine', Carnot had pointed out 'the proper objects to be kept in view in the practical construction and working of such machines'. An ideal of economy had thus been set for the practical engineer.[56]

In an additional section comparing the relative advantages of the air- and steam-engines, read to the RSE on 30 April 1849, Thomson stressed the vast advantage of the former on account of the much greater 'fall' of temperature which could be utilized. He stressed that the 'beautiful engine invented by Mr Stirling of Galston may be considered as an excellent beginning for the air-engine; and it is only necessary to compare this with Newcomen's steam-engine, and consider what Watt has effected, to give rise to the most sanguine anticipations of improvement'. The Thomson brothers, Gordon and Forbes in particular had become increasingly optimistic that practical air-engines might indeed be constructed to replace steam-engines whose economy was limited by the dangerous pressures produced by steam at comparatively low boiler temperatures (ch.8). Gordon for instance had told Thomson in 1847 that he was 'laying a trail for getting up a substantial Company for the manufacture of the [Stirling] Engine'. Over the next few years, Joule, Rankine and Napier would all take up the challenge set by Carnot's theory to find a workable air-engine which could bring fame and fortune to its designer.[57]

As we have seen, the Thomson brothers' reflections on the Stirling air-engine had prompted James to predict, on the basis of Carnot's theory, a depression of the freezing point of water under pressure (ch.3). William communicated his brother's prediction to the RSE at the same meeting which received the 'Account'. Early in the following year, William and his Glasgow College assistant Mansell took Carnot into the nascent 'laboratory' or apparatus room, hitherto used principally for the preparation of classroom demonstrations, and began experimental measurements of the depression of the freezing point of ice under pressures. To this end Mansell had constructed a very accurate ether thermometer: 'This thermometer is assuredly the most delicate that ever was made, there being 71 divisions in a single degree of Fahr.,' the professor told his class.[58] In a remarkable letter to Forbes on 10 January 1850, he recorded the course of experimental work almost as it happened. At first there appeared to be problems with standardizing the new thermometer:

The ether thermometer w^h Mansell has made for me is so sensitive that more than 2 i[nches] (nearly three I think) correspond to 1° Fahr. We have divided the tube very roughly, & we find that somewhere about 70 divisions corre-spond to 1° Fahr. I have found it quite impossible as yet however, & with the means I have at my command to get anything but the rudest estimate of the value of my divisions in Fahr. degrees. The estimate I mention was made by comparing our ether therm[r] with one of Crichton's, but there are great difficulties, I find, in the way of making any comparison at all, & it will be impossible, without other means of comparison, to attain to any satisfactory accuracy.[59]

Later in the letter Thomson admitted that at this point he was not obtaining the hoped-for agreement between James's theoretical prediction and the experimental results: hence he attempted to cover himself with the option of explaining away the discrepancies in terms of the problems of comparison. However, everything changed in the course of penning the letter: 'As soon as we got the thermometer (hermetically sealed in a glass tube) into Oersted's apparatus [for demonstrating the compressibility of water, purchased from the Parisian instrument maker Pixii] everything was satisfactory. The column of ether remained absolutely stationary until pressure was applied'. Then:

> When 9 atmospheres of pressure was applied by the piston, the column of ether sank very rapidly, and appeared to settle about 7½ divisions lower than previously. After that we gave the mass of ice & water in the appar-atus a pressure of 19 atmospheres, and the column of ether sank again very rapidly, until it stopped as nearly as possible 17½ divisions below the primitive position. If 70 of these divisions correspond to 1° Fahr., the temperature of the mixture of ice & water would have been lowered by .250 of a Fahr. degree, or by .139 of a degree cent. Now according to my brother's theory & calculations, 19 atmospheres ought to lower the freez-ing point by .1425°, and therefore the agreement is wonderfully satisfactory.

Thomson admitted 'The fact is when I commenced writing to you I was afraid the agreement was not satisfactory, because the result of my brother's, w[h] I remembered was, I thought, adapted to Fahrenheit's scale'. He thus confessed frankly that he had only just 'become aware of the agreement w[h] really surprises me by being so close'. Here indeed is a revealing private confession: only with the perceived agreement are the doubts about the 'rudest estimate of the value of my divisions' and the possibility of attaining 'to any satisfactory accuracy' suddenly forgotten amid the desire 'as soon as possible to communicate a notice to the Royal Society [of Edinburgh] and to have something published (a few lines would do) in the Philosophical Magazine, so that people may repeat the experiment before the frost goes'.

Just seven days later Thomson was communicating the news to his class. In one experiment, '0.246 was the lowering according to experiment; 0.2295 according to theory'. In another, '0.106 is the lowering by experiment; 0.109 by theory'. 'This is within 1/3000 of a Fahr. degree', he concluded jubilantly.[60] From his private uncertainties, confided to and initially resolved in his letter to Forbes, he now staked his credibility on a semi-public, classroom audience. Were he to have to retract his conclusion, he stood to lose face with his most important local audience with whom he had had to establish himself as a very young successor to an ailing and inactive predecessor in the chair of natural philosophy. For their part, Glasgow students were now witnessing natural philosophy as they had never seen it before. Earlier generations had been used to receive established scientific truth. The latest student body was seeing natural knowledge in the making. Out of the apparatus room issued not merely new demonstration apparatus purchased from the best instrument makers of Paris and London, but new scientific truths en route to their public trial before the RSE.[61]

Clausius reforms Carnot

Born in Prussia and receiving his secondary education at the Gymnasium in Stettin, Clausius entered the University of Berlin in 1840 where he spent over three years studying mathematics and the natural sciences. Qualifying as a Gymnasium teacher, he remained in Berlin throughout the 1840s and cultivated his scientific research in the University. Inspired particularly by his association with the physicist Magnus, the subject of his earliest publication (1847) and of his doctoral dissertation (1848) considered 'the light-dispersing and luminous effects of the atmosphere through theoretical considerations'.

In 1849 he turned to a study of equations of motion of elastic bodies, drawing on both the experimental researches of Regnault and the mathematical investigations of Navier, Poisson, Clapeyron and others. In these investigations elasticity and heat were often very closely connected. For example, heat was measured by the expansive properties of gases and liquids, properties which depended upon pressure and therefore elasticity. Following appointment both as physics teacher in the Berlin Artillery and Engineering School and as *Privatdozent* at Berlin University in 1850, his research focused almost wholly upon heat.[62]

Clausius's 1850 paper for Poggendorff's *Annalen*, 'On the Moving Force of Heat, and the Laws Regarding the Nature of Heat which are Deducible Therefrom', was occasioned by Thomson's 1849 presentation of the conflict between Joule and Clapeyron. Claiming that 'we must look steadfastly into this theory which calls heat a motion', Clausius expressed his opinion that the difficulties were not so great as Thomson considered them to be. Thus we did not have to reject Carnot for 'we find that the new theory is opposed, not to the real fundamental principle of Carnot, but to the addition "no heat is lost"'. In other words, 'it is quite possible that in the production of work both may take place at the same time; a certain proportion of heat may be consumed, and a further portion transmitted from a warm body to a cold one; and both portions may stand in a certain definite relation to the quantity of work produced'.[63]

Explaining that he did not at present wish to discuss the 'nature of the motion

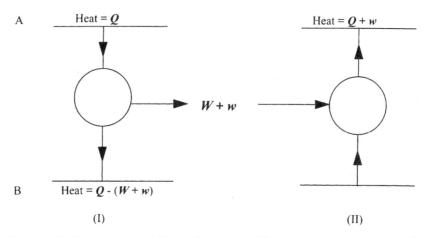

Figure 5.4 Clausius invited his readers to consider two arrangements operating between a warm body A and a cold body B. He then supposed that the working substance of arrangement (I) produced more work than that of arrangement (II). If, for instance, II produced work W from heat Q falling from A to B, then (I) might produce work $W + w$ from Q. Although Clausius offered no visual representations, and indeed avoided any reference to 'engines', this figure illustrates his argument. Work produced by (I) is being used to drive (II) in reverse. But because (I) produces an amount of work w greater than that needed to restore (II) to its original state, w would be manifest in the form of a greater transmission of heat in (II) from B to A than from A to B. By repetition of this process, any quantity of heat might be transferred from a cold to a warm body 'without expenditure of force or any other alteration whatever'. Since this conclusion contradicted the general tendency of heat to pass from warmer to colder bodies, Clausius reasoned that Carnot's claim (that the work done was independent of the nature of the working substance) was vindicated (Clausius 1851: 102–3).

which may be supposed to exist within a body', Clausius assumed 'generally that a motion of the particles does exist, and that heat is the measure of their *vis viva*'. More generally still, he laid down a 'maxim', or 'principle of the equivalence of heat and work', founded upon this assumption of heat as *vis viva*:

> In all cases where work is produced by heat, a quantity of heat proportional to the work done is expended; and inversely, by the expenditure of a like quantity of work, the same amount of heat may be produced.[64]

In the second part of the paper, Clausius reformed what he now termed 'the assumption of Carnot' – that the production of work was 'the equivalent of a mere transmission of heat from a warm body to a cold one, the quantity of heat being thereby undiminished' – into a second maxim to be applied in connection with the first. Thus the latter part of the assumption, that the quantity of heat remained undiminished, had to be rejected as contradictory to the first maxim, but the former portion of Carnot's assumption could 'remain substantially as it is'. Clausius made clear, however, that it was the *maximum* of work which must be considered, for 'a transmission of heat may take place by conduction without producing any mechanical effect'. In the ideal case, then, 'the work depends solely upon the quantity of heat transmitted, on the temperatures [of the hot and cold bodies A and B] . . . and not upon the nature of the substance which transmits it'.

In order to establish this last point, Clausius invited his readers to suppose that two substances existed such that the first (I) could produce more work by the transmission of a certain quantity of heat from A to B than the second (II) (Figure 5.4). We might then apply (I) to produce work from heat and (II) to produce heat from work. If we brought (I) and (II) to their original states, 'the work expended and the work produced would exactly annul each other, and thus, in agreement with our [first] maxim also, the quantity of heat would neither be increased nor diminished'. But with regard to the *distribution* of the heat 'more heat would be brought from B to A than from A to B, and thus on the whole a transmission from B to A would take place'. By this means, and with no expenditure of force or other alteration whatever, 'any quantity of heat might be transmitted from a *cold* body to a *warm* one'. This inference contradicted the general behaviour of heat 'which everywhere exhibits the tendency to annul differences of temperature, and therefore to pass from a *warmer* body to a colder *one*'. Hence, Clausius concluded, 'it would appear that we are *theoretically* justified in retaining the first and really essential portion of the assumption of Carnot'.[65]

News of the publication of Clausius's paper in the prestigious *Annalen* was for Thomson simultaneously welcome and troublesome. On the one hand, Clausius had acknowledged his specific debt to Thomson's 'Account' and thereby raised its credibility on an international, rather than simply Scottish, stage. Furthermore, his reform of Carnot's theory of the motive power of heat gave greatly added public significance to the work of Joule just at the moment Joule was finally achieving his much sought after goal of Royal Society recognition. But on the other hand, Clausius's publication threatened to do for Thomson what Mayer had done for Joule: to rob him of some at least of the credit for formulating a new theory of the motive power of heat. Just how Thomson and his growing number of allies sought to represent Clausius and Mayer will be a recurring theme in subsequent chapters.

CHAPTER 6

'Everything in the Material World is Progressive'

Self-missioned Leader through Creation's maze!
   Dost thou interpret *thus* God's mighty scheme –
Weaving the cobweb fancies of a dream
   O'er each grey vestige of His mystic ways?
When thus 'mid chaos thou didst blindly grope,
   Gathering new links for matter's heavy chain,
Dwelt there not in thy soul the secret hope
   That some strong truth would rend the bond of pain
Which fixed thee to Progression's iron wheel?
   Oh teach not suffering earth such hopeless creed:
For heavy were her curse if doomed to feel
   That, in her frequent hour of bitter need,
Her lifting eye of prayer could only see
   Necessity's stern laws, graven on eternity.
      Sonnet 'To the Author of the "Vestiges of Creation"' (1845)[1]

By 1850 a sustained correspondence network held together the shared enthusiasms of Gordon, Joule, and the Thomson brothers concerning the problems of motive power and the nature of heat. With private scientific practice located mainly in William Thomson's Glasgow classroom and Joule's Manchester laboratory, presentations organized at the Royal Society of Edinburgh by J.D. Forbes provided the principal forum for the construction of public credibility. The RSE performances functioned to secure priority, to permit the promotion of mutual credibility, and to win new recruits to the cause. They also served, unlike the Royal Society of London, to allow debate of unresolved and even controversial issues in an atmosphere of gentlemanly discussion. Moreover, RSE membership, though select, was neither exclusive of engineers and commercial gentlemen, nor was it limited to a senior generation of learned men. It was therefore within this context that early in 1850 the growing network gained another young recruit, Macquorn Rankine, engaged in a similar quest for scientific credibility after several years of involvement in practical engineering.

Following in-depth discussions with Rankine over a 12-month period, William Thomson read the first instalment of his 'Dynamical Theory of Heat' to the RSE on 17 March 1851. He had finally been persuaded by the recent investigations of Clausius and Rankine that a dynamical theory of heat, allowing for the actual conversion of heat into work, offered the best way of reforming the theory of the motive power of heat. But still a fundamental problem remained: the conversion of heat into mechanical effect could be admitted for heat engines operating with a fall from high to low temperatures, while the *reconversion* or recovery of mechanical effect 'lost' as the heat of conduction or friction in less-than-perfect engines could not. In the weeks prior to his public presentation, therefore, Thomson struggled in a draft manuscript, revealing of his private convictions and uncertainties, to resolve this problem and simultaneously to construct a new demonstration of Carnot's criterion for a perfect engine.

His resolution rested upon a fundamental conviction that 'Everything in the material world is progressive'.[2] The directional flow of energy through space offered human beings the opportunity of directing, though not restoring, those mighty gifts of the Creator, the energies of nature. But such an irreversible tendency was not 'loss' of energy *in* the material world. Human beings had a duty to employ engines for the benefit of mankind and in aid of its commercial and moral 'progress'. Failure properly to direct and harness those gifts of energy was therefore only a waste, and in that sense a sin of 'dissipation', with respect to human beings rather than in nature.

Yet Thomson's 1848 confession that 'as yet much is involved in mystery with reference to these fundamental questions of natural philosophy' (ch.5) remained strikingly apt. The evangelical Chalmers, with whom the Thomson family was well acquainted, had spoken again and again in the 1830s of the tendency of nature's designed collocations towards derangement and thus seemed to retain a strong Calvinist sense of an universe cursed by depravity, imperfect in its parts and destined for inevitable decay (ch.2). In contrast, Thomson's 1851 phrase 'Everything in the material world is progressive' was loaded with potentially explosive meaning. Although most probably inspired in part by the Cambridge directionalist geology of Sedgwick and Hopkins, the phrase could also be read as aligning him with academic radicals such as John Pringle Nichol, Glasgow professor of astronomy and one major inspiration for the infamous *Vestiges of the Natural History of Creation*.

The new energy vision, however, never really offered the kind of onwards and ever-upwards progression espoused by developmental and evolutionary theorists like Robert Chambers or Herbert Spencer. Choosing not to speak publicly in terms of 'progress' for another ten years (ch.7), Thomson instead opted in 1852 for a very different terminology with which to capture a sense of the transitoriness of the visible creation. Thus the doctrine of 'the universal dissipation of mechanical energy' set temporal limits to the 'present arrangements' of the solar system and so to human progress and advancement. Presbyterian students, then, could read the doctrine as fully compatible with a traditional vision of a fallen world which was subject, like unregenerate souls, to depravity and death.

An unresolved mystery, however, was the question of whether or not God had written a principle of dissipation into nature. Thomson's terms 'progression' (private) and 'dissipation' (public) shared the notion of directionality or flow in nature. In that sense, energy carried the connotation of a 'fall' in nature and a sense of the 'ageing' of natural systems such as those of the earth and sun. Furthermore, Thomson articulated a doctrine, complementary to evangelical presbyterian theology, that only God could 'regenerate' that energy. But the issue of nature's perfection or imperfection was a fundamental one. Was the creation really subject to similar imperfections as those which rendered man a sinful, fallen creature? Or was the downward directionality everywhere apparent a mark of the progressive perfection of a dynamic creation within which human beings needed to acknowledge in humility before God their mental and moral limitations? These were questions which would pervade debates within and beyond the science of energy for another three decades (chs 9–12).

The Rankine–Thomson dialogue (1850–1)

Characterized as 'a Scot of Scots' by his colleague P.G. Tait, Macquorn Rankine was born in Edinburgh in 1820. His father, a former Rifle Brigade Lieutenant, became increasingly involved with practical engineering during the railway booms of the 1830s and 1840s. Daughter of an eminent Glasgow banker, his mother was related to the chemist Thomas Graham who in 1837 moved from the chair of chemistry at Anderson's College, Glasgow, to that at University College, London, and whose *Elements of Chemistry* (1842) had critically examined the material and undulatory theories of heat.[3]

Although Macquorn briefly attended Ayr Academy (1828–9) and Glasgow High School (1830), his schooling was largely informal on account of poor health. At the end of 1834 a maternal uncle presented him with a Latin copy of Newton's *Principia*. As he recorded in his private journal, a careful reading of this classic text 'was the foundation of my knowledge of the higher mathematics, dynamics and physics. My knowledge of the higher mathematics was obtained chiefly by private study'. Another journal entry, according to P.G. Tait, recorded his lasting obligations to his parents 'for early instruction in the fundamental principles of the Christian religion and the character of its Founder'.[4] Although the journal is no longer extant, these remarks, placed alongside his close friendship with Elder and Napier (ch.8), suggest that Rankine's religious position differed but little from the new Moderate theology of Norman Macleod with its emphasis on the person of Christ as God Incarnate and 'perfect man'.

Entering Edinburgh University in late 1836, Rankine left at the end of his second session without taking a degree. During his two years at the University, he attended classes on chemistry, botany, natural history and natural philosophy, but not, it seems, mathematics. As natural philosophy professor Forbes was then at the height of his experimental investigations into radiant heat. Eager to find favour with Cambridge mathematical reformers and advocates of the undulatory theory of light (notably Airy, Whewell and Herschel), Forbes claimed his results showed that both light and radiant heat manifested wave-like properties, results which suggested that both sets of phenomena derived from transverse undulations in an etherial medium.[5]

As part of his campaign to enhance the quality of Scottish university education along Oxbridge lines, Forbes created a Gold Medal prize which Rankine won for an essay on the undulatory theory of light in 1837. Attending the natural philosophy class for a second time, he won another prize for an essay on 'Methods in Physical Investigation' the following year. Both essays related closely to his subsequent 'hypothesis of molecular vortices' which he would use to explain in mechanical terms the identity of light and radiant heat and whose legitimacy he would often seek to justify with appeals to scientific method.[6]

Rankine's first scientific paper, 'On the Laws of the Conduction of Heat and on Their Application to Geothermal Problems', appeared in the short-lived *Edinburgh Academic Annual* (1840). Embodying the recent interests both of Forbes (underground temperatures) and Kelland (heat conduction), it aimed to establish the 'true' law of heat conduction and the 'true' distribution of temperature through the body of the earth.[7] Limited to a very local audience, its content nevertheless reflected issues which were currently high on the agenda of BAAS meetings.

Sympathetic to the industrial applications of natural philosophy in general, Forbes had included some 17 lectures on heat and the steam engine in the natural philosophy course for 1836–7. Rankine's choice of a career in practical engineering did not therefore represent a sharp break with natural philosophy. At the same time any thoughts which Forbes might have entertained of encouraging this star pupil to embark on a Cambridge mathematical education quickly evaporated. Constrained by his father's limited financial resources, Macquorn abandoned university study in favour of industry.[8]

Having assisted his father in superintending the construction of the Leith branch of the Edinburgh and Dalkeith Railway in 1837–8, Rankine crossed to Ireland in 1838 to become a pupil of John Macneill. Employed on surveys for civil engineering schemes such as harbour and river improvements, Rankine's principal interest now focused on practical railway engineering. Upon his return to Scotland in the early 1840s, he complemented employment on Scottish railway projects with a series of papers submitted to the Institution of Civil Engineers.[9]

Between 1845 and 1848, Rankine was employed on various schemes connected to the Caledonian Railway Company's ambitious goal of linking Scotland and England. By early 1848 the celebrated West Coast route between London and Glasgow had been opened, thereby uniting directly the first and second cities of the Empire. At almost the same time, the great railway mania of the mid-1840s collapsed. Projects everywhere, including a proposal to dispose of the old Glasgow College to a railway company in exchange for the financing of a new university, were halted.[10]

Coinciding with these traumatic events, Rankine shifted his interests from railway engineering to other engineering schemes (especially water supply) and back to natural philosophy. This phase of his career, marked by frequent appearances on the British Association stage and by new scientific dialogues with William Thomson and J.R. Napier on the subject of thermo-dynamic (most notably air) engines, culminated in appointment to the Glasgow chair of engineering in 1855 as successor to Gordon (ch.8). It was a position which he developed with characteristic zeal, building up the engineering classes and publishing a classic series of engineering treatises until his death in 1872, when he was succeeded by James Thomson.[11]

In June 1849 Rankine was elected to the Royal Society of Edinburgh. His sponsor was none other than Professor Forbes, the Society's secretary. By early October, Rankine placed in the hands of his former teacher an entire paper on the hypothesis of molecular vortices which consisted of, first, the details of the hypothesis 'which ascribes the elasticity connected with heat to the centrifugal force of small revolutions of the particles of bodies' (the statical part) and second, the application of the model to the mechanical action of heat (the dynamical part). On 4 February 1850, Rankine read both parts, 'On the Centrifugal Theory of Elasticity, as Applied to Gases and Vapours' and 'On the Mechanical Action of Heat, especially in Gases and Vapours', to the Society. Around the same time, Forbes despatched the draft to his colleague William Thomson who produced a critical report. As a result the first part of Rankine's original paper initially appeared in summary form as an introduction to the second part, though a full and revised version of the first part was later published in the *Phil. Mag.* (1851).[12]

In the summary introduction Rankine explained that the 'ensuing paper forms part of a series of researches respecting the consequences of an hypothesis called that of Molecular Vortices, the object of which is, to deduce the laws of elasticity, and of heat as connected with elasticity, by means of the principles of mechanics, from a physical supposition consistent and connected with the theory which deduces the laws of radiant light and heat from the hypothesis of undulations'. Here the debt to Forbes was very clear. But Rankine also claimed that his 'researches were commenced in 1842, and after having been laid aside for nearly seven years, from the want of experimental data, were resumed in consequence of the appearance of the experiments of M. Regnault on gases and vapours'. As he told Thomson in 1850, his 'first attempt to apply mathematical reasoning to the subject arose from my seeing the translation . . . of Clapeyron's paper on the opposite [material] theory'.[13] Clapeyron's memoir, together with Regnault's investigations, therefore constituted major resources for Rankine's paper.

Rankine acknowledged that his molecular model had many conceptual resemblances to the models of earlier natural philosophers, including Aepinus and Davy. Each atom of matter consisted of a nucleus surrounded by an elastic atmosphere, self-repulsive but retained by the attraction of the nucleus. As with Joule quantity of heat was the *vis viva* of revolutions or oscillations among the particles of the atmospheres which Rankine supposed to constitute vortices about the nuclei (ch.4). Apart from attempting to develop the mathematical consequences of the vortex hypothesis, Rankine's claim to originality was that the luminiferous medium transmitting light and radiant heat consisted of the nuclei of the atoms, which vibrated independently, or nearly so, of their atmospheres. His model thus avoided the problems of a separate etherial medium filling the space between nuclei, a medium which required the seemingly contradictory properties of great rigidity to permit transverse undulations *and* minimal resistance to planetary motion.[14]

Offering a detailed account of how a 'dynamical theory of heat' could explain the conversion of heat into work, Rankine took as a principal problem the calculation of the variations in heat in a body deriving from variation of volume and temperature with the aim of determining the effect of cyclic processes in a heat engine. Employing the vortex hypothesis, he argued that the variation of heat would derive from three components: change of volume (external work), change of molecular distribution dependent upon change of volume, and change of molecular distribution dependent upon change of temperature. The last two 'internal changes' were represented as changes in a function U, determined in any given case from the principle of conservation of *vis viva*. He also noted that 'the process followed in ascertaining the nature of the function U is analogous to that employed by M. Carnot in his theory of the motive power of heat'.[15]

At the same time, however, Rankine contrasted his conclusions with those of Carnot's theory:

> According to the theory of this [Rankine's] essay, . . . and to every conceivable theory which regards heat as a modification of motion, no mechanical power can be given out in the shape of expansion, unless the quantity of heat emitted by the body in returning to its primitive temperature and volume is *less* than the quantity of heat originally received; the excess of the latter quantity above the former disappearing as heat, to appear as expansive power, so that the sum of the *vis viva* in these two forms continues unchanged.[16]

Shortly after Rankine's presentation to the RSE, he began direct correspondence with Thomson whose report, though critical, had spoken positively of the author's 'Good remarks on Joule' and 'Good principles of *vis viva*'. Anxious to please both Forbes and Thomson, Rankine modified his paper to accommodate Thomson's objections prior to publication. Thomson's report had also noted that Rankine claimed 'to be the first person who applies Math[ematical] Anal[ysis] to the theory of molecular vortices w[h] was first intelligibly stated by S.H.D. [Sir Humphry Davy]'. Indeed, as Rankine later explained to Thomson, he had no wish to relinquish ownership of the molecular vortex hypothesis: 'although it undoubtedly rests on a much less firm basis than that which is founded on the general law of the mechanical convertibility of heat, and although I believe you did my paper essential service by inducing me to make it less prominent & less detailed than it was originally, still I conceive it may lead to some useful results'.[17] The issue was an important one, for in the subsequent promotion of the new sciences of energy and thermodynamics Rankine's role would often be played down on account of its 'hypothetical' and thus 'controversial' nature compared to the more general formulation of Thomson (ch.7).

On 15 August 1850, Thomson recorded in his personal mathematical notebook that he had 'just written to Rankine telling him of Clausius' paper in Poggendorff (incompl. in the Number [April] I have seen) on the Motive Power of Heat &c'.[18] Thomson had thus become aware of the *first* part of Clausius's 1850 paper which focused on 'deductions from the principle of the equivalence of heat and work'. The second part, on the modified Carnot principle, he had yet to see. Within days Rankine replied to 'thank him for calling my attention to the paper by Clausius, in Poggendorff's Annalen, on the Mechanical Theory of Heat. I approve of your suggestion to send a copy of my paper either to Clausius or Poggendorff'.[19]

Early next month Rankine wrote again to Thomson to inform him that the first part of Clausius's paper 'consisting entirely of deductions from the law of convertibility of Heat and Power, agrees, as far as it goes, with my own investigations'. He stated that he had now also looked over the second part which: 'consists of deductions from the same law [of the convertibility of heat and work], taken in conjunction with a *portion* of the principle of Carnot. . . . Clausius gives a sort of *a priori* proof of this second law, which so far as I have yet been able to consider the subject, seems to me very unsatisfactory'.[20]

Rankine later stressed that he had 'always thought the principle of Clausius . . . had an appearance of probability; but I was not satisfied with his mode of proving it'.[21] The 'proof' which caused Rankine to feel uneasy was Clausius's attempt to establish the modified Carnot principle on what he saw as a widely agreed assumption, namely that the transference of heat from a cold to a hot body was impossible without compensation (chs 5 and 8). By April 1851 Rankine had presented his own 'proof', derived from his molecular vortex theory, to the RSE.[22]

From Thomson's point of view, the dialogue with Rankine highlighted two key issues. First, the separate papers of Rankine and Clausius, both drawing on Joule as a common resource, showed Thomson the importance of a principle of conversion of heat into work. Rankine's detailed molecular vortex hypothesis in particular offered a possible model for the conversion of heat into work. And second, Rankine's reservations about the 'sort of *a priori*' proof of a modified Carnot principle in the second part of Clausius' paper convinced Thomson of the need for a new foundation and a new derivation of that principle.[23]

A proposition stated in Rankine's paper excited both Thomson and Joule on account of its relevance to the mutual convertibility of heat and work. At first sight Rankine's proposition seemed mundane:

> If vapour at saturation is allowed to expand, and at the same time is maintained at the temperature of saturation [which decreases along with the pressure], the heat which disappears in producing the expansion is greater than that set free by the fall in temperature, and the deficiency of heat must be supplied from without, *otherwise a portion of the vapour will be liquified in order to supply the heat necessary for the expansion of the rest.*[24]

What interested Thomson here was the striking difference between steam issuing from an orifice in a high pressure and a low pressure boiler: in the former case the steam was 'dry' and non-scalding; in the latter case 'wet' and scalding. Somehow there had, in the former case, to be 'a supply of heat from without'. Rankine's result, Thomson thought, could 'be reconciled with known facts [of the non-scalding property of steam from a high pressure boiler] only by means of your discovery, that heat is evolved by the friction of fluids in motion . . . Either then Mr Rankine's conclusion is opposed to the facts, or *some heat must be acquired by the steam as it issues from the boiler*'.[25] Thus far, the phenomenon appeared to provide further dramatic evidence of the conversion of work into heat by fluid friction.

Thomson, however, also saw here possible experimental conditions for the conversion of heat into work, a process of which he had long been sceptical. Still wary of Rankine's reasoning from a specific molecular vortex hypothesis, Thomson now accepted instead that, combining a recent experimental result of Regnault (that the 'total heat' of saturated steam increases slowly with the temperature) with Joule's mechanical equivalent, 'we may demonstrate Mr Rankine's remarkable theorem without any other hypothesis than the convertibility of heat and mechanical effect'.[26] With these remarks, then, Thomson had at last come very close to accepting in full Joule's claim for the mutual convertibility of heat and work.

Thomson's 'Dynamical Theory of Heat'

In February 1851 Joule informed Thomson that he had just 'read with very great pleasure your letter received this morning by which you seem to have completely solved the difficulty which before seemed to render the results of Carnot's theory and what we must consider the true theory irreconciliable'. He therefore urged Thomson 'to lose no time in following it out to all its legitimate deductions and send it to the RSE'.[27] 'Following it out to all its legitimate deductions' produced Thomson's massive series of papers 'On the Dynamical Theory of Heat' published between 1851 and 1855 and occupying over 150 pages in his collected *Mathematical and Physical Papers*. In this section, however, I focus on his enunciation of two propositions, attributed respectively to Joule and to Carnot and Clausius, the second of which required for its foundation Thomson's resolution of the problem of the irrecoverable nature of mechanical effect 'lost' as heat.

For his first proposition Thomson adopted Joule's principle of mutual convertibility of heat and mechanical effect:

> PROP. I. (Joule). – When equal quantities of mechanical effect are produced by any means whatever from purely thermal sources, or lost in purely thermal effects, equal quantities of heat are put out of existence or are generated.[28]

This proposition, Thomson claimed, 'is shown to be included in the general "principle of mechanical effect"'. His demonstration also depended on the assumption of a dynamical theory of heat. According to the dynamical theory, he reasoned, 'the temperature of a substance can only be raised by working upon it in some way so as to produce increased thermal motions within it, besides effecting any modifications in the mutual distances or arrangements of its particles which may accompany a change of temperature'. By the principle of mechanical effect, then, 'the work which any external forces do upon it, the work done by its own molecular forces, and the amount by which the half *vis viva* of the thermal motions of all its parts is diminished, must together be equal to the mechanical effect produced from it; and consequently to the mechanical equivalent of the heat which it emits'. Were the temperature and physical condition of the substance to be restored after a cycle of operations to their original condition, the second and third parts of the work disappear, and 'we conclude that the heat which it emits or absorbs will be the thermal equivalent of the work done upon it by external forces, or done by it against external forces; which is the proposition to be proved'.[29]

Thomson's unqualified commitment to this first proposition did not, therefore, rest on experiment: while he had been persuaded empirically of the conversion of work into heat, the converse process still remained experimentally questionable. Thus in an early draft he noted: 'the author considers that as yet no experiment can be quoted which directly demonstrates the disappearance of heat when mechanical effect is evolved; but he considers that the fact has only to be tried to be established experimentally, having been convinced of the mutual convertibility of the agencies by Mr Joule's able arguments'.[30]

Mr Joule's 'able arguments' centred on the assertion that *if* a dynamical theory of heat were true, *then* a principle of *mutual* convertibility necessarily followed. In Thomson's words: 'it is only by the general confirmation of the [dynamical] theory that heat is motion afforded by his [Joule's] experiments that this assertion [of mutual convertibility] is supported'.[31] Since Thomson did not appear to have been persuaded by such arguments when first put to him by Joule in 1847–8, and retained considerable doubts in the following two years, we must suppose that in this draft he was in the process of rehearsing the arguments for a dynamical theory in readiness for his public appearance on the subject at the Royal Society of Edinburgh in March 1851.

Further support for this interpretation derives from Thomson's presentation of Sir Humphry Davy's experiment of melting two pieces of ice together. In a notebook entry of 1848 James Thomson had remarked that 'Davy's exp[eriment] of melting 2 pieces of ice by their mutual friction w^d be incompatible with my view of the subject as it w^d show that *vis viva* or work actually produced *quantity of heat*. (Is the exp. to be trusted?)'.[32] In the 1851 draft, however, William began to portray Davy as laying the foundations of the dynamical theory of heat. Nevertheless, Thomson's account was still riddled with doubts and deletions:

> Sir Humphry Davy, in his first published work [1799], laid down the following proposition. 'The phenomena of repulsion [heat] are not dependent on a peculiar elastic fluid [caloric] for their existence, or caloric does not exist'. <Even if the form of demonstration by which he supported this proposition

be not considered satisfactory . . . > and in support of it he describes an experiment in which two pieces of ice in circumstances so arranged as to prevent the possibility of any heat being received by either from external sources, were melted by being rubbed together. Whatever may be thought of the reasoning founded on this . . . <practically it has not been convincing> (it has certainly not been generally *convincing* perhaps because not generally known) this one experiment must be admitted as sufficient in the present state of science to establish the proposition. In the train of reasoning which follows Sir H.D. concludes that heat consists of a motion excited among the corpuscles of bodies . . . The Dyn^l T of H thus laid down by Sir H.D. has not met with any efficient support <that has contributed to its advancement> until Mr Joule of Manchester commenced the valuable series of researches[33]

In contrast Thomson's public presentation contained no such reservations. Davy, through his ice melting experiment, had decisively demolished the caloric theory and thus 'established' the dynamical theory of heat. The more recent researches of Mayer and Joule (ch.4) on fluid friction and heat generated by electric currents would, furthermore, 'be sufficient to demonstrate the immateriality of heat, and would so afford, if required, a perfect confirmation of Sir Humphry Davy's views'.[34]

Thomson's public articulation then reached a more delicate stage. In the original version, read to the RSE., he appeared to retreat from his opening proclamation by acknowledging that Davy had only 'established beyond all doubt the fact that heat may be created by mechanical work'. He therefore attempted to explain that:

> the converse proposition, that heat is lost when mechanical work is produced from thermal agency, appears to have been first enunciated by Mayer in 1841. In 1842 the same proposition was enunciated by Joule, and a number of most admirable experiments illustrating the mutual convertibility of heat and mechanical effect . . . are described in his paper on Magneto-electricity[35]

Realizing that the conversion of heat into mechanical effect was not self-evident in such experiments, Thomson altered the passage entirely for the full and final version of his paper, published in the Society's *Transactions*. There he resorted once more to the argument that, given the truth of the dynamical theory, *mutual* convertibility followed: 'Considering it as thus established [by Davy, Mayer and Joule], that heat is not a substance, we perceive that there must be an equivalence between mechanical work and heat, as between cause and effect'.[36]

Such remarkable convolutions bear witness to the enormous behind-the-scenes difficulties that Thomson confronted as he sought to make the dynamical theory of heat credible. Public audiences, in contrast, would see little but the apparently automatic logic of the theory, based upon seemingly incontrovertible evidence piled up by a series of contributors to the physical truth. Once the credibility of the dynamical theory was accepted, however, the reconciliation with Carnot also became plausible. Acknowledging the priority of Clausius and Rankine in enabling that reconciliation, Thomson laid down a second proposition:

> PROP.2. (Carnot and Clausius). – If an engine be such that, when it is worked backwards, the physical and mechanical agencies in every part of its motions are all reversed, it produces as much mechanical effect as can be produced by any thermo–dynamic engine, with the same temperatures of source and refrigerator, from a given quantity of heat.[37]

Essentially the same as his 1849 statement of Carnot's criterion for a perfect engine, Thomson now required a new 'demonstration' of this proposition.

'The earth shall wax old &c.'

On 1 March 1851, Thomson privately recorded that the 'difficulty which weighed principally with me in not accepting the theory so ably supported by Mr Joule was that the mechanical effect stated in Carnot's Theory to be *absolutely lost* by conduction, is not accounted for in the dynamical theory otherwise than by asserting that *it is not lost*; and it is not known that it is available to mankind'.[38] Unlike Carnot's theory, then, the dynamical theory was consistent with the belief in a 'principle of mechanical effect' (that is, that no mechanical effect was ever lost in nature). Thomson, however, confessed here that the dynamical theory was no more than an 'assertion', a belief, which offered no solution to the puzzle of whether or not the heat 'lost' in conduction could be 'recovered' in the form of mechanical effect 'available' to mankind for the production of useful work and thus the creation of wealth.

'The fact is', Thomson continued, 'it may I believe be demonstrated the work is *lost to man* irrecoverably [when conduction occurs]; but not lost in the material world'. Reintroducing the word 'energy' for the first time since his 1849 footnote (ch.5), Thomson expressed his analysis in theological and cosmological terms: 'Although no destruction of energy can take place in the material world without an act of power possessed only by the supreme ruler, yet transformations take place which remove irrecoverably from the control of man sources of power which, if the opportunity of turning them to his own account had been made use of, might have been rendered available'.[39] In other words, God, as 'supreme ruler', had established a fundamental law of 'energy conservation' by which only He could create or destroy energy. Nevertheless, the energies of nature (a waterfall, for instance) were continually being transformed in ways which offered human beings the opportunity of harnessing these energies to the creation of wealth. By implication, mankind had a moral duty to 'make use of' such opportunities: failure so to do amounted to the sin of 'waste' (ch.2).

Nature's transformations of energy had a *direction* which only God could reverse: 'Everything in the material world is progressive. The material world could not come back to any previous state without a violation of the laws which have been manifested to man, that is, without a creative act or an act possessing similar power'. Thus God had ordained for nature two basic laws of energy: its conservation and its 'progressive' transformation. Human beings could neither create nor destroy energy, and could not reverse the progressive transformation of energy in nature. They could, however, direct those energies to beneficial and productive ends such as the driving of a vortex turbine from a fall of water which, though in itself aesthetically pleasing, would otherwise be wasted to mankind.

This 'progressive' tendency, however, also amounted to a gradual 'running down' of the visible material world, or at least of the solar system, in which the sources of useful work were never wholly replenished:

> I believe the tendency in the material world is for motion to become diffused, and that as a whole the reverse of concentration is gradually going on – I believe that no physical action can ever restore the heat emitted from the sun, and that this source is not inexhaustible; also that the motions of the earth & other planets are losing vis viva w^h is converted into heat; and that although some vis viva may be restored for instance to the earth, by heat received from the sun, or by other means, that the loss cannot be *precisely* compensated & I think it probable that it is under-compensated. What many writers, for instance Pratt, say, that volcanoes & other sources of mechanical effect are found to compensate the losses, is (I believe) nonsense; since it first ought to be shown that the losses ... could have produced any appreciable effect on the rotation ... of the earth within the short period during w^h man has lived on it – 'The earth shall wax old &c'. The permanence of the present forms & circumstances of the physical world is limited.[40]

By quoting from the 102nd Psalm here, Thomson made his vision consistent with a Biblical perspective of change and decay. The passage read in full:

> 25. Of old hast thou laid the foundation of the earth: and the heavens are the work of thy hands.
>
> 26. They shall perish, but thou shalt endure: yea, all of them shall wax old like a garment; as a vesture shalt thou change them, and they shall be changed.
>
> 27. But thou art the same, and thy years shall have no end.[41]

In this remarkable creed, the Glasgow professor was privately trying out his fundamental cosmological convictions. His was a directional, not a cyclical, economy of nature in which no 'compensation' or 'balance' was conceivable in practice. The compensatory mechanisms advocated by writers such as Archdeacon J.H. Pratt (c.1808–71), author of a well-known Cambridge mathematical tripos text on mechanics, were dismissed as nonsensical.[42] Instead, the Glasgow professor, drawing upon several resources with which he was already familiar, constructed simultaneously a new basis for Carnot's criterion *and* laid the foundation for a new cosmological synthesis.

The first resource was Joseph Fourier's *Théorie analytique de la chaleur* (1822). It is well known that in May of 1840, Professor Nichol introduced the young

Thomson to Fourier's treatise, that William mastered the text in a couple of weeks while on a visit to Germany soon after, and that, at the age of sixteen, he launched his scientific career in the *Cambridge Mathematical Journal* with a powerful defence of Fourier's mathematics against the Cambridge-educated professor of mathematics in the University of Edinburgh, Philip Kelland. While a Cambridge undergraduate himself in the early 1840s, William contributed to the *Journal* a whole series of papers related to Fourier. The founders and supporters of the *Journal* (D.F. Gregory and R.L. Ellis, for example) had done much to promote Fourier. Thomson, for his part, saw Fourier's approach as practical, direct, elegant and simple in contrast to the hypothetical and complex action-at-a-distance force physics of Laplace's followers, notably S.D. Poisson (1781–1840).[43]

In the 'Preliminary discourse' to his *Théorie analytique*, Fourier had drawn attention to three possible sources of the heat in the body of the earth, observed to increase generally with depth in mines: primitive heat in the earth's mass, the sun's heat, and the natural heat of the heavens. While he treated only the effects of the sun's heat in the *Théorie*, he analysed the 'secular cooling' deriving from a primitive store of heat in the earth in papers of 1820 and 1824.[44] This work provided the basis for a strong Cambridge interest in the 'geological' dimension to Fourier's physics, beginning with William Whewell's praise for Fourier's approach in his 'Report' to the British Association in 1835. Led principally by Adam Sedgwick, the Cambridge school had been eager to oppose the geological doctrines of Charles Lyell (1797–1875) who emphasised 'steady-state averages' or 'uniformitarianism': that geological agencies acted with more or less the same intensity over all time. Sedgwick in particular believed that this doctrine of equal intensity did not conform to geological facts:

> Volcanic action is essentially paroxysmal; yet Mr Lyell will admit no greater paroxysms than we ourselves have witnessed – no periods of feverish spasmodic energy, during which the very framework of nature has been convulsed and torn asunder. The utmost movements that he allows are a slight quivering of her muscular integuments.[45]

Lyell had particularly argued against Fourier's doctrine of primitive or central heat by advancing an alternative theory to account for the earth's thermal store in terms of a compensatory mechanism with no discernible direction or 'progression' from beginning to end. From the mid-1830s Cambridge mathematical coach William Hopkins began to publish a series of studies concerned with the internal constitution of the earth. His support for the central heat doctrine against Lyell was apparent throughout these papers. Hopkins argued for an originally fluid earth which cooled, leaving cavities of still-molten matter in the solid mass as the source of elevation and volcanoes. In opposition to Lyell's essentially fluid earth, Hopkins could claim that the present temperature distribution over the earth's surface and within its mass was consistent with Fourier's model of a solid cooling by conduction. As one of Hopkins's star pupils of the mid-1840s, William Thomson's ardent support for the Fourier model must be set within this context of the physical geology William Hopkins.[46]

Although all of Thomson's *Cambridge Mathematical Journal* articles were directed towards a specialist mathematical audience, the second of two papers employing Fourier's theory of heat conduction to investigate thermal distributions over time provided the basis of Thomson's inaugural dissertation as Glasgow professor of natural philosophy. This paper of 1844, 'Note on Some Points in the Theory of Heat', opened with the claim that:

> In problems relative to the motion of heat in solid bodies, the initial distribu-
> tion, which is entirely arbitrary, is usually one of the data. When this is the
> case, and the circumstances in which the body is placed are known, the
> distribution at any subsequent period is fully determined, and if our analysis
> had sufficient power, would become known in every case.[47]

Such problems depended for their solution upon accurate, easily measurable data such as temperatures and conductivities, and upon the mathematical law of heat conduction. The solution 'would be an expression for the temperature of any point in the body in terms of the co-ordinates of the point, and the times measured from the instant at which the distribution is given'. Everything was therefore directly measurable without reference to controversial hypothetical entities or speculative theories. Thomson had typically chosen to investigate a problem from the perspective of maximum consensus and minimum controversy, that is, as a 'latitudinarian' in natural philosophy.[48]

'It is in many cases an interesting investigation to examine what this expression becomes when negative values are assigned to the time', Thomson continued. He explained that in cases where, for a particular negative value of time, 'we find that the expression gives an actual arithmetical value for the temperature of every point in the body ... the given initial distribution would have been produced by the spontaneous motion of heat'. However, he pointed out that 'the arbitrary initial distribution may be of such a nature that it cannot be the natural result of any previous possible distribution, or that it cannot be any stage except the first in a system of varying temperatures'. Distributions with cusps or angular points, he claimed, could not result from previous distributions: 'all such abrupt transitions or angles which may exist in an initial distribution, will disappear instantaneously after the motion has commenced'.[49]

With respect to negative values of time, Thomson assumed that a diverging Fourier series represented impossible physical distributions. He considered for convergent series three cases of initial distribution. First, he examined the case in which divergence for all negative time occurred whereby 'the distribution will be impossible, and therefore the given distribution cannot be any stage but the first in a system of varying temperatures'. Second, he outlined the case in which con-vergence occurred for finite negative time leading to 'an essentially primitive dis-tribution'. Hence 'a finite *age* ... may be assigned to the given initial distribution'. And third, in the case of convergence for all negative time, 'no limit can be assigned to the *age* of the initial distribution'.[50]

Here, for the first time, Thomson had introduced the word 'age' into his mathe-matical discourse. Not until his inaugural dissertation two years later, however, did he explicitly refer to the question of the earth's age. Entitled 'On the Distribution of Heat through the Body of the Earth', very little of the dissertation has in fact survived. Nonetheless, several key points are clear. First, he had chosen to mark his triumphant return to Glasgow with a dissertation which implicitly paid homage to J.P. Nichol, the Glasgow professor who had probably done most to inspire William's love for natural philosophy, who had introduced him to Fourier's text six years earlier, and who had been one of William's three most ardent supporters in the campaign for the election of the new professor. That the subject of the dissertation should have identified so closely with Nichol's 'progressionist' cosmology is therefore hardly surprising.[51]

Second, while Thomson had been warned before and during the election campaign to emphasize that he was no mere '$x + y$' man lest he be thought too advanced for the 'popular' nature of the natural philosophy class, he could now demonstrate simultaneously his skills as a mathematical and experimental natural philosopher. Here was no 'high priest' seeking to guard the mysteries of the universe against the uninitiated and ignorant masses, but a 'minister' of natural philosophy who had acquired the skills by which he could display to his fellow-professors, and soon to the students of the university, the *true* workings of the universe in terms which they, with a little effort, could then *see* for themselves in simple, practical terms. In presbyterian and democratic fashion, the new professor of natural philosophy could present the laws of nature not on the strength of his authority, not as *dogmas*, but as *principles* which every right-thinking student of nature could reach for himself. His technique thus effectively removed at a stroke all disputes about the age of the world and the literalness of the scriptural account and placed the issue of the earth's age into the sphere of mathematics and experiment.[52]

Thus far Thomson was fulfilling the expectations of those Glasgow College reformers who had engineered his election and prevented the appointment of a traditional presbyterian 'moderate'. But Scottish culture was in political and religious turmoil (ch.2). Reform threatened to turn to revolution as the spectre of Chartism stalked the streets of Glasgow. The great railway mania had collapsed into economic destitution. The Irish Famine was spreading disease and disorder to the West of Scotland. William's younger brother had perished from typhus in 1847 and his father from cholera two years later. The Thomson family, once regular attenders at Chalmers's former parish of St John's, resisted following their present minister, Dr Thomas Brown, into the Free Kirk but showed no enthusiasm for continuing with the remnants of the established Kirk.[53]

For the most part they chose to worship with the 'secessionist' congregation of Rev. David King who had married Elizabeth Thomson, eldest sister of William. James, however, turned openly to the far more radical Unitarianism after his father's death. And William, more familiar with Anglicanism from his Cambridge years, adopted 'the habit of regularly conforming to the [Scottish] Episcopal Church & not appearing more than once or twice or three times in the course of a session at an Established Church'. Such defiance of orthodoxy prompted a Cambridge friend to congratulate William on acquiring the title of 'latitudinarian' in

Glasgow: 'there I suppose it is equivalent to being a member of ye Episcopal Church'. Norman Macleod had indeed written privately in 1841 of the possible dawn of an 'episcopal era' in Scotland's ecclesiastical history, not least because of those 'souls who love to eat their bread in peace, and who, weary of the turmoil of our [presbyterian] Church, flee to the peace of the Church of England, which seems to reflect the unchangeableness of the Church invisible'.[54] But the reforming young professor could not, by virtue of his position in one of the country's premier sites for the education of ministers of the kirks, wholly retreat into Anglicanism.

At war with itself, Scotland's Christian inheritance was at risk too from 'infidels' within and without. The new Free Church College in Edinburgh, inaugurated in 1851, had set up a chair of natural science specifically to provide an 'armour of defence' against named infidels such as David Hume, James Hutton, Laplace, Cuvier and others, but including also the anonymous *Vestiges*, first published just one year after the Disruption as though to take advantage of the plight of the Kirk.[55]

Aimed at a popular rather than elite scientific audience, *Vestiges* eventually went through some twenty editions in America and thirteen in Britain. As James Secord has shown, Chambers had developed a deep distaste for 'those dogs of clergy', notably of an 'evangelical' caste, in the years between the Reform Act of 1832 and the publication of *Vestiges*. Abandoning membership of the presbyterian Church of Scotland after attacks from the pulpit on his (and his brother's) *Chambers's Edinburgh Journal* for its secular style, Chambers developed a specific dislike of 'evangelical' presbyterians (later associated with the Free Kirk) who insisted on strict adherence to scriptural doctrine, particularly as set out in the *Westminster Confession*. References to God in *Vestiges* were therefore little more than a means of rendering its avowedly secular message more palatable to its middle-class readers.[56]

Writing to a friend in 1845, Thomson's older brother James had perceived the secular tone of *Vestiges*. James's reaction appeared highly critical:

> I do not like the development theory as given in *The Vestiges*. It appears to me that the author substitutes for the Creator, what he calls *Law*; and that , if he gives his assent to the existence of God as a First Cause, he, at least supposes Him to be now infinitely removed from all the Works of Nature, and that everything goes on now of itself just as a clock goes after its weights have been wound up. Now I am strongly impressed with the idea that a law is in itself nothing and has no power; and I can view what we call the Laws of Nature in no other light than merely as expressions of the will of an Omnipresent and Ever Acting Creator.[57]

James highlighted here what he saw as the principal defect of *Vestiges*, namely that it had apparently treated God as a 'deistic' First Cause without any continuing role in the government of nature. Significantly, however, he did not reject the development theory as such: 'I do not like the development theory as given in *The Vestiges*'. Nor did he offer criticism in terms of a conflict with the literal interpretation of *Genesis*, that is, with the strict doctrine of creation expressed in *The Larger Catechism* (1648) of the Church of Scotland. We should not, however, be surprised by James's lack of real hostility towards a development theory. After all, the Thomson brothers' most revered teacher at Glasgow College had been none other than Nichol, a key inspiration for Chambers's perspective of progressive development. Indeed, the whig values of 'progress', 'utility', 'improvement' and 'modernity' which Chambers professed, following a rapid rejection of toryism, were also those which the Thomson family had long espoused.[58]

More generally other reviewers also perceived a dangerously secular message in *Vestiges*. In an article printed in the American *Broadway Journal* in 1845, for example, a certain Rudolph Hertzman believed that 'in modern days, the habit of generalization tends decidedly to materialism':

> Take, for instance, a book recently published, which for lucid arrangement and admirably sustained generalization, is unsurpassed by any other work on the same subject: I mean 'Vestiges of the Natural History of Creation' . . . The author is no materialist; on the contrary, he takes great pains to disclaim all such tendencies, yet what a storehouse of materialism would the book afford to one who doubted every truth which did not come through the intellect. His own faith in his theories adds an irresistable charm to his arguments, and it requires a most determined examination of truth to detect in many instances the workings of his imagination from the action of his reason. His system of progression has no limit short of Deity.[59]

The reviewer, however, extracted from the text of *Vestiges* a frightening message of a heaven 'filled only by an infinite Intelligence to which we are but as atoms of dust on the rolling wheel of progression'. He therefore penned the sonnet (quoted in full above) which uttered a cry against 'Necessity's stern laws, graven on eternity'.

Within the context of British science in the 1840s, the senior generation of 'Gentlemen of Science' who had led the British Association for over a decade represented *Vestiges* as an unacceptable deviation from the standards of orthodoxy. Chambers was not part of this ruling elite in British science. His work carried all the wrong connotations: in its scope it was all-embracing and speculative rather than specialist and based upon sound and sober 'inductive' method; it made errors obvious to the geological and other specialists of the time; and it did not conform to the theological assumptions of the Gentlemen who regarded species as exempt from any law of development *within* the natural order and who supposed man, especially Cambridge man, as the highest of God's special creations.[60] In all these respects, the Cambridge professor of geology, Adam Sedgwick, summed up the differences between *Vestiges* and its orthodox opponents in an 1845 letter to Charles Lyell:

> [*Vestiges* begins] from principles which are at variance with all sober inductive truth. The sober facts of geology shuffled, so as to play a rogue's game; phrenology (that sinkhole of human folly and prating coxcombry); spontaneous generation; transmutation of species; and I know not what; all to be swallowed, without tasting or trying, like so much horse-physic!! Gross

credulity and rank infidelity joined in unlawful marriage, and breeding a deformed progeny of unnatural conclusions![61]

Just how dangerous might *Vestiges* be to impressionable young Cambridge men can be seen from the enthusiasm with which the poet Alfred Tennyson embraced parts of the developmental theory in the closing sections of his long poem *'In memoriam'* (1833–50). As a solution to his early doubts about the traditional 'providential' role of God in a nature 'red in tooth and claw', Tennyson offered his readers a highly unorthodox vision of a universe in a grand progression which promised the development of man into a 'crowning race':

> . . . They say
> The solid earth whereon we tread
> In tracts of fluent heat began
>   And grew to seeming-random forms,
>   The seeming prey of cyclic storms,
> Till at the last arose the man;
> . . .
> And, moved through life of lower phase,
>   Result in man, be born and think,
>   And act and love, a closer link
> Betwixt us and the crowning race . . .[62]

As Richard Yeo has well argued, *Vestiges* proved extremely unsettling to the British scientific establishment. The book challenged the authority of elite men of science by direct appeal to a public, popular readership. Just as in the United States, Alexis de Tocqueville pointed to the rejection of the authority of learned elites by a public imbued by an egalitarian ideology, so too in Scotland, if not in Britain as a whole, would 'the people' respond enthusiastically to this 'democratic' strategy by which the worth of the product would be decided by popular demand. Chambers could thus direct his own criticism towards the 'gentlemanly specialists' who dominated the British Association and whose increasing narrowness of vision prevented them from appreciating the grandeur of his own synthetic vision. No wonder then that the gentlemanly Roderick Murchison should wish for 'a real *man of armour*' to slay *Vestiges* and thereby 'do infinite service to *true* science'.[63]

Vestiges thus presented a formidable set of challenges to anyone seeking to sustain a Christian perspective on man and nature. On the one hand Sedgwick's personal criticisms of the author and his (or her) ignorance of science would scarcely stem the popular appeal of the developmental theory. On the other hand a retreat to the authority of scripture, literally interpreted, would not win influence among the BAAS's elite gentlemen of science who worked comfortably with a progressionist (directionalist) account of the physical world over millions of years. In private, therefore, the Glasgow professor of natural philosophy had begun to sketch a new energy perspective which, avoiding both the narrowness of biblical literalism and the dangers of unadulterated naturalism, might begin the task of building a new scientific authority with both national and local appeal.

One such Christian perspective appeared as a popular text, *The Principles of Geology Explained, and Viewed in Their Relation to Revealed and Natural Religion* (1850). The author was none other than Thomson's brother-in-law, Rev. David King. King admitted the evidence of geology that the earth was much older than had once been thought, but upheld the first verse of Genesis as affirming the original creation of all things by God. Drawing on the work of Chalmers (ch.2), he suggested that there might have been an indefinite time between this original creation and the special creations described from the second verse onwards:

> Our best expositors of scripture seem to be now pretty generally agreed that the opening verse in Genesis has no necessary connexion with the verses which follow . . . On this principle of interpretation, the bible recognises, in the first instance, the great age of the earth, and tells us of the changes it underwent at a period long subsequent, in order to render it a fit abode for the family of man. The work of the six days was, according to this view, not creation in the strict sense of the term, but a renovation – a remodelling of pre-existent materials. Some difficulty, however, remains in explaining the transactions of these days, so as to establish their accordance with geological discoveries.[64]

King was above all concerned with a reconciliation of scripture and geological science, without damage to either. As a minister who frequently discussed scientific issues with his brothers-in-law, his 1850 publication accords with William Thomson's non-literal but scriptural economy of nature drafted in the following months.

David King's unpretentious little book, however, provoked the wrath of one reviewer, P. McFarlane, who sought to show 'the utter incompatibility of these principles [of modern geology] with both science and sacred history' in a book-length text published in Edinburgh in 1852. In a note, McFarlane cited a recent favourable review of King's work by Hugh Miller: 'Were we asked, where, in the course of a few hours' reading, a man might best acquaint himself with the actual state of the question between geologists and theologians, we would at once refer him to the work of Dr King'. He also quoted another reviewer's endorsement that 'Beginning with Hugh Miller . . . down to critics of minor note, a similar verdict has everywhere been given'. But McFarlane's own verdict was very different, reflecting a severely Calvinist view of the Fall not only of man but of the physical world:

> If we compare the universe, or rather that portion of it forming the different members of the solar system, to a splendid fleet of new and elegantly furnished and rigged vessels just launched, and sailing from port on a summer morning, this world . . . would resemble one of these barks, whose inexperienced and rash master, seduced into the wrong course, soon ran his charge against a sunken rock, although that rock was distinctly laid down in his chart, and moreover beacon-marked as a most dangerous spot, to be carefully avoided – that by this grave error, himself, his crew, his vessel, and all

its contents, forthwith became a melancholy wreck. That sometime after this catastrophe, an immense sea or wave swept over the already much damaged vessel, overwhelming most of its now drunken and dissipated crew; and left the crashed hull we now behold; or, to drop the figure, that at the Fall this previously magnificent world was fearfully convulsed and shattered, physically as well as morally.

McFarlane thus preached that geological phenomena testified to the 'solemn fact (one of the Bible's great truths) of the presently fallen PHYSICAL state of nature', a truth which 'harmonizes in melancholy strains with the still more awfully solemn corresponding fact ... of the present appalling lapsed MORAL condition of this sin-blighted planet'. As a result of the Fall of man, then, 'nature in whole ... appears fearfully fallen throughout ALL her domains'.[65]

If Thomson, like his brother-in-law, aimed to distance himself from such traditional literalism, he also sought to avoid the implications of a steady-state, eternal world. We have seen that Thomson's 1851 draft adopted as a basic principle the belief that only God could create or destroy 'energy', a consensus already fully articulated by both Thomson and Joule in private and in public (chs 4 and 5). Joule's position, however, had been that this principle of perfectly reciprocal exchanges (*vis viva* converting to and from mechanical force measured as work done) alone governed nature's economy. This atemporal view was consistent with Enlightenment conceptions of a perfectly balanced series of natural economies which, interpreted by liberal Anglicans or Scottish moderates in accordance with standard arguments from design, testified to the wisdom and goodness of God.

Everything in recent Scottish presbyterian culture on the other hand pointed to the inadequacy of such an exchange economy. Like the dynamic stability and self-sustaining character of Laplace's solar system, such an economy would offer mankind little or no awareness of its debt to the ultimate creator and sustainer of that energy. Nor would there be much immediate sense of man's moral duty to make use of these exchanges for the advancement of mankind in what seemed like a vast perpetual motion machine without a role for human free will.

Rejecting all such steady-state economies of nature, Thomson shared with his mentor Nichol a belief that natural laws were adequate, *pace* Chalmers, to account for nature's present arrangements. But Nichol's progressionism had been tainted by its association with *Vestiges*. Thomson therefore combined Nichol's view that the present arrangements of nature were the result of natural law with Chalmers's frequent assertion that only the Almighty could *renew* nature's arrangements as they followed a natural tendency to the progressive transformations of energy from concentrated to diffused states. There was thus a directional tendency in nature that only God could reverse. The energies of nature obeyed laws which accounted for the present physical arrangements, but those arrangements could not be restored or renewed by any natural laws, still less by any human agency or effort. While Nichol had allowed for an ultimate cycle of renewal within the natural order, it followed from Thomson's intense commitment, like that of Chalmers, to absolute directionality from beginnings to endings in nature that there had to be a Creator whose grace could regenerate the natural order in a manner analogous to the regeneration of fallen human beings. God, then, was not some optional and abstract conception, but the ultimate provider of all the sources of power in nature.[66]

Although Thomson did not engage directly with the question of ultimate beginnings for another three years (ch.7), his reflections upon Fourier's theory had already brought him to the problem of initial distributions of heat. As S.P. Thompson recorded: 'Lord Kelvin used to declare that it was this mathematical deduction [from Fourier] which convinced him that there must have been an origin to the natural order of the cosmos; that therefore natural causes could not be deduced backwards through an infinite time. There *must* have been a beginning'.[67] Citing and paraphrasing Thomson, Maxwell adopted an identical stance in his presidental address to Section A of the BAAS in 1870:

> if we reverse the process [of diffusion of energy], and inquire into the former state of things by causing the symbol of time to diminish, we are led up to a state of things which cannot be conceived as the result of any previous state of things, and we find that this critical condition actually existed at an epoch, not in the utmost depths of a past eternity, but separated from the present time by a finite interval.
>
> This idea of a beginning is one which the physical researches of recent times have brought home to us more than any observer of the course of scientific thought in past times would have had reason to expect.[68]

The energy principles fitted perfectly a presbyterian economy of nature. The energies of nature formed a vast reservoir which enabled man to seize 'the opportunity of turning them to his own account'. Nature's energies were thus structurally analogous to other divine gifts to mankind: the divine gifts constituted a downward flow in which human beings, having received the gift and being unable to reciprocate, had a duty to make use of that gift in what appeared to be an exchange economy, though one which depended for its continuation upon the inward flow of energy if it were to avoid stasis (ch.2). Over the next decade, as a new 'Moderatism' took root in the College and in the City, the increasing public authority of the 'science of energy' would be deployed to undermine the credibility of all sorts of pretenders to true knowledge of nature (ch.9). Nor, indeed, would the new science be confined to a merely local context, for the BAAS provided the means by which the 'science of energy' could aspire to be the science of Britain. For the moment, however, Professor Thomson had more immediate problems to resolve.

Demonstrating Carnot's criterion

Committed both to the conservation of energy and its progressive transformation with no possibility of full restoration in nature, Thomson then sought for a new demonstration of Carnot's criterion. He approached the problem by posing two complementary questions: 'Is it possible to continually get work by abstracting heat from a body till all its heat is removed? Is it possible to get work by cooling a body below the temperature of the medium in which it exists?' The first of these questions simply captured his long perplexity over the 'recovery' of useful work from the heat derived from conduction or fluid friction. The second question related to the possibility of obtaining useful work from the heat of a body at a temperature *lower* than the surroundings. Thomson's reply was immediate and decisive: 'I believe we may consider a negative answer as axiomatic.[69]

In the published paper, Thomson transformed the second question into a firm axiomatic statement:

> It is impossible by means of inanimate material agency to derive mechanical effect from any portion of matter by cooling it below the temperature of the coldest of the surrounding objects.

In a footnote, he reformulated the first question to explain that 'If this axiom be denied for all temperatures, it would have to be admitted that a self-acting machine might be set to work and produce mechanical effect by cooling the sea or earth, with no limit but the total loss of heat from the earth and sea, or, in reality, from the whole material world'.[70] Here then was Thomson's axiomatic assertion that an unqualified acceptance of the mutual convertibility of heat and mechanical effect would not do. In a 'progressive' universe not all the heat could be recovered as useful work: the production of useful work from heat depended on the existence of a difference of temperature levels between 'boiler' and 'condenser'.

Demonstration of the 'second proposition' in the dynamical theory of heat followed a pattern different from that of Clausius (ch.5). In his new 'proof', Thomson imagined a compound arrangement of two engines in which engine A was supposed more efficient than a reversible engine (that is, A violated Carnot's criterion for a perfect engine). Engine B, on the other hand, was taken to be a perfectly reversible engine running in reverse, as a refrigerator utilizing work to restore a temperature difference (Figure 6.1). The net effect of this imagined arrangement, therefore, would not only be the restoration of the original temperature difference by B, but the production of net work (due to A's greater efficiency than B). The production of this surplus work would, according to the dynamical theory, mean that A delivered less heat to the surroundings (condenser) than B extracted, with the result that such an arrangement would eventually extract all the heat from the material world in the form of useful work. Hence, to conform to the axiom, no engine could be more efficient than a perfectly reversible one.[71]

On 15 December 1851 Thomson presented to the RSE the fifth part of his 'Dynamical Theory' entitled 'On the Quantities of Mechanical Energy contained in a Fluid in Different States, as to Temperature and Density'. Here he developed publicly the new language of energy. He explained that the 'total mechanical energy of a body might be defined as the mechanical value of all the effect it would produce, in heat emitted and in resistances overcome, if it were cooled to the utmost, and allowed to contract indefinitely or to expand indefinitely according as the forces between its particles are attractive or repulsive, when the thermal

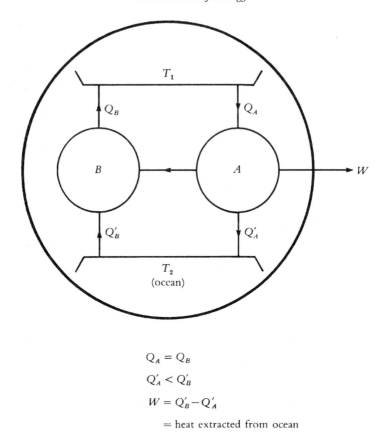

$$Q_A = Q_B$$
$$Q'_A < Q'_B$$
$$W = Q'_B - Q'_A$$
= heat extracted from ocean

Figure 6.1 In Thomson's new 'demonstration', engine *A* (supposed to be more efficient than a reversible engine) drives engine *B* (reversible). The combination produces 'surplus' work *W*, with no other effect than the extraction of heat from the ocean, contrary to his second proposition (Smith and Wise 1989: 331).

motions within it are all stopped'. However, he pointed out that 'in our present state of ignorance regarding perfect cold, and the nature of molecular forces, we cannot determine this "total mechanical energy" for any portion of matter, nor even can we be sure that it is not infinitely great for a finite portion of matter'.[72]

Thomson aimed instead to present an interpretation of 'mechanical energy' which, in its practicality, could command maximum assent from all parties:

> [In order to avoid such problems] it is convenient to choose a certain state, as standard for the body under consideration, and to use the unqualified term, *mechanical energy*, with reference to this standard state; so that the 'mechanical energy of a body in a given state' will denote the mechanical value of the effects the body would produce in passing from the state in which it is given,

to the standard state, or the mechanical value of the whole agency that would be required to bring the body from the standard state to the state in which it is given.[73]

A couple of months later Thomson read a brief, three-part series of papers to the RSE on sources available to man for the production of mechanical effect. In the first part he committed himself to a dynamical theory of radiant heat and light and offered the following example of the possible economical employment of the sun's heat:

> Mechanical effect of the statical kind [raising weights, for instance] might be produced from solar radiant heat, by using it as the source of heat in a thermo-dynamic engine. It is estimated [from data provided by the French experimental physicist C.S.M. Pouillet (1790–1868) on the amount of heat radiated from the sun in any given time and Joule's mechanical equivalent of a thermal unit] that about ... the work of 'one horse power' might be produced by such an engine exposing 1800 square feet to receive solar heat, during a warm summer day in this country; but the dimensions of the moveable parts of the engine would necessarily be so great as to occasion practical difficulties in the way of using it with œconomical advantage that might be insurmountable.[74]

In the second part of the series, Thomson considered the power of living creatures over matter, and in particular the theory which attributed animal heat and motion to chemical action in the light of the new thermo-dynamic principles. As an example, he estimated that 'perhaps even as much as 1/4 of the work of the chemical forces may be directed to the overcoming of external resistances by a man exerting himself for six hours a day in such operations as pumping'. He suggested, however, that the animal body did not act as a thermo-dynamic engine but it was 'very probable that the chemical forces produce the external mechanical effects through electrical means'. And he maintained, against materialistic determinism, that 'Whatever be the nature of these means, consciousness teaches every individual that they are, to some extent, subject to the direction of his [own] will'. As in the draft of his 'Dynamical Theory', human beings had been granted the *choice* of directing the 'energies' of nature to beneficial ends.[75]

The final part of the series classified the range of 'stores' available to man for the production of work: the food of animals, natural heat, solid matter at a height, the natural motions of air and water, natural combustibles such as wood or coal, and artificially produced combustibles such as hydrogen. Thomson then investigated the 'sources' from which these stores had derived their 'mechanical energies'. Agreeing with John Herschel's views, Thomson concluded that 'Heat radiated from the sun ... is the principal source of mechanical effect available to man'. Another important source was to be found in the motions of the earth, sun, and moon (tide-mills and trade-winds, for example), while the third class (mainly terrestrial sources such as hot springs) counted for comparatively little of the mechanical effect obtained by human beings. Overall, therefore, the series presented a concise survey of the economy of nature (nature's stores of mechanical effect) in relation to the human economy (mankind's capacities to direct the energies of nature for practical purposes through engines of various kinds). Only once, however, did Thomson use the term 'mechanical energies'.[76]

Not until a meeting of the RSE on 19 April 1852 did Thomson publicly present the full cosmological picture which he had begun to paint in his 1851 draft. In this public presentation, however, he spoke not of nature's 'progression' but of 'dissipation'. The shift, as we have noted, was significant. Distancing himself publicly from the onwards and ever-upwards progression of *Vestiges*, Thomson now made the new science especially appealing to anyone, not least his own Scottish audiences, familiar with presbyterian theology. Now ranked alongside such terms as 'depravity' and 'decay', 'dissipation' carried connotations of moral and financial waste. As his father had written to him in 1842, a Glasgow student, son of a unitarian minister, had been found 'to have been attending the theatre and other amusements from night to night – to have been indulging in habits of dissipation'.[77]

'On a Universal Tendency in Nature to the Dissipation of Mechanical Energy' opened with a statement that its aim was to 'call attention to the remarkable consequences which follow from Carnot's proposition, established as it is on a new foundation, in the dynamical theory of heat':

> that there is an absolute waste of mechanical energy available to man, when heat is allowed to pass from one body to another at a lower temperature, by any means not fulfilling his [Carnot's] criterion of a 'perfect thermo-dynamic engine'. As it is most certain that Creative Power alone can either call into existence or annihilate mechanical energy, the 'waste' referred to cannot be annihilation, but must be some transformation of energy.

In an unpublished sketch of the same paper, he put the point in more specific economic terms, that 'when heat is diffused by conduction there is an economical loss as regards its value as a source of mechanical effect'.[78] But in the published version, he effected a transition from particular 'engineering' contexts to a 'universal', cosmological one.

In order to elucidate the nature of the 'transformation of energy', Thomson divided 'stores' of mechanical energy into two classes, statical and dynamical, which echoed traditional Cambridge textbook divisions of mechanics. Statical stores of energy included weights at a height, electrified bodies, and quantities of fuel. Dynamical stores of energy included masses of matter in motion, bodies having thermal motions among their particles, and undulations of light or radiant heat in a volume of space. He then laid down four propositions 'regarding the *dissipation* of mechanical energy from a given store, and the *restoration* of it to its primitive condition'. Each proposition, he asserted without elaboration, was a necessary consequence of the axiom which provided the demonstration of Carnot's criterion. The first proposition referred to reversible processes while the others treated of 'unreversible' processes (friction, conduction and absorption of radiant heat or light) in which 'perfect restoration' to a primitive condition was impossible because of a dissipation of mechanical energy. Embodying previous perplexities over 'waste' and 'recovery' of useful work, these 'conclusions' were now presented as consequences of a general and universal principle of energy 'dissipation' throughout nature.[79]

Thomson drew three further conclusions from these propositions and from 'known facts with reference to the mechanics of animal and vegetable bodies' which up to this point he had excluded from direct discussion. First, he proclaimed that 'There is at present in the material world a universal tendency to the dissipation of mechanical energy'. Though presented here simply as another deduction, this conclusion in fact expressed Thomson's fundamental belief in the 'irreversible' nature of the created universe, the very belief which grounded his demonstration of Carnot's criterion.

Second, 'Any *restoration* of mechanical energy, without more than an equivalent of dissipation, is impossible in inanimate material processes, and is probably never effected by means of organised matter, either endowed with vegetable life, or subjected to the will of an animated creature'. By denying the possibility of reversing the tendency towards dissipation (at least in inanimate nature), this conclusion complemented the first. It also initiated a debate about the role of living beings (vegetable and animal) in a universe of energy.[80]

Finally, 'Within a finite period of time past the earth must have been, and within a finite period of time to come the earth must again be, unfit for the habitation of man as at present constituted, unless operations have been, or are to be performed, which are impossible under the laws to which the known operations going on at present are subject'.[81] Expressed in formal legalistic style (almost certainly to establish his rights to intellectual property), this gratifying conclusion seemed to follow from the first two: given the dissipation of energy in nature, and the impossibility of its full restoration, the earth existed very much as a finite abode for mankind.

Thus far the principal forum both for Thomson's 'Dynamical Theory' and now for 'energy' physics and cosmology had been the RSE. But from July 1852, reprints of his 'Dynamical Theory' began to appear in the pages of the *Phil. Mag.* By the end of the year, a new cosmology of energy had been well and truly aired in its pages.[82] When in 1852 Thomson employed the term 'energy', he no longer concealed the term in a footnote. Instead, he offered an energy synthesis, in terms of dual principles of energy conservation and energy dissipation, to a very wide audience. The dynamical theory of heat, placed on a firm 'experimental' basis, was now taken for granted. And the universal, cosmological primacy of energy opened up new questions about the origin, progress and destiny of the solar system and its inhabitants.

CHAPTER 7

'The Epoch of Energy': the New Physics and the New Cosmology

But the scientific importance of the principle of the conservation of energy does not depend merely on its accuracy as a statement of fact, nor even on the remarkable conclusions which may be deduced from it, but on the fertility of the methods founded on this principle. . . . .It gives us a scheme by which we may arrange the facts of any physical science as instances of the transformation of energy from one form to another. It also indicates that in the study of any new phenomenon our first inquiry must be, How can this phenomenon be explained as a transformation of energy? What is the original form of the energy? What is its final form? and What are the conditions of transformation?

Maxwell judges the significance of the principle of energy conservation for scientific practice (1876–7)[1]

On 20 January 1852 William Thomson saw for the first time Hermann Helmholtz's 'admirable treatise on the principle of mechanical effect' published nearly five years earlier as *Über die Erhaltung der Kraft*. In a footnote added to the 1852 *Phil. Mag.* reprint of part one of his paper 'On the Dynamical Theory of Heat', Thomson stated that had he 'been acquainted with it in time', he should have had occasion to refer to it with respect to thermo-electric questions and 'on numerous other points of the dynamical theory of heat, the mechanical theory of electrolysis, the theory of electro-magnetic induction, and the mechanical theory of thermo-electric currents in various papers communicated to the Royal Society of Edinburgh' and to the *Phil. Mag.*[2] Far from seeing Helmholtz as a threat to British priorities, however, Thomson would rapidly appropriate the German physiologist's essay to the British cause, deploying it ultimately as a means of enhancing the international credibility of the new 'epoch of energy'.

For his part Helmholtz too derived dramatic gains in credibility from Thomson's enthusiastic recognition of the value of the 1847 essay which had hitherto received rather mixed reactions from German physicists. John Tyndall, whom Helmholtz first met in August 1853, translated the essay for *Scientific Memoirs. Natural Philosophy* (edited by Tyndall and the publisher William Francis) in the same year. Also in 1853, Helmholtz travelled to England for the Hull meeting of the British Association where he met William Hopkins whose presidential address did much to promote, especially on Thomson's behalf, the new doctrines of heat. He became acquainted with other members of Thomson's circle, notably Stokes and the Belfast chemist Thomas Andrews (1813–85), though it was not until 1855 that he met Thomson in person. By 1853 he could therefore write that his *Erhaltung der Kraft* was 'better known here [in England] than in Germany, and more than my other works'. As a result, Helmholtz's essay acquired the status of a major contribution to physical science.[3]

Reviewing Helmholtz's career in 1877, James Clerk Maxwell claimed that in the 1847 essay Helmholtz had shown 'if the forces acting between material bodies were equivalent to attractions or repulsions between the particles of these bodies, the intensity of which depends only on the distance, then the configuration and motion of any material system would be subject to a certain equation which, when expressed in words, is the principle of the conservation of energy'.[4] Maxwell's appraisal vividly illustrates the nature of the British re-reading of Helmholtz's essay from 1852, a re-reading which would render Helmholtz as an ally in the establishment of the new 'science of energy':

> [Helmholtz] is not a philosopher in the exclusive sense, as Kant, Hegel . . . are philosophers, but one who prosecutes physics and physiology, and acquires therein not only skill in discovering any desideratum, but wisdom to know what are the desiderata, e.g., he was one of the first, and is one of the most active, preachers of the doctrine that since all kinds of energy are convertible, the first aim of science at this time should be to ascertain in what way particular forms of energy can be converted into each other, and what are the equivalent quantities of the two forms of energy.[5]

So powerful has this re-reading been that Helmholtz scholars have tended to interpret the 1847 essay in terms of the extent to which it measured up to a British 'energy' perspective rather than as a treatise firmly grounded in German traditions of *Kraft*, understood as separation in space.[6]

In this chapter I begin with Helmholtz's *Erhaltung der Kraft* in its original context of German physiology and physics. Re-read by Thomson as the author of a treatise on the 'principle of mechanical effect', Helmholtz found himself feted in North Britain as a key contributor to the principle of energy conservation, now construed as a basic and independent doctrine whose truth rested not on any rational deduction from more fundamental principles but upon Joule's experimental measurements. Their confidence boosted by this latest and not unwilling recruit, the North British scientists of energy moved quickly from local to national contexts via the BAAS. At successive British Association meetings throughout the 1850s, they and their allies among the 'old lions' promoted a new physics and a new system of the world which together would constitute a powerful but by no means unproblematic 'science of energy'.

The second section of this chapter tracks the promotion of Scottish-centred energy physics at annual meetings of the British Association. Transforming the science of work from its initial industrial contexts, the promoters of the science of energy gave it universal character and universal marketability transcending local customs and conventions. The science of energy became the intellectual property of the elite scientific practitioner, especially in Section A (Mathematics and Physics) of the British Association. These practitioners presented themselves as responsible for interpreting and directing the grand economy of nature upon which the wealth of nations ultimately depended. From the mid-1850s, the new scientific elite also brought quantitative estimates of the age of the sun and earth to bear upon debates about the origin of the solar system and life on earth. Adherents to the new energy doctrines thus sought to police the theoretical and empirical scope of the geological and biological sciences. In particular, the fundamental assumptions of Charles Lyell and Charles Darwin were subjected to critical scrutiny and their claims to have laid bare the natural history of the earth and its inhabitants challenged. What Thomson, addressing Section A at the fiftieth annual meeting of the British Association in 1881, called 'the epoch of energy' therefore forms the principal subject of the present chapter.

Physics and physiology: Hermann Helmholtz's Erhaltung der Kraft

As with Mayer, Helmholtz's professional career lay largely in medical science rather than physics. Unlike Mayer, however, much of Helmholtz's scientific work took place firmly within the institutional context of German universities, first as professor of physiology at Königsberg (1849–55), then in the chair of anatomy and physiology at Bonn (1855–8), and thirdly as professor of physiology at Heidelberg (1858–71). Only at the age of fifty did he become professor of experimental physics in Berlin (1871–88) and finally president of the new *Physikalisch-technische Reichsanstalt*, founded to promote science in the service of German industry and empire, until his death in 1894. Physiology, rather than physics, thus stood at the core of his professional career until the 1870s.[7]

Helmholtz and Mayer also differ strikingly in the degree and manner of credibility afforded to their research. The difference was in large part due to what might be termed Helmholtz's 'scientific cosmopolitanism'. Through his early association with the Berlin Physical Society (*Physikalische Gesellschaft*), through his succession of academic locations in the German-speaking world, and especially through his British acquaintances, Helmholtz exploited to the full the advantages offered by such social contexts for the promotion of his researches. Yet the initial reception of his 1847 essay had inauspicious. Rejected by Poggendorff's *Annalen*, Helmholtz published locally with no guarantee of a mainstream scientific readership. Furthermore, open hostility over the assumptions of his essay soon broke out between Helmholtz and his fellow-German Rudolf Clausius in the early 1850s.[8]

Born in Potsdam not far from the great Prussian capital, Berlin, Helmholtz grew up in a family limited in wealth but rich in culture. In due course, he attended the Potsdam Gymnasium (1832–8) where he studied subjects from mathematics to classics. Given his father's limited income, he required external funding if he were to enter university. The Royal Friedrich-Wilhelm Institut of Medicine and Surgery offered just the kind of state support which Hermann needed, in return for eight years service as an army surgeon.

In his first term he reported to his parents that he had used a spare interval to read Homer, Byron, Biot and Kant. Here indeed was a wide range of texts, from the classical Homer to the 'romantic' Byron and from the French physics of J.B. Biot (1774–1862) to the critical philosophy of Immanuel Kant. Biot's four-volume physics text, with its commitment to Laplace's style of physics, had been translated into German in 1824–5. In his second term Helmholtz apparently devoted more spare time to studies of Kant, whose writings served as a major philosophical resource for the introduction to *Erhaltung der Kraft*.[9]

It was in this second term, however, that he first became inspired by the physiological lectures of Johannes Müller. Although he had been introduced to French physiology from François Magendie's textbook (*Précis élémentaire de physiologie*) prior to his four-year course at the Institut, it was Müller who was to provide one of the most enduring social and intellectual foundations for Helmholtz's scientific career. By 1841–2 he 'lived entirely in the circle of Müller's pupils, since he had already formed a friendship with the physiologists [Ernst] Brücke and [Emil] du Bois-Reymond, who were two years senior to him, and like him devotedly attached to their teacher'.[10]

Appointed an army surgeon in 1843 and conveniently stationed in neighbouring Potsdam, Helmholtz continued to move in the Müller circle throughout the 1840s and beyond. His first publication, 'On the Nature of Fermentation and Putrefaction' appeared in Müller's *Archiv für Anatomie und Physiologie* (1843). It was in part based on work carried out by Helmholtz in his master's laboratory where Liebig's latest investigations of the physiology of respiration and nutrition were being appraised.[11]

This paper demonstrated his early opposition to Müller's views on *Lebenskraft* as an expression of purposive organization. Arguing from experiments in which he claimed to have eliminated the presence of micro-organisms, Helmholtz concluded that putrefaction was a purely chemical process which involved the breakdown of organic materials. Consequently, he argued, chemical processes alone, without any notion of *Lebenskraft*, were sufficient to replicate processes previously thought to require living organisms. Fermentation, on the other hand, was 'a form of putrefaction bound to and modified by the presence of an organism'. Helmholtz's demonstration that the laws of chemistry and physics could account for *all* the phenomena of life was therefore far from complete.[12]

Two years later Helmholtz targeted what he called 'One of the most important questions of physiology, one immediately concerning the nature of the *Lebenskraft* . . . [that is], whether the life of organic bodies is the effect of a special self-generating, purposive force or whether . . . the mechanical force and heat generated in the organism can be completely deduced from the process of material exchange or not'.[13] He chose to attack the problem through a specific investigation of muscle activity.

As a practising physiologist he had a particular enthusiasm for froglegs. He thus explained to du Bois Reymond early in 1847, the year of his *Erhaltung der Kraft*, that he was waiting 'impatiently for the spring and the frogs'.[14] Helmholtz's experiments involved the removal of the pairs of legs from several frogs. Taking one leg from each pair, weighing them, and finally arranging the legs end to end in series, he connected this set of froglegs to a Leyden jar which subjected the set to electric discharges until the legs showed no further 'irritability'. The other set, after weighing, was not subjected to electrical action. Both sets then had the muscles removed from the bones. Helmholtz found the difference in total weight between the two sets to be negligible but, after quantitative chemical analysis, he found a significant difference between the electrified and non-electrified sets: a material, chemical exchange had apparently taken place in the electrified set of froglegs. The analysis seemed to confirm that an increase in one kind of constituent was balanced by the decrease in another kind. Thus he concluded that muscle action involved chemical exchange or transformation within the muscle.[15]

Published by the Medical Faculty of the University of Berlin, the *Encyclopaedia of Medical Knowledge* (1845) contained a contribution from Helmholtz on 'Physiological Heat'. Here for the first time he considered rival theories of heat. Rejecting the caloric (material) theory on account of its inability to explain the continuous production of heat by friction, he explicitly supported a mechanical theory:

> the relationships of free and latent heat set forth in the language of the materialistic theory remain the same if in place of the quantity of matter we put the constant quantity of motion in accordance with the laws of mechanics. The only difference enters where it concerns the generation of heat through other motive forces and where it concerns the equivalent of heat that can be produced by a particular quantity of a mechanical or electrical force.

Bringing heat under the rule of mechanics entailed, in Helmholtz's argument, that 'the forces present in an organised body can only produce a particular quantity of heat motion, and ... that a particular quantity of a motive force [*bewegende Kraft*] can only produce the same definite quantity [of heat] whatever the complication of its mechanism might be'.[16] Given that animals liberated heat, the issue was whether the chemical and physical forces arising from the intake of food and oxygen were alone sufficient to account for this animal heat or whether a special *Lebenskraft* with the capacity to supply force indefinitely was needed.

Helmholtz then engaged directly with Liebig's views that animal heat, a purely chemical process, was due to the oxidation of organic substances containing carbon and hydrogen in the lungs and was equivalent to the heat generated by combustion in a free state. Instead of dismissing earlier experimental work by Dulong and Despretz that the total quantity of animal heat exceeded such generation by respiration, Helmholtz widened the site of heat production from respiration in the lungs to the animal tissues, including the muscles. While confessing that more accurate quantitative determinations were necessary, he claimed that 'we must be temporarily satisfied with the fact that at least very nearly as much heat is generated through chemical processes in the animal as we find in it and its output, and that we must regard it as experimentally demonstrated that by far the greatest part of organic heat is a product of chemical forces'. Clearly, though, there remained the possibility that at least some of the animal heat derived not from chemical processes but from (for example) the activity of the nervous system which could be linked to *Lebenskraft*.[17]

The Berlin Physical Society, founded in 1845 by du Bois Reymond, Brücke and others including the electrical engineer Werner Siemens, counted among its members a new generation of German physiologists desirous of presenting their researches in purely physical terms, free from all remnants of *Geist* and *Lebenskraft*. Here, then, was a forum of like-minded young researchers with ambitions for the radical advancement of physiology along mechanistic lines and whose careers were being launched in the context of the growth of the German university system and the decline of *Naturphilosophie*. By 1847, Helmholtz was presenting his research before the Physical Society, reporting on physiological heat to the Society's *Progress of Physics (Fortschritte der Physik)* from its first volume and arguing for new and rigorous foundations for physiological science.[18] Those new foundations were to be provided by his *Erhaltung der Kraft*.

In February 1847 Helmholtz wrote to his friend du Bois Reymond inviting private criticism of a draft of the introduction to his forthcoming memoir *'Über die Erhaltung der Kraft'*: 'I want to know if you think the style in which it is written [is] one that will go down with the physicists. I pulled myself together at the last reading, and threw everything overboard that savoured of philosophy, wherever it was not absolutely essential'. Helmholtz's private uncertainties reveal his concerns about the kind of reception which his ambitious memoir could expect, not least from physicists like Gustav Magnus (1802–70) in whose Berlin physical laboratory Helmholtz had recently carried out some physiological research. Because such physicists wielded considerable patronage, power and prestige, Helmholtz was especially desirous of winning their approval.[19]

When he read the memoir before the Physical Society on 23 July 1847, the younger generation of Helmholtz's friends and contemporaries were enthusiastic. The physicists were less impressed. Magnus, to whom Helmholtz sent the essay in the hope of obtaining support for its publication in Poggendorff's *Annalen*, was only persuaded to provide that support through the intervention of du Bois Reymond. In Poggendorff's opinion, however, the work was not sufficiently experimental to justify publication in his *Annalen*: 'The *Annalen* is necessarily dependent above all on experimental investigations'.[20] The largely-theoretical memoir, with no new experimental results, had thus to find publication outside the official scientific journals. Backed by strong support from his Berlin circle, Helmholtz not only persuaded a local publisher, G.A. Reimer, to produce the work, but was even paid an honorarium for the 71-page text.[21]

Helmholtz's *Erhaltung der Kraft* opened with the statement that 'The principal contents of the present memoir show it to be addressed to physicists [*Physiker*] chiefly'. He thus stressed that he was going to lay down the 'fundamental principles' of the memoir 'purely in the form of a physical premise, and independent of metaphysical considerations [*philosophischen Begründung*]'. He would then 'develope the consequences of these principles, and submit them to a comparison with what experience has established in the various branches of physics'.[22]

In spite of his denial of 'philosophical foundations', the 'Introduction' has often been interpreted in its philosophical, notably 'Kantian', context. What, then, did he mean by its independence from 'philosophical foundations'? The answer lies in the physiological context discussed in the previous section: that Helmholtz was above all concerned to dispose of all traces of *Geist* and *Lebenskraft* from natural science and to establish a mechanical approach to physiology on persuasive rational and empirical foundations. In this quest, of course, we see him pursuing a 'metaphysical' and philosophical goal which transcends purely physical and empirical investigation. But for Helmholtz and his contemporaries 'philosophical' carried the disreputable connotations of 'speculative', 'imaginary' and ultimately sterile *Naturphilosophie*.[23]

At the outset Helmholtz made clear that he was concerned to demonstrate that the propositions contained in his memoir were based upon and deducible from either of two maxims or assumptions. First, 'that it is not possible by any combination whatever of natural bodies to derive an unlimited amount of mechanical force', that is, perpetual motion is impossible. And second, 'that all actions in nature can be ultimately referred to attractive or repulsive forces, the intensity of which depends solely upon the distances between the points by which the forces are exerted'.[24]

Helmholtz further promised that in due course he would show the identity of these maxims. But in his 'Introduction' he set out to consider their bearing on what he termed 'the true aim of the [physical] sciences'. Arguing that the final goal of theoretical physical science was to 'refer back the phænomena of nature to unchangeable final causes', Helmholtz now transformed this requirement into a search for 'unchangeable forces'. He then invoked a conception of motion and force, rooted in German scientific culture, in terms of 'spatial relationship' [*räumliche Verhältnisse*] and synthesized it with a Laplacian perspective on the forces of attraction and repulsion acting centrally between material points.[25] First, 'motion is the alteration of spatial relationship'. From experience we know that motion can only appear as a change in the spatial relationship of at least two material bodies. Second, force, as the cause of motion, can 'only be conceived of as referring to the relation of at least two material bodies towards each other'. Force was therefore 'to be defined as the endeavour of two masses to alter their relative position':

> But the force which two masses exert upon each other must be resolved into those exerted by all their particles upon each other; hence in mechanics we go back to forces exerted by material points. The relation of one point to another, as regards space, has reference solely to their distance apart: a moving force, therefore, exerted by each upon the other, can only act so as to

cause an alteration of their distance, that is, it must be either attractive or repulsive.

The problem of physical science therefore became simply that of referring 'natural phænomena back to unchangeable attractive and repulsive forces, whose intensity depends solely upon distance'.[26]

Helmholtz also explained that many of the general mechanical principles applicable to systems of bodies (the principle of virtual velocities, the conservation of motion of the centre of gravity, and the conservation of *vis viva* found in most of the existing treatises and textbooks on rational mechanics) were 'only valid for the case that these bodies operate upon each other by unchangeable attractive or repulsive forces'. Indeed, he claimed that virtual velocities was just a special case of *vis viva* which 'therefore must be regarded as the most general and important consequence of the deduction we have made'.[27]

His 'Introduction' complete, Helmholtz had thus far endeavoured to arrive at a 'rational' foundation for physical science in the form of a claim that all natural phenomena ultimately depended upon unchangeable attractive and repulsive forces. It was not the intensity of these forces which remained constant, however, for he had made their intensity depend upon the distance apart of the two or more material points between which the forces acted. Already he had hinted that it was to be *vis viva* which would be the principal measure of the conserved *quantity* of force. Helmholtz, however, was not presenting a formulation of the principle of conservation of energy as an independent, fundamental principle. Rather, his foundations rested very decisively on a commitment to a system of attractive and repulsive forces, from which the principle of conservation of *vis viva* followed as the most important consequence, a measure of 'conservation of force' [*Erhaltung der Kraft*].

In the first principal section of his essay Helmholtz began from the assumption of the impossibility of perpetual motion, that is, that 'it is impossible, by any combination whatever of natural bodies, to produce force continually from nothing'. He cited in particular Carnot and Clapeyron's deductions from this assumption of a series of laws regarding the specific and latent heats of natural bodies. Clapeyron's 1834 memoir had been partly translated for Poggendorff's *Annalen* in 1843 (ch.3). Now Helmholtz wanted to render the same assumption applicable throughout the whole range of physical phenomena.

Generalizing the Carnot–Clapeyron cycle of operations for reversible processes far beyond its previous application to gases, Helmholtz argued that if a system of natural bodies, acted upon by forces 'mutually exerted among themselves', were thereby moved to a new position with the net production of external work, the same quantity of work would be required to restore the system to its original position: 'For were the quantity of work greater in one way than another, we might use the former for the production of work and the latter to carry the bodies back to their primitive positions, and in this way produce an indefinite amount of mechanical force [work]'. In other words, we would have constructed 'a *perpetuum mobile* which could not only impart motion to itself, but also to exterior bodies'.[28]

The mathematical expression of this principle was, Helmholtz stated, to be found in the 'known law of the conservation of *vis viva*' but followed the comparatively recent (particularly French engineering) practice of restyling the traditional *vis viva* as ½*mv²* (ch.3). He then devoted the remainder of this section to showing that the principle 'is alone valid where the forces in action may be resolved into those of material points which act in the direction of the lines which unite them, and the intensity of which depends only upon the distance', that is, that the principle was not applicable to all possible kinds of forces, but only to central forces.[29]

In the second section of the memoir, Helmholtz sought to transform the principle of conservation of *vis viva* into a much more general law, 'the principle of the conservation of force', for mechanical systems subject to the condition of centrally acting forces. Thus if ϕ were the intensity of the force acting in the direction r (positive for attraction; negative for repulsion), then he showed that for a material point of mass m and when Q and q and R and r represent corresponding tangential velocities and distances:

$$\tfrac{1}{2}mQ^2 - \tfrac{1}{2}mq^2 = -\int_r^R \phi \, \mathrm{d}r$$

This equation, Helmholtz explained, expressed on the left-hand side the difference of *vires vivae* of m at two different distances. The quantity on the right-hand side had to be interpreted as 'the sum of the intensities of the forces which act at all distances between R and r'. He therefore named the forces which tended to move m, before the motion actually occurred, *tensions* [*Spannkräfte*] by contrast to *vires vivae* [*lebendige Kräfte*]. Thus the law of the conservation of force would be stated as: 'The increase of *vis viva* of a material point during its motion under the influence of a central force is equal to the sum of the tensions which correspond to the alteration of its distance'.[30] The choice of the term *Spannkraft*, consistent with the notion of spatial relationships, carried the meaning of increased tension or intensity of force with increased distance, analogous to the stretching of an elastic string.

Helmholtz moved to a still more general mathematical statement of the law for any number whatever of material points. Expressed verbally, the law became:

> In all cases of the motion of free material points under the influence of their attractive and repulsive forces, whose intensity depends solely upon the distance, the loss of [quantity of] tension is always equal to the gain in *vis viva*, and the gain in the former equal to the loss in the latter. Hence *the sum of the existing tensions and vires vivae is always constant*. In this most general form we can distinguish our law as *the principle of the conservation of force*.

He immediately proceeded to show that the most general law of statics, the principle of virtual velocities, followed from his principle of conservation of force, that is, that all the laws of statics may be deduced from the equations he had set forth.[31]

Having established his foundations, Helmholtz began a wide-ranging and systematic application of the principle of conservation of force [*Das Princip von der Erhaltung der Kraft*]. Initial applications were simple: for example, he considered briefly the application of the principle (as conservation of *vis viva*) to the motions of heavenly and terrestrial bodies under the influence of gravitational force. Provided friction or inelastic collision did not occur, the *vis viva* acquired by a body falling through a height would be sufficient to carry it back to the same height.[32]

Following a preliminary discussion of the nature of inelastic collision and friction, Helmholtz claimed in the fourth part of his memoir that the conventional representation of friction as a force which acts against existing motion had been only made to facilitate calculation and was a very inadequate expression of the complex processes of action and reaction of molecular forces. As such it led to the inference that by friction (and by inelastic collision) '*vis viva* was absolutely lost'. But the heat developed also represented 'a force by which we can develope mechanical actions; the electricity developed, whose attractions and repulsions are direct mechanical actions, and the heat it excites, an indirect one, has also been neglected'.[33]

The problem then to be solved was 'whether the sum of these forces always corresponds to the mechanical force which has been lost'. In cases which avoided molecular changes and electrical effects, the question became: 'whether for a certain loss of mechanical force a definite quantity of heat is always developed, and how far can a quantity of heat correspond to a mechanical force'. In response to the first part of this question, the development of heat from work, Helmholtz pronounced upon the value of Joule's experiments (ch.4):

> For the solution of [this] first question but few experiments have yet been made. Joule has measured the heat developed by the friction of water in narrow tubes, and that developed in vessels in which the water was set in motion by a paddle-wheel; in the first case he found that the heat which raises 1 kilogramme of water 1°, was sufficient to raise 452 kilogrammes through the height of 1 metre; in the second case he found the weight to be 521 kilogrammes. His method of measurement however meets the difficulty of the investigation so imperfectly, that the above results can lay little claim to accuracy. Probably the above numbers are too high, inasmuch as in his proceeding a quantity of heat might have readily escaped unobserved, while the necessary loss of mechanical force in other portions of the machine is not taken into account.[34]

Although Helmholtz questioned the accuracy of Joule's method and results, he nevertheless placed considerably more reliance upon Joule's investigations as evidence against a material theory, and in favour of a mechanical theory of heat, than has generally been recognized.[35]

Turning to the converse process, the development of work from heat, Helmholtz set out the Carnot–Clapeyron theory of the motive power of heat and its connection with a caloric theory:

The material theory of heat must necessarily assume the quantity of caloric to be constant; it can therefore develope mechanical forces only by its effort to expand itself. In this theory the force-equivalent of heat can only consist in the work produced by the heat in its passage from a warmer to a colder body; in this sense the problem has been treated by Carnot and Clapeyron, and all the consequences of the assumption, at least with gases and vapours, have been found corroborated.[36]

Having rejected the material theory in favour of a mechanical theory in his 1845 encyclopaedia article, he now highlighted the principal problem which the material theory continued to face: that of accounting for the heat developed by friction. In particular he dismissed an attempt to explain frictional heating by conduction: 'if it were true, then in the neighbourhood of the rubbed portions a cold proportionate to the intense heat often developed must be observed'.

Helmholtz also put considerable emphasis on the development of heat from electricity. He examined two cases of the production of electricity by mechanical means, those of electrostatic induction and of electromagnetic machines. In the first case, an electrophorus, as an insulated conductor, is brought near an insulated charged body and thus acquires an induced charge which may be discharged into a Leyden jar which can then be itself discharged to produce heat without the production of corresponding cold. In this manner we 'have consumed a certain amount of force, for at each removal of the negatively-charged conductor from the inducing body the attraction between both is to be overcome'. Similarly, in magneto-electric machines 'heat *ad infinitum* may be developed by the bodies constituting the machine, while it nowhere disappears. That the magneto-electric current developes heat instead of cold, in the portion of the spiral directly under the influence of the magnet, Joule has endeavoured to prove experimentally'.[37]

Such cumulative evidence appeared decisive against all caloric theories: 'it follows that the quantity of heat can be absolutely increased by mechanical forces, that therefore calorific phænomena cannot be deduced from the hypothesis of a species of matter, the mere presence of which produces the phænomena, but that they are to be referred to changes, to motions, either of a peculiar species of matter, or of the ponderable or imponderable bodies already known, for example of electricity or the luminiferous aether'. Under this interpretation, quantity of heat would express firstly 'the quantity of *vis viva* of the calorific motion' (corresponding to the old term 'free heat') and secondly 'the quantity of those tensions between the atoms which, by changing the arrangement of the latter, such a motion can develope' (corresponding to 'latent heat').[38]

Instances of collision, friction, and chemical processes, however, tended to focus on the production of heat rather than on the 'disappearance' of heat. Thus 'Whether by the development of mechanical force heat disappears, which would be a necessary postulate of the conservation of force, nobody has troubled himself to inquire' – with the possible exception of Joule:

> I can only in respect to this cite an experiment by Joule, which seems to have been carefully made. He found that air while streaming from a reservoir with

a capacity of 136.5 cubic inches, in which it was subject to a pressure of 22 atmospheres, cooled the surrounding water 4°.085 Fahr. when the air issued into the atmosphere, and therefore had to overcome the resistance of the latter. When, on the contrary, the air rushed into a vessel of equal size which had been exhausted of air, thus finding no resistance and exerting no mechanical force, no change of temperature took place.[39]

Helmholtz made no mention of his little-known countryman Mayer (nor indeed of Gay-Lussac) with regard to this experiment (ch.4). Once again, in spite of reservations of the accuracy of Joule's results, Helmholtz placed considerable credibility upon the Englishman's researches.

In the final part of this section, Helmholtz summarized Clapeyron's theory of the motive power of heat (ch.3). Noting that Clapeyron's deduction of the 'law' that the maximum mechanical effect is the same for all bodies 'can only be admitted when the quantity of heat is regarded as unchangeable', Helmholtz nevertheless accepted that Clapeyron's specific formulae for gases were supported by experiment as well as following from K.H.A. Holtzmann's more recent formula.

Holtzmann, on the other hand, had begun from the assumption that a quantity of heat entering a gas either increases the temperature or causes an expansion in volume of the gas: 'The quantity of work thus produced by the heat he [Holtzmann] assumed to be the mechanical equivalent of heat; he calculated from the experiments of Dulong upon sound, that the heat which raises the temperature of 1 kilogramme of water 1° Centigrade, would raise a weight of 374 kilogrammes 1 metre'. Accordingly, Helmholtz noted that the free expansion of a gas must exhibit no change of temperature, an inference which 'indeed appears to follow from the above-mentioned experiment of Joule'. Thus 'the increase and diminution of temperature by compression and dilatation would, under ordinary circumstances, be due to the excitation of heat by mechanical force, and *vice versa*'. Joule's own experimental results, however, for the mechanical equivalent (the 'force-equivalent of heat') were in the range from 452 to 521, considerably higher than Holtzmann's value of 374. Apart from commenting upon the fact that the latter figure, unlike those of Joule, had been calculated 'from the experiments of others', Helmholtz made no attempt to account for the discrepancy.[40]

Sections five and six of the memoir ranged widely and systematically across the whole of electrical science, beginning with statical electricity and continuing through galvanism, thermoelectricity, and magnetism to conclude with electromagnetism. In these investigations he was guided by the conservation principle throughout an impressive range of electrical conversion processes which brought together chemical, thermal, mechanical and electrical phenomena. He again drew extensively on recent experimental research, including that of Joule, Faraday, Ohm and Lenz. His stated aim was 'to draw conclusions regarding laws which are as yet but imperfectly known . . . and thus to indicate the course which the experimenter must pursue'.[41]

Helmholtz's *Erhaltung* ended with brief reference to the physiological context which had generated it. In that original context the memoir had been constructed to function as a theoretical and experimental foundation which would demonstrate the fallacy of any doctrine involving *Lebenskraft*. This 'policing role' had become vastly extended in the course of the memoir itself. His strategy had been to show that the principle was applicable not simply within rational mechanics but had universal applicability to all the forces of nature. He had therefore assembled a substantive weight of physical, chemical and physiological evidence drawn from a wide range of experimental resources in support of that claim. Now the theoretical physicist could assert mastery over all the branches of physical and chemical science.[42]

Within a very few years this role was strengthened when *Erhaltung* ceased to be the locally published property of a German physiologist. Having 'discovered' Helmholtz's treatise on 20 January 1852, Thomson began reflecting on the theoretical and experimental foundations for his paper 'On a Universal Tendency in Nature to the Dissipation of Mechanical Energy'. In a draft for the opening paragraphs he recognized that in the past the principle of mechanical effect had been restricted to 'conservative' cases, that is, where the effects are independent of path between initial and final states:

> The principle of mechanical effect first stated in all its generality by Newton at the conclusion of the scholium to his 3^d Law of Motion, and enunciated more or less <completely> explicitly by subsequent writers in the two propositions commonly called 'the principle of virtual velocities' and 'the principle of the conservation of vis viva', is first deduced from the fundamental axioms of mechanics as a theorem applicable to cases in which either working or resisting forces may be regarded as arbitrarily applied.[43]

Thomson continued, however, that 'as soon as it is established in Natural History that all the working or resisting forces of inanimate matter, as well as all the mechanical actions of living Creatures either are due to the inertia of matter, or are mutual forces between material particles which, when overcome through any spaces are always ready to restore the work spent, by working backwards through the same spaces, the postulate . . . [that a potential function exists for every force] assumed in the theorem of the "conservation of vis viva" becomes known as a Universal Truth'. In this laden sentence Thomson expressed his conviction that the empirical establishment of 'conservation of energy' *throughout* nature would guarantee the existence of a potential function for all the forces of nature, including apparently non-conservative ones such as friction. Thus the effect of every force would be independent of path and the principle of mechanical effect raised to the status of a completely general variational equation from which the principle of virtual velocities and other equations of motion could be derived.[44]

In his draft Thomson was in effect reworking Helmholtz's arguments in the light of his own fundamental convictions. At first sight the analyses appear identical. But Helmholtz's commitment to a basic physics of attractive and repulsive forces acting at a distance contrasted strikingly with Thomson's early preference for continuum approaches to physical agencies such as electricity and magnetism.

Erhaltung der Kraft as conservation of force, whose quantity is measured in terms of *vis viva* and whose intensity is expressed in terms of attractive or repulsive forces acting at a distance, was now being read as an independent 'Universal Truth', 'conservation of mechanical energy', whose quantity is measured as mechanical effect and whose intensity is understood in terms of a potential gradient.[45]

Reworking Helmholtz further, Thomson drafted his own range of empirical evidence in support of this 'Universal Truth' that mechanical energy could not be altered in quantity by any natural agency:

> This state of certainty we may regard as now reached, in consequence of the recent advancement of science in the establishment of the dynamical theory of heat and light; in the discoveries of Rumford, Davy, and Joule regarding the thermal effects of the friction of solids, fluids, and electric currents, and the aggregate thermal and non-thermal mechanical effects of electric currents employed in electrolysis or in raising weights [as in an electromagnetic engine]; in the discoveries of Priestley, Sennebier, and Davy regarding the influence of sunlight on vegetation; and in the ample confirmation of the mechanical energy of animals afforded by the researches of Dulong, Dumas, and Liebig. *We may consequently regard it as certain that, neither by natural agencies of inanimate matter, nor by the operations arbitrarily effected by animated Creatures, can there be any change produced in the amount of mechanical energy in the Universe*; and the belief that Creative Power alone can either call into existence or annihilate mechanical energy, enters the mind with perfect conviction.[46]

Very soon afterwards Rankine formally restyled the 'principle of mechanical effect' as 'the law of the conservation of energy', that 'the sum of the actual and potential energies in the universe is unchangeable'.[47] The new language, developed by Thomson and Rankine, signified their concern not merely to avoid ambiguities in speaking about 'force' and energy' in physics and engineering, but to reinforce a whole new way of thinking about and doing science.

Rankine's 'On the General Law of the Transformation of Energy' (originally communicated to the Glasgow Philosophical Society on 5 January 1853 and reprinted in the *Phil.Mag.*) explained that 'the term *energy* is used to comprehend every affection [state] of substances which constitutes or is commensurable with a power of producing change in opposition to resistance, and includes ordinary motion and mechanical power, chemical action, heat, light, electricity, magnetism, and all other powers, known or unknown, which are convertible or commensurable with these'. As with Thomson's 1851 axiom, Rankine's language here read like that of a legal document which in a sense it was. Associated through personal and family connections with lawyers, commercial gentlemen and inventors, Rankine and Thomson were quick to apply the language of patents to scientific property which could then be marketed to a scientific public in return for increased credibility. Like any product in a capitalist economy, the science of energy would be gauged by the extent of its market appeal. Again in legalistic style, Rankine distinguished all conceivable forms of energy into two kinds, 'actual' or 'sensible', and 'potential' or 'latent'.[48]

In the preceding article in the same issue of the *Phil. Mag.*, Thomson added a note referring to the 'admirable terms "potential" and "actual" introduced by Mr Rankine in his paper "On the Transformation of Energy" . . . to designate the two kinds of energy which I had previously distinguished by the inconvenient adjectives of "statical" and "dynamical"'. Joule too firmly endorsed the new perspectives. As he told Thomson in February 1853: 'The term energy employed by you seems admirably suited as the expression of anything which might ultimately by proper transformations be exhibited in the form of, say, heat'. Furthermore, 'The adjectives potential and actual employed by Mr Rankine seem to deserve the approbation you express in reference to them'.[49] From now on, the new terminology of energy would replace Thomson's 'principle of mechanical effect' and Helmholtz's 'law of conservation of force' as well as various older versions of conservation of *vis viva*.

Yet the North British scientists of energy did not always present a wholly united public image. During the preparation of their *Treatise on Natural Philosophy* (1867), Thomson and Tait substituted 'kinetic' for 'actual' in 1862 (ch.10). Rankine himself publicly criticized the change as undesirably restrictive and seized upon Maxwell's remark in his *Theory of Heat* (1871) that 'We cannot even assert that all energy must be either potential or kinetic, though we may not be able to conceive any other form'. In Rankine's view 'this was the very reason which induced me in 1853 to propose the word "actual" for denoting energy that is not potential, rather than any word expressly denoting motion, and which still induces me to prefer the word "actual" to the word "kinetic" for that purpose'.[50]

The British Association and the advancement of energy

By the mid-1840s the annual meetings of the British Association for the Advancement of Science had become a familiar ritual of Victorian life. But leadership of the Association was slowly changing. Many members of the founding generation of Oxbridge-dominated Anglicans – polymaths such as William Whewell, geologists such as Adam Sedgwick and William Buckland, and astronomers such as Sir John Herschel – no longer played such a central role after 1845. And although Section A could boast a distinctive style of scientific practice (modelled upon a mythical 'Newtonian' inductive approach of data, laws and general theory), the Association had never constructed a universal science of its own to rival, for example, that of Laplacian physics. For a younger, more entrepreneurial generation there thus existed a gap in this large scientific market which a credible new physics might readily fill.[51]

A contemporary spectator at the Ipswich meeting (1851), however, would scarcely have discerned significant changes in the Association's hierarchical character. The Cambridge-educated Astronomer Royal, George Biddell Airy, delivered the presidential address. Astronomy, hitherto 'Queen of the sciences', predominated. But Airy referred briefly to 'the investigations which have lately been made by able engineers regarding the Mechanical Equivalent of Heat. The subject, in this form, is yet new; but I think that the importance of an accurate determination cannot be over-rated'.[52] Unnamed, these 'able engineers' (probably Joule and Rankine) were not accorded the status of the Cambridge-trained mathematicians, natural philosophers and astronomers of Section A. Yet Airy, sympathetic to engineering concerns, had nonetheless highlighted here the need for 'accurate determination' (the hall-mark of his own astronomical practice at the Royal Greenwich Observatory) of the mechanical equivalent of heat.

For the 1852 Belfast meeting the BAAS president was a very different figure. A particularly astute soldier of science, Colonel Edward Sabine (1788–1883), senior member, treasurer and vice-president of the Royal Society, was best known for his role in promoting the so-called 'Magnetic Crusade' in the late 1830s. Once held up by Babbage as an embodiment of the Royal Society's failings, Sabine had become one of the most powerful patrons of young Cambridge mathematicians, such as Archibald Smith and William Thomson, who had little liking for the older generation of Oxbridge lions.[53]

Almost certainly well primed by Thomson, Sabine pronounced from the presidential chair a highly favourable verdict on the new theory of heat in general and upon Joule in particular. Given the Royal Society's reluctance to accord him a place in its *Transactions* until 1850, Joule may well have felt that Sabine's judgment had come somewhat late in the day. Nevertheless Joule was now being represented within the BAAS as a model gentleman of science who, employing all the correct procedures of experimental and numerical accuracy, had placed the new theory of heat beyond controversy.[54] Sabine also drew attention to the 'Mathematical developments of the theories of heat and electro-dynamics, in accordance with these [Joule's] principles ... given in various papers by MM. Helmholtz, Rankine, Clausius and Thomson, published principally within the last two years'. He noted with special satisfaction that Section A 'will have a great advantage in being presided over by the last-named of these gentlemen, a native of Belfast, who at so early an age has attained so high a reputation, and who is taking a leading part in the investigations to which I have referred'.[55]

The 28-year-old Glasgow professor of natural philosophy, elevated for the first time to the presidency of the prestigious Section A, commanded unrivalled authority in the field. His series of papers 'On the Dynamical Theory of Heat' had been appearing both in the *Transactions of the Royal Society of Edinburgh* and in the *Phil. Mag.*, as had his recent paper 'On the Universal Tendency in Nature to the Dissipation of Mechanical Energy'. He was receiving presidential patronage from Sabine. And his now-frequent public usage of the term 'energy' linked together the investigations of a formidable range of practitioners: some, like Joule, Rankine and his brother, part of a well-established personal network; others, like Clausius and Helmholtz, a means to elevate the significance of the new science from merely local relevance to international and universal importance.

Although Thomson read at least four papers to Section A during the 1852 meeting, his dissipation paper was not one of them. Instead, Rankine presented a summary of the new physics and cosmology under the provocative title 'On the Reconcentration of the Mechanical Energy of the Universe'. He began by observing that:

> it has long been conjectured, and is now being established by experiment, that all forms of physical energy, whether visible motion, heat, light, magnetism, electricity, chemical action, or other forms not yet understood, are mutually convertible; that the total amount of physical energy in the universe is unchangeable, and varies merely its condition and locality, by conversion from one form to another, or by transference from one portion of matter to another.[56]

These transformations and transferences, he added to his *Phil. Mag.* version, constituted 'the phænomena which are the objects of experimental physics'. Thus experimental physics could be re-read and redefined as the study of energy conversions (ch.9).

Rankine noted that Professor Thomson had pointed out that 'in the present condition of the known world there is a preponderating tendency to the conversion of all the other forms of physical energy into heat, and to the equable diffusion of all heat; a tendency which seems to lead towards the cessation of all phænomena'.[57] Here Rankine was taking further Thomson's original third conclusion which had merely referred to the finite character of the earth as a place fit for human life (ch.6). Two years later, Helmholtz also broadened Thomson's original claim when he reflected that 'we must admire the sagacity of Thomson, who, in the letters of a long-known little mathematical formula which speaks only of the heat, volume, and pressure of bodies, was able to discern consequences which threatened the universe, though certainly after an infinite period of time, with eternal death'.[58]

While not disputing Thomson's claim 'to represent truly the present condition of the universe, so far as we know it', Rankine refused to accept the pessimistic conclusion. He therefore speculated that radiant heat – 'the ultimate form to which all physical energy tends' – might be totally reflected at the boundaries of the very interstellar medium through which the radiation had been transmitted and diffused. This energy might then be 'ultimately re-concentrated into foci; at one of which, if an extinct star arrives, it will be resolved into its elements, and a store of energy reproduced'. Rankine therefore concluded:

> Thus it appears, that although, from what we can see of the known world, its condition seems to tend continually towards the equable diffusion, in the form of radiant heat, of all physical energy, the extinction of the stars, and the cessation of all phenomena; yet the world, *as now created*, may possibly be provided within itself with the means of reconcentrating its physical energies, and renewing its activity and life.[59]

Although we as yet know too little of Rankine's specific religious position within presbyterian culture, the key phrase 'as now created' is consistent with the evidence that he placed great store on the teachings of the Bible. Outside the growing liberal presbyterianism of Glasgow College until 1854, Rankine may have felt that Thomson's temporal vision of the universe came too perilously close to the deistic undertones in *Vestiges* (ch.6) and that what was needed instead was a universe of recent creation functioning as a perfectly reversible thermo-dynamic engine. It is therefore possible that Rankine sought to embody Thomson's notion of a perfect thermo-dynamic engine in a universe 'as now created' in the image of a God of perfection. 'Dissipation' once again related to human beings' inefficient design of engines rather than to any ultimate imperfections in nature, the apparent tendency towards 'the cessation of all phænomena' being in reality only one phase of a larger reversible cycle of the created world.[60]

Wielding vastly greater scientific authority and shamelessly recruiting BAAS presidential allies in the guise of William Hopkins to his cause (below), Thomson ensured that it was his, not Rankine's vision, which held sway. Indeed, Rankine was never again to air such a 'cyclical' speculation in public. So strong was Thomson's commitment to a directional universe that he was apt to dismiss all such hypotheses of 'compensation' and 'balance' as wholly unwarranted by the evidence of nature and inconsistent with a liberal reading of scripture (ch.6). As he clarified his position in 1862:

> The result would be a state of universal rest and death, [only] if the universe were finite . . . But it is impossible to conceive a limit to the extent of matter in the universe; and therefore science points rather to an endless progress, through an endless space, of action involving the transformation of potential energy into palpable motion and thence into heat, than to a single finite mechanism, running down like a clock, and stopping for ever.

Thomson's version of the new energy cosmology, then, ruled out both the possibility of reconcentration *and* a view of the universe as a simple mechanical clock winding down to an inevitable cessation of motion. Furthermore, his public phrase 'an endless progress', echoing his private claim in 1851 that 'Everything in the material world is progressive', was symptomatic of the authority of the new science of energy. In particular, Thomson could now speak confidently of the universe in terms of an endless flow, implying a perfect infinite creation rather than an imperfect finite mechanism. Equally, he and his allies could attempt to break the monopoly on 'progress' held by evolutionary theorists such as Herbert Spencer who in 1857 had enunciated a fundamental natural law: 'that in which Progress essentially consists, is the transformation of the homogeneous into the heterogeneous'.[61] Such doctrines of onwards and ever-upwards evolutionary progress were indeed anathema to the North British scientists of energy (ch.9).

Very conveniently, the BAAS president at the Hull meeting (1853) was none other than Thomson's former Cambridge mathematical coach, William Hopkins, who had recently served as president of the Geological Society of London. Respected for his researches in physical geology, he was known as the 'wrangler-maker' on account of his ability to produce many of the top-ranking Cambridge mathematical graduates classed from 'Senior Wrangler' downwards. Ideally placed to give credibility to the new doctrines of heat, he devoted at least half of his BAAS address to a wide-ranging review of the subject.[62]

His enthusiasm for geological science originally fired by the Cambridge professor of geology, Adam Sedgwick, around 1833, Hopkins had read to the Cambridge Philosophical Society his first major scientific memoir within two years. He there introduced a new subject, 'Researches in Physical Geology', which would bring the power and prestige of mathematical analysis to bear on geological phenomena. Geological science would thereby be raised to the same high status as physical astronomy. As with the mythologized history of physical astronomy, the new geology would have three stages: first, a geometrical description of the relevant motions (analogous to Kepler's laws of planetary motion); second, postulation of a very general force that would cause the motions (analogous to gravitational force); and third, derivation of the actual motions from the general force according to dynamical principles.[63]

The formulation of geometrical laws of faults, fissures, mineral veins and the like constituted the goal of this memoir. In subsequent investigations, Hopkins sought the very general cause, an 'elevatory force', responsible for all such phenomena wherever and whenever they occurred. He found that cause in the widely held doctrine of central heat whereby the earth, assumed to have been formed as a hot fluid mass that subsequently solidified, had undergone, and continued to experience, a progressive (directional) cooling in time. Like the doctrine of universal gravitation in its generality and simplicity, central heat became for Hopkins the fundamental agency of geological dynamics.[64]

Hopkins's preferred model was of a largely solid earth consisting of cavities. Hot vapours or fluids forced into those cavities from below would produce elevatory pressures in local regions. This model clashed directly with the steady-state (non-progressionist) geological theory of Lyell (ch.6) which denied the doctrine of primitive heat while upholding the notion of a largely liquid interior supporting a thin terrestrial crust less than 100 miles thick. Most of Hopkins's subsequent investigations were directed towards justifying the adequacy of his model and extending its explanatory power. Between 1838 and 1842 a series of papers for the Royal Society argued on mathematical grounds that the observed behaviour of the terrestrial planet around its axis (precession and nutation) was consistent with a solid, but inconsistent with a liquid, interior. A long report to the BAAS (1847) treated the phenomena of earthquakes and volcanoes within the same framework.[65]

Hopkins's 1853 presidential address enlisted the new theory of heat in support of geological theory. A scientific lifetime's credibility invested in geological researches had brought increasing returns to Hopkins: presidency of the Geological Society in 1851, presidency of the BAAS in 1853, and now seemingly irrefutable support from the new physics of his one-time pupil. He had indeed emerged as the most scientifically active of the older generation of Cambridge lions, an outsider who had entered the University relatively late in life and who had been excluded by marriage from a College fellowship. His alliance with the rising generation had been quickly cemented by his institution of a series of experiments, supported by Royal Society grants, guided by Thomson's thermo-dynamic expertise and aided in practice by Joule and Fairbairn, in Manchester to determine the effects of enormous pressures on the melting points of substances, the results of which he interpreted as supporting the solidity of the earth.[66]

In his address, Hopkins deployed the new theory of heat in the Cambridge war against Lyellian geology. He defined the controversy as one of 'progression' *versus* 'non-progression' (rather than, as Whewell had done earlier, as 'catastrophism' versus 'uniformitarianism') in geological theory. Progressionists held that 'the matter which constitutes the earth has passed through continuous and progressive changes from the earliest state in which it existed to its actual condition at the present time'. The earliest state (fluid or gaseous) was one of 'enormous primitive heat of the mass', the gradual loss of which caused progressive change. On the other hand, non-progressionists recognize 'no primitive state of our planet differing essentially from its existing state'. The only changes they recognized were 'those which are strictly periodical, and therefore produce no permanent alteration in the state of our globe'.[67]

The theory of non-progression, however, remained in serious doubt: 'the theory [cannot] derive present support . . . by an appeal to any properties of inorganic matter or physical laws, with which we are acquainted'. In contrast, Hopkins presented Thomson's cosmological views as resting upon well-founded physical laws. The new theory of heat asserted 'the exact equivalence of heat and motive power; and that a body [such as the sun], in sending forth heat, must lose a portion of that internal motion of its constituent particles on which its thermal state depends'. Furthermore, 'no mutual action of these constituent particles can continue to generate motion which might compensate for the loss of motion thus sustained'. This 'simple deduction from dynamical laws and principles' was 'independent of any property of terrestrial matter which may possibly distinguish it from that of the sun'. In the absence of adequate external supplies of 'thermal energy', Hopkins concluded, 'the heat of the sun must ultimately be diminished, and the physical condition of the earth therefore altered, in a degree inconsistent with the theory of non-progression'.[68]

Hopkins' indictment here of non-progressionists extended also to Rankine whom he damned with faint praise for having 'ingeniously suggested an hypothesis according to which the reconcentration of heat is conceivable'. He rejected Rankine's 'very ingenious, though, perhaps, fanciful hypothesis' as 'affording a sanction to the theory of *non-progression*'. By contrast, he stated that his own convictions 'entirely coincide with those of Prof. Thomson'. Thus, Hopkins concluded, 'If we are to found our theories upon our knowledge, and not upon our ignorance of physical causes and phænomena, I can only recognize in the existing state of things a passing phase of the material universe'.[69]

Hopkins's reading of the new theory of heat itelf bore a distinctive Cambridge stamp. Invoking no appeal to unobservable particles, he defined a 'dynamical theory of heat' as that which 'proposes to explain the thermal agency by which motive power is produced, and to determine the numerical relations between the *quantities* of heat and the *quantity* of mechanical effect produced by it'. He even included Carnot's original theory under this 'macroscopic' definition: 'Carnot was the first to give such a theory a mathematical form'. Such 'dynamical theories', macro-scopic in character, were characteristic of the new generation at Cambridge, exemplified particularly in Stokes's hydrodynamical and optical memoirs of the period.[70]

With respect to the experimental basis for the new theory, Hopkins acknow-ledged Rumford's 'rough attempt' to determine a mechanical equivalent of heat. But, he stressed, 'it was reserved for Mr Joule to lay the true foundation of this theory by a series of experiments, which, in the philosophical discernment with which they were conceived, and the ingenuity with which they were executed, have not often, perhaps, been surpassed'. Once again Joule had been unreservedly endorsed as a model gentleman of science within the Association's elite. Hopkins even reminded his audience of Joule's links with one of the BAAS's early heroes: 'we have in Mr Joule a pupil, a friend, and fellow-townsman of Dalton'.[71]

Hopkins also proclaimed that the new dynamical theory was 'in perfect har-mony with the opinions now very generally entertained respecting *radiant heat*'. It had been 'established beyond controversy' that 'light is propagated through space by the vibrations of an exceedingly refined ethereal medium . . . and it is now supposed that radiant heat is propagated in a similar manner' whereby the par-ticles of a heated body are in a state of vibration.

In this context he referred to Mr. Rankine's 'ingenious paper on a molecular theory of heat'. Professor Thomson, on the other hand, had 'given a clear and compendious mathematical exposition of the new dynamical theory of heat, founded on Mr. Joule's principle of the exact equivalence of heat and mechanical effect', a theory which was not, 'like Mr. Rankine's, a *molecular* theory, but one which must henceforth take the place of Carnot's theory'.[72] Although Hopkins here referred approvingly to the 'ingenuity' of Joule's experimental technique as well as to that of Rankine's theoretical manipulations, he reserved the highest accolade for his own former pupil, *Professor* Thomson, whose academic status, Cambridge mathematical training, and critical attitude towards speculative hypotheses together qualified him for leadership of Section A, alongside the likes of Stokes and Hopkins himself. Concomitantly, while Rankine would remain a central member of the energy network on account of his engineering expertise, his ventures into molecular physics and speculative cosmology earned him less credi-bility from among his scientific peers. As Thomson told Tait privately in 1864, 'Rankine . . . gave a somewhat magnificent but very imperfect & lame molecular theory (read & judge). Only less lame than all its predecessors'.[73]

With this presidential address Hopkins had set up Thomson as the BAAS's 'real man of armour' for whom the geologist Roderick Murchison had once called as a bulwark against the kind of dangerous and populist speculations embodied in *Vestiges* (ch.6). Although Thomson himself was not at the Hull meeting, the presi-dential address had opened the way for his fuller account of energy cosmology to Section A at the Liverpool meeting (1854). He also read the same paper, 'On the Mechanical Antecedents of Motion, Heat and Light', to the French Academy of Sciences, prompting the Abbé Moigno to comment critically on the views of the popular young British *savant*, the 'spoilt child of English science'.[74]

Reminding the BAAS audience of his 1852 discussion of animal heat and its origins (ch.6), Thomson recalled his verdict that, with minor exceptions, 'every kind of motion . . . that takes place naturally, or that can be called into existence through man's directing powers on this earth, derives its mechanical energy either from the sun's heat or from motions and forces among bodies of the solar system'. In a series of papers read to the RSE in the spring of 1854, he had further discussed the nature and cause of the sun's heat. The conclusions owed much to intensive discussions with his Cambridge friend Stokes.[75]

Thomson rejected two possible ways of accounting for the sun's heat: that of a 'primitive' store of heat in the sun and that of intrinsic (chemical) combustion. Neither hypothesis, he argued, could supply the required quantity of heat even over a 6,000-year period, that is, over the period of human history. Instead, he claimed to have 'shown that the sun's heat is probably due to friction in the atmosphere between his surface and a vortex of vapours, fed externally by the evaporation of small planets, in a region of very high temperature round the sun, which they reach by gradual spiral paths, and falling in torrents of meteoric rain, down from the luminous atmosphere of intense resistance, to the sun's surface'.[76]

While acknowledging that J.J. Waterston (1811–83), one-time civil engineer and now naval instructor for the East India Company in Bombay, had put forward such a 'meteoric' view at the Hull meeting, Thomson argued that, unlike Waterston's 'extra-planetary' origin for the meteoric matter, his own view was consistent with planetary dynamics in supposing the meteoric matter to lie within the earth's orbit. Indeed, he believed this matter was visible as a 'tornado of dust' or 'Zodiacal Light' whirling around the sun. He also estimated the sun's age as some 32,000 years and that 'Sunlight cannot last as at present for 300,000 years'. This dramatic vision, of meteors retarded in their orbits by an aetherial medium and progressively spiralling inwards towards the sun, was analogous to James Thomson's designs for vortex turbines with maximum economy, though in the one case the result was heat, in the other work (ch.3).[77]

Thomson, however, now asked a further question: 'from what source do the planets, large and small, derive the mechanical energy of their motions?' He regarded this question as one which could be answered by the application of mechanical reasoning 'for we know that from age to age the potential energy of the mutual gravitation of those bodies is gradually expended, half in augmenting their motions, and half in generating heat'. This kind of action could be traced either backwards or forwards in time:

> backwards for a million of million of years with as little presumption as forwards for a single day. If we trace them forwards, we find that the end of this world as a habitation for man, or for any living creature or plant at present existing in it, is *mechanically inevitable*; and if we trace them backwards according to the laws of matter and motion . . . we find that a time must have been when the earth, with no sun to illuminate it, the other bodies known to us as planets, and the countless smaller planetary masses at present

seen as the zodiacal light, must have been indefinitely remote from one another and from all other solids in space.

But Thomson now checked his unbridled enthusiasm for 'mechanical' reasoning backwards in time. 'All such conclusions', he emphasized, 'are subject to limitations, as we do not know at what moment a creation of matter or energy may have given a beginning, beyond which mechanical speculations cannot lead us'. Because the omnipotent God could have chosen to create matter or energy out of nothing at any time, human reasoning from present to past time was by no means infallible.

A striking instance of this point occurred with respect to the earth's history. Mechanical reasoning showed that the earth had been once without tenants. In Thomson's view science could not point to any antecedent or past state for the appearance of life on earth 'except the Will of a Creator', a view which he modified in 1871 to allow for the arrival by meteors of primitive life on earth. With these limitations in mind, Thomson nevertheless urged that 'we may legitimately push [our speculations] . . . into endless futurity, and we can be stopped by no barrier of past time' unless we arrived at a state of matter in some finite past epoch 'derivable from no antecedent by natural laws'.

Thomson concluded 'that the bodies now constituting our solar system have been at infinitely greater distances from one another in space than they are now' or that '*the potential energy of gravitation may be in reality the ultimate created antecedent of all the motion, heat, and light at present in the universe*'. He used this conclusion to attack the 'ordinarily stated' nebular theory (promoted and popularized by his old Glasgow teacher, John Pringle Nichol) which assumed primitive matter in a gaseous state. In contrast the new view showed evaporation of matter into a gaseous state to be 'a necessary consequence of heat generated by collisions and friction'.[78]

At this stage Thomson referred to Helmholtz's recent Königsberg lecture, 'On the Interaction of Natural Forces' (February 1854), in which the particles constituting the sun's present mass were supposed drawn together by gravitational attraction from a state of infinite diffusion, though not from an originally gaseous state. Helmholtz, who had attended the Hull meeting, would have been well acquainted with Hopkins's presentation of the new 'North British' views. Against this context, he had now constructed his own cosmological theory. His estimate of the heat generated, Thomson noted, yielded some 20 million times the amount of heat at present radiated by the sun in one year. But Thomson claimed that most of this heat would have been radiated off immediately, leaving not enough to account for the sun's present store of heat. He therefore reiterated preference for a meteoric theory of compensation.

Although maintaining here an independent position from that of Helmholtz, Thomson largely abandoned his meteoric theory in the early 1860s on the grounds that an adequate supply of such meteors within the earth's orbit seemed too much at variance with the observed motion of comets passing close to the sun. By 1861 he had adopted instead Helmholtz's 'contraction' model of the sun's heat in which the progressive shrinkage of the sun under gravitational attraction released vast quantities of heat over immense periods of time. Thomson therefore informed the 1861 Manchester meeting of the BAAS that 'the sun is probably an incandescent liquid mass, radiating away heat without any appreciable compensation by the influx of meteoric matter'.[79]

Taken together, these British Association meetings in the period 1851–4 transformed the new energy physics and cosmology from hitherto largely local contexts in Manchester, Glasgow and Edinburgh to national and even international contexts. With the patronage of Sabine, representing the Royal Society and a powerful imperial strand in British science, and Hopkins, representing the latest 'progressionist' crusade of Cambridge geometrical geology, Thomson had staged little short of a *coup d'etat*. Having appropriated in turn the insights of Carnot and Helmholtz, among other continental *savants*, Thomson and his allies had launched their promotion of energy physics and cosmology at a time when the older generation of British Association lions had failed to construct and sustain a distinctively British natural philosophy. With a new science of energy in the making, Section A's 'men of armour' could now effectively discipline other, often very active sections such as geology, and police wayward popularizers such as the author of *Vestiges*. Fresh opportunities for such disciplinary authority were not long in coming (ch.9).

CHAPTER 8

The Science of Thermodynamics

Mechanical action may be derived from heat, and heat may be generated by mechanical action, by means of forces either acting between contiguous parts of bodies, or due to electric excitation; but in no other way known, or even conceivable in the present state of science. Hence thermo-dynamics falls naturally into two divisions, of which the subjects are respectively, *the relation of heat to the forces acting between contiguous parts of bodies*, and *the relation of heat to electric agency*.

William Thomson coins the term 'thermo-dynamics' at a meeting of the RSE, 1 May 1854[1]

Adapted from William Thomson's phrase 'thermo-dynamic engine' (ch.5), the word 'thermodynamics' entered the scientific language through Macquorn Rankine's *Manual of the Steam Engine and Other Prime Movers* (1859), constructed in four parts ('Muscular Power', 'Water and Wind Power', 'Steam and Other Heat Engines' and 'Electro-magnetic Engines'). The third part of the *Manual* contained a 150-page chapter entitled 'Principles of Thermodynamics' which Maxwell later billed 'the first published treatise on the subject' (ch.12). There Rankine consolidated the terminology of 'First' and 'Second Laws of Thermodynamics', introduced by him in an article for John Pringle Nichol's *Cyclopaedia of the Physical Sciences* (1857). Concepts of 'work' and 'energy' also unified the *Manual* as a whole into one of the best selling engineering texts of the nineteenth century. By 1908, indeed, long after the author's death, the *Manual* had reached its seventeenth edition.[2]

The *Manual*, however, was only one end-product of several years' demanding labour by Rankine and his associates directed to a complete reconstruction of the principles and practice of the motive power of heat. This chapter examines the stages in the construction of this new science of thermodynamics. In name, style, and character the new science bore all the hallmarks of Clydeside engineering. Informed by engineering practice, Rankine in particular drew heavily upon the concerns of his marine engineering friends, John Elder (1824–69) and James Robert Napier (1821–79). In turn, Rankine sought to offer a credible new science which would serve as an infallible guide to the designing of more efficient and compact prime movers, especially for marine use. Its principles represented in heat engines, thermodynamics would thus draw credibility from heat engines, and heat engines would conversely derive credibility from thermodynamics. In contrast, '*die mechanische Wärmetheorie*' of Clausius was the work of a Prussian theoretical physicist whose ultimate goal was to probe the molecular and mechanical foundations of physical phenomena such that even the elusive entropy concept might eventually be reduced to mechanical explanation (ch.12). We begin, however, with the Rankine story.[3]

Macquorn Rankine and 'the science of thermodynamics'

On a September day in 1869 the Rev. Norman Macleod conducted a funeral service for a member of his Glasgow congregation. No stranger to such sombre occasions, Macleod nevertheless recorded that this city funeral 'was one of the most impressive sights I ever witnessed'. On the south bank of the River Clyde the bustling shipbuilding and engineering works had closed, 'forge and hammer at rest, and silent as the grave'. Along the River itself, the forest of masts in Glasgow's crowded harbour was 'draped in flags, lowered half-mast in sign of mourning'. Following the coffin was a 'very army of workmen, dressed like gentlemen . . . column after column', upwards of 4,000 employees of one firm. Respectful crowds lined the streets, 'as if gazing on the burial of a prince'.[4] Everyone, it seemed, had cause to mourn the death, at the age of 45, of John Elder (1824–69), shipbuilder and marine engine builder whose steam compound engines had begun to make Glasgow and the Clyde a maritime legend.

One-time pupil of Lewis Gordon at Glasgow College, Elder had been part of a close-knit community of marine engineers and academics which included William Thomson, Macquorn Rankine and James Robert Napier, son of Robert and nephew of David, the 'founding fathers' of marine engineering and iron shipbuilding on the Clyde. Elder, Napier and Rankine had all attended at various times the Glasgow High School before entering upon engineering apprenticeships. They also shared a strong presbyterian heritage. Robert Napier himself had, as was the Scottish custom for an eldest son, been intended for the Kirk. His brother Peter eventually became a well-known minister of the Blackfriars Church in Glasgow. Rankine's own 'earliest recollection is that of my mother teaching me the Lord's Prayer, next my father explaining to me the character of Jesus Christ'. In later life Thomas à Kempis's *Imitation of Christ* would be a major inspiration. Another of Elder's friends, Rev. W.G. Fraser, recorded:

> When speaking with me on religious subjects, in his [Elder's] own quiet, clear, flowing, and forcible way, about translating the *facts* of Christ's life into our own lives, the unmistakeable impression was left on my mind that he was

actually making this part of his own religion, in endeavouring to improve the temporal condition of those around him. Whatever he did for the bodily comfort of those under him, flowed, I have no doubt, from this living principle rising from the centre of his own spiritual being – a God-given and Christ-implanted principle in the soul, leading to the imitation of Christ in doing good to the bodies of men.[5]

'Imitation of Christ' was symptomatic of the shift from the harsh evangelicalism, centred upon individual sin and redemption through atonement, to the much softer 'incarnationalism' centred on the perfect humanity of Christ (ch.2). Moral and physical 'imperfections' thus lay not with God or nature but in man whose duty it was to seek out the causes of those imperfections, whether in himself or in his works, in order better to approach the ideals of perfection embodied in Christ and in the creation.

Under the direction of his father who had been manager of Robert Napier's engineering works since 1821, Elder served a five-year apprenticeship. After gaining further practical experience in England, he returned to Napier's works as chief draughtsman in 1849. Indeed, such a passage through the stables of the Napier cousins was wholly characteristic of the second generation of innovative Clyde shipbuilders and engineers: William Denny, David Tod and John Mac-Gregor, J. and G. Thomson, A.C. Kirk, and William Pearce. In 1850 for example Tod and MacGregor evoked widespread interest among shipbuilders and owners with the launch to their own account of the 1600 ton *City of Glasgow* for a Glasgow to New York service. Unlike the usual wooden-hulled, paddle-driven Atlantic mail steamers of the 1840s, the ship boasted an iron hull and screw propulsion. Within the year the *City of Glasgow* had been bought by the new Liverpool & Philadelphia Steam Ship Company of Liverpool (later popularly known as the Inman Line). In contrast, the rival American-owned Collins Line and the Liverpool-based British & North American Royal Mail Steam Packet Company of Samuel Cunard, both with wooden paddle steamers, looked decidedly outdated. With the *City of Glasgow*, Glasgow's shipbuilders quickly made ocean-going iron steamers, hitherto exemplified by rather 'experimental novelties' such as Brunel's troubled and troublesome *Great Britain* (1845), credible to prospective investors.[6]

Relinquishing his position with Napier, Elder entered into partnership with a firm of millwrights from 1852. As Randolph, Elder and Co., the firm began marine engine construction and two years later fitted a compound steam engine, designed and patented by Elder, to the 764-ton iron screw steamer *Brandon*, intended for a new service between Limerick and New York. The *Brandon* made only one round voyage from Europe to New York before being taken up by the British Government as a Crimean war transport. But at the end of 1854 the *North British Daily Mail* reported that it was 'gratifying to learn that the result, both as to speed and economy, has equalled the expectations of her engineers'.[7]

The *Brandon's* demonstrable reduction in fuel consumption quickly prompted orders for engines to suit two ships of the Pacific Steam Navigation Company whose services were concentrated along the western seaboard of South America. In the absence of local supplies of fuel, coal had to be shipped out from Britain by sailing vessels. When, therefore, the Crimean War put a premium on such tonnage, Pacific Steam consulted Elder on the prospects for greater engine economy. The outcome was the coastal paddle steamer *Inca* (290 gross tons) with a recorded fuel consumption of only 2.5 lbs of coal per indicated horse-power per hour and the larger *Valparaiso* (1060 gross tons) burning 3 lbs compared to a low of 4 to 4.5 lbs in steamers prior to that time.[8]

The consequent rise in Elder's engineering credibility was dramatic. In Rankine's partisan verdict this initial 30–40 percent saving 'rendered it practicable to carry on steam navigation on the Pacific Ocean with profit'. But such a view was shared by the joint managing director of Pacific Steam, William Just, who declared in 1870 that Elder's 'promised results were always more than realised; and . . . such was the progress in improvement in the double-cylinder engines, that the last-delivered vessels surpassed the *Valparaiso* in the economy of fuel as far as she surpassed the ordinary type of machinery'. During his remaining lifetime, Elder supplied some 30 sets of engines and built 22 ships for Pacific Steam, including four large steamers of 3,000 tons designed for a new direct mail service from Europe to Chile, a 19,000 mile round trip representing the longest 'steam-line' in the world. In Just's words, those vessels 'have been so remarkable as regards regularity in performance of the voyage . . . and economical in the consumption of coal, that the attention of many large steam-ship owners, who had long remained sceptical, has been more particularly attracted to the merits of the compound engine'.[9]

Back in the early 1850s, however, no such practical demonstrations were on view. But on Clydeside there was a rising confidence in the prospects for more economical forms of motive power. By now firmly located in Glasgow, Rankine was building an impressive network of scientific and engineering contacts. On the one hand, links forged with William Thomson and J.D. Forbes had enabled him to promote his investigations of the mechanical action of heat at the Royal Society of Edinburgh. On the other hand, the Glasgow Philosophical Society, together with his existing engineering contacts, provided the opportunity to develop formal and informal collaboration with two Glasgow engineers, J.R. Napier and John Elder. With Napier, Rankine would attempt to patent and develop a new form of air-engine embodying the latest thermo-dynamic insights. And, in Lewis Gordon's words, besides Napier, John Elder 'was the intimate friend of Rankine' such that Elder's 'bold improvements' in marine steam machinery 'were constantly discussed with Rankine, whose scientific aid in insuring success was gracefully and munificently acknowledged by Elder's widow, by the gift of a large endowment to increase the emoluments of the chair of Civil Engineering and Mechanics'.[10]

The Rankine–Napier–Elder triangle provided the basis for the embodiment of the new dynamical theory of heat into engineering practice. Mutually concerned with the economy of thermo-dynamic engines, the two ostensible marine engineering rivals in fact worked closely together: Napier in some cases constructing the hulls and Elder supplying the engines. For his part, Rankine offered state-of-the-art scientific expertise derived from his own contributions to the theory of the motive power of heat and from his close associations with the Scottish natural philosophers, most notably Thomson. Fashioning for himself a role as mediator between natural philosophy and practical engineering, Rankine would soon seize the opportunity for establishing that role institutionally when appointed as Gordon's successor to the Glasgow chair of Civil Engineering and Mechanics in 1855. But in the four years or so prior to that appointment, Rankine's enthusiasm for thermo-dynamic engines in theory and in practice focused on two specific projects: a collaboration with Napier to patent a workable air-engine; and the discussions with Elder over the development of a marine compound engine.[11]

These practical projects were not simply informed by a few specific features of the new theory of heat. To Rankine and his associates, both projects were the very embodiment of the science of energy. As we have seen, mechanical effect or energy had very recently been represented as having its source in the sun's energies, themselves the result of the original potential energies of gravitation established at creation as the gift of God to His universe (ch.7). Mankind had but limited access to this vast cosmological progression through the energies available to man for the production of useful work. As the great driving force of the industrial age, coal represented the most readily available store of potential energy, easily transformable into actual energy through thermo-dynamic engines.

The great goal of Glasgow engineers and natural philosophers alike was therefore to minimize waste and maximize useful work, that is to attempt to approach the economic goal of a perfect thermo-dynamic engine through an understanding of the causes of waste. As Rankine later remarked of Elder's goal, he, unlike previous engineers, 'took a more correct view of the real advantages of the compound engine' by regarding it 'as a means, not so much of increasing the indicated power produced by a given expenditure of steam, as of diminishing that waste of power which causes the effective power to fall short of the indicated power'. Thus reform of the marine compound engine was 'not to be expected . . . except by an engineer who had studied and understood the principles of the then almost new science of thermodynamics'.[12]

Patents by Craddock in the early 1840s had suffered from excessive strain on the crankshaft and bearings due to the piston-rods acting in the same direction with very nearly the sum of their forces. There had also been problems with steam condensing within the cylinders themselves. In Rankine's opinion, Elder had designed an engine to obviate both these defects. First, his compound arrangement with a pair of cranks exactly opposite each other enabled the two driving forces to balance one another 'as regards pressure on the bearings'. Second, Elder had solved the practical problem which had recently dogged high pressure engines:

> He knew, in common with other practical men, the fact that when high rates of expansion were used with a view to economy of fuel, their economical action was defeated by the gathering in the cylinders of large quantities of liquid water, which evaporated when the exhaust-port opened, and carried away heat to the condenser; but he learned also – what was known to very

few practical men fifteen years ago – that the formation of that liquid water originated in the disappearance of heat during the performance of work by the expansion of the steam, and that the remedy was to supply the cylinder with additional heat to replace that which so disappears, by returning to the practice of Watt and the Cornish engineers, and resuming the use of the steam-jacket.[13]

This specific solution to a practical problem of waste related to remarks made by Rankine in a short paper, 'On the Economy of Single-acting Expansive Steam Engines, and Expansive Machines Generally' presented to the RSE in April 1851. Computing the action of Cornish pumping engines and comparing the results with experimental measurements made on a large Cornish engine at Old Ford, Rankine noted that the 'results of experiment generally shew a somewhat less expenditure of steam for a given duty that theory indicates'. He attributed this anomaly to 'the cylinder being heated by a jacket communicating with the boiler, in which the temperature is much higher than the highest temperature in the cylinder', thereby providing a theoretical legitimation for the introduction of steam jackets and offering a solution in practice to the problem confronted by Elder.[14]

Two months after the trial trip of Elder's S.S. *Brandon*, Rankine delivered a timely address 'On the Means of Realizing the Advantages of the Air-Engine' to Section G (engineering) of the Liverpool meeting of the British Association. The presentation hailed the embodiment of 'thermo-dynamics' in steam- and air-engine practice and the embodiment of steam- and air-engine practice in thermo-dynamics. The opening section defined a 'thermo-dynamic engine' as 'any body, or assemblage of bodies, which produces mechanical power from heat'. All our systematic knowledge of the mutual relations of heat and motive power was deducible from two laws: the 'law of the mutual convertibility of heat and mechanical power' (involving the quantity 'known by the name of *Joule's equivalent*') and the law of 'the efficiency of thermo-dynamic engines' or the 'ratio which the available power bears to the mechanical equivalent of the whole heat expended'.[15]

Although both laws bore a close resemblance to the two laws stated by Thomson in his 1851 'Dynamical Theory of Heat' (ch.5) as well as to contemporary formulations by Clausius and by Rankine himself, Rankine was concerned that the second law in particular should be framed in a manner 'most readily applicable to the performance of engines worked by heat'. To achieve this goal, he classified the efficiency of thermo-dynamic engines into two types: 'theoretical maximum efficiency' and 'actual efficiency'. The former term was defined as the ratio of the theoretical maximum value of the heat permanently converted into mechanical power (that is, the proportion of heat remaining after deduction of 'the necessary loss of heat' during the condensation or refrigeration part of the cycle) to the whole heat. The latter term resulted from the engine not fulfilling exactly the conditions prescribed by theory, resulting in a *waste* of heat and power. It was concerns with 'actual efficiency' which brought together Rankine, Elder and Napier.[16]

Treating not of general causes of loss of power (such as the friction of the machinery), Rankine's 1854 address focused specifically on those causes which 'affect the relation between the expenditure of heat and the action of the working elastic substance upon the piston, – in other words, the *indicated* power of the engine'. Thus restricted, five classes of such waste could be identified: first, 'Imperfect communication of heat from the burning fuel to the working substance'; second, 'Imperfect abstraction of the heat'; third, 'Communication of heat to or from the working substance at improper periods of the stroke'; fourth, 'Any expenditure of the heat in elevating the temperature of the working substance'; and fifth, 'Imperfect arrangement of the series of changes of volume and pressure undergone by the working substance during the stroke'.[17]

Rankine noted that in actual steam-engines the various means of diminishing the first three causes were already being practised to a great extent. Tubular boilers presenting a large surface to the products of combustion helped reduce dramatically the first kind of waste. Improved condensation assisted in the reduction of the second. And elimination of conduction and radiation from cylinder, steam-passages and boiler reduced the third. But the closely connected fourth and fifth classes of waste required an understanding of the action of steam in a perfect engine and an assessment of how far short actual engines fell from the ideal. In Rankine's account, imperfections occurred at two points in the cycle constituting the stroke (Figure 8.1).

First, maximum theoretical efficiency would be possible if, after condensation of a portion of the expanded steam (process D), compression of the remaining uncondensed steam resulted in liquefaction of that portion and its simultaneous elevation of temperature (and that of the condensed steam) to the original temperature and conditions at the beginning of the cycle (process A). In such a case, the only heat expended at the commencement of the next cycle would be the latent heat of evaporation, that is, the additional heat required to convert the now-liquified working substance back to its vapour state. But no practical mechanism existed by which to fulfil these ideal conditions and therefore a certain amount of heat was wasted in raising the temperature of the feed-water from the point of condensation to that of evaporation. Estimates showed, however, that unless the methods of heating the feed-water were 'unskilful', the actual waste was comparatively unimportant in practice, lowering the efficiency from its theoretical maximum by only about one-sixteenth.[18]

Second, Rankine explained that maximum theoretical efficiency would be possible if the steam, its supply from the boiler cut off, expanded without receiving or losing heat. During this expansion the temperature fell by the conversion of heat into mechanical power to the point of condensation (process C). But in order to achieve this result in practice, the cylinder and engine would be 'bulky, heavy, and costly, and the action of the steam upon the piston, during the latter portion of the stroke, would be so feeble as to cause unsteadiness of motion unsuitable for the driving of machinery'. On the other hand, an absence of expansive-working reduced efficiency to as little as a quarter of its theoretical maximum. Thus single-acting pumping engines, in which the expanding steam acted on one side of the piston only, were well suited to expansive working and had apparently approached

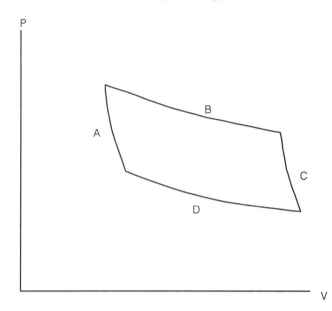

Figure 8.1 Rankine's 'diagram of energy' for 'perfect' heat engines. During stage (B) the working substance expands at a constant elevated temperature producing power and 'losing' heat. During stage (C) the substance may also expand and the temperature falls to a low level such as that of the condenser. During (D) the substance is compressed at the constant low temperature. Heat is produced but must be abstracted to avoid elevation in temperature since increased pressure would impede the return of the piston. Alternatively, but very imperfectly, the substance may be expelled to the atmosphere as in a locomotive. During (A) the substance is raised in temperature by compression alone to its high level. Avoiding elevation of temperature by expenditure of heat is the means to theoretical maximum efficiency.

the theoretical maximum. But double-acting engines, where the full range of expansive-working was problematic, tended to have efficiencies of less that half the theoretical maximum. In Rankine's estimates, such engines, with an average coal consumption of 4 lbs per horse-power per hour, might be improved by attention to the above causes of waste to about three-quarters of their theoretical maximum or 2.5 lbs per horse-power per hour: a target lower than that of Elder's recent *Brandon* but one met by his *Inca* a couple of years later.[19]

In general the possibility of utilizing a larger 'fall' of temperature motivated the search for workable air-engines in which the working substance carried few of the attendant risks of boiler explosions from high pressure steam. Enthusiasm among the North British natural philosophers and engineers had been fired some years earlier by the doyen of Scottish air-engine inventors, Robert Stirling (ch.3). Lewis Gordon, for example, had announced to William Thomson in April 1847 that 'Mr

Stirling is in Town – I am laying a trail for getting up a substantial Company for the manufacture of the Engine'. Eighteen months later he reported to the Glasgow professor the good news that his business partner, R.S. Newall, was to provide £500 to 'enable Regnault to complete his experiments on Steam and Effects of gases' in the face of funding cuts threatened by the 1848 Revolution on the streets of Paris. As to 'The true *Motive* – the Sterling [sic] disinterestedness of the offer are manifest'.[20]

Taking his cue from Carnot, Clapeyron and his brother James, William Thomson had waxed lyrical in his 1849 'Account' of Carnot's theory: 'The beautiful engine invented by Mr Stirling of Galston, may be considered an excellent beginning for the air-engine; and it is only necessary to compare this with Newcomen's steam-engine, and consider what Watt has effected, to give rise to the most sanguine anticipations of improvement'.[21] Others like Joule were initially less sanguine. At the end of 1850, the Manchester natural philosopher echoed his earlier pessimism regarding electro-magnetic engines (ch.4) by telling his Glasgow mentor that 'neither the air-engine, nor any other will be made to supersede the steam-engine'. But by the next spring he had been persuaded: 'The more I look into the subject the more I feel that air will supersede steam'.[22]

The new dynamical theory of heat apart, one of the reasons for a fresh upsurge of air-engine enthusiasm in the early 1850s was John Ericsson's ambitious project centred on the construction of a 1902-ton paddle-driven ship. Launched at New York on 15 September 1852, the wooden *Ericsson* was being fitted with a hot-air or caloric engine consisting of four working cylinders, over 4 metres (14 feet) in diameter and nearly two metres (6 feet) in stroke. These cylinders were apparently suspended 'like enormous camp-kettles over the furnace fires'.[23] Above them were installed four air-compressing cylinders of the same stroke and 3.5 metres (11 feet) diameter. Although preliminary trials took place in early 1853, the ship was not ready for full trials until the spring of the following year.

News of the Ericsson project almost certainly prompted Rankine's initial approach to J.R. Napier early in February 1853. As Ben Marsden has shown, in a formal letter Rankine called attention to both Stirling's and Ericsson's engines. The former boasted 'a maximum duty equivalent to about 1,800,000 foot-pounds for each lb. of coal' (compared to about 1,400,000 foot-pounds for the Fowey Consols engine). At the core of commercial resistance to the adoption of air-engines over steam engines lay the 'economizer' or 'regenerator' 'for storing up and using over and over again a considerable portion of the heat which produces the alternate rise and fall of temperature'. Echoing James Thomson's 1848 assessment (ch.3), Rankine told his friend that both inventors had believed 'if they could carry to perfection the operation of storing up & using over again the heat of *change of temperature*, their engines would produce power without expenditure of heat; that is to say, power out of nothing; a conclusion opposed not only to science, but to common sense, and tending to produce distrust of the alleged efficiency of the engines'.[24]

In order to build up the credibility of the air-engine, Rankine would have to act on two complementary fronts. On the one hand, he would have to explain the functioning of the 'regenerator' in accordance with 'scientific principles'. But such scientific principles themselves had only just been formulated and indeed had scarcely found widespread acceptance (ch.7). It was clearly not enough to form part of a relatively private network of North British natural philosophers and engineers. Rather, he needed simultaneously to ensure both the authoritative assent of the Royal Society of London *and* to promote a new science of 'thermodynamics' among practical engineers themselves. On the other hand, the latter audience, together with its commercial backers, demanded a practical demonstration. Rankine thus sought to win Napier's commitment by appeal to just such a practical demonstration now in the offing: 'If Captain Ericsson's ship, however, should succeed in crossing the Atlantic, of which there seems to be no reason to doubt, it is evident that public opinion will turn in favour of air-engines, and that a demand for them may be expected to arise at once; and it is not unlikely that for marine purposes, and in all districts where either fuel or water is scarce, and perhaps also for locomotives, they will ultimately supplant the steam-engine altogether'.[25]

Persuading Napier with such an appeal was the first step in a grand strategy. After all, there was no guarantee that Ericsson's project would ultimately prove any more viable than Stirling's more modest efforts. Rankine began a scheme to unite scientific and financial capital in a Glaswegian partnership. From Rankine would come primarily the theoretical expertise focused on the principal issue that Stirling, Ericsson and Joule had ignored: 'to reduce the waste of heat to a minimum, and secure the nearest approach to the theoretical duty'. From Napier would come primarily a knowledge of the market for ocean 'steam' navigation: '240 tons of coal would carry a ship of 500 horse-power to Australia'. Collaboration could lead to untold symbolic and financial returns on capital invested. The scene was set for the launch of a new Clydeside enterprise.[26]

Instructing Napier in the principles of the new science of heat, Rankine made much use of the pressure–volume or indicator diagram which traced a four-part cycle of isothermal expansion (B), decreasing pressure at constant volume (C), isothermal compression (D), and increasing pressure at constant volume (A) (Figure 8.3). In the case of the proposed air-engine, a higher temperature of some 600°F was cited, implying a vastly greater fall of temperature than for steam-engines. Drawing on his 1851 paper 'On the Economy of Heat in Expansive Machines', efficiencies were to be calculated by the ratio of the difference between the higher and lower temperatures to the higher temperature (expressed in terms of an absolute scale). These analytical tools enabled a clear assessment of the sources of waste in all sorts of heat engines, including of course both the Stirling and Ericsson engines. Refined for publication, they soon found public expression in Rankine's 1854 'Air-engine' paper as well as in a much longer presentation to the Royal Society which coincided with his election as FRS in June 1853.[27]

The hallmarks of the proposed Napier and Rankine air-engine were to be its compactness and its economy, rendering it ideal for ships and locomotives. Ericsson's engine was both 'enormous' and inefficient, while Stirling's, although compact, involved a 'great waste of heat' because 'the air on its way to be cooled passes over the hot surface, and carries off heat to the refrigerator which produces no useful effect'. In contrast, the compactness of the Napier and Rankine engine would be facilitated by working at high pressure and temperature, while its efficiency would be enhanced by 'screening the air from the heating surface except when it is expanding' and through increasing the heating surface by the use of tubes. Guided by Rankine's principal thermo–dynamic tool, the indicator diagram, all these features were to be embodied in the patent specification.[28]

Formal agreement between the partners had been reached by the beginning of June 1853. Rankine was to be recognized as the original inventor of the improvements 'after long study of the principles which govern the action of machines in which the motive power is derived from heat, and of the facts which have been ascertained relative to the subject by experience and observation'. For his part, Napier's 'skill and experience in the construction of engines, and especially of marine engines, and the well-known eminence of the firm with which you are connected, are essential to the practical carrying out of these improvements in an efficient and profitable manner'. Napier would provide the capital for development and would receive reimbursement after which any subsequent returns would be divided equally between the partners. Little over a week later the provisional patent, including a detailed provisional specification, was deposited at the Patent Office in London. Around the same time Rankine sought the active support of an uncle to secure a patent in North America.[29]

The actual design was relatively straightforward. Arranged as a double-acting engine, one receiver to the left side of a vertical cylinder exerted expansive power in a downward direction against the cylinder piston while the other receiver exerted its power in an upward direction against the same piston. Tubes at the lower and upper ends of each receiver provided the means of supplying heat to or extracting heat from the air. Each receiver also contained two moving components. Designed to 'regulate the transmission of heat' to the air during expansion but to inhibit such transmission at other times, a 'heat screen' operated at the lower end of the receiver. The 'plunger' (equivalent to the 'regenerator') functioned at the upper end of the receiver, both 'heat screen' and 'plunger' working in a co-ordinated manner (Figure 8.2).[30]

Co-ordination of all the operations in such a double-acting air-engine, always likely to be a major practical challenge, would produce an effect 'of a similar nature to that which takes place in all engines driven by the action of heat on an elastic substance viz., that the substance is alternately expanded at a higher temperature, and compressed at a lower, so that the power developed by the expansion is greater than the power consumed by the compression, and a surplus of power remains to drive the machinery'.[31]

By the autumn of 1853 plans were in hand for the construction of both a model (non-functional) and an experimental (fully working) air-engine. The model, to be built by the Glasgow instrument maker James White, was required in America to satisfy the patent office there. The experimental engine would be used as a proto-type to generate data and thereby convince prospective customers of the economy and feasibility of the design. By the end of the year the complete specification, fleshing out the earlier features, had been filed in the Great Seal Patent Office.

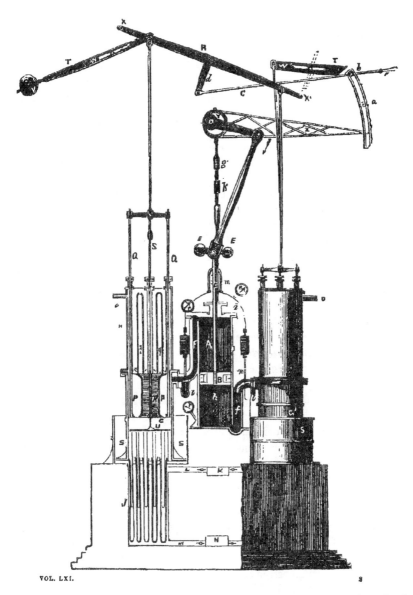

Figure 8.2 Rankine and Napier's patent hot-air engine as illustrated on the front page of the *Mechanics' Magazine* for 21 October 1854. The receivers to the right and left of the central cylinder exert power on the piston in upward and downward directions respectively. This cutaway drawing also gives some impression of the complexities of the engine compared to simple steam-engines. The illustration was located by Ben Marsden to whose work on the history of the air-engine this chapter is much indebted.

Around the same time, Rankine was formally admitted to the Royal Society where on 19 January 1854 he presented his first major paper to the Society, 'On the Geometrical Representation of the Expansive Action of Heat, and the Theory of Thermo-dynamic Engines'. In line with the Royal Society's professed distaste for controversial theories, the paper avoided all reference to Rankine's hypothesis of molecular vortices and formulated its message concerning the motive power of heat very much in terms of the indicator diagram.[32]

Rankine introduced his subject with a discussion of Watt's indicator (pressure–volume) diagram which he now renamed the 'diagram of energy'. Referring back to his 1851 paper 'On the Economy of Heat in Expansive Machines', he explained that 'the area of the diagram represents at once the potential energy or motive power which is developed at each stroke and the mechanical equivalent of the actual energy, or heat, which permanently disappears'. He further claimed that 'the principles of the expansion are capable of being presented to the mind more clearly by the aid of diagrams of energy than by means of words and algebraical symbols alone'. With them, he wished above all to solve new questions for 'all classes of engines, whether worked by air, or by steam, or by any other material'.[33] Grounded in Rankine and Napier's project to construct a practical air-engine, this paper took the new 'energy' perspective out of the relatively local contexts of Glasgow and Edinburgh to the very core of elite British science. Simultaneously, Rankine's authority as a scientific engineer was enormously enhanced.

Such diagrams of energy served a number of simultaneous functions. For the engineer they allowed the engine quite literally to chart its own performance by tracing directly on paper the work done in each cycle. From such inscriptions the engineer could therefore read messages of various kinds: whether the particular engine needed fine-tuning in order to improve its output of useful work, or whether the type of engine compared unfavourably with rival types of engine, for example. Although such functions were scarcely new (ch.3), the 'diagram of energy' now had the full authority of the new theory of heat. Conversely, for the North British natural philosophers Rankine was ensuring that the 'science of energy' was no mere abstract system, nor one loosely connected to possible engineering applications, but was instead one which was fully and firmly embodied in engineering practice. Just as Napier had been persuaded by such inscriptions of the advantages of the air-engine, so too would 'diagrams of energy' serve direct pedagogical roles for students of engineering science when Rankine took over the Glasgow University engineering chair in 1855.[34]

Rankine quickly deployed his enhanced scientific authority to provide an explanation of the hitherto controversial 'regenerator' within the framework of the new theory of heat. He explained that since the 'latent heat of increase of volume at an elevated temperature' was 'the direct source of the power of a thermodynamic engine', it followed that the more we minimized 'the heat which is expended in elevating the temperature of the working substance, the more nearly we shall attain to the maximum theoretical efficiency of the engine'. Elevation of temperature was, he argued, theoretically possible without any expenditure of heat'. There were two distinct ways of approaching this goal in practice: first, by compression during

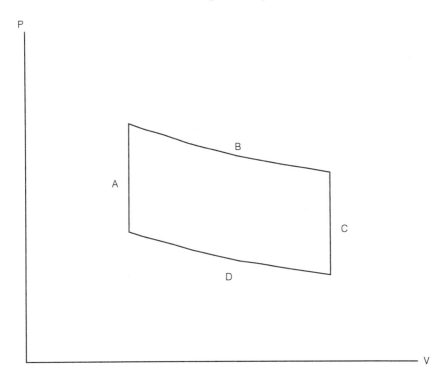

Figure 8.3   Rankine's 'diagram of energy' for the 'perfect' air-engine. In contrast to Figure 8.1. the stages (C) and (A) involve no expansion or compression. The respective fall and restoration of temperature is achieved by means of the regenerator storing up or regenerating the heat lost during (C) and gained during (A). (A) and (C) are then designated 'curves of equal transmission'. Once again avoiding elevation of temperature by expenditure of heat is the route to maximum theoretical efficiency.

the process A of the cycle, the power to do so having been obtained by lowering the temperature of the substance entirely by expansion in process C; and second, by storing up in a mass of some solid conducting material (called an economizer or regenerator) the heat given out by the working substance, while its temperature is being depressed, during the process C, and employing the heat, so stored up, to produce the required elevation of temperature during the process A (Figures 8.1 and 8.3).[35]

It followed that the principal role of the regenerator was to permit an approach to maximum efficiency with much less expansion and thus to enable air-engines to be designed in accordance with the criterion of compactness so much in demand for ships and locomotives but so little fulfilled by the Ericsson project. In contrast, the Napier and Rankine air-engine was to embody an approach to this goal: instead of 'curves of no transmission' (adiabatics) during processes C and A, they aimed at 'curves of equal transmission' (shown as parallel lines of constant volume but decreasing pressure) (Figure 8.3). In both cases the maximum theoretical efficiency would be the same, but in the latter case there would be no need for large increases and decreases in volume.[36] Either way Rankine had used his authority to crush all extravagent claims for the notorious regenerator.

Prospects for a practical air-engine suddenly waned in the spring of 1854. White had not delivered the model and the experimental engine had not yet been made to work.[37] Worst of all disaster overtook Ericsson's embodied vision of a bright air-engine future as the *Ericsson* at last put to sea:

> we attained a speed of from 12 to 13 turns of our paddle wheels, equal to full 11 miles an hour, without putting forth anything like our maximum power. All went magnificently until within a mile or two of the city (on our return from Sandy Hook), when our beautiful ship was struck by a terrific tornado on our larboard side, careening the hull so far as to put completely under the lower starboard ports, which unfortunately the men on the freight deck had opened to clear out some rubbish, the day being fine.[38]

Lying in shallow water the *Ericsson* was subsequently raised but put into service as a trans-Atlantic steamer. By 1867 the humiliation was complete: the *Ericsson* was converted to sail only and as such survived another 25 years.[39]

Without such practical demonstration the Scottish air-engine enthusiasts could scarcely hope to gain the financial backing which the project demanded. But the practical failures neither weakened the theoretical prospects for such an engine nor undermined the authority of thermodynamics. Indeed, the project had served Rankine well. His credibility among Clydeside marine engineers had increased dramatically. By the beginning of 1855 Rankine was being employed as a substitute for Gordon as professor of civil engineering and mechanics at the University. Having long sought to consolidate the seemingly complex formulations of 'energy' and 'thermodynamics' into concise 'sciences' on firm and unchallengeable axiomatic foundations, he thus took the opportunity in May of that year to present to the Glasgow Philosophical Society just such a science framed in suitably scholarly language with strong echoes of the *Principia*.[40]

'Outlines of the Science of Energetics' set out methodological principles which distinguished mechanical hypotheses from an 'abstractive method' of framing physical theories. The latter method aimed 'at a body of principles, applicable to physical phenomena in general, and which, being framed by induction from facts alone, will be free from the uncertainty which must always attach, even to those mechanical hypotheses whose consequences are most fully confirmed by experiment'. Not wishing to disown his own molecular vortex hypothesis, however, Rankine argued that a hypothetical theory was a necessary, preliminary step towards framing an abstractive theory. Extending his 1853 paper (ch.7), Rankine proposed to formulate, as an abstractive theory, the 'science of energetics'. Thus 'Energy, or the capacity to effect changes, is the common characteristic of the various states of matter to which the several branches of physics relate; if, then, there be general laws respecting energy, such laws must be applicable . . . to every branch of physics, and must express a body of principles as to physical phenomena in general'.[41]

Publication of this high-sounding, scholarly memoir coincided with, and probably facilitated, his permanent appointment to the Glasgow engineering chair in the autumn of 1855. That appointment may not have marked the complete abandonment of the air-engine project, but Rankine's primary concerns were now pedagogical. Thus talk of air-engines was more likely to relate to classroom and textbook teaching of the new thermodynamics than to the pursuit of commercial patents.[42] Thermodynamics itself, however, needed to be propagated to wider audiences and by 1857 the first edition of Nichol's *Cyclopaedia of the Physical Sciences* gave Rankine a fresh opportunity to write about 'Heat, Theory of the Mechanical Action of, or Thermo-dynamics', this time explicitly in terms of a 'First' and 'Second Law of Thermo-dynamics'. The article concluded with a forceful statement of Rankine's goals for securing the authority of the new 'sciences' of thermodynamics and energy, as well as consolidating the relationship between them:

> Although the mechanical hypothesis [of molecular vortices] . . . may be useful and interesting as a means of anticipating laws, and connecting the science of thermo-dynamics with that of ordinary mechanics, still it is to be remembered that the science of Thermo-dynamics is by no means dependent for its certainty on that or any other hypothesis, having been now reduced to a system of principles, or general facts, expressing strictly the results of experiment as to the relations between heat and motive power. In this point of view, the laws of Thermo-dynamics may be regarded as particular cases of more general laws applicable to all such states of matter as constitute ENERGY, or the capacity to perform work, which more general laws form the basis of the SCIENCE OF ENERGETICS, a science comprehending as special branches, the theories of motion, heat, light, electricity, and other physical phenomena.[43]

Although Rankine's *Manual* (published two years later) was also written in the new language of energy for the education of engineers, Rankine's closest colleagues in natural philosophy had reservations about his foundations for the science of thermodynamics. From the beginning, Thomson had urged him to reduce his dependence upon the molecular vortex hypothesis (ch.6). But Maxwell and Tait, far less imbued with the heady industrial spirit of Clydeside, became more critical, especially after Rankine's death. Maxwell acknowledged that in the *Manual* Rankine had 'disencumbered himself to a great extent of the hypothesis of molecular vortices, and builds principally on observed facts'. Nevertheless, Maxwell cautioned, he 'makes several assumptions, some expressed as axioms, others implied as definitions, which seem to us anything but self-evident'. Most strikingly, 'when we come to Rankine's Second Law of Thermodynamics we find that . . . its actual meaning is inscrutable'.[44]

Finding Rankine's formulation of thermodynamics unsuitable for their student audiences, and frustrated by the lack of progress with the epic *Treatise on*

Natural Philosophy (ch.10), leading advocates of the science of energy took action to supply the perceived need in the form of textbooks such as Tait's *Thermo-dynamics* (1868) and Maxwell's *Theory of Heat* (1871) (ch.12). Spurred by an increasing distaste for Clausius's use of theoretical concepts which frequently seemed too far removed from concrete physical reality, Tait in particular felt strongly 'that both Clausius & Rankine are about as obscure in their [thermo-dynamic] writings as anyone can well be'.[45] And Maxwell too expressed a clear preference for the style of Thomson over that of his co-founders of the science of thermodynamics:

> He has always been most careful to point out the exact extent of the assump-tions and experimental observations on which each of his statements is based, and he avoids the introduction of quantities which are not capable of experimental measurement. It is therefore greatly to be regretted that his memoirs on the dynamical theory of heat have not been collected and reprinted in an accessible form, and completed by a formal treatise, in which his method of building up the science should be exhibited in the light of his present knowledge.[46]

Not until 1882 did Thomson's papers appear together, and then only as the first volume of his collected *Mathematical and Physical Papers* rather than as a separate treatise.[47] Thomson had simply too many projects under way to assume the mantles of treatise or textbook writer. Yet his characteristic style, captured above by Maxwell, was to permeate many of the debates over the foundations and formulation of thermodynamics (ch.12).

Rudolf Clausius and 'Die mechanische Wärmetheorie'

In his 'Dynamical Theory' (part I) Thomson had credited Clausius with 'the merit of first establishing the [second] proposition upon correct principles'. But Thomson's reading of Clausius went much further. He claimed that Clausius's 'demonstration' was founded on the axiom that 'It is impossible for a self-acting machine, unaided by any external agency, to convey heat from one body to another at a higher temperature'. Not stated anywhere in Clausius's 1850 paper, the axiom had been constructed by Thomson in such a fashion that although this and Thomson's own axiom were different in form nevertheless 'either is a con-sequence of the other'. Clausius's 1850 paper had indeed shown how Carnot's theory of the motive power of heat could be reconciled with the convertibility of heat and work: in accordance with Carnot, whenever work was done in a cyclic process, some heat was transferred from a hotter to a colder body; while in accordance with Joule (and others) the rest was converted into work (ch.5). But Clausius's scientific style, characterized by logical and linguistic consistency, dif-fered so radically from the North British engineering perspectives that it would only be a matter of time before mutual credibility yielded to bitter controversy (ch.12). In this section, however, I consider Clausius's own distinctive reformulations of a mechanical theory of heat in the period prior to the outbreak of those hostilities.[48]

By 1854 Clausius himself admitted that his original expression of 'Carnot's theorem' or 'second fundamental theorem in the mechanical theory of heat' was 'incomplete, because we cannot recognize therein, with sufficient clearness, the real nature of the theorem, and its connexion with the first fundamental theorem'. He aimed to remedy the defect by offering more general and concise foundations for the mechanical theory of heat in terms of two theorems: that of the 'equivalence of heat and work' and that of 'the equivalence of transformations'. In contrast to his 1850 paper which had referred only to 'the general deportment of heat ... to pass from a *warmer* body to a *colder* one', Clausius stated as a fundamental principle: '*Heat can never pass from a colder to a warmer body without some other change, connected therewith, occurring at the same time*'.[49]

The second theorem now expressed 'a relation between two kinds of transformations, the transformation of heat into work, and the passage of heat from a warmer to a colder body, which may be regarded as the transformation of heat at a higher into heat at a lower temperature'. Its analytical expression became:

$$\int dQ/T = 0$$

Clausius then examined three further issues. First, he derived a relation between T and 'Carnot's function' (denoted by C) which showed C to be a function of temperature only (ch.3). Second, he briefly considered *non-reversible* processes. Here he modified the second theorem to the statement that 'The algebraical sum of all transformations occurring in a circular process can only be positive', with reversible processes forming the limit (in which the algebraic sum was zero). Non-reversible transformations were *uncompensated* transformations: for example, the sudden expansion of air from one vessel into an empty one produced no corresponding work; yet restoration of the air to its original pressure and volume required work to be done. Finally, on the assumption that a perfect gas expanding at constant temperature only absorbs as much heat as is consumed in the external work produced, Clausius concluded that in all probability T was simply the absolute temperature.[50]

Reinforced by his recent researches into gas theory (ch.12), Clausius was soon openly permitting molecular considerations to enter his memoirs on the mechanical theory of heat. By 1862 he explained his view that the processes 'by which heat can perform mechanical work ... always admit of being reduced to the alteration in some way or another of the arrangement of the constituent molecules of a body'. Inferring that 'the effect of heat always tends to loosen the connexion between the molecules, and so to increase their mean distances from one another', he then expressed 'the degree in which the molecules are dispersed' in terms of a magnitude called the *disgregation* of the body. Thus the effect of heat was redefined as simply '*tending to increase the disgregation*'.[51]

A ninth memoir in his series of investigations on the mechanical theory of heat appeared in Poggendorff's *Annalen* in 1865. It was to form the concluding chapter in his treatise on *The Mechanical Theory of Heat* (1865). Although he there introduced 'entropy', the paper itself had been little concerned with the promotion of a radically new concept. Recognizing that earlier versions of the mechanical theory of heat had sometimes led to difficulties of calculation in its application, Clausius simply aimed 'to render a service to physicists and mechanicians by bringing the fundamental equations . . . from their most general forms to others which, . . . being susceptible of direct application to different particular cases, are accordingly more convenient for use'.[52]

Clausius now designated as positive an element of heat *dQ absorbed* (rather than emitted) by a changing body and thus the expression of the second fundamental theorem became:

$$\int dQ/T \leqslant 0$$

For reversible changes, he denoted dQ/T as the complete differential of a magnitude S depending only on the present condition of the body, and not on the way by which it reached this condition. He therefore wrote:

$$dS = dQ/T$$

He then explained that the 'physical meaning' of the magnitude S had already been provided in his 1862 memoir. There $\int dQ/T$ was expressed as the sum of two integrals, $\int dH/T$ (depending solely upon temperature) and $\int dZ$ (the disgregation which depended on the arrangement of the particles of the body). Clausius now termed the first integral the 'transformation-value' of the heat present in the body and the second integral the transformation value of the existing arrangement of the particles of the body. The sum of these two transformation values yielded S.[53] Only at this late stage in his memoir did Clausius introduce a new nomenclature:

> We might call S the *transformational content* of the body . . . But as I hold it to be better to borrow terms for important magnitudes from the ancient languages, so that they may be adopted unchanged in all modern languages, I propose to call the magnitude S the *entropy* of the body, from the Greek word τροπή, *transformation*. I have intentionally formed the word *entropy* so as to be as similar as possible to the word *energy*; for the two magnitudes to be denoted by these words are so nearly allied in their physical meanings, that a certain similarity in designation appears to be desirable.[54]

Clausius concluded his memoir with a summary of the cosmological implications of the second theorem. Acknowledging that Thomson had first drawn attention to the application of this theorem to the universe, he expressed 'the fundamental laws of the universe which correspond to the two fundamental theorems of the mechanical theory of heat':

1. *The energy of the universe is constant.*

2. *The entropy of the universe tends to a maximum.*[55]

These neat statements, of course, could be seen as lending further authority to the new thermodynamics by virtue of their seeming identity to North British perspectives on conservation and dissipation of energy. But consensus among the protagonists was much more apparent than real. Thus the logical and linguistic similarities should not be allowed to mask the striking conceptual differences between Clausius and his North British contemporaries. Prussian theoretical physicists who gave primacy to logical and linguistic consistency contrasted with Scottish engineers and natural philosophers steeped in cultures of industry and presbyterianism. Such cultural disparities were to have profound consequences for interpretations of the laws of energy and thermodynamics in the following two decades (ch.12).

CHAPTER 9

North Britain *versus* Metropolis: Territorial Controversy in the History of Energy

> Just as, in the eye of the chemist, every chemical change is merely a rearrangement of indestructible and unalterable matter; so to the physicist, every physical change is merely a transformation of indestructible energy; and thus the whole aim of natural philosophy, so far as we yet know, may be described as the study of the possible transformations of energy, with their conditions and limitations; and of the present forms and distribution of energy in the universe, with their past and future.
>
> *P.G. Tait, 'Rede Lecture' to a distinguished academic audience in the Senate House of the University of Cambridge (1873)*[1]

Professor Peter Guthrie Tait, occupant of the Edinburgh chair of natural philosophy, could scarcely have chosen a more appropriate audience before which to put on record the pedigree of the new science of energy. According to the speaker it was none other than Cambridge's own Isaac Newton who, 'in a short Scholium a couple of centuries ago', had pointed out the 'grand idea of the conservation, or indestructibility, of energy' which formed the very 'groundwork of modern physics'. Established for heat by Rumford and Davy, and 'extended to all other forms of energy by the splendid researches of Joule',[2] the energy doctrines owed nothing, it seemed, to continental actors. It was as though the names of Leibniz, Helmholtz, Mayer and Clausius had never existed.

Tait's goal in this 'Rede Lecture' was to assist in preparing the ground for Maxwell's introduction of experimental physics into the mathematical and moral culture of Cambridge University during the early 1870s. Following his appointment to the new Cavendish chair in 1871, Maxwell had explained to the mathematical physicist J.W. Strutt (1842–1919, later Lord Rayleigh) that it would 'need a great deal of effort to make Exp. Physics bite into our university system which is so continuous and complete without it'. And he added with typical irony that 'if we succeed too well, and corrupt the minds of youth till they observe vibrations and deflexions ... we may bring the whole university and all the parents about our ears'.[3] Cambridge was a university imbued far more with the spirit of mathematical symbols and moral syllogisms on paper than with alien apparatus on the bench. As Maxwell warned Tait just prior to the lecture:

> The Senate House is a place to write in, to graduate in and to vote in. The Public Orator I believe can speak in it provided he employs the Latin tongue. . . . If you have a good audience there will not be much echo from [statues of King] Geo II or [prime minister] Pitt and if you erect a lofty platform the light spot on the screen, and the under side of your table may be seen by all. . . . you will do very well always remembering that to speak familiarly of a 2nd Law [of thermodynamics] as of a thing known for some years, to men of culture who have never even heard of a 1st Law, may arouse sentiments unfavourable to patient attention.[4]

With missionary zeal Tait would preach the gospel of the 'science of energy' before the assembled ranks of Cambridge dons. Invoking visible experimental demonstration within those sacred walls, he would dispel the clouds of ignorance which still beset the ancient institution. Neither Thomson nor Maxwell, Rankine nor Helmholtz could have acted such a part. Tait had taken upon himself the role of leading propagandist for energy.

In the decade following publication of Darwin's *Origin*, Huxley and his eight fellow members of the 'X-Club' (founded in 1864) seized upon Darwin's evolutionary theory as one of their most powerful weapons with which to promote a gospel of 'scientific naturalism'. Adrian Desmond in particular has shown that a prime target of Huxley's 'New Reformation' was Anglican Cambridge whose clerical, gentlemanly dons could be represented as the out-moded relics of pre-professionalized science. In this chapter I introduce John Tyndall, another member of the X-Club circle. Tyndall would seek to place energy conservation alongside evolution as one of the key doctrines of the new creed. In so doing he unleashed a massive extension of hostilities to North Britain.[5]

As the context for Tait's 'Rede Lecture' suggests, the science of energy offered a powerful rival reform programme to the University of Cambridge, competing directly with scientific naturalism but with the significant advantage that it posed no threat to the traditional public harmony between science and theology within the ancient institution. Conservative dons of a generation such as Stokes could therefore rest easy with the 'science of energy'. But the rivalry between North British and metropolitan reformers was not simply confined to a Cambridge battleground. It was rather the peripatetic British Association and the pages of the *Phil. Mag.* that provided principal national forums for the vigorous debates.[6]

I focus on the controversy between P.G. Tait and John Tyndall, fought ostensibly over the historical claims of Joule and Mayer. I argue that the Tait–Tyndall controversy should be construed as a contest for scientific authority. Their respective groups, informal social networks, were competing intensely for a monopoly on 'truth', once the preserve of the old Oxbridge Anglican lions. Tyndall's group mobilized a rhetoric of professionalization in its attack on the clerical dons and to serve its cause of self-aggrandisement. More specifically, Tyndall took upon himself the task of attempting to appropriate the new conservation of energy doctrine and deploy it, as Huxley was doing with Darwinian evolution, in the service of scientific naturalism. Tyndall, however, needed to wrest control from the North Britishers, whose 'ownership' had already been well entrenched through the British Association's recognition accorded to the experimental credentials of Joule. With a particularly shrewd knowledge of German physics Tyndall found an alternative hero to Joule in the person of Julius Robert Mayer.

The claims of the scientific naturalists to speak authoritatively for science in general, and for the new science of energy in particular, was challenged less by the Anglican dons and more by the North Britishers who were also perfectly willing and able to mobilize a rhetoric of expertise against Tyndall and friends. The clash over the claims of Mayer can thus be seen as expressing some of the fundamental differences between scientific naturalism (united by a stand against Christian doctrine) and the science of energy (promoted as a natural philosophy in harmony with, though not subservient to, Christian belief).

Dissipating Darwinism

Since its publication in 1859 Charles Darwin's *The Origin of Species by Means of Natural Selection* had been the subject of public controversy. Whether or not the legendary confrontation between Huxley and Bishop Samuel Wilberforce at the 1860 Oxford meeting of the BAAS provided a catalyst, Thomson certainly prepared himself to enter the debate sometime prior to the following year's meeting in Manchester. His aim was to offer an authoritative ruling from the perspective of the new energy physics and cosmology. As Joule wrote:

> I am glad you feel disposed to expose some of the rubbish which has been thrust on the public lately. Not that Darwin is so much to blame because I believe he had no intention of publishing any finished theory but rather to indicate difficulties to be solved. . . . It appears that now a days the public care for nothing unless it be of a startling nature. Nothing pleases them more than parsons who preach against the efficacy of prayer and philosophers who find a link between mankind and the monkey in the gorilla – certainly a most pleasing example of what *muscular* Christianity may lead us to.[7]

Notebook drafts during the summer of 1861 make clear, however, that Thomson's strategy would be to undermine the indefinitely long time-scale maintained by geologists such as Lyell rather than to attack Darwin directly. By showing that 'the Sun was sensibly hotter a million years ago than he is now', Thomson hoped to persuade geologists present that 'speculations assuming somewhat greater extremes of heat, more violent storms and floods, more luxuriant vegetation and hardier and coarser grained plants and animals in remote antiquity are more probable than those of the extreme quietist school', that is than those of Lyell's followers who held that geological agents acted with the *same* intensity throughout all past time.[8]

Order-of-magnitude estimates of the probable ages of the sun and earth were to be the decisive weapons in Thomson's hands with which to slay the Darwinian dragon of evolution by natural selection. In his 1861 address to the BAAS, 'Physical Considerations Regarding the Possible Age of the Sun's Heat', Thomson first invoked quantitative estimates provided by John Herschel and Pouillet of the annual rate at which heat was radiated from the sun. Assuming the specific heat of the sun to be the same as that of liquid water (the highest of all known terrestrial liquids and solids) and knowing the mass of the sun, Thomson calculated the annual lowering of the sun's temperature to be 1.4° C. Although he believed there were persuasive reasons to suppose the sun's matter to be very much like the earth's, Thomson argued that the sun's actual specific heat was probably, on account of the enormously greater pressures involved, between ten and ten thousand times greater than the specific heat of water. The result was therefore to reduce the rate of fall of the sun's temperature and to extend the age of the sun backwards and forwards in time. Working from Helmholtz's estimate of the amount of heat generated by shrinkage, Thomson concluded:

> It seems therefore, on the whole, most probable that the sun has not illuminated the earth for 100,000,000 years, and almost certain that he has not done so for 500,000,000 years. As for the future, we may say with equal certainty that inhabitants of the earth cannot continue to enjoy the light and heat essential to their life for many million years longer, unless new sources, now unknown to us, are prepared in the great storehouse of Creation.[9]

Consistent with his theology of nature, which emphasized the contingency of all the energies of nature upon God's will, Thomson kept open the possibility that God could endow nature with new gifts of available energy to allow continuation of the solar system beyond the time that the present laws allowed.

In a longer version for *Macmillan's Magazine* Thomson confronted Darwin directly over the quantitative time-scale put forward in *The Origin of Species*. 'What are we to think of such geological estimates as 300 million years for the "denudation of the Weald"?' Thomson asked his readers rhetorically. He was referring to Darwin's example of the Weald of Kent, the wide valley between the North and South Downs of southeast England. The Weald, according to Darwin, had been caused by the encroachment of the sea upon a line of chalk cliffs at a rate of about one inch per century. Darwin had provided an easy hostage to Thomson's fortune, for Thomson could simply argue that it was far more plausible to believe that the sea had acted with greater intensity in past times than that 'the physical conditions [properties] of the sun's matter differ 1000 times more than dynamics compel us to suppose they differ from those of matter in our laboratories'.[10]

As a sequel Thomson read 'On the Secular Cooling of the Earth' to the RSE on 28 April 1862. The full version appeared in the *Phil. Mag.* early the following year. Again the specific target was Lyell's steady-state theory of the earth which 'violates the first principles of natural philosophy in exactly the same manner and to the same degree, as to believe that a clock constructed with a self-winding movement may fulfil the expectations of its ingenious inventor by going for ever'. Thomson's approach was grounded upon some of his earliest investigations in mathematical physics (ch.6): 'Fourier's great mathematical poem' now provided a 'beautiful working out of a particular case belonging to the general doctrine of the Dissipation of Energy'. Not only, however, did Thomson's estimates of 50–500 million years for the age of the earth approximate to those derived from his recent calculations on the sun's age, but they also coincided remarkably closely with those of the geologist and leading figure in the BAAS, John Phillips (1800–74), with whom Thomson had been corresponding over the measurement of underground temperatures.[11]

By 1862, then, Thomson's strategies meant that many of the leading figures in the British Association felt able to participate in the promotion of the new doctrines or at least to feel that here was a new perspective which, unlike some of the rival cosmologies on offer, supported their own most deeply held convictions concerning man's place in nature. In particular, Cambridge men such as William Hopkins and Adam Sedgwick, together with BAAS stalwarts such as John Phillips, could feel secure in the knowledge that their commitment to progressionist cosmology and geology was fully consistent with the new energy views. Equally, traditional British Association values of accurate experimentation and measurement were upheld in the work of Joule, by now the figurehead of the new energy physics. And again, although some of the 'raw materials' of the energy doctrines had been imported from the Continent (notably in the papers of Fourier, Carnot, Helmholtz and Clausius), the 'finished product' would be exported to the world as a distinctively British artefact, superior in every way to any foreign imitation.

The making of Professor Tait

Whereas Thomson's father had come from a family of yeomen farmers, Tait's father had been secretary to the fifth Duke of Buccleuch, head of one of the wealthiest and most powerful aristocratic families in the land. While Thomson's mother came from a Glasgow commercial background, Tait's mother had strong Edinburgh banking connections. Between 1841 and 1847 Tait's schooling took place at the gentlemanly Edinburgh Academy where Fleeming Jenkin (ch.13) and Maxwell (ch.11) were among his contemporaries.[12]

Following one session at Edinburgh University (1847–8) during which he attended the classes of Kelland (mathematics) and Forbes (natural philosophy), Tait entered St Peter's College, Cambridge. Coached by William Hopkins, he emerged as Senior Wrangler and First Smith's Prizeman in 1852, with a former pupil of Thomson's at Glasgow University, W.J. Steele, as Second Wrangler. Tait quickly took up the offer of a Cambridge fellowship with credentials which surpassed even those of his future collaborator, Thomson, whom he seems to have first met in person around this time. 'I could coach a coal scuttle to be Senior Wrangler', Tait apparently once remarked with characteristic modesty. Within two years he was offered the mathematics chair at Queen's College, Belfast.[13]

While in Belfast from 1854 until 1860 Tait completed the foundations for his subsequent career. In mathematics he allied himself with the distinguished Irish mathematician and astronomer William Rowan Hamilton (1805–65) whose quaternion methods were to remain a source of great pleasure to Tait, and a cause of much irritation to Thomson, for many years after the commencement of the Thomson–Tait collaboration (ch.10). In experimental science Tait exploited to the full the laboratory facilities and expertise of the chemist Thomas Andrews, former pupil of Thomson's father and now Vice-principal of the College. Another colleague was Thomson's brother James who had begun teaching at Queen's in the same year as Tait, finally becoming professor of engineering in 1857. On a personal level, Tait married a sister of the Porter brothers, former pupils of the Thomsons at Glasgow and students at St Peter's College.[14] Tait had shrewdly consolidated his mathematical base while simultaneously cultivating a wider sphere of influence through a personal and scientific network centred on Belfast.

In a further career move Tait competed against Maxwell for a vacant chair of natural philosophy at Marischal College, Aberdeen in 1856. Though his rival obtained the chair, it was a Pyrrhic victory for the gentlemanly Maxwell. After his chair had been abolished by a fusion of Marischal with King's College into one University of Aberdeen, Maxwell soon found himself competing again with Tait in a bid to succeed Forbes in the more prestigious Edinburgh chair of natural philosophy. Obtaining that chair in 1860, Tait remained Edinburgh professor of natural philosophy until a short time before his death in 1901.[15]

Unlike the long-neglected Glasgow professorship to which Thomson had succeeded in 1846, Tait's chair had been associated with one of the most distinguished of British experimental philosophers. Forbes had also been a leading figure in the early years of the British Association.[16] Secure in his academic career, Tait could therefore build upon the high social and scientific status enjoyed by his predecessor, but with the advantage of outstanding Cambridge mathematical credentials of his own. Over the next decade, Tait acquired space for a physical laboratory along the lines of Thomson's Glasgow laboratory established in the course of the 1850s.[17] But his principal concern was to secure and develop a framework of physics which would bring together the mathematical and experimental components of natural philosophy. As promoted in Thomson and Tait's *Treatise on Natural Philosophy*, the science of energy would stand as the hallmark of British physics (ch.10).

Tait's first book had been *Dynamics of a Particle* (1856), begun as a collaboration between first and second wranglers but completed by Tait after Steele's premature death in 1855. Written in a tradition of textbooks for Cambridge mathematical students, 'Tait and Steele' did not employ either of the terms 'work' or 'energy' until the second edition (1865), that is, only well after Tait's collaboration with Thomson had commenced.[18] Not until Tait's public inaugural lecture at Edinburgh on 'The Position and Prospects of Physical Science' in 1860, do we find signs of enthusiasm for 'energy'.

The inaugural lecture set out Tait's position right at the beginning of his career as Edinburgh professor and just before the collaboration with Thomson. From one perspective the lecture shows the new professor carefully surveying his territory and firmly marking the disciplinary boundaries of natural philosophy. Early on he staked a dramatic claim by asserting that chemistry was 'in reality a branch of Natural Philosophy', but later assured his audience that with respect to chemistry he had 'no *right*, even if I had the will and the qualifications, to encroach upon the territory of a colleague'.[19]

Tait presented natural philosophy as a science which treated of matter and its *forces*, where force was '*any* cause which produces, or tends to produce, a change in a body's state of rest or motion'. Its principal branches were mechanics, physical astronomy, optics and hydrodynamics (including sound, tides and waves), all of which readily boasted a Newtonian pedigree and constituted a traditional core of the Cambridge mathematical syllabus. Having committed almost two-thirds of his lecture thus, Tait by-passed the subject of the 'modern dynamical theory of heat' on the grounds of pressure of time and moved quickly through electricity and magnetism to 'the great question' of the mutual relations of these 'natural forces'. Many people, he told his audience, 'suspect that these natural forces are merely modifications or transmutations one of another . . . In this we have what has been called the CORRELATION OF THE PHYSICAL FORCES'.[20]

William Robert Grove (1811–96), professor of experimental philosophy at the commercially minded London Institution from 1841 until 1846, had delivered a course of lectures 'On the Correlation of Physical Forces' at the Institution in 1843. His book of the same title first appeared in 1846, reaching its third edition by 1855 and its sixth edition by 1874. These later versions tended to assimilate some of the recent conclusions of Thomson and others with respect to heat, though the references to 'Mr. Thompson' can scarcely have impressed the Glasgow professor. Grove was an influential figure in metropolitan scientific circles, and especially in moves to reform the Royal Society from the 1840s. Tait himself took Grove seriously enough in his inaugural lecture to devote space to illustrations of the doctrine, emphasizing its basis in the claim that 'the amount [of one force] which can be derived from a given amount of another has a definite value'.[21] Only later would he dismiss Grove with contempt:

> Do you T[homson] try to read Grove's Book, and tell me whether you feel all along the impression of humbug. I certainly do, and cannot manage more than a few pages without falling asleep.
>
> I believe Grove is not a bad fellow, but he is (in my opinion) woefully loose and unscientific. He is not a Tyndall, as of course you know, since the latter is unique; but he is as vague and cloudy.[22]

As yet, however, Tait had not introduced the term 'energy'. Like Hopkins in 1853 he first did so in reference to Thomson's views on the sun's heat as the principal source of mechanical work available to man and on the finite duration of the sun on account of a gradual dissipation of energy (ch.7). 'What then', he asked, 'supplies the sun, or is he steadily drawing upon limited resources, and diminishing his own capital to increase ours?'

The language of political economy here seemingly resonated with the engineering concerns of Rankine and Thomson. Tait, however, did not yet align himself fully with his Glasgow brethren in ascribing a new, fundamental status to 'conservation of energy'. In attempting to give the 'somewhat vague principle of the Correlation of Forces' a *'precise statement'* as 'the great principle of CONSERVATION OF FORCE', he followed Helmholtz's *derivation* of 'conservation of force' from assumptions of forces acting centrally at a distance between particles (ch.7). A few pages later he again echoed Helmholtz in showing how the principle could be demonstrated from an assumption of the impossibility of perpetual motion (that is, that *'by no arrangement of apparatus can work be procured from nothing'*).[23]

Only in the final quarter of his lecture did the new professor explain that in the principle of conservation of force 'we do not mean force in the ordinary acceptation of the word'. He told his audience that 'the principle is now better known as the CONSERVATION OF ENERGY'. Accepting the divisions of energy into its 'actual' and 'potential' forms, Tait stated the principle 'as it has been put by Rankine, to whom the terms are due: *"In any system of bodies, the sum of the potential and actual energies is never altered by their mutual action"'*.[24]

Expressed in this inaugural lecture, Tait's position with respect to energy conservation at the very beginning of his Edinburgh professorship is revealing. He had devoted much time to the traditional and secure territories of mechanics, optics and hydrodynamics. Not content with these limited goals, however, he had seized Grove's 'correlation of forces' as a means of both greatly widening and powerfully unifying the conquests of his subject while grounding the new doctrines on a traditional central force, action-at-a-distance foundation. Very soon, however, Tait would cast himself in the role of promoter and defender of the new faith, simultaneously propagating the new doctrines and denouncing heretics with all the zeal of a fresh convert.

The making of John Tyndall

At first sight the social origins of John Tyndall might be thought to qualify him for membership of the group of North British natural philosophers clustered around a Glasgow–Edinburgh axis. Born into an Irish Protestant family in County Carlow, Tyndall had comparatively humble origins. His father, a former shoemaker, earned his living as a police officer of the Irish Constabulary, while his mother was a local farmer's daughter. Benefiting from the new state system of National Schools in Ireland, Tyndall was introduced to mathematics of a very practical kind through the school textbooks of James Thomson, father of William. From his Roman Catholic schoolmaster he also learnt the basic skills of land surveying.[25]

Lacking the resources for a college or university education at this time, Tyndall embarked in 1840 upon a career as civil assistant in the military-run Ordnance Survey office in Youghal, County Cork, moving to Preston in Lancashire two years later. Dismissed for insubordination (he and eighteen other civil assistants had dared to petition the authorities for better working conditions), Tyndall found more lucrative employment in Manchester with a firm engaged in various surveys for projected railway lines. 'Railway mania' had reached a peak of competitive intensity and these boom years enabled Tyndall to accumulate a store of capital which would provide him with the funds required for a university education not in Britain but in Germany.[26]

Tyndall gained a year's teaching experience in 1847–8 at the Quaker-run Queenwood College in Hampshire, an institution which emphasized science in relation to industry and agriculture. Also on the staff was Edward Frankland (1825–99) who had spent three months in the small German University of Marburg under the chemist Robert Bunsen (1811–99). Frankland communicated his enthusiasm for German scientific education to Tyndall and the two men set out for Marburg in the autumn of 1848.[27]

One of some nineteen German universities active in the mid-nineteenth century, and not far from Liebig's Giessen, Marburg was the only university in the state of Hessen and as such was the privileged site, with about 300 students, for the training of the state's elite professionals and bureaucrats. In physics (under C.L. Gerling) as in chemistry (under Bunsen), students had an opportunity for laboratory practice. Under the newly appointed *Privatdozent* (most junior) professor of physics, Hermann Knoblauch (1820–95) Tyndall began his experimental investigations into the magnetic (and diamagnetic) properties of bodies, researches which stemmed from the work of Michael Faraday at the Royal Institution. Tyndall, meanwhile, completed a dissertation on a mathematical subject (screw surfaces) and was awarded the degree of PhD.[28]

By the early 1850s he had forged connections with many of the leading German men of science, notably those associated with the Berlin Physical Society, including Magnus (in whose laboratory Tyndall worked for a time), Poggendorff, Du Bois-Reymond, and Clausius. While in Berlin in 1851 Tyndall even called on the veteran Alexander von Humboldt (1769–1859) from whom he was to convey greetings to Faraday. At the same time he received from Du Bois-Reymond a copy of Helmholtz's *Erhaltung der Kraft* and subsequently translated it for *Scientific Memoirs* (ch.7). In his own words of 1873:

> In 1851, through the liberal courtesy of the late Professor Magnus, I was enabled to pursue my scientific labours in his laboratory in Berlin. One evening during my residence there my friend Dr Du Bois-Reymond put a pamphlet [*Erhaltung der Kraft*] into my hands, remarking that it was 'the production of the first head in Europe since the death of Jacobi', and that it ought to be translated into English. Soon after my return to England I translated the essay and published it in 'Scientific Memoirs', then brought out under the joint-editorship of Huxley, Henfrey, Francis, and myself.[29]

In Britain Tyndall established equally-impressive connections: in particular with Colonel Edward Sabine (who was instrumental in Tyndall's election as FRS in 1852), and with Richard Taylor and William Francis (proprietors of the *Phil. Mag.* for which Tyndall translated several German memoirs, beginning with those of Clausius on the mechanical theory of heat). Of equal importance to his ambitions Tyndall appeared on the BAAS stage at the Belfast meeting of 1852 when

Thomson (as president of Section A) and Rankine were launching their new energy physics and cosmology (ch.7).[30]

It was around 1851 that Tyndall first became acquainted with Faraday. Although differing in their views of the subject, both men had a common research interest in the new phenomenon of diamagnetism, identified and named by Faraday. But since Faraday did not write testimonials in support of candidates for academic posts, Tyndall's acquaintance with the celebrated natural philosopher at first played little role in advancing his career. Indeed, in 1851–2 he unsuccessfully competed for three university chairs of physics (Toronto, Cork, and Galway). But in February 1853 he delivered a lecture on diamagnetism as one of the RI's Friday Evening Discourses (initiated by Faraday in 1826 to promote the dissemination of scientific knowledge). A few months later Tyndall was offered the chair of natural philosophy (once held by Thomas Young) at the RI. His friendship with Faraday, as well as with other savants with RI connections (notably Sabine and Henry Bence Jones), had begun to bear fruit.[31]

Tyndall had now placed himself at the very heart of elite metropolitan science. He increasingly identified himself with the living legend of Faraday and cultivated an association with ambitious promoters of professional science: Lyon Playfair and T.H. Huxley, for example. He had links with Sir David Brewster and Sir Charles Wheatstone from the old guard of British natural philosophy. Through the Royal Society, he benefited from acquaintances as influential as Sabine and Stokes. He worked closely with the proprietors of the *Phil. Mag.*, becoming one of its honorary editors in 1854. And he continued to develop his powerful connections with German physicists.[32]

By the late 1850s Tyndall had expanded his personal research territory into an area of far greater sensitivity than that of diamagnetism. Investigations into the nature of glaciers had been the occasion for an intense debate between two well-known gentlemen of science in the 1840s, Professor J.D. Forbes of Edinburgh and Mr William Hopkins of Cambridge. Publication of Tyndall's *The Glaciers of the Alps* (1860) called forth similar hostility from Forbes who viewed Tyndall as an intruder into his personal domain. Following Forbes's retirement from the Edin-burgh chair in 1860, Brewster and Playfair apparently urged Tyndall to apply for the lucrative post but he chose to remain 'close to the heart of England and in the midst of my personal friends'.[33] With Tait as Forbes's successor the scene was set for controversy between the successful candidate for the chair in the Scottish capital and the Irishman who sought to command the centre of Imperial science.

Although Tyndall had irritated Thomson at a number of British Association meetings, Thomson had accumulated more than sufficient scientific credibility not to feel threatened by Tyndall. After all, Tyndall had few of the skills in mathe-matical physics which the Cambridge-trained Thomson displayed with impressive effect at such meetings. Nor had Tyndall much interest in the mighty telegraphic projects from which Thomson would shortly gain his fortune and his knighthood. For his part Tyndall viewed Thomson with severely qualified admiration, believ-ing that Thomson was prone to employ the rather ungentlemanly tactic of claim-ing that he had been misunderstood rather than admitting to a change of mind.[34]

With Tait much more was at stake. Entering into full collaboration with Thomson in late 1861 (ch.10), he quickly took upon himself the task of promoting the science of energy as the *true* natural philosophy, complete with the purest Newtonian pedigree and free from continental influences, to an Empire-wide and world-wide public. From his point of view the threat to his authority and territory came not from foreign men of science in themselves, but from a rival British natural philosophy professor. Tyndall did not have the credentials of a senior wrangler from Newton's *alma mater*, but he had been attempting to acquire the mantle of the greatest living English natural philosopher, Faraday. He also wielded an unsurpassed knowledge of German physics, particularly that which related most directly to energy and thermodynamics. And he stood as the closest competitor in the market for accessible texts in physics. The spark that lit the fuse of conflict between the rival natural philosophers came with Tyndall's Friday Evening Discourse 'On Force' in June 1862.

That spring Tyndall had been delivering extempore a course of lectures 'On Heat, Regarded as a Kind of Motion' at the RI, while employing a short-hand writer with a view to subsequent publication. In the same period the RI was devoting a series of Friday Evening Discourses, coinciding with the 1862 London International Exhibition, to 'the various agencies on which the material strength of England is based'. Subjects such as iron, coal, and cotton preceded Tyndall's own contribution 'On Force'.[35]

Most of Tyndall's lecture was devoted to an *apparently* uncontroversial account of the principles of 'force', illustrated by appeals to direct experience and experiment. Informing his audience, for example, that the amount of mechanical force represented by a pound of coal 'would be equivalent to the work of 300 horses', Tyndall explained that we would need '108 millions of horses working day and night with unimpaired strength, for a year [in order that] their united energies would enable them to perform an amount of work just equivalent to that which the annual produce of our coal-fields [84 million tons] would be able to accomplish'.[36]

Turning to cosmological perspectives, Tyndall's exposition seemed uncontroversial because apparently familiar. His account of the sun's heat as the principal source of mechanical effect on earth, for example, read like a popular account of one of William Thomson's early energy papers (ch.6). Phrases containing the word 'energy' even appeared throughout the lecture, though specific terms in the language of the 'science of energy' such as 'actual energy', 'potential energy', and 'dissipation of energy' were conspicuously absent.

In a calculated rhetorical move, however, Tyndall ascribed all of the physical and cosmological features of 'force' to none other than Mayer, the 'man of genius' who arrived 'at the most important results some time in advance of those whose lives were entirely devoted to Natural Philosophy'. By comparison to the priority now afforded to Mayer, Joule's paper on the mechanical value of heat (1843) and Waterston's paper on the meteoric theory of the sun's heat (1853), though independently formulated, took second place. And for his part Thomson had merely applied his 'admirable mathematical powers to the development of the theory'.[37]

To strengthen his claims Tyndall invoked the authority of two leading German men of science, 'both particularly distinguished in connexion with the Dynamical Theory of Heat'. Although unnamed in the lecture, they were readily identifiable as Clausius and Helmholtz. Tyndall quoted Clausius thus: 'I must here retract the statement in my last letter, that you would not find much matter of importance in Mayer's writings: I am astonished at the multitude of beautiful and correct thoughts which they contain'. Helmholtz, who had himself addressed the RI 'On the Application of the Law of Conservation of Force to Organic Nature' a year earlier, was praised as having repeatedly cited Mayer's work, though without apparently knowing of his essay on celestial dynamics.[38]

Tyndall's promotion of Mayer as the neglected genius who had anticipated most if not all the key features of the new theory of heat clearly omitted those ingredients, particularly the absence of a dynamical theory of heat, in the Mayer corpus which stood at variance with the new views. But our present interest is with the image of Mayer which Tyndall was attempting to construct: that of a romantic man of genius, working alone, ignored by society, and capable of the most profound imaginative insights into the workings of nature ahead of those lesser mortals with the advantages of a lifetime devoted to the practice of physics. It was indeed no coincidence that Tyndall was a lifelong friend and admirer of the celebrated 'Romantic' writer, critic and historian, Thomas Carlyle, author of *On Heroes and Hero Worship*, which espoused just such an image.[39]

Haunted still by the spectre of Mayer which threatened his own recently acquired hero status (ch.4), Joule wrote to Thomson on 9 July to inform him that 'Tyndall names you, myself, and Waterston only, and with a view to how we were anticipated by Mayer'. He particularly denounced Tyndall for betraying the heritage of the RI: 'I think the walls of the Royal Institution might be almost expected to cry out against the neglect by the present Professor, of Davy and Thomas Young. To the former, the ablest man who ever belonged to the Institution, the merit surely belongs of making an experiment decisively in favour of the dynamical theory'. Joule 'therefore felt constrained and much against my will to write to the *Phil. Mag.* as few lines as temperately as I can to set the history in the Magazine aright'.[40]

Joule publicly resorted to the tactic of placing Mayer in the context of a long history of the dynamical theory of heat beginning with John Locke in the seventeenth century. In this way Mayer's claims to originality would be undermined. Unaware of Mayer's own dismissal of such mechanical theories of heat, Joule quoted from Marc Séguin's *De L'Influence des Chemins de Fer* (1839) which had used the same 'hypothesis' (that the heat evolved in compressing an elastic fluid is exactly the equivalent of the compressing force) to argue against a caloric theory. Joule, however, insisted that 'at the time Séguin and Mayer wrote, there were no known [experimental] facts to warrant the hypothesis they adopted'. They had not established the dynamical theory: 'To do this required experiment; and I therefore fearlessly assert my right to the position which has been generally accorded to me by my fellow physicists as having been the first to give a decisive proof of the correctness of this theory'.[41]

Responding to Joule in August 1862 Tyndall protested that his RI lectures on heat *had* acknowledged that it was 'to Mr Joule . . . that we are almost wholly indebted for the experimental treatment of this subject'. He thus professed that he had no wish to question Joule's 'claim to the honour of being the experimental demonstrator of the equivalence of heat and work'. But he strongly defended the stand he had taken on behalf of 'a man of genius', including Mayer's adoption of the hypothesis of which Joule and especially Thomson had been sceptical: 'Mayer, I submit, was perfectly warranted in assuming that the molecular attractions were insensible, and that the quantity of heat . . . was entirely expended in raising the weight, and had its true mechanical equivalent in the weight so raised'. And he concluded with the news that he had been urging one of the *Phil. Mag.* editors to publish in translation Mayer's original papers: 'the true court of appeal in connexion with this subject'.[42]

As we shall see in the next section, however, the issues did not simply concern matters of personal priority. At successive BAAS meetings since 1852, Tyndall had been witness to the growing power and prestige of energy physics. In the aftermath of Darwin, as metropolitan *savants* seized evolutionary theory as a means to enhance their credibility over the old generation of naturalists represented by Richard Owen and Adam Sedgwick, Tyndall was quick to perceive the value of conservation of energy in the armoury of scientific naturalism. But in order to appropriate the doctrine for these ends, he needed to break any perceived North British monopoly on physical truth, now grounded upon the experimental work of James Prescott Joule. Familiar with Helmholtz's earlier doubts about the accuracy of Joule's experiments, Tyndall himself owed no debts to the Manchester natural philosopher. He was thus not constrained in his construction of a wholly different kind of 'pioneer' of energy, the neglected but prophetic genius of a German physician named Mayer who, by his anticipation of this new truth, could be used to shatter that North British monopoly.

North British defenders of the faith

In January 1862 Thomas Henry Huxley drew large numbers of working men to Edinburgh's Queen Street Hall. His now-standard linkage between man and ape was apparently well received. As the Free Church *Witness* remarked, when 'their kindred to the brute creation was most strongly asserted, the applause was the most vigorous'.[43] The unholy alliance of Huxley and Darwin, far more threatening than the discredited *Vestiges*, had launched a crusade among Scotland's working classes. Christian opposition remained divided. On the one hand the Free Kirk had now three Colleges strategically placed in Edinburgh, Glasgow and Aberdeen, as well as a number of periodical publications, including the *Witness* and *North British Review*, with which to defend their faith against the infidel. On the other hand the Church of Scotland, under the leadership of a new generation of liberal evangelicals like Norman Macleod, could deploy the intellectual resources of the Scottish universities against the metropolitan agnostics.

Reformers within ancient northern universities, especially in Edinburgh and Glasgow, had ensured that their institutions had neither degenerated into 'sectarian' theological colleges, finishing schools for the rump of the established Church, nor suffered catastrophically from Free Church College competition (ch.2). Indeed, while Glasgow could boast of its Cairds and its Thomsons, embodying the liberal and practical traditions of the West, Edinburgh could point to the principalship of the celebrated experimentalist and Free Churchman Sir David Brewster just as St Andrews could feel proud of its episcopalian Principal Forbes, living symbols of the Universities' non-sectarian character.[44]

Over the next 20 years or so liberal intellectual values eroded still further the old sectarian divisions, pushing even the Free Kirk into a relaxation of the Calvinistic rigour of the Westminster Confession by 1892. But Scottish reformers could not take a triumph of liberal Christian values for granted. Their academic and ecclesiastical dominance was not therefore a self-evident position, but one which had to be constructed and defended within a highly competitive market place offering to student and public audiences, post-1860, every cultural artefact from strict calvinism to scientific naturalism.

In 1868, for example, William Thomson wrote to his brother of the need to secure a 'moderate and liberal' Member of Parliament for Glasgow and Aberdeen Universities in opposition to the candidate put forward by a 'religiously most illiberal' section 'of free church partisans'.[45] In the same year Tait took on as his assistant a talented trainee for the Free Kirk, William Robertson Smith (1846–94), who, appointed to the professorship of Hebrew at Aberdeen's Free Church College in 1870, was subsequently tried for heresy on the basis of an 1875 contribution on the 'Bible' (promoting historicist, non-literal readings) to the 9th edition of the *Encyclopaedia Britannica* and deprived of his professorship. In Tait's words this trial was effected by an 'alliance between ignorant fanaticism and cultivated jesuitry which deplored the "unsettling tendency" of his articles!'[46]

In this context of continuing Scottish cultural turmoil Thomson and Tait were teaching their students how to deploy the energy doctrines to police the rise of scientific naturalism in general and Darwinian evolution in particular by placing checks on geological time. Most of Thomson's Glasgow students, for example, continued to be destined for ministry in the Scottish churches. Under a dual system of prizes offered from around 1850, 'The Cleland Gold Medal' would be awarded for the best theological essay in the first of two sessions and for the best essay in natural philosophy in the second, while 'The University Silver Medal' would similarly go to the best natural philosophy essay in the first and to the best theology essay in the second session. Natural philosophy topics shared the theme of 'energy' in all its diverse relations, including, in the mid-1860s, essays entitled 'The Estimation of Time in Geological and Cosmical History' and 'The Doctrine of Uniformity in Geology' (which students were doubtless expected to refute).[47]

Tyndall's 'Force', however, threw down a fresh and potentially very serious challenge from the metropolis. As Huxley's staunchest ally he was preparing to deploy 'conservation of energy' in the cause of scientific naturalism. Hitherto uninvolved in the priority dispute between Joule and Mayer but well aware of

Huxley and Tyndall's crusading ambitions, Tait first saw a copy of Tyndall's lecture while staying with Thomson in Arran during the summer of 1862. According to Tait, Thomson 'had been repeatedly asked by the Editor of "Good Words" to contribute a scientific article to its columns' and so 'we seized the opportunity of distributing among its 120,000 readers a corrective to the erroneous information which we saw was stealing upon them through the medium of *popular* journals'.[48]

The article 'Energy' (1862), co-authored by Thomson and Tait, was written in Tait's polemical style as a popular statement of 'authentic' energy physics and cosmology in opposition to any rival 'substitutes' such as that just propounded by Tyndall. That it should have appeared in the 'latitudinarian' Church of Scotland magazine *Good Words*, edited by Thomson's friend Rev. Norman Macleod (ch.2), was in part due to the authors' concern to promote the new natural philosophy in opposition to both biblical literalists at home and the scientific naturalists south of the border.

As editor of *Good Words*, Macleod aimed 'to avoid, on the one hand, the exclusively narrow religious ground – narrow in its choice of subjects and in its manner of treating them – hitherto occupied by our religious periodicals; and, on the other hand, to avoid altogether whatever was antagonistic to the truths of Christianity, and also as much as possible whatever was calculated to offend the prejudices, far more the sincere convictions and feelings, of fair and reasonable "Evangelical" men'.[49] Thomson and Tait's 'Energy' exactly matched this theological stance. Offering neither a cosmology constrained by 'exclusively narrow religious ground', nor one 'antagonistic to the truths and spirit of Christianity', its harmony of 'science' and 'scripture' was well targeted at 'the sincere convictions and feelings of fair and reasonable "Evangelical" men'.

'Energy' presented a popular account of the meaning and history of the 'ONE GREAT LAW of Physical Science, known as the *conservation of energy*'. The climactic cosmological vision reaffirmed the Helmholtz-Thomson contraction model. The authors therefore had no reservations about asserting an age of the sun vastly greater than any literal six-day creation such as that expressed in the *Larger Catechism* and upheld by a substantial section of the Free Kirk in its commitment to the undiluted *Westminster Confession* (ch.2).

Thomson and Tait also offered a view of 'the possible origin of energy at creation as excessively instructive'. Created 'simply as the difference of position of attracting masses, the potential energy of gravitation was the original form of all the energy in the universe'. Given that 'all energy tends ultimately to become heat, which cannot be transformed without a new creative act into any other modification, we must conclude that . . . the result will be an arrangement of matter possessing no realizable potential energy, but uniformly hot . . . chaos and darkness as "*in the beginning*" [Genesis 1:1–2]'. This indeed was a 'latitudinarian' re-reading of Genesis. Thomson and Tait's remarks here on the transitory nature of the *solar system* should be compared with Thomson's view (also in 1862) of the *universe* as 'an endless progress, through an endless space' (ch.7).[50]

In the meantime the solar system will have endured 'tremendous throes and convulsions, destroying every now existing form'. Thus, 'As surely as the weights of a clock must run down to their lowest position, from which they can never rise again, unless fresh energy is communicated to them from some source not yet exhausted, so surely must planet after planet creep in, age by age, towards the sun' and a fiery end. Thomson and Tait therefore emphasized the consistency between their cosmological vision and the reinterpreted scriptural texts:

> we have the sober scientific certainty that heavens and earth shall 'wax old as doth a garment [Psalm 102:26]'; and that this slow progress must gradually, by natural agencies which we see going on under fixed laws, bring about circumstances in which 'the elements shall melt with fervent heat [2 Peter 3:10]'. With such views forced upon us by the contemplation of dynamical energy and its laws of transformation in dead matter, dark indeed would be the prospects of the human race if unillumined by that light which reveals 'new heavens and a new earth [2 Peter 3:13]'.

Though not addressed to the BAAS or even to the scientific community as such, 'Energy' provided a concise statement of all the new physical and cosmological doctrines fundamental to the 'epoch of energy'. Those doctrines set definite limits to the earth as a habitation for human beings and explained the origin and nature of the great source of energy available to mankind from the sun. But the article went further in announcing the utmost primacy of the conservation of energy as the 'ONE GREAT LAW of physical science'. As such it served as a flyer for the vast *Treatise on Natural Philosophy* then under construction (ch.10).

'Energy' was strongly didactic in style. The authors spoke as ministers of natural philosophy rather than high priests of knowledge, that is, they addressed their audience not as guardians of some hidden truths known only to the anointed few, but as teachers whose task it was to let the presbyterian people see the truths of nature for themselves and shield the common man from the seductions of defective and erroneous systems:

> Of late several attempts have been made, with various success, to impart to the great mass of the interested but unscientific public an idea of the ONE GREAT LAW of Physical Science, known as the *Conservation of Energy*, and it is on account of the defects, or rather errors, with which most of these attempts abound, that we have aimed primarily at preparing an article, which shall be at all events *accurate*, as far as human knowledge at present reaches.[51]

In particular the commonly applied phrase 'conservation of force' was both inconvenient and erroneous. Force, in general defined as 'that which produces, or tends to produce, motion', was absolutely lost if no motion be produced. Force in this sense was not conserved. Finding an even worse example of such confusion of 'force' and 'energy', the authors cited an unnamed popular magazine which carried the statement that ' "The sum-total of the Forces in the Universe is Zero" – a statement meaningless if it be applied to Force in its literal sense, and untrue if it refer to Energy'. From the outset, then, Thomson and Tait prepared to expose those natural philosophers (such as Mayer and Tyndall) who employed an ambiguous language of 'force' rather than an 'accurate' and 'precise' language of energy.

The authors proceeded to a popular but 'definite' exposition of that language, employing a range of every-day examples. Thomson and Tait now replaced Rankine's 'actual' with their own preferred 'kinetic' energy 'which indicates motion as the form in which the energy is displayed'. According to the energy conservation law, then, *'Energy is never lost'*, that is, *'the sum of the potential and kinetic energies does not vary'*. Practical, concrete illustrations of potential and kinetic energy followed, as did examples of the transformations of visible energy into heat (fluid friction) and heat into visible energy (heat engines). Indeed, their account ranged from causes of 'waste' in long-distance telegraphy, through a brief consideration of 'free will directing the motions of matter in a living animal' to the nature of the sun's heat and the fate of the solar system.[52]

'Accuracy' in scientific language complemented 'accuracy' in the historical record:

> We were certainly amazed to find in a recent number of another popular magazine, and in an article specially intended for popular information, that one great branch of our present subject, which we have been accustomed to associate with the great name of Davy, was in reality discovered so lately as twenty years ago by a German physician. Such catering for the instruction of the public requires careful looking after; and we therefore propose to place on a proper basis the history of the discovery . . . And it especially startles us that the recent attempts to place Mayer in a position which he never claimed, and which had long before been taken by another [i.e. by Joule], should have found support within the very walls wherein Davy propounded his transcendent discoveries.[53]

Thomson and Tait's scorn was thus directed at the unnamed Tyndall for his multiple sins of ignoring, even violating the heritage of, Davy, and attributing the dynamical theory of heat not only to a German but above all to a mere physician, an 'outsider' without authority in such matters.

In their 'accurate' historical account it was Locke who had conjectured that heat was motion, Davy who (in his melting of ice experiment) had 'proved' that heat could not be a substance and must therefore be motion, Rumford who had 'arrived at a very approximate answer to the question, "How much heat can be produced by the expenditure of so much work?" or, in other words, with the modern phraseology, *"What is the Dynamical Equivalent of Heat?"'* But it was of course Joule who stood as 'The founder of the modern dynamical theory of heat, an extension immensely beyond anything previously surmised'.

According to Thomson and Tait, Joule had in 1843 'published the results of a well planned and executed series of experiments, by which he ascertained that a pound of water is raised one degree Fahrenheit in temperature by 772 foot-pounds of mechanical work done upon it'. Furthermore, 'The actual method which he first employed was to force water through small tubes'. Other methods produced numerical results which 'agreed so well with each other in very varied experiments, that the definite transformation of work into heat was completely established, and the "dynamical equivalent of heat" determined with great accuracy'.[54] Moreover, as Tait added a few months later in the *Phil. Mag.*, this research formed 'but a *very small* part of what he [Joule] has done for the science of energy'.[55]

By contrast Mayer was credited with merely stating the results obtained by 'preceding naturalists' (including the fundamental result of Davy), with suggesting new experiments, and with propounding a method for finding the dynamical equivalent of heat. As though to underline the limited competence of the German physician, Thomson and Tait again subjected 'Mayer's Hypothesis' to detailed scrutiny and found it wanting:

> The experimental investigations of subsequent naturalists [Joule and Thomson!] have shown that this hypothesis is altogether false, for the generality of fluids, especially liquids, and is at best only *approximately* true for air; whereas Mayer's statements imply its indiscriminate application to all bodies in nature, whether gaseous, liquid or solid, and show no reason for choosing air for the application of the supposed principle to calculation, but that at the time he wrote air was the only body for which the requisite numerical data were known with any approximation to accuracy (ch.4).[56]

By February 1863 Tyndall had discovered Thomson and Tait's 'Energy' in *Good Words*. Stung into a rapid response in the *Phil. Mag.* a month later, he stressed that as a busy man of science the 'time I am able to devote to unscientific reading is so stinted as to leave me in almost total ignorance of the general periodical literature of the country'. In particular, he singled out articles from *Good Words* which combined a Christian theme with a soft and sentimental family tone, a combination which contrasted with the 'manly' world of the true Victorian men of science, not least the members of the Alpine Club.[57]

Pursuing the high moral ground, Tyndall claimed that Thomson and Tait had acted against the 'interests' and 'dignity' of science itself by 'taking difficult and disputed points, which apparently involve imputations on individual character, into such a court' as that formed by the readers of *Good Words*. Thus 'these respectable persons are placed in a false position when they are virtually called upon to decide between the rival claims of Joule and Mayer, and to form an opinion as to the scientific morality of myself'. Remarking that 'with such an audience authority is, of course, decisive', Tyndall noted the tactic of employing *two* authors by which to sway this court of the people. Contemptuous of such dubious 'democratic' appeals, Tyndall therefore attempted to transfer the case to a quite different court, 'that of instructed men of science'.[58]

Tyndall's strategy here was an expression of the underlying contest between the North British natural philosophers and the metropolitan naturalists over scientific authority. Raising questions of the correct forum for discussion of such scientific matters in general, and for the establishment of facts in historical cases in particular, Tyndall had given a decisive answer that such matters were not to be decided in the pages (or through the audience) of general periodical literature but by means of the consensus of a legitimate (and legitimating) group of 'instructed men of science'. His answer was thus a means of publicly undermining the authority of his North British rivals who had demonstrably broken with these unwritten conventions of the new scientific world.[59]

Tait's private response was to seek to crush Tyndall 'at once ... *writing to Brewster* who is Editor of the *Phil. Mag* ... and render him [Tyndall] an object of scorn to every man of any sense of honour'. The Edinburgh professor then assumed the role of public health officer superintending the health of the scientific community and rooting out potential sources of disease: 'Such a nuisance must be abated, even at the risk of becoming a Commissioner of Sewers; and I consider that Sir David [Brewster] is just the man for us – as we shall thus be saved from appealing to Joule and others'. Brewster, then Principal of Edinburgh University and one of the *Phil. Mag.* 'honorary editors', would appear as the perfect elder statesman of science, respected by all sides as an experimental natural philosopher, to preside over the dispute and ensure that justice be seen to be done. Above all, Joule's 'heroic' status would be preserved by keeping him aloof from any personality war. And Brewster, readily accessible to Edinburgh's natural philosophy professor and respected Free Churchman with no sympathies for scientific naturalism, was being used to ensure that Tait's reply to Tyndall would be inserted in the very next number of the *Phil. Mag.*[60]

As Tait had privately told Thomson, Tyndall had 'cooked' his account of *Good Words* as a journal which appeared to print sentimental religious articles. In his published letter to Brewster, Tait therefore claimed that in nearly every number there appeared a scientific paper by men such as Brewster himself, Sir John Herschel, or Forbes. Conversely, Tyndall's highly controversial lecture 'On Force' had been addressed to a lay audience at the RI and was subsequently reported in publications such as the *Illustrated London News* and the *Engineer*.[61]

As to Mayer's claims Tait refuted Tyndall's allegation that at the time of writing 'Energy' Thomson and Tait had been 'unacquainted with the real merits of Mayer' by pointing out Thomson's crediting of Mayer in 1851 (ch.6). Tait, however, attempted to weaken Mayer's claims still further by asserting that his 1842 paper 'has *no claims to novelty or correctness at all*, saving this, that by a lucky chance he got an approximation to a true result from *an utterly false analogy*; and that even on this point he had been anticipated by Séguin, who, [in 1839] had obtained and published the same numerical result from the same hypothesis'. With strong sarcasm Tait challenged his opponent to expose the errors in Thomson and Tait's critique of 'Mayer's hypothesis' whereupon 'we shall be happy ... to hail the additions to scientific knowledge ... (involving at least a reconstruction, if not a destruction of thermo–dynamics)'.

In response, finally, to Tyndall's demand for a clarification of the personal charges against him, Tait asserted that there was in Tyndall's lecture 'little about the general principle of Conservation of Energy'. Thus Thomson and Tait had not been accusing Tyndall of putting forward Mayer 'as having any claims to this great generalization'. Rather, Tyndall had appeared to claim for Mayer 'a succession of results regarding transformations of energy which had been elaborated by mathematicians and naturalists from Galileo to Davy'. 'Depreciation' and 'suppression', then, were appropriate words by which to accuse Tyndall of neglecting almost everyone but Mayer.[62]

Tyndall adopted new tactics in his reply of May 1863. This time he chose to address William Thomson personally through the *Phil.Mag*: in 'Energy', Thomson's name had preceded that of Tait 'not by right of alphabet'; Tait had admitted that the article had been written in compliance with external requests made to Thomson; and the Glasgow professor was 'the older and more famous man'. Tyndall thus attempted to divide his opponents one from the other in the expectation of soliciting a direct response from Thomson which he could then set against the stated position of Tait. Attempting to out-manoeuvre his opponents by drawing Thomson alone into a duel, Tyndall trusted that 'Prof. Tait will see that simple chivalry makes it my duty to decline entering into any contest with him at present; and seeing this, he will, I doubt not, have the grace and modesty to stand aside and allow you and me to settle this affair between ourselves'.[63]

The main thrust of Tyndall's counterattack, however, centred on the question in Tait's April letter asking him if he was aware that Mayer's 1842 paper had '*no claims to novelty or correctness at all*'. Tyndall proceeded to show that his alleged ignorance was shared by two leading European men of science who had more than a passing interest in such matters: Helmholtz and Emile Verdet (1824–66). He quoted from the former's Königsberg lecture (1854): 'The first man who correctly perceived and rightly enunciated the general law which we are here considering, was a German physician, J.R. Mayer, of Heilbronn, in the year 1842'. And he quoted from Verdet's recent lectures on the mechanical theory of heat to the Chemical Society of Paris: 'These researches [upon which the mechanical theory is founded] are the exclusive work of three men [Mayer, Joule and Colding], who without concert, and even without knowing each other, arrived simultaneously in almost the same manner at the same ideas. The priority in the order of publication belongs without any doubt to the German physician Jules [sic] Robert Mayer'.[64]

By this stage, if not earlier, Tyndall appeared less as defendant and more as prosecutor. Indeed, his next move was to place Thomson and Tait in the dock for their misrepresentation of Joule's researches. In particular, they had *inaccurately* claimed that Joule in 1843 had ascertained by 'the results of a well-planned and executed series of experiments' a mechanical equivalent of 772 foot-pounds:

> You here prove yourself to be as ill-informed regarding the labours of Joule as you are regarding those of Mayer. It was in 1849, and not in 1843, that Mr Joule proved the mechanical equivalent of heat to be 772 foot-pounds. His determinations of the mechanical equivalent of heat, published in 1843, varied from 1040 to 587 foot-pounds. They were so discordant that nobody attached any value to them. It was with reference to Mr Joule's earlier experiments that Helmholtz expressed himself thus, in his celebrated essay ... [(ch.7)]. But to write as you write regarding his first experiments is simply to betray a want of acquaintance with the requirements of refined experimental inquiry.

And, contrary to Thomson and Tait's dismissal of Mayer's hypothesis, Tyndall reminded Thomson that 'the approximation is so close that it cost Mr Joule six years of labour [until 1849], according to his own methods, to arrive at the degree of accuracy attainable by the method of Mayer'.[65]

Tyndall's letter ended on a note of high rhetorical drama. Ranging widely over Mayer's work, he claimed that all the ideas contained in 'Energy' had been employed or anticipated by Mayer: in particular, the illustration of heat from fluid friction and the meteoric theory of the sun's heat. But Mayer had gone further with his account of tidal friction, a feature 'eminently illustrative of the clear glance which this hard-working physician had obtained into the system of nature'. Indeed, 'Here was a phenomenon which had been for centuries the subject of observation, measurement, and calculation; still no astronomer, no mathematician, no natural philosopher perceived its inevitable mechanical effect':

> Name, if you can the 'mathematician or naturalist, from Galileo to Davy', who ever 'elaborated' these things. You cannot do so; and while science lives, the name of Mayer will be associated with these questions. In the presence of such facts, it ill becomes you to talk to me of suppression and depreciation. You may send your statements into the world labelled 'Good Words', but the world before which you and I now stand will see that the 'trade mark' is incorrect; that if to be pitiful, if to be courteous, if to cherish that charity which thinketh no evil, be marks of goodness, these utterances of yours are *not* good words, but the reverse. Judged of by the facts, and apart from your own uninformed convictions, they are not even words of truth.

Professor Thomson deliberately ignored Tyndall's bait, choosing instead to adopt an even higher moral ground. Believing that Tyndall's tone in addressing him was 'unprecedented in scientific discussion', Thomson pursued a tactic (characteristic of Faraday in similar circumstances) of declining 'to take part personally in any controversy' with Tyndall. Thomson nevertheless affirmed his support for 'the correctness of the opinions and information' respecting 'all the scientific questions' touched upon in his and Tait's article on 'Energy'.[66] Significantly he said nothing as to the historical record. In the division of labour between the two Scottish professors, it fell to Tait to resume the offensive.

Towards the close of 1863 Tait had been invited by the editor of the *North British Review* to contribute an article on Tyndall's 'theory of heat'. Hitherto very much a Free Church periodical, the *North British* editor now assured Tait that it 'will no longer confine itself to the leavings of Sir D. Brewster's portfolio' and that 'A serious attempt will be made to keep it up to the highest point; & it w^d be a great thing if we c^d have a Scotch Review first-rate in writing & thoroughly liberal'. A new liberal and latitudinarian trend was thus evident even in parts of the Free Kirk during the 1860s, paralleling the rise of the new 'Moderatism' in the established Church of Scotland. This trend, contrary to the sectarian fragmentation of Scottish presbyterianism two decades earlier, meant that at least in Edinburgh and Glasgow the Universities and their professors could claim to be participants in a broad Scottish, even British, Christian culture embodying all that was progressive and universal in opposition to that which was narrowly doctrinal and local.[67]

Responding to the invitation, Tait obliged with two historical articles which were published in the spring of the following year. Both articles, one on the dynamical theory of heat and the other on energy, formed a large part of Tait's *Sketch of Thermodynamics* (1868), in which the second article, somewhat revised, became 'Historical Sketch of the Science of Energy'. More than in the earlier controversies over the history of the dynamical theory of heat and conservation of energy, Tait constructed an image of Mayer as the very embodiment of illegitimate scientific practice. By so doing, Tait was in effect reinforcing the authority of the North British 'scientists of energy' whose science, in contrast, utilized best practice:

> But when we find, in modern times, conclusions, however able, drawn without experiment on such a [metaphysical] text as '*Causa æquat effectum*' [cause equals effect], we feel that the writer and his supporters are, as regards method, little in advance of the science of the dark ages. This is one of the fundamental characteristics of all the writings of Mayer. . . . Mayer, therefore, and others who have followed a course similar to his, cannot be considered as having any claims to the credit of securely *founding* the science of Energy; though their works have become of great value as developments and applications, since the science has been based upon correct reasoning and rigorous experiments.[68]

By calling into question Mayer's competence in experimental physics, by labelling him as a physician rather than a physicist, by pointing to his reliance on metaphysical principles, and by associating him with the science of the pre-Reformation 'dark ages', Tait sought to secure not only the reputation of Joule but the whole authoritative basis of a North British science of energy against the very real threats of its appropriation by the scientific naturalists.

Tyndall, on the other hand, promoted Mayer as a neglected genius, whose insights into nature's workings transcended and anticipated the conclusions of ordinary men of science. Such insights would then be brought for judgement before the bar of scientific experts. At a time when the role of the 'professional' scientist was being defined in Britain, these controversies served to show, not a simple conflict between 'amateur' and 'professional', but an intense battle to secure and shape scientific authority. Above all, Thomson and Tait needed to protect the vast amounts of intellectual capital which from 1861 they had begun to invest in a grand project to develop a complete science of energy (ch.10).

CHAPTER 10

Newton Reinvented: Thomson and Tait's *Treatise on Natural Philosophy*

> The world of which they give the Natural Philosophy is not the abstract world of Cambridge examination papers – in which matter is perfectly homogeneous, pulleys perfectly smooth, strings perfectly elastic, liquids perfectly incompressible – but it is the concrete world of the senses, which approximates to, but always falls short alike of the ideal of the mathematical as of the poetic imagination ... Nowhere is there actual rest; nowhere is there perfect smoothness; nowhere motion without friction.
>
> *Anonymous review of Thomson and Tait in the* Scotsman *newspaper,*
> *6 November 1868*[1]

Over the first three decades of the British Association, geologists had given their Section C both disciplinary focus and international prestige. In contrast, Section A (Mathematics and Physics), once the preserve of Cambridge lions such as Herschel and Whewell, remained at the head of the BAAS hierarchy, but appeared to lack the disciplinary focus of its younger relative. The scientists of energy, on the other hand, prided themselves on the coherence of their natural philosophy, unified through the doctrines of energy, but their programme had been sketched and articulated only in a range of papers scattered in the scientific and popular periodicals of the mid-Victorian period.[2]

As Tait, latest convert to the crusade, recognized early in the 1860s, the new science of energy needed to be represented in disciplinary space, either in the form of a treatise or textbook, or both, if it was to resist appropriation by ambitious rivals such as Tyndall. The *Treatise on Natural Philosophy* was just such an attempt to embody North British energy physics in canonical form, capable of being literally and metaphorically translated around the world. Popular textbooks, notably Balfour Stewart's *Conservation of Energy*, would follow. But above all, the *Treatise* promised to raise natural philosophy to a level of academic and cultural prestige above that of, and with jurisdiction over, those 'softer' branches of science which now included Darwin's 'natural history' alongside geology. That prestige would be further enshrined in the Cambridge Tripos as former wranglers such as Thomson and Maxwell returned to examine the new generations of mathematical undergraduates.[3]

Printed in proof form at irregular intervals between late 1862 and mid-1867, the first volume of Sir William Thomson and Peter Guthrie Tait's *Treatise on Natural Philosophy* (1867) heralded a complete, state-of-the-art account of physical science in four volumes. In style, as in structure and in content, the *Treatise* (familiarly known as T&T' on account of the way in which Thomson (T) and Tait (T') addressed one another in their correspondence) represented all the values of the science of energy. A strong preference for engineering and geometrical modes of expression and a powerful dislike of symbolic abstractions characterized the work. So also did a closely related concern with 'democratic' values which reflected a presbyterian academic and economic culture emphasizing direct, personal experience over Catholic authoritarianism and Anglican hierarchies.[4]

Taking the *Principia* as the sacred text of the natural philosopher, Thomson and Tait claimed to expose the errors of previous scholars and restore to the text its 'original' meaning, uncorrupted by the interpretations of the profane. Founded on the doctrine of conservation of energy and expounded by these Scottish professors of natural philosophy, the 'true' Newtonian gospel would henceforth be accessible to everyone rather than to a privileged priesthood of Cambridge dons. Maxwell dramatized the new 'democratic' dynamics in his review for *Nature* (1879) of the second edition:

> The credit of breaking up the monopoly of the great masters of the spell [Lagrange, Hamilton and others], and making all their charms familiar in our ears as household words, belongs in great measure to Thomson and Tait. The two northern wizards were the first who, without compunction or dread, uttered in their mother tongue the true and proper names of those dynamical concepts which the magicians of old were wont to invoke only by the aid of muttered symbols and inarticulate equations. And now the feeblest among us can repeat the words of power and take part in dynamical discussions which but a few years ago we should have left for our betters.[5]

In the end, however, only one of the projected four volumes was ever written. The protracted struggles of the two authors to complete even that volume became so severe as to make construction of the remaining volumes well nigh impossible. Much of the blame lay with Thomson, whose other interests, combined with his dislike of writing, inevitably frustrated Tait. 'I am getting quite sick of the great Book', Tait told his co-author in 1864: 'if you send only scraps and these at rare intervals, what can I do?'[6] But that single volume, expanded into two volumes for a second edition (1879–83), prompted Maxwell to comment in 1879 that for twelve years 'we have been exploring the visible part of the work, marking its bulwarks and telling the rising generation what manner of palace that must be, of which these are but the outworks and first line of defences'.[7]

Writing the Treatise

As Tait moved into his second session with the natural philosophy class at Edinburgh University, he approached the publisher Macmillan with a proposal for a text, suitable for his own teaching purposes and devoted largely to experimental physics. At this stage, William Thomson offered to join the project which Tait then presented to Macmillan in revised form around early December 1861:

> I proposed 3 volumes, each about the size of my Dynamics [Tait and Steele] – and I expressly stipulated that *one* of these (or ⅓ of the work, if the Exp^l & Math^l parts were to be *mixed*) should be devoted to Math^l Physics – including all sorts of Potentials, Dynamical Theory of Heat &c &c. I said also that as we were both (you and I, that is) at present *in training* we could turn out a volume by August next, another in Jany or Feb^r /63 & the third in the ensuing summer. The Professors here, to whom I have mentioned the affair, are particularly enthusiastic, & I have little doubt of our increasing our students by 30 per cent (at least) if we do the business well.[8]

The original joint proposal was thus for a three-volume work to be completed by mid-1863 and aimed specifically at Scottish students of natural philosophy with a view to boosting professorial income from student fees.

Based very much upon their respective lectures, the publication would secure the experimental and mathematical territory of natural philosophy at home and abroad: 'I fancy that we might easily give in three moderate volumes a far more complete course of Physics, Experimental and Mathematical, than exist (to my knowledge) either in French or German. As to English, there are NONE'. Indeed, Tait compared their prospects to the success of his erstwhile presbyterian colleague at Queen's, the theologian James McCosh (later of Princeton): 'we shall make a great hit – besides being translated into Russian & perhaps Hindi & Chinese – like the immortal McCosh whom I daresay you know'. More seriously Tait expressed an anxiety, again underlining his concern with securing income, that all necessary precautions be taken against the pirating of the work in America and elsewhere.[9]

Tait acknowledged that Thomson's special contribution would be to the third volume on mathematical physics, 'the *unique* one as I call it in my note to Macmillan, and which I could never have ventured on alone'. But very soon the proposal showed signs of further change. By the end of December 1861 Tait was proposing to the publisher at least two volumes on experimental physics and probably two on mathematical physics. Suggesting a plan for the contents of the experimental volumes, he proposed some twelve chapters beginning with 'General reflections on Matter, Force, Motion, Measures, Energy, Work, and Experiment', proceeding through such subjects as sound, light and heat, and ending with conservation of energy.[10]

Tait, however, sensed that Thomson was already proving difficult to control. He therefore suggested a change of direction: 'What we want *at once* is not the fame of authorship, but the supply of a want in elementary teaching'. By early January 1862 the authors seemed to have agreed on the need to separate the project into that which would supply immediate teaching needs and that which would secure the scientific credibility of the authors for posterity. Tait confided to Thomson that 'We may gain considerable credit, and perhaps profit, by the present [popular] undertaking; but the other will go over Europe like a statical charge'. He urged Thomson to assert possession against any imposters: 'Don't you think it would be prudent to warn the profane off such ground by a timely notice?' What he called 'our *Great* work', Thomson and Tait's own '*Principia Mathematica*', 'A Mathematical Treatise on Natural Philosophy', would be '*unique*' such that no 'living man could attempt [it] alone, not even Helmholtz'.[11]

By the end of January something like the final plan had been agreed. 'No mathematics will be admitted', Tait told his old Belfast colleague Thomas Andrews, '(except in notes, and these will be more or less copious throughout the volume, being printed in the text but in smaller type)'. In the end, however, the idea of restricting mathematics to a minimum of notes in anticipation of the 'great work' was abandoned as Thomson expanded the notes into the 'small print' part of the *Treatise*, threatening to engulf the whole work. Tait desperately fought a rearguard action to maintain the democratic aim of a popular text. As he wrote to Andrews in September 1862 concerning a draft chapter: 'I hope to make the large type part of it intelligible even to savages or gorillas'.[12]

Throughout the collaboration Tait was ever-anxious to impose temporal order upon his discursive and disordered Glasgow colleague. 'I shall start at 2pm on Friday & be with you at the College about half past 4. I have then a choice between 7^h20 & 9^h15 for my return', Tait informed Thomson of their first meeting. 'I shall lunch heavily in Princes' Street [Edinburgh] before starting so that I shall be independent of all creature comforts (but a pipe, w^h is easily managed on the street) till I get back'. Tait's style was proverbial: all he seemed to need was 'a few hours' contemplation of a subject, with a pot of beer, and the lurid glare from my pipe showing in the darkness; then I can sit down and write you off a chapter in double quick time'. Thomson, by contrast, could do no such thing, feeling 'a repugnance' to writing 'which is not common'. A letter from Macmillan in 1865 demonstrates that the problems faced by Tait in controlling his zealous Glasgow colleague were of no common order:

> I have seen corrections run rather high before, but never anything approaching what you have done. You will of course send me a cheque for all over a fair allowance – 10/- a sheet is what we ordinarily reckon. This would have been £17 – yours is about £117. This point you will think over quietly and talk with Thomson about. But what I want to have as clear as may be about how many more pages your volume I [one] may be. Make as accurate a guess as possible, like a dear professor.

Eventually produced by Macmillan as 'publishers to the University of Oxford', the *Treatise* formed part of a Clarendon Press Series of educational, rather than commercial works. Selling rapidly but failing to make profits for authors or publisher, the *Treatise's* commercial fate prompted Tait to remark in 1870: 'We do look like a couple of sold gorillas'.[13]

Assigning a 'Newtonian' pedigree

Volume one had two principal divisions. 'Preliminary Notions', divided into two long chapters on 'Kinematics' and 'Dynamical Laws and Principles' and two short chapters on 'Experience' and 'Measures and Instruments', constituted over half the volume. 'Abstract Dynamics', devoted entirely to 'Statics', formed the second principal division. To anyone familiar with older textbooks on mechanics, which tended to separate statics from dynamics, the inclusion of statics under abstract dynamics would have appeared strange. But the authors had agreed a distinctive classification. 'Kinematics' (adopted from Ampère) was defined as 'the purely geometrical science of motion in the abstract' without reference to matter and force.[14] As Maxwell explained Thomson and Tait's usage in his review for *Nature* (1879):

> The guiding idea, ... now for the first time boldly and explicitly put forward, is that geometry itself is part of the science of motion, and that it treats, not of the relations between figures already existing in space, but of the process by which these figures are generated by the motion of a point or a line. ... Our knowledge, therefore, of whatever kind, may be compared to that which a blind man acquires of the form of solid bodies by stroking them with the point of his stick, and then filling up in his imagination the unexplored parts of the surface according to his own notions about continuity and probability.[15]

Kinematics was to be studied prior to 'Dynamics', defined 'in its truest sense as the science which treats of the action of *force*, whether it maintains relative rest, or produces acceleration of relative motion'. Thus dynamics itself divided into 'statics' and 'kinetics'. Statics, then, had been made a branch of dynamics.[16]

From the beginning the authors made clear that 'Conservation of Energy' was to play a fundamental role in the vast edifice. Their 'Preface' announced:

> One object which we have constantly kept in view is the grand principle of the *Conservation of Energy*. According to modern experimental results, especially those of JOULE, Energy is as real and indestructible as Matter. It is satisfactory to find that NEWTON anticipated, so far as the state of experimental science in his time permitted him, this magnificent modern generalization.
>
> We desire it to be remarked that in much of our work, where we may appear to have rashly and needlessly interfered with methods and systems of proof in the present day generally accepted, we take the position of Restorers, and not of Innovators.[17]

Thomson and Tait proclaimed their re-reading of Newton in a way which neither Newton nor his self-styled followers would have acknowledged as 'Newtonian'. The authors of the *Treatise* clothed their radical reforms of natural philosophy in a hitherto unrecognized interpretation of Newton's *Principia*. It was a daring strategic move aimed at reducing or even destroying the rights of many of the nineteenth-century claimants to discovery of conservation of energy. Simultaneously it asserted a British pedigree for the doctrine. Accompanying these benefits went a reinforcement of the ideological values of the Scottish natural philosophers who sought to promote accurate measurement against metaphysics as the primary stamp of their discipline. Newton and Joule, not Leibniz and Mayer, would serve these ends (ch.9).

Tait's ardent campaign to establish a Newtonian pedigree for conservation of energy began in 1863 with a lecture to the Royal Society of Edinburgh and continued in communications to the *Phil. Mag.* Historians of science have attributed this reading to Tait alone, and condemned him for it.[18] But Thomson had been following much the same line in his natural philosophy classes since at least the 1849–50 session at Glasgow College. At that time Thomson had expounded 'Newton's law of dynamical effect' and 'the great principle of work spent in dynamics'. In a hydrodynamic example he showed that the action (*actio agentis* from Newton's original Latin), interpreted as work done by a piston compressing water in a cylinder, was equal to the reaction (*actio resistentis*), interpreted as the *vis viva* of a jet of water issuing from an orifice in the side of the cylinder.[19] Furthermore, Thomson subsumed the principle of virtual velocities in statics under this principle of mechanical (or dynamical) effect:

> Every problem in nature of which we have forces in equilibrium may be brought under this equation. Newton lays it down in a very general way . . . I will give you a view that will bring it under the law of dynamical motion . . . The condition necessary & sufficient for equilibrium is that for every infinitely small displacement the mechanical effect produced by one set of forces is precisely = to the work spent in the others.[20]

Already, then, Thomson was reversing Lagrange's attempt to subsume dynamics under statics through the principle of virtual velocities. Maxwell put the issue in terms of a 'whig' historical perspective in 1877: 'Lagrange and Virtual Velocity . . . is the germ of the method of energy which was fully developed in mathematical form in the *Mécanique analytique*, but very little appreciated outside the inner circle of mathematicians till the physical theory of energy became generally known'.[21] As we have seen, the physical theory of energy, based not on Helmholtz's theoretical assumption of attractive and repulsive forces but upon the empirical claims of Joule and others, had been privately and publicly formulated around 1852 by Thomson and Rankine (ch.7). And by 1862 'Conservation of Energy' had been elevated into the 'ONE GREAT LAW OF PHYSICAL SCIENCE' (ch.9).

As part of this process of 'elevation' Tait delivered his lecture 'On the Conservation of Energy' to the RSE in April 1863. Echoing Newton's wish expressed in the 'Preface' to the *Principia* that we should derive all the phenomena of nature from mechanical principles, Tait told his audience that everything needed for such a goal 'has been most distinctly laid down by Newton in his Axiomata, and the Scholia appended to them'. A summary of these features would 'lead us easily and naturally to the Conservation of Energy – though stated for visible motions only, and without reference to the energies of heat, electricity, &c'.[22]

Tait then argued that in order to extend our investigations from a single particle to a *body*, or a system of bodies, 'we require the additional [third] law'. Newton had shown that there were *two* ways to measure action, leading to 'two classes of important dynamical theorems'. First, 'Mutual pressures, tensions of rods and cords, attractions, stresses in solids or liquids, &c.&c., form one class of Actions and Reactions'. Well-known mechanical principles such as 'conservation of momentum' and 'conservation of areas' followed as consequences. So also did 'the general statement of the equations of equilibrium and motion', known as 'D'Alembert's Principle', which yielded 'the mathematical expression of the circumstances of any dynamical problem'.

Second, Newton had pointed out '*another* kind of action and reaction, [also] ruled by the third law'. Tait here quoted Newton's Latin scholium to the third law which he and Thomson later translated in their own words:

> If the Action of an agent be measured by its amount and its velocity con-
> jointly; and if, similarly, the Reaction of the resistance be measured by the
> velocities of its several parts and their several amounts conjointly, whether
> these arise from friction, cohesion, weight, or acceleration; – Action and
> Reaction, in all combinations of machines, will be equal and opposite.

According to Tait's exegesis of this passage from the Newtonian scriptures, the Action 'here spoken of, the product of a force by the rate of motion of its point of application, is now known as the *rate of doing work*, or the horse-power of the prime mover'.[23] The corresponding Reaction, in various forms, was 'the *rate of losing work* by the resistances, such as friction, cohesion, and weight', as well as by 'the resistance due to the *acceleration* of the various parts of the system'.[24] This statement, though restricted to machines and their visible motions, constituted the 'Conservation of Energy'.

The *Treatise* itself credited James Watt with introducing 'the practical unit of a "*Horse-power*", or the rate at which an agent works when overcoming 33,000 times the weight of a pound through the space of a foot in a minute; that is, producing 550 foot-pounds of work per second' (ch.3). Thus the name of one of Clydeside's most celebrated sons had been added to the roll of honour for 'conservation of energy'.[25]

The legend of Watt, embodying the progressive and democratic values of presbyterian culture, flourished in mid-Victorian Glasgow. Speeches made at the opening of the new University of Glasgow in 1870 highlighted the perceived connections between 'science' and 'wealth', represented by the mythology of James Watt. With unintended irony the Duke of Montrose acclaimed Watt as 'a citizen of their own . . . lowly in origin, but strong in intellect and perseverence, with small means to carry out his own inventions, but still becoming in later life the father and practical inventor of the great machine which had brought fortune and prosperity to this city, and was now the great means of . . . universal locomotion by land and by sea'. He trusted that the example of Watt 'would stimulate them all . . . and that this day would be the first day of a new era of regeneration and prosperity (loud cheers)'. Mr E.S. Gordon, MP for the Universities of Glasgow and Aberdeen, put the matter even more forcefully in response to the rhetorical question 'To what are we to attribute the wealth and extensive commerce of this country?':

> it was due mainly to the great inventions of our countryman Watt (Applause). That noble man did not appear to have been a student at the Glasgow College; but when he was excluded from following his occupation as a mathematical instrument maker in the city, owing to his not having served his apprenticeship with the trades there, he was taken under the shelter of the University; and, if he did not attend classes, at all events he was indebted to the scientific advice of those great men, Black and Robertson [Robison], in bringing to perfection his invention of the steam engine (Applause). It would thus be seen that science, as taught in our Universities, had much to do with the production of the national wealth. In other words, science was the great promoter of invention, and invention was the great promoter of national wealth.[26]

No matter, then, that Watt had been neither student nor professor at the University. No matter either that his days of prosperity had been in Birmingham rather than Glasgow. Glasgow College and its reformers were too shrewd to allow historical accuracy to stand in the way of their claim to Watt, so useful in the promotion of their own goals in those heady days of Glasgow's prosperity in the 1860s.

Democratic dynamics: Chapter II of T&T'

As construction of the mighty *Treatise* began to fall increasingly behind schedule in the early 1860s, Tait wrote to Thomson in October 1862:

> I would far rather that you spent your time upon Chapter II, Force, Time, Energy, Work, Laws of Motion &c &c., for the want of it is keeping me back in my work at Statics & Kinetics – and besides II *ought* to be a long chapter. It is there that the shorthand writer would be invaluable. Import him to Brodick [Isle of Arran] for 3 days.[27]

Just as Tyndall had attributed much of the responsibility for 'Energy' to the Glasgow professor, so Thomson's role in the writing of the *Treatise* was no nominal one. At the core of the single volume published in 1867 stood Chapter II on 'Dynamical Laws and Principles' which aimed to make the 'one great law of physical science', conservation of energy, the foundation for the whole of dynamics. As Tait's letters show, it was Thomson who had the primary responsibility for this central chapter.[28]

In the first chapter, on kinematics, the *Treatise* 'considered as a subject of pure geometry the motion of points, lines, surfaces and volumes . . . and the results we there arrived at are of course altogether independent of the idea of *matter*, and of the *forces* which matter exerts'. In the second chapter, however, they engaged the question, not 'of how we *might* consider such motion . . . to be produced, but of the *actual* causes which in the material world *do* produce them'. Thus the results enunciated in this chapter had to be regarded as 'due to actual experience in the shape either of observation or experiment'.[29]

'We cannot, of course, give a definition of *Matter* which will satisfy the metaphysician'. With this characteristically caustic remark T&T′ introduced matter in an empirical, practical manner as '*that which can be perceived by the senses*, or as *that which can be acted upon by, or can exert, force*'. Both definitions involved the idea of force which 'is a direct object of sense; probably of all our senses, and certainly of the "muscular sense"'.[30] Maxwell responded quickly but privately to what he evidently saw as an extraordinarily pragmatic, indeed philosophically crude, account of matter: 'Matter is *never* perceived by the senses. According to Torricelli, quoted by Berkeley "Matter is nothing but an enchanted vase of Circe fitted to receive Impulse and Energy, essences so subtle that nothing but the inmost nature of material bodies is able to contain them"'.[31] But Thomson especially, immersed in a world of Victorian engineering rather than in a world of classical mythology, had no time for the finer aspects of traditional Scottish academic culture.

In similarly practical manner T&T′ claimed to follow Newton in defining 'quantity of matter' or mass as proportional to volume and density. Although the definition really assigned meaning to density rather than mass, the *Treatise* claimed that for matter of uniform density the mass was proportional to the volume or space it occupied. For his part, Maxwell commented that 'Newton's statement is meant to distinguish matter from space or volume, not to explain either matter or density'.[32] Once again, T&T′ were scarcely interested in such questions. Far more significant to them were the issues of *measurement* of mass and other physical quantities to which they devoted a substantial part of their second chapter.

The *Treatise* then set down definitions of momentum or quantity of motion (proportional to mass and velocity) and *vis viva* or kinetic energy (proportional to the mass and the square of the velocity), as well as their respective rates of change. They explained inertia, proportional to the quantity of matter in a body, as matter's 'innate power of resisting external influences'. It therefore followed that 'some *cause* is requisite to disturb a body's uniformity of motion, or to change its direction from the natural rectilinear path'. That cause was called 'force': 'Force is wholly expended in the *Action* it produces; and the body, after the force ceases to act, retains by its inertia the direction of motion, and the velocity which were given to it'.[33]

According to the *Treatise* three elements were required to specify a force. First, we needed to know its place of application. In this respect there were two distinctive kinds of force: one kind whose point of application was a surface (material objects never touched at a mere geometrical point); and the other kind whose place of application is a solid (such as a weight acting by gravity). Thus, 'When a heavy body rests on the ground . . . force of the second character, acting downwards, is balanced by force of the first character acting upwards'. Second, we needed to know the direction of a force, that is, the line in which it acts. And third, we needed to know the magnitude of the force. But 'Before measuring anything, it is necessary to have a unit of measurement, or a standard to which to refer' (ch.13).[34] Mathematical points thus yielded to material particles and surfaces; philosophical inquiries into the nature of matter and force gave way to definitions with everyday, practical expression; and metaphysical speculation was extinguished by a relentless promotion of accurate measurement.

T&T' based measurement of force on Gauss's absolute system (ch.13) which now received a Newtonian pedigree. They told their readers, 'accustomed to speak of a force of so many pounds', that 'Newton gives no countenance to such expressions' because 'the weight of a given quantity of matter differs in different latitudes'. To drive the point home to a commercial audience, they provided a practical and democratic, rather than abstract and elite, illustration of the issue:

> Whereas a merchant, with a balance and a set of standard weights, would give his customers the same quantity of the same kind of matter however the earth's attraction might vary, depending as he does upon *masses* for his measurement; another, using a spring balance, would defraud his customers in high latitudes, and himself in low, if his instrument (which depends on forces and not on masses) were correctly adjusted in London.

Under the absolute system 'accuracy' went hand in hand with such *moral* epithets as 'integrity', 'honesty' and 'commercial fairplay'. In this system, argued the *Treatise*, we begin with a unit of mass (the imperial pound or international standard weight such as the gramme) and from it derive the unit of force according to Newton's definition. Thus in the absolute system which Gauss 'first introduced practically' we have the unit force as 'that force which, acting on a national standard unit of matter during the unit of time, generates the unit of velocity'. This standard force was then 'independent of the differing amounts of gravity at different localities'.[35]

The *Treatise* moved quickly from the definition and measurement of 'force' to that of 'work'. Thus 'A force is said to *do work* if its place of application has a positive component motion in its direction; and the work done by it is measured by the product of its amount into this component motion'. The unit for the measurement of work 'adopted in practice by British engineers' was the 'foot-pound' while that in 'purely scientific measurements' was the 'kinetic unit force acting through unit of space'. Discussion of definitions concluded with an explanation of work done in relation to kinetic and potential energy.[36]

In a private draft (November 1862) Thomson had interpreted the work–energy relation as an absolute kinetic measure for work, and thus for force. Independent of location, the kinetic unit was 'a universal unit *of force*'. Its advantage among scientific people was analogous to the 'same [advantage] as men of letters formerly found in using Latin as their language, practically a universal language'. In other words the 'expression of force & work in terms of the absolute kinetic unit has now & must always continue to have the same advantage (or greater) that said use of Latin formerly had'. But the comparison no longer held good. Latin was nowadays 'practically retained by only one branch of the old stock & for quite a different reason (preventing people from acquiring their learning &c &c'.[37] Indicting the Catholic Church here for its abuse of power over the faithful, Thomson had laid bare his presbyterian credentials. Knowledge had to be accessible to everyone and not merely to the privileged priesthood. Truth had to be a matter of direct personal experience rather than a matter of taking the Church authorities on trust.

In this respect, however, the absolute kinetic unit also had the disadvantage of the old Latin in not being the vernacular:

> We keep it & shall keep it forever because it alone is or can ever be a universal language: but with truly X^{ian} [Christian] benevolence (how diff^t from the papistic spirit above alluded to) we translate every scientific result out of a language which not one of the 1000 intelligent mechanics of Glasgow (each with as good a head as any mathematician) would understand into terms perfectly appreciated by every inhabitant of this great prosperous & influential city. We do so by dividing by 32.2[38]

The problem of 'democratic' accessibility was therefore solved by dividing the absolute kinetic unit by the acceleration due to gravity at Glasgow to give a gravitational unit equal to the weight of an imperial pound, a unit instantly comprehended by the intelligent mechanics who, like James Watt, had helped turn Glasgow into the thriving Second City of the Empire.

Such unguarded remarks serve to reveal much about the manner in which the *Treatise* was embedded in mid-Victorian Scottish culture. The authors' construction of a dynamics centred on work and energy was radically contingent upon Scottish academic, religious and industrial culture. There was thus nothing self-evident or essential about the authors' choices: many other formulations of dynamics existed, but these were consciously rejected as unsuited to the wants of the modern age. The values of that age were embodied in the iconography of Watt: a man who, with his artisan origins, virtues of work and economy, independence and inventiveness, exemplified all that was best in a democratic presbyterian culture. But in order to make the new natural philosophy serve that culture and make it universal, Thomson and Tait now sought to render the Newtonian scriptures accessible to all men.

Claiming to have adhered thus far to many, though not all, of Newton's definitions, T&T' turned to statements of Newton's laws in his own words by offering first the Latin text and then their translation. They particularly drew attention to a later tendency exemplified in Cambridge texts on mechanics by Whewell, Pratt and Samuel Earnshaw (1805–88) published in the 1830s and 1840s for the use of mathematics undergraduates. The tendency in these texts was 'to split the second law into two, called respectively the second and third, and to ignore the third [law] entirely, though using it *directly* in every dynamical problem' and introducing instead D'Alembert's principle (below). This principle, they asserted, was really 'Newton's rejected third law in another form'. Indeed, they aimed to show that 'Newton's own interpretation of his third law points out not only D'Alembert's principle, but also the modern principles of Work and Energy'.[39]

Consideration of Newton's first two laws yielded a definition and a measure of force. But it was the third law which enabled us to understand completely 'the more complex cases of motion, especially those in which we have mutual actions between or amongst two or more bodies; such as, for instance, attractions, or pressures, or transference of energy in any form'. In his natural philosophy class lectures of 1862, Thomson had made the same point: 'The action of a force . . . is . . . a transformation . . . of Energy'. 'Force' was thus rapidly losing status to the fundamental terms of 'mass' and 'energy'.[40]

The *Treatise* then introduced its readers to the 'true' reading of Newton's scholium to his third law, stated in Tait's 1863 lecture. D'Alembert's principle followed from this re-reading:

> if we consider any one material point of a system, its reaction against acceleration must be equal and opposite to the resultant of the forces which that point experiences, whether by the actions of other parts of the system upon it, or by the influence of matter not belonging to the system. In other words, it must be in equilibrium with these forces. . . . all the forces acting on points of the system form, with the reactions against acceleration, an equilibrating set of forces on the whole system. This is the celebrated principle first explicitly stated, and very usefully applied, by D'Alembert in 1742. . . . [His] principle is found practically most useful in showing how we may write down at once the equations of motion for any system for which the equations of equilibrium have been investigated.[41]

Everything was now in place for an elucidation of what the *Treatise* called 'the abstract theory of energy', that is, for a theory of energy which would largely exclude considerations of dissipation. The 'physical theory of energy', inclusive of dissipation, would await future publication. Foundations for the abstract theory of energy had been laid by Newton 'in an admirably distinct and compact manner' in the scholium to his third law of motion. Newton's *actio agentis* or action of an agent was 'evidently equivalent to the product of the effective component of the force, into the velocity of the point on which it acts', that is, 'in modern English phraseology, the rate at which the agent works'.[42]

T&T' performed a further transformation (or what they termed a 'logical conversion') of Newton's words in the light of this interpretation of the action of an agent:

> Work done on any system of bodies (in Newton's statement, the parts of any machine) has its equivalent in work done against friction, molecular forces, or gravity, if there be no acceleration; but if there be acceleration, part of the

work is expended in overcoming the resistance to acceleration, and the additional kinetic energy developed is equivalent to the work so spent.

Work done against molecular forces (bending a spring) or gravity (raising a weight) could be reproduced without loss. In general, if the mutual forces between the parts of a limited system of bodies yielded or consumed 'the same amount of work during any motion whatever' when the system passed from one configuration to another, the system was '*dynamically conservative*'.

The *Treatise* then laid down a proposition, defining a dynamically conservative system, upon which 'the whole theory of energy in physical science is founded': 'If the mutual forces between the parts of a material system are independent of their velocities, whether relative to one another, or relative to any external matter, the system must be dynamically conservative'. Demonstration rested on consideration of two paths leading from one configuration to another. If more work were done in passing between the two configurations by the first than by the second path, then by passing in one direction by the first and returning by the second path we would have 'a continual source of energy without any consumption of materials [fuel], which is impossible'.[43]

In such a conservative system the amount of work done during the passage from one configuration to another was the potential energy of the system, usually fixed at zero for the initial configuration. This approach provided a simple understanding of the notion of a potential function:

> The potential energy of a conservative system, at any instant, depends solely on its configuration at that instant, being, according to [the above] definition, the same at all times when the system is brought again and again to the same configuration. It is therefore, in mathematical language, said to be a function of the co-ordinates by which the positions of the different parts of the system are specified.[44]

Examples, promised for a later section of the *Treatise*, included gravitational, electrical and magnetic potentials (drawing on the memoirs of George Green and C.F. Gauss) with respect to the mutual action between two spheres.

In nature, however, the 'hypothetical condition' of a conservative system was '*apparently* violated in all circumstances of motion'. Indeed, for Newton and his successors 'it was supposed that work was *absolutely lost* by friction'. There was an 'inevitable loss of energy of visible motions' due to one of several classes of resistance: (1) 'mutual friction between solids sliding upon one another'; (2) the viscosity of fluids and the imperfect elasticity of solids; (3) the resistance of electric currents; and (4) imperfect magnetic retentiveness. Thus 'No motion in nature can take place without meeting resistance due to some, if not all, of these influences' which 'impede the action of all artificial mechanisms' as well as the movements of all bodies in the solar system (as in tidal friction). The ultimate tendency of the solar system would therefore be towards 'the falling together of all [bodies] into one mass, which, although rotating for a time, must in the end come to rest relatively to the surrounding medium'.[45]

As a pointer to later (never completed) sections of the *Treatise* T&T' explained that 'The theory of energy cannot be completed until we are able to examine the physical influences which accompany loss of energy in each of the classes of resistance'. Readers would there see that 'in every case in which energy is lost by resistance, heat is generated' as a 'perfectly definite equivalent for the energy lost'. In general, conservation of energy entailed the conclusion that 'if any limited portion of the material universe could be perfectly isolated, so as to be prevented from either giving energy to, or taking energy from, matter external to it, the sum of its potential and kinetic energies would be the same at all times: in other words, that every material system subject to no other forces than actions and reactions between its parts, is a dynamically conservative system'.[46]

T&T', however, went considerably further and stated a conviction that nature as a whole constituted a conservative system:

> But it is only when the inscrutably minute motions among small parts, possibly the ultimate molecules of matter, which constitute light, heat, and magnetism; and the intermolecular forces of chemical affinity; are taken into account, along with the palpable motions and measurable forces of which we become cognizant by direct observation, that we can recognise the universally conservative character of all natural dynamic action, and perceive the bearing of the principle of reversibility on the whole class of natural actions involving resistance, which seem to violate it.

This belief in the 'universally conservative character of all natural dynamic action' related specifically to Thomson's enduring commitment to continuum theory, expressed in his contemporaneous enthusiasm for a vortex theory of matter (ch.12). As an ultimate goal, therefore, a complete 'physical dynamics' would account for all the motions, visible and invisible, which produced the phenomena of nature.[47]

As a 'provisional' goal T&T' introduced into their abstract dynamics (concerned with finite systems) what they termed 'a special reckoning for energy lost in working against, or gained from work done by, forces not belonging palpably to the conservative class'. They thus brought into abstract dynamics certain kinds of action and reaction, most notably that of friction between solids sliding on solids and, much less conspicuously, the effects of viscosity of fluids, imperfect elasticity of solids, imperfect electric conduction and imperfect magnetic retentiveness. The law of conservation of energy, which they claimed comprehended the whole of abstract dynamics by allowing the conditions of equilibrium and motion in all cases to be derived from it, now became:

> The whole work done in any time, on any limited material system, by applied forces [that is, by forces applied arbitrarily from outside the system], is equal to the whole effect in the forms of potential and kinetic energy produced in the system together with the work lost in friction.

Considering this law of energy in the case of a material system 'whose relative motions are unresisted by friction', the *Treatise* stated that the system would be in equilibrium in any particular configuration if and only if 'the work done by the applied forces is equal to the potential energy gained, in any possible small displacement from that configuration'. This statement was 'the celebrated principle of virtual velocities which Lagrange made the basis of his *Mécanique analytique*'. The condition of equilibrium followed from the fact that 'the system cannot possibly move away from any particular configuration except by work being done upon it by the forces to which it is subject'.[48]

Instead of Newton's third law (the equality of action and reaction), Lagrange had employed 'D'Alembert's principle' which applied to many-body systems. The principle began from a consideration of forces applied to each of several bodies. Because the bodies formed a system of connection and constraint, they could not in practice acquire the motions which these forces would produce in the absence of such constraints. The principle then assumed the fictitious motions (and forces) to consist of the motions (and forces) which the bodies actually acquire together with the motions (and forces) destroyed by constraint. But the latter forces produced no motion and were thus in equilibrium. Consequently, the actual forces, if reversed in sign, must always be in equilibrium with the applied forces. In this sense the principle represented a generalized statement of the action–reaction law. In its most common form the sum of the applied forces minus the actual forces (mass times acceleration) was zero.[49]

In Lagrange's words, D'Alembert's principle yielded 'a direct and general method for resolving, or at least for putting into equations, all the problems of dynamics that one can imagine'. This powerful tool reduced 'all the laws of motion of bodies to those of their equilibrium, and thus brings dynamics back to statics'. To achieve this goal, Lagrange combined D'Alembert's principle with a traditional principle of statics, the principle of virtual velocities.[50]

'Virtual velocities' represented possible changes in the configuration of a system, consistent with the constraints on the system. A 'virtual velocity' was thus not an actual velocity in so far as it did not involve time and was not due to actual forces in the system. In the simple balance, for instance, any small displacement of one of the weights expressed a virtual velocity. Lagrange stated the principle thus:

> if any system of arbitrarily many bodies or points, each acted on by any force, is in equilibrium; and if one gives to this system any small movement, as a result of which each point traverses an infinitely small space called its virtual velocity, the sum of the forces, each multiplied by the space which its point of application traverses in the direction of this same force, will always be equal to zero; regarding as positive the small spaces traversed in the direction of the pressure and as negative the spaces traversed in the opposite sense.[51]

In the case of the simple balance, therefore, the sum of the forces (weights) multiplied by their respective displacements (virtual velocities) would be zero.

Combining the two principles yielded Lagrange's most general relation (the sum of the applied forces minus the actual forces multiplied by the displacement equals zero). From this relation he could deduce the equations of motion of any mechanical system as well as several conservation laws (especially *vis viva*) and the principle of least action. Conservation of *vis viva* was used by Lagrange not as a physical or metaphysical foundation, but as an important derivative principle, useful for solving particular mechanical problems. Above all he articulated what became known as the 'Lagrangian' formulation of the equations of motion in generalized coordinates.[52]

In order to deduce the equations of motion of any system, Thomson and Tait returned to D'Alembert's principle that 'forces acting on the different points of a material system, and their reactions against the accelerations which they actually experience in any case of motion, are in equilibrium with one another'. Lagrange's principle of virtual velocities, subsuming D'Alembert's principle, then became:

> in any actual case of motion, not only is the actual work done by the forces equal to the kinetic energy produced in any infinitely small time, in virtue of the actual accelerations; but so also is the work which would be done by the forces, in any infinitely small time, if the velocities of the points constituting the system, were at any instant changed to any possible infinitely small velocities, and the accelerations unchanged. This statement, when put in the concise language of mathematical analysis, constitutes Lagrange's application of the 'principle of virtual velocities' to express the conditions of D'Alembert's equilibrium between the forces acting, and the resistances of the masses to acceleration. It comprehends ... every possible condition [stable, unstable, neutral equilibrium] of motion. The 'equations of motion' in any particular case are, as Lagrange has shown, deduced from it with great ease.[53]

What Thomson and Tait had done was to reverse Lagrange's derivation of dynamics from statics and to make Lagrange's principle of virtual velocities derivative from the principle of energy conservation. Upon this dynamical basis, made more secure by the introduction of a principle of least action, would rest the whole of natural philosophy.

Least action

The effect of moving the principle of conservation of energy to centre stage in the *Treatise*, and of relegating force to the status of a derivative concept, left T&T''s system of dynamics open to the criticism that they had removed the very notion of a causal agent (force) from physical science and had thus destroyed the primary and traditional goal of natural philosophy. In the course of Chapter II, however, they introduced the notion of 'extremum principles', notably that of least action from Sir William Rowan Hamilton, which would replace force as the causal agent with the tendency of energy to transfer in such a manner that the action between two states or configurations was a minimum.[54]

In eighteenth-century dynamics the principle had been employed by Maupertuis and Euler to replace the summing up of individual forces with an optimization condition, that of 'least action' or maximum economy. Eager to suppress all suggestions of purposive behaviour in science, Laplace in particular had reduced the status of least action to that of a principle derivative from force. But the sophisticated mathematical tool for handling extremum conditions, the variational calculus, became in the hands of Lagrange and Laplace a highly potent device for the solution of mathematical problems, one which reappeared in the works of Hamilton.[55]

The abstract and metaphysical character of Hamilton's mathematical physics at first sight held little attraction for the scientists of energy. But Tait, enamoured of Hamilton's quaternion methods in mathematics, was also an enthusiast for Hamilton's extension of the principle of least action. Failing to persuade Thomson of the value of quaternions, why did Tait succeed with a principle tainted by much that his pragmatic colleague despised?[56]

Since the mid-1840s Thomson had thought of a physical system (involving physical agents such as electricity or heat, for example) as an engine (ch.13). An engine or a physical system produced mechanical effect or work by the 'fall' of a physical agency from a higher to a lower state. The tendency was thus towards the lowest possible state: a river flowed towards the sea and a body cooled towards the temperature of its surroundings. At the lowest possible state, equilibrium would be attained: the river could flow no further and a body could lose no more heat.[57]

Thomson then utilized this perspective in a whole complex of electrical, magnetic, thermal and hydrodynamical problems. The key condition was that the equilibrium state is always produced by the least possible expenditure of mechanical effect, that is, in the most economical manner. As early as 1847 Thomson had viewed this condition as a special case of the least action principle. In hydrodynamics, for example, equilibrium will require that the 'action', resulting from the application of forces at the boundary of a perfect fluid and taken as the *vis viva* or kinetic energy integrated over time, be a minimum. Thomson was thus generally well prepared to accept the utility of a principle of least action when Tait apparently talked him into admitting it into the *Treatise*.[58]

'Action' in the *Treatise* meant the rate of working of an engine or the rate of production of kinetic energy. Thomson especially felt strongly that 'the *action of a steam engine* is a very intelligible phrase, and may be used without any vagueness to express the energy with which the engine is working'. Lagrange's 'action', on the other hand, meant total *vis viva* integrated over time between two states or average kinetic energy multiplied by time. In Thomson's opinion, this version offered 'no definite idea to the mind'. Nevertheless, because Thomson's preferred 'action' did not properly generalize to the principle of least action and because Lagrange's 'action' was now in any case readily linked to a principle of working through energy considerations, the authors of the *Treatise* accepted Lagrange's 'action'.

Reinterpreted in this manner as a principle of working or of production of kinetic energy, the least action principle together with the energy conservation principle yielded the principle of virtual velocities (seen by Thomson as a principle of virtual work). Thus for every possible displacement of a dynamical system, the work done by the applied forces would equal the change in kinetic energy.

Hamilton had given least action a further formulation as a principle of varying action. Here the initial and final states, together with the *vis viva* for different natural motions, were varied, yielding Hamilton's 'characteristic function' for the action in terms of *vis viva* and initial and final states alone. In the *Treatise* this function provided a single minimized energy function which determined the entire finite change in the system between any two states.[59]

Maxwell's comments on the new and democratic dynamics in his *Nature* review (1879) included remarks on the slowness of natural philosophers to make use of the dynamical theorems developed by Lagrange and Hamilton in particular. With characteristically subtle words Maxwell, however, observed how much had changed since the publication of Thomson and Tait's *Treatise*: 'Now, however, we have only to open any memoir on a physical subject in order to see that these dynamical theorems have been dragged out of the sanctuary of profound mathematics in which they lay so long enshrined, and have been set to do all kinds of work, easy as well as difficult, throughout the whole range of physical science'.[60]

Maxwell was especially concerned to highlight the northern wizards' exposition in the second edition 'of a very potent form of incantation, called by our authors the Ignoration of Coordinates'. Here Maxwell remarked that 'the co-ordinates of Thomson and Tait are not the mere scaffolding erected over space by Descartes, but the variables which determine the whole motion', variables which we may picture as 'so many independent driving-wheels of a machine which has as many degrees of freedom'. He then explained to readers of *Nature* that the approach was a mathematical illustration of the methodological principle that, 'in the study of any complex object, we must fix our attention on those elements of it which we are able to observe and to cause to vary, and ignore those which we can neither observe nor cause to vary'. He drove home his claim with a mechanical illustration drawn not from an industrial but from a religious context familiar to many Cambridge undergraduates:

> In an ordinary [church] belfry, each bell has one rope which comes down through a hole in the floor to the bellringers' room. But suppose that each rope, instead of acting on one bell, contributes to the motion of many pieces of machinery, and that the motion of each piece is determined not by the motion of one rope alone, but by that of several, and suppose, further, that all this machinery is silent and utterly unknown to the men at the ropes, who can only see as far as the holes in the floor above them.[61]

'What', Maxwell then asked rhetorically, 'is the scientific duty of the men below'?

These scientific bell-ringers had 'full command of the ropes, but of nothing else'. In particular they could 'give each rope any position and any velocity'. They could also estimate the momentum 'by stopping all the ropes at once, and feeling what sort of tug each rope gives'. Furthermore, they could 'ascertain how much work they have to do in order to drag the ropes down to a given set of positions'. By expressing the work in terms of these positions 'they have found the potential energy of the system in terms of the known co-ordinates'. Finally, if 'they then find the tug on any one rope arising from a velocity equal to itself or to any other rope, they can express the kinetic energy in terms of the co-ordinates and velocities':

> These data are sufficient to determine the motion of every one of the ropes when it and all the others are acted on by any given forces. This is all that the men at the ropes can ever know. If the machinery above has more degrees of freedom than there are ropes, the co-ordinates which express these degrees of freedom must be ignored. There is no help for it.[62]

Maxwell's *Nature* review did not simply confer credibility on the *Treatise* by way of lavish praise from a close colleague. He instead demonstrated how it could empower mathematical physicists to avoid speculations about the hidden machinery of nature and yet to produce fruitful results. Maxwell himself had made much use of this approach in his own *Treatise on Electricity and Magnetism* (1873) which in many respects filled much of the massive space left by Thomson and Tait's incomplete project (ch.11). Meanwhile, Helmholtz translated the *Treatise* into his own language in 1871, insisting in the preface that the deficiencies of its formal analytical exposition were more than outweighed by its consistency to physical fact. If the *Treatise* did not quite 'go over Europe like a statical charge', it had certainly presented the definitive dynamical foundations for the science of energy.[63]

CHAPTER 11

Gentleman of Energy: the Natural Philosophy of James Clerk Maxwell

. . . and so much to do for science before we all go.
Sir William Thomson to G.G. Stokes on hearing of Maxwell's fatal illness, 1879[1]

The main object of the *Treatise on Electricity and Magnetism* (1873), wrote Tait in a contemporary review for *Nature*, 'besides teaching the experimental facts of electricity and magnetism . . . is simply to upset completely the notion of *action at a distance*'.[2] In the mid-1840s Weber had constructed a major new unifying theory of electricity based on the interaction of electric charge at a distance. But between 1854 and his death a quarter of a century later, James Clerk Maxwell made relentless efforts to depose Weber's theory from its pre-eminent position as the most powerful and persuasive interpretation yet on offer.

Locating Maxwell in opposition to continental action-at-a-distance theories and in alignment with Faraday's 'field' theories, however, reveals only part of the historical picture. In this chapter I present Maxwell as a core member of the North British group. Shaped by a distinctive presbyterian culture, Maxwell's deeply Christian perspective on nature and society became inseparable from his central commitment to the science of energy. Yet the science of energy was in a state of *construction* rather than a finished edifice. It provided the cultural and conceptual framework around which Maxwell would build credibility for himself and for his controversial electromagnetic theory. To that end he would depend heavily on private critical discussions with his closest scientific colleagues, Thomson and Tait, and would attempt to tailor his successive investigations to specific public audiences.

'More power of useful work': the culture of James Clerk Maxwell

Maxwell's father, John Clerk, had added the name Maxwell after inheriting a neglected estate in Kirkcudbrightshire, south-west Scotland, from Maxwell ancestors. James would later describe this North British region as 'perhaps the least Episcopal part of Scotland by reason of the memory of the [English] dragoons . . . most of the persecutors' families are now Presbyterian and Whig'.[3] By profession a lawyer, John married Frances Cay, daughter of a one-time Judge of the Admiralty Court in Scotland. Together they organized the construction of a new home, Glenlair, which became the focal point of family life on the estate. An elder in the parish kirk of Parton, John Clerk was noted above all for 'a persistent practical interest in *all useful processes*'. Thus he would apparently remark upon a 'sad waste of work' in relation to 'the folly of coachmen in urging a horse to speed as soon as they saw the top of a hill, when, by waiting half a minute until the summit was really attained, they might save the animal'.[4]

Born in Edinburgh on 13 June 1831, James would be brought up, the only surviving child, in a genteel culture which combined Christianity and practicality. Lacking commercial and competitive values, it was a world away from that of William Thomson, seven years his senior. With their intensely competitive life-style, the Thomsons contrasted markedly with the conservative yet eccentric lairds of Glenlair. Frances Clerk Maxwell, from episcopalian rather than presbyterian stock, died when James was eight, apparently from the same form of cancer that would end his own life forty years later. After private tutoring James attended Edinburgh Academy, a prestigious institution that promoted an English-style liberal education which, with an emphasis on disciplining the student's mind through classical languages and geometry, was well suited to the tastes of aspiring Scottish gentlemen. With his eccentric mannerisms and broad dialect, however, James was not easily assimilated to the conventions of a refined Anglo–Scottish culture.[5]

James's schooling (1841–7) coincided with the period of the Disruption. At the end of his life he recollected:

> The ferment about the Free Church movement had one very bad effect. Quite young people were carried away by it; and when the natural reaction came, they ceased to think about religious matters at all, and became unable to receive fresh impressions. My father was so much afraid of this, that he placed me where I should be under the influence of Dean Ramsay, knowing him to be a good and sensible man.[6]

The Cambridge-educated Dean of Edinburgh, Edward Bannerman Ramsay (1793–1872), hailed from a gentrified Scottish culture very much like that of the Maxwells. While staunchly Episcopalian on matters of Church government, Dean Ramsay's low church evangelical views allowed for a cross-denominational outlook as, for example, when he supported Dean Stanley in a move to open the pulpit of Westminster Abbey to clergymen from outside the Church of England. Like many of his contemporaries who disliked the Free Kirk ferment, Ramsay was none-theless an ardent admirer of the evangelical Chalmers. A liberal conservative in politics, Ramsay appeared doctrinally very close to Norman Macleod, opponent of extreme evangelicals, religious enthusiasts and biblical literalists (ch.2).[7]

While in Edinburgh the young Maxwell thus learnt to feel equally at home with presbyterian and episcopal forms of protestantism. His episcopalian friend (and later biographer) Lewis Campbell recorded that on Sundays James generally attended the presbyterian St Andrews Church with his father in the forenoon, while in the afternoon, at the behest of his maternal aunt, Jane Cay, he would go to St John's Episcopal Chapel where he was also for a time a member of Dean Ramsay's catechetical class. Back in the comparative tranquillity of Glenlair he worshipped devoutly in the new parish church of nearby Corsock. As Lewis Campbell recalled, James 'always sat preternaturally still . . . not moving a muscle, however long the sermon might be. Days afterwards he would show, by some remark, that the whole service, whether good or bad, had been, as it were, photographed upon his mind'.[8]

Entering Edinburgh University at the age of sixteen, James enjoyed above all the lectures of J.D. Forbes, professor of natural philosophy, and Sir William Hamilton, professor of logic and metaphysics. As with Rankine and Tait, Maxwell's natural philosophy owed much to Forbes's direct influence and patronage.[9] In contrast to Forbes's qualified admiration for Cambridge standards, Hamilton represented traditional Scottish philosophical values by which students examined the foundations of knowledge rather than learning the techniques of problem solving.[10]

Maxwell's cousin, Jemima Wedderburn, had married Hugh Blackburn, professor of mathematics at Glasgow College from 1849 in succession to William Thomson's father who had died of cholera in January of that year. Together with a friendship between the Edinburgh and Glasgow professors of natural philosophy (Forbes and Thomson), these close family networks brought the young Maxwell firmly into Glasgow University scientific circles. Moreover, Blackburn, since Cambridge undergraduate days, had been Thomson's closest personal friend and had accompanied him to Paris after graduation from Cambridge in 1845. Not surprisingly it was Blackburn who 'insisted that the mathematical discipline of Cambridge would enable him [Maxwell] to exercise his genius more effectively'.[11] Dean Ramsay and Professor Forbes also urged that James should travel south. But there were doubts: 'certain floating prejudices about "the dangers of the English universities", [High Church] Puseyism, infidelity, etc., had then considerable hold, especially on the Presbyterian mind'.[12]

By the summer of 1850 James was thoroughly immersed in the scientific world of Thomson and his friends, including Stokes, especially at the time of the BAAS Edinburgh meeting and in the weeks prior to his own admission to St Peter's College, Cambridge. He explained to Lewis Campbell that he had received an important commission from Professor Thomson:

> Professor W. Thomson has asked me to make him some magne-crystallic preparations which I am now busy with [at Glenlair]. . . . Not that I am turned chemist. By no means; but common cook. My fingers are abominable with glue and chalk, gum and flour, wax and rosin, pitch and tallow, black oxide of iron, red ditto and vinegar. By combining these ingredients, I strive to please Prof. Thomson, who intends to submit them to Tyndall and

Knoblauch, who, by means of them, are to discover the secrets of nature, and the origin of the magne-crystallic forces.[13]

Acting like a laboratory assistant to Thomson, Maxwell was here being introduced to some of the latest features of Faraday's experimental research into the magnetic properties of matter.[14] But above all he was acquiring practical skills, already highly prized by his father, in matters of experimental science.

In October 1850 the young Scot entered St Peter's College, but soon transferred to the larger Trinity College where the prospects for a fellowship after graduation were thought to be better. Yet the formal discipline of a Cambridge undergraduate education, reflected in one of his many poems, had no great appeal:

> In the grate the flickering embers
> Served to show how dull November's
> Fogs had stamped my torpid members,
>   Like a plucked and skinny goose.
> And as I prepared for bed, I
> Asked myself with voice unsteady,
> If of all the stuff I read, I
>   Ever made the slightest use.[15]

Nevertheless Cambridge offered its Scottish students mathematical skills unavailable in North Britain. In due course Maxwell graduated second wrangler and first-equal Smith's prizeman (1854), obtained a Trinity College Fellowship (1855) and taught mathematics to undergraduates. Soon after he published his first substantial study of electromagnetism and obtained the chair of natural philosophy at Marischal College, one of Aberdeen's two university colleges.[16]

Maxwell's Cambridge years (1850–6) allowed him not only to hone his mathematical skills but also allowed him to engage critically with some of the most controversial cultural issues of the time, including *Vestiges of Creation* and F.D. Maurice's *Theological Essays*. His own participation in the meetings of the elite Apostles' Club bear witness to the breadth of his philosophical interests. A common thread, however, was a desire to distinguish two kinds of human pursuit, the one false and the other true. On the one hand, 'theories are elaborated for theories' sake, difficulties are sought out and treasured as such, and no argument is to be considered perfect unless it lands the reasoner at the point from which he started'. This kind of disembodied activity seemed exemplified by John Henry Newman's reasonings on the development of Christian doctrine which were characterized by 'impatience of facts and fascinated by the apparent simplicity of some half-apprehended theory'. On the other hand there was that pursuit 'which in its lowest form is called the useful, and has for its ultimate object the extension of knowledge, the dominion over Nature, and the welfare of mankind'.[17]

This latter pursuit, defined in opposition to such 'theoretical dogmas' as those of High Church Anglicanism, Maxwell made his own. It was a pursuit fully sanctioned by the presbyterian culture of his native land and represented in the iconography of James Watt (chs 3&10). It was also a pursuit which received the imprimatur of his father who in 1853 issued detailed instructions to the Cambridge undergraduate concerning the importance of seeing for himself industrial and craft practices:

> View, if you can, [Birmingham's] armourers, gunmaking and gunproving – swordmaking and proving – *Papier-maché* and japanning – silver-plating by cementation and rolling – ditto, electrotype – Elkington's Works – Brazier's works, by founding and by striking up in dies – turning – spinning teapot bodies in white metal, etc. – making buttons of sorts, steel pens, needles, pins, and any sorts of small articles which are curiously done by subdivision of labour and by ingenious tools – glass of sorts is among the works of the place, and all kinds of foundry work – engine-making – tools and instruments (optical and philosophical) both coarse and fine.[18]

Such interests in craft practices not only aided Maxwell in his developing friendship with Joule, Rankine and the Thomsons but also provided a grounding in workshop practices which would stand him in excellent stead when he undertook experimental investigation on behalf of the British Association (ch.13).[19]

In contrast to the tranquillity of Glenlair and Cambridge, and lacking the intellectual sophistication of Edinburgh, Aberdeen had a reputation for a severe northern climate and an equally harsh presbyterianism. 'No jokes of any kind are understood here', Maxwell informed Lewis Campbell. 'I have not made one for two months, and if I feel one coming I shall bite my tongue'. He was writing a couple of days before Christmas, not a time for celebration in Calvinist Scotland. Yet the religious character of Aberdeen was surprisingly diverse. Although in the past 'the Presbyterians persecuted more than the Prelatists', the result of such persecution had been to strengthen the persecuted such that 'there are three distinct Episcopal religions in Aberdeen, all pretty lively'.[20]

During this period of religious revivals in Scotland, especially northern Scotland, Maxwell sharpened and developed his own interpretation of Christianity. This interpretation, developed in a number of substantial letters to close friends, coalesced perfectly with, and would serve to shape further, the new science of energy (esp.ch.12). At the core of his analysis was the concept of free will, considerations of which brought into focus the relationships between God, Christ, nature, human society and the individual human being. Underlying his analysis were deeply entrenched presbyterian concerns with sin and bondage, grace and freedom, faith and work. While still at Cambridge he came close to espousing for himself the notion of a depraved will: 'I maintain that all the evil influences that I can trace have been internal and not external . . . that I have the capacity of being more wicked than any example that a man could set me, and that if I escape, it is only by God's grace helping me to get rid of myself, partially in science, more completely in society – but not perfectly except by committing myself to God as the instrument of His Will'. While at Aberdeen, Maxwell would subject this traditional presbyterian perspective, scarcely distinguishable from the original *Westminster Confession of Faith* itself, to his own powers of refinement.[21]

To begin with he differentiated sharply between 'the fallen world' of human society, characterized by 'worldliness', and the Kingdom of Christ. Upon Lewis Campbell taking holy orders, Maxwell thus wrote of the dangers of a clergyman becoming immersed in 'a clerical world' which differed 'from the world of business or of fashion only in the general colouring of its scenery' and that it had 'always seemed to me that men who have fallen into this "religious world" have completely failed in getting into the Church, seeing that the Church professes to be an escape from the world, and the only escape'. Furthermore, he stressed that he believed in this distinction 'not only as a theory, but as a fact, that a man will find the thing so if he will try it himself'. He therefore drew close parallels between the role of the natural philosophy professor and that of the minister of Christ as teacher and guide, leading their respective flocks to see the truth for themselves rather than imposing upon those flocks the mere worldly authority of professor or priest:

> I have to tell my men that all they see, and their own bodies, are subject to laws which they cannot alter, and that if they wish to do anything they must work according to those laws, or fail, and therefore we study the laws. You have to say what men are and the nature of their actions depends on the state of their wills, and that by God's grace, through union with Christ, the contradictions and false actions of those wills may be settled and solved, so that one way lies perfect freedom, and the other way bondage under the devil, the world, and the flesh, and therefore you entreat them to give heed to the things which they have heard.[22]

A couple of months later Maxwell provided his Cambridge friend R.B. Litchfield with a refined treatment of the will in relation to human freedom and order. Indeed, he went so far as to propose his own, distinctively Christian notion of liberty in opposition to the liberty of all manner of Enlightenment proponents, including political economists and utilitarians. 'All the time I have lived and thought', he began, 'I have seen more and more reason . . . to hold that all want of order, caprice, and unaccountableness results from interference with liberty, which would, if unimpeded, result in order, certainty, and trustworthiness (certainty of success in predicting)'. Thus far the claims seemed uncontroversially in line with *laissez faire* doctrines of nineteenth-century Britain. But he immediately issued a caution: 'Remember I do not say that caprice and disorder are not the result of free will (so called), only I say that there is a liberty which is not disorder, and that this is by no means less free than the other, but more'.[23]

Maxwell then presented three stages 'in the evolution of the idea of liberty'. In the first phase 'human actions are the resultant (by par^m of forces) of the various attractions of surrounding things, modified in some degree by internal states, regarding which all that is to be said is that they are subjectively capricious, objectively the "RESULT OF LAW", – that is, the wilfulness of our wills feels to us like liberty, being in reality necessity'. Thus free will in this phase was merely an illusion.

In the second, related phase, 'the wilfulness is seen to be anything but free will, since it is merely a submission to the strongest attraction, after the fashion of material things'. From many traditional Christian perspectives this state implied that 'a man's will is the root of all evil in him, and that he should mortify it out till nothing of himself remains, and the man and his selfishness disappear together'. With respect to Roman Catholicism, for example, 'People get tired of being able to do as they like, and having to choose their own steps, and so they put themselves under holy men, who, no doubt, are really wiser than themselves'. Such a solution was for Maxwell both morally wrong and intellectually impossible as a treatment for human sin:

> it is not only wrong, but impossible, to transfer either will or responsibility to another; and after the formulae have been gone through, the patient has just as much responsibility as before, and feels it too. But it is a sad thing for any one to lose sight of their work, and to have to seek some conventional, arbitrary treadmill-occupation prescribed by some sanitory jailors.[24]

At the end of his life Maxwell also affirmed with regard to Calvinist theology that he could not 'ascribe so much to a depraved will as some people do, though I do to a certain extent believe in it'. Rather, much 'wrong-doing seems to be no more than not doing the right thing; and that finite beings should fail in that does not seem to need the supposition of a depraved will'.[25]

It was therefore a third state to which Maxwell turned, a phase 'in which there appears a possibility of the exact contrary to the first state, namely, an abandonment of wilfulness without extinction of will, but rather by means of a great development of will, whereby, instead of being consciously free and really in subjection to unknown laws, it becomes consciously acting by law, and really free from the interference of unrecognised laws'.[26] Will as 'wilfulness' and will as 'willingness' were therefore different categories for human action. The former came close to the old sense of 'depraved will' which Maxwell had all but abandoned in favour of the limitations and failings of finite human beings to do right. For the latter to be truly free required a union with Christ, not a freedom from the straightjacket constraints of Christian dogma as Enlightenment philosophers had held. Only with that union would 'the world', recovered from the bondage of sin, become perfect once more as part of Christ's Kingdom.

Soon after his engagement in February 1858 to Katharine Dewar, daughter of the Principal of Marischal College, Maxwell developed his Christian perspective still further. In a letter to Katharine of 6 May he regretted that some early Christians, by adopting certain Judaic practices, had thereby 'altered the character of the whole of their religion by making it a thing of labour and wages, instead of an inward growth of faith working by love, which purifies the heart now, and encourages us to wait for the hope of righteousness'. Faith thus took priority over works insofar as divine favours were not to be earned or returned by mere worldly 'labour and wages', that is, the exchange economy of the political economists was not appropriate to the relations between God and man. Instead, God had provided finite beings with certain free gifts, just as He had endowed the living creation with the free gifts of light, air and water (ch.2):

Our flesh is God's making, who made us part of His world; but then He has given us the power of coming nearer to Himself, and so we ought to use the world and our bodies as means towards the knowledge of Him, and stretch always as far as our state will permit towards Him ... the fruit of the spirit comes when, like good trees, we stretch our best affections upwards till we see the sun, and breathe the air and drink the rain, and receive all free gifts, instead of sending our branches after our roots, down among things that once had life but now are decaying, and seeking there for nourishment that can only be had from above.[27]

A few days later he told Katharine that Lewis Campbell, preaching on the text 'Ye must be born again', had shown that all the world needed to be 'born *from above*'. Acceptance of such a gift of spiritual birth meant entry into union with Christ such that 'the more we enter together into Christ's work He will have the more room to work His work in us'.[28] He would subsequently elaborate on the divine division of spiritual labour: 'Now let us read ([1] Cor[inthians]) chapter xii, about the organisation of the Church, and the different gifts of different Christians, and the reason of these differences that Christ's body may be more complete in all its parts'. A more distinct awareness of our union with Christ would mean that 'we would know our position as members of His body, and work more willingly and intelligently along with all the rest in promoting the health and growth of the body, by the use of every power which the spirit has distributed to us'.[29]

Maxwell's creed articulated only with greater finesse many of the themes which have appeared throughout this study of the North British science of energy: the emphasis on divine gifts, spiritual and physical; the high premium placed on useful work, again both spiritual and material, made possible through a willingness to accept and then direct those gifts to beneficial ends; and a stress on the parallels between spiritual perfection (Christ as 'perfect humanity') and physical perfection (a perfect thermodynamic engine), both impossible for finite human beings to achieve but both set as goals by which to improve the flawed condition of the human world (chs 2 and 8). The same creed informed Maxwell's academic philosophy. As natural philosophy professor he considered that Marischal College combined 'more discipline and more liberty, and therefore more power of useful work' and looked forward to the best students 'choosing' to return to form 'a select class for the higher branches. They have all great power of work'.[30] It was to be a creed that also underpinned his whole commitment to a developing science of energy in general and to statistical thermodynamics in particular (ch.12).

Constructing electromagnetic credibility: 1855–62

A few days after completing the Mathematical Tripos, Maxwell despatched to Thomson a letter whose informal tone indicated that its author regarded himself both as a personal friend of the Glasgow professor and as a prospective apprentice to a master craftsman of electrical science. 'Suppose a man to have a popular knowledge of electrical show experiments and a little antipathy to Murphys Electricity', he wrote, 'how ought he to proceed in reading & working so as to get a little insight into the subject wh[ich] may be of use in further reading?' In particular, 'If he wished to read Ampère Faraday &c how should they be arranged, and at what stage & in what order might he read your articles in the Cambridge [Mathematical] Journal?'[31]

These apparently-innocuous remarks could scarcely have been better aimed at their intended target. Thomson's interest in the mathematical theory of electricity extended back to the early 1840s. On the basis of Fourier's treatment of heat conduction, Thomson had constructed a mathematical analogy between electrostatic induction and heat conduction. Instead of forces acting at a distance over empty space, he viewed electrical action mathematically as represented by a series of geometrical lines or 'surfaces of equilibrium' intersecting at right angles with the lines of force. These surfaces would later be called equipotential lines or surfaces. At each stage he correlated the mathematical forms in thermal and electrical cases, but avoided any physical inferences about the nature of electricity as an actual contiguous action like fluid flow.[32]

The analogy enabled Thomson to reformulate the action at a distance mathematical theory of electricity (developed by Poisson and employed in Robert Murphy's Cambridge textbook on electricity) into Faraday's theory of contiguous action, though without Faraday's quantity–intensity distinction. In the analogy, force at a point was analogous to temperature gradient while specific inductive capacity of a dielectric was analogous to conductivity. Over the next decade or so Thomson would search for the mechanism of propagation, perhaps in terms of an elastic solid model such as that used to explain the wave nature of light, or in terms of a hydrodynamical model which would show not only electricity, magnetism, and heat, but ponderable matter itself, to result from the motions of an all-pervading fluid medium or ether. This quest for a unified field theory acquired special urgency once Thomson adopted a dynamical theory of heat around 1850, and would remain with him until his death in 1907.[33]

Most of Thomson's early electrical papers had been published in the *Cambridge Mathematical Journal* which he himself edited from 1845 until 1852. His strenuous efforts to expand the *CMJ* into a national journal for mathematical sciences were marred by the stubborn preponderance of contributions from pure mathematicians and correspondingly few papers on physical subjects.[34] With few converts to his own style of electrical science, Thomson was inspired by Maxwell's enthusiasm to respond with a long letter symptomatic of his own depth of interest. Just as Joule, Thomson and Rankine had initially traded credibility on the topic of heat engines, so now with respect to electrical science was the mutual credibility of Thomson and Maxwell on the increase.

In return Maxwell expressed his appreciation: 'As I wish to study the growth of ideas as well as the calculation of forces, and I suspect from various statements of yours that you must have acquired your views by means of certain conceptions which I have found great help, I will set down for you the confessions of an electrical freshman'. First, Maxwell explained that he had 'got up the fundamental principles of tension [electrostatics] easily enough' and had been 'greatly aided by the analogy of the conduction of heat, wh: I believe is your invention at least I never found it elsewhere'.[35] So enamoured of this method was Maxwell that twelve months later he opened his first electrical paper, 'On Faraday's Lines of Force', by stating:

> The method pursued in this paper is a modification of that mode of viewing electrical phænomena in relation to the theory of the uniform conduction of heat, which was first pointed out by Professor W. Thomson in the Cambridge and Dublin Mathematical Journal . . . Instead of using the analogy of heat, a fluid, the properties of which are entirely at our disposal, is assumed as the vehicle of mathematical reasoning.[36]

Second, Maxwell confessed privately to Thomson that he had been reading Ampère's investigations on the attractions between current-carrying wires. Unlike Laplacian physics these inverse square forces acted between current elements rather than point atoms. Ampère had formulated a new law for electromagnetic force and a new theory in terms of electrical fluid circulating around the molecules in magnetic substances (ch.4). Maxwell 'greatly admired' these investigations but thought there was a showy quality about them 'got up, after Ampère had convinced himself, in order to suit his views of philosophical inquiry, and as an example of what it ought to be'.[37]

Third, he noted that Thomson had spoken of 'magnetic lines of force'. By contrast to people who 'seem to prefer the notion of attractions of elements of current [acting] directly', Faraday 'seems to make great use' of these lines of force. Maxwell therefore 'thought that as every current generated magnetic lines & was acted on in a manner determined by the lines thro wh: it passed that something might be done by considering "magnetic polarization" as a property of a "magnetic field" or space and developing the geometrical ideas according to this view'. By the word 'polarization' he aimed to express 'the fact that at a point of space the south pole of a small magnet is attracted in a certain direction with a certain force'.[38] Here he had begun employing the new language of 'magnetic field' (publicly introduced by Faraday in 1846), 'field of force' (used by Thomson in a letter to Faraday in 1852), and 'lines of magnetic force' (introduced by Thomson and Faraday in 1851–2). As with the language of energy, these unorthodox terms embodied a new way of thinking about electrical phenomena.[39]

By the spring of 1855 Maxwell was telling his father that he was 'working away at electricity again' and that he had been 'working my way into heavy German writers'. To Thomson he wrote of his specific reading of Weber's electrodynamic measurement determinations (*Electrodynamische Maasbestimmungen*): 'I have been examining his mode of connecting electrostatics with electrodynamics, induction &c & I confess I like it not at first'. By September his attitude towards Weber's theory had hardened: 'I have . . . been working at Weber's theory of Electro Magnetism as a mathematical speculation which I do not believe but which ought to be compared with others and certainly gives many true results at the expense of several startling assumptions'. While he would later acknowledge that the value of

Weber's researches, 'both experimental and theoretical, renders the study of his theory necessary to every electrician', Maxwell and his 'North British' colleagues would never feel at home with Weber's electrical doctrines on account of their seeming violation of energy conservation.[40] With repect to action-at-a-distance in general, he explained to Mark Pattison (rector of Lincoln College, Oxford) in 1868:

> I cannot admit any theory which considers matter as a system of points which are centres of force acting on other similar points, and admits nothing but these forces. For this does not account for the perseverance of matter in its state of motion and for the measure of matter . . . there is a desire among men to explain action apparently at a distance by the intermediate action of a medium and then to explain the action of the medium as much as possible by its motion, and so to reduce Potential Energy to a form of Kinetic Energy.[41]

As we shall see, Maxwell's participation from the mid-1850s in the British trend away from action-at-a-distance was strongly shaped by his growing commitment to the science of energy.

Three months prior to reading 'On Faraday's Lines' to the Cambridge Philosophical Society on 10 December 1855, he told Thomson that 'I would be much assisted by your telling me whether you have not the whole draught of the thing lying in loose papers and neglected only till you have worked out Heat or got a little spare time'. He suggested that Thomson no doubt had the mathematical part of the theory in his desk and that all he had to do was to explain his results for electricity. And he also playfully professed not to 'know the Game laws & Patent laws of science' but expressed his intention 'to poach among your electrical images, and as for the hints you have dropped about the "higher electricity", I intend to take them'.[42] Pre-empting the possibility that his Cambridge peers might accuse him of lifting his ideas too directly from Thomson, the future laird of Glenlair moved quickly to the first public presentation of his own electrical views.

Published eventually in the *Cambridge Philosophical Transactions* in 1856, 'On Faraday's Lines' was directed largely at winning scientific credibility from his Cambridge mathematical peers. 'The present state of electrical science seems peculiarly unfavourable to speculation', he began. Prefacing the paper with a methodological introduction, he emphasized that we must 'discover some method of investigation which allows the mind at every step to lay hold of a clear physical conception, without being committed to any theory founded on the physical science from which that conception is borrowed, so that it is neither drawn aside from the subject in pursuit of analytical subtleties, nor carried beyond the truth by a favourite hypothesis'.[43] Maximum consensus and minimum controversy would be achieved through a geometrical and analogical style, with an absence of speculative theorizing.

Such methodological introductions usually tell historians of science less about the path by which scientific actors arrive at their theories, and much more about the manner in which they wish to have their arguments interpreted by specific audiences. Written by a young Trinity College don for an audience representing (since the foundation of the Cambridge Philosophical Society in 1819) the University's mathematical and scientific establishment, 'On Faraday's Lines' was designed to appeal to an older generation of Cambridge mathematical reformers, notably William Whewell, who had advocated geometrical reasoning over analytical subtleties as the pedagogical core of the University's 'liberal education'. Anxious to clear a distinctive space for his own work amid the jungle of contemporary electrical science, Maxwell complained that until now the student had to 'make himself familiar with a considerable body of most intricate mathematics, the mere retention of which in the memory materially interferes with further progress'.[44]

Furthermore, this career-making paper belonged to a strong Cambridge 'kinematical' research tradition (exemplified by the hydrodynamical and optical papers of Stokes and the physical geology of William Hopkins) which regarded the formulation of geometrical laws as the prerequisite to mathematical dynamical theory. Indeed, it was a style which had come to be regarded as the hallmark of the mathematical (Section A) elite of the BAAS whereby accurate geometrical mapping of the 'field' was a necessary preliminary to unifying mathematical theory.[45]

Endeavouring 'to lay before the mind of the geometer a clear conception of the relation of the lines of force to the space in which they are traced', Maxwell claimed to 'have found the geometrical significance of the "Electrotonic State" [Faraday's term for an electrical state of tension], and [to] have shown how to deduce the mathematical relations between the electrotonic state, magnetism, electric currents, and the electromotive force, using mechanical illustrations to assist the imagination, but not to account for the phenomena'.[46] Thus 'On Faraday's Lines' presented a 'kinematical' point of view, with reference only to the geometry of motion of an imaginary fluid and without reference to causal mechanisms. For a young don it was a safe and sure way of gaining credibility from his elite Cambridge audience, a credibility that he would very soon exchange for a Scottish chair of natural philosophy.

In their respective investigations which now formed the core of North British opposition to continental theorists, both Thomson and Maxwell were showing how action-at-a-distance forces could be translated into the preferred *flow* conceptions. As Maxwell noted in the introduction to 'Faraday's Lines', with respect to the 'laws of the conduction of heat in uniform [continuous] media . . ., [the] word *force* is foreign to the subject'. Again, both men understood very well how the Laplace-Poisson treatment of problems in gravitational and electrical attraction had been conceptually transformed into the 'potential function' by Green and Gauss, a move which represented to Thomson and Maxwell a distinct shift from microscopic point atoms acting at a distance to macroscopic geometrical 'fields' mapped by equipotential lines and potential gradients.[47] With kinematical foundations secure, Maxwell could now prepare for a more adventurous 'dynamical' stage of investigation in which action-at-a-distance forces would yield to energy distributions in the field.

At the conclusion of his extensive series 'On the Dynamical Theory of Heat', Thomson had submitted a short paper, 'Dynamical Illustrations of the Magnetic and the Helicoidal Rotatory Effects of Transparent Bodies on Polarized Light', to the Royal Society of London in June 1856. His principal concern was to offer a dynamical illustration of Faraday's 'magneto-optic effect', namely that a beam of polarized light directed along lines of magnetic force rotated in a right-handed manner, while a beam in the opposite direction rotated in a left-handed manner. Thomson inferred that these magnetic rotations depended, not upon spiral structures within the transparent body, but 'on the direction of the moving particles' about the lines of force.[48]

Thomson insisted that the Faraday effect demonstrated that magnetism comprised actual rotations of matter which in turn offered 'a definition of magnetization in the dynamical theory of heat'. Magnetism was thus to be understood simply as a net alignment of the rotational motions which constituted heat. Furthermore, he issued a challenge to all investigators of electromagnetic phenomena:

> The explanation of all phenomena of electro-magnetic attraction or repulsion, and of electro-magnetic induction, is to be looked for simply in the inertia and pressure of the matter of which the motions constitute heat. Whether this matter is or is not electricity, whether it is a continuous fluid interpermeating the spaces between molecular nuclei, or is itself molecularly grouped; or whether all matter is continuous . . . it is impossible to decide, and perhaps in vain to speculate, in the present state of science.[49]

Fired by this challenge, Maxwell told an old Cambridge friend, C.J. Monro, in May 1857 that he had been 'grinding at many things and lately during this letter at a Vortical theory of magnetism & electricity which is very crude but has some merits, so I spin & spin'. But Maxwell was not simply drawing upon Thomson's theoretical speculations. In a letter to Thomson at the end of 1856, he had referred to his cousin William Cay who had been working in Belfast with James Thomson, enthusiast for horizontal waterwheels or vortex turbines, adding that he had himself 'made a splendid vortex lately quite smooth and 7 inches deep in the middle'. He also told Monro of the latest projects for a mill race and drainage system at Glenlair. And in November 1857 he informed his Aunt Jane that he had received 'a letter from Willy [Cay] to-day about jet pumps to be made for real drains, but not saying anything about the Professorship of Engineering' at Queens College, Belfast, which James Thomson was soon to secure.[50]

Uniting these threads of speculative theory and practical engineering, the doctrines of energy increasingly attracted Maxwell's attention. In late 1856 he told Thomson that he had 'been looking at Rankine's Thermodynamic diagrams [the diagrams of energy]' (ch.8). By June 1857 he informed Monro that he had been 'practising my hand of write [sic] at exposition of fundamentals . . . Theory of Angular Momentum & Couples of Work done & Vis Viva, of Actual & Potential Energy [Rankine's terms], with continual jaw on the doctrine of measurement by units all through'.[51] One such draft opened with the announcement: 'Sketch of an Introduction to the Higher Parts of Mechanics, being the Science of Energy as it relates to the Motion and Rest of a Single

Particle, of Two Particles or of a system of Particles, this system being either of invariable form, elastic or fluid'.[52]

Five months later Faraday sent Maxwell a copy of his 'On the Conservation of Force'. In a lengthy and positive reply Maxwell urged that we 'keep our words for distinct things more distinct'. Thus 'Energy is the power a thing has of doing work arising either from its own motion or from the "tension" subsisting between it and other things' while 'Force is the tendency of a body to pass from one place to another and depends upon the amount of change of "tension" which that passage would produce'.[53] Faraday, however, denied that he was employing 'force' in this sense, insisting that 'What I mean by the word is the *source* or *sources* of all possible actions of the particles or materials of the universe: these being often called the *powers* of nature'.[54] But Maxwell had already seen in Faraday's perspective a credible framework for a science of energy in opposition to a science of action-at-a-distance forces:

> Now as far as I know you [Faraday] are the first person in whom the idea of bodies acting at a distance by throwing the surrounding medium into a state of constraint has arisen, as a principle to be actually believed in. We have had streams of hooks and eyes flying around magnets, and even pictures of them so beset, but nothing is clearer than your description of all sources of force keeping up a state of energy in all that surrounds them, which state by its increase or diminution measures the work done by any change in the system. You seem to see the lines of force curving round obstacles and driving plump at conductors and swerving towards certain directions in crystals, and carrying with them everywhere the same amount of attractive power spread wider or denser as the lines widen or contract.[55]

The letter concluded with a direct reference to 'questions relating to the connexion between magneto electricity and certain mechanical effects which seem to me opening up quite a new road to the establishment of principles in electricity and a possible confirmation of the physical nature of magnetic lines of force. Professor W. Thomson seems to have some new lights on this subject'.[56] In the same month Maxwell discussed with Thomson the question of irreversibility in a perfect fluid which also related closely to molecular vortices: 'I do not see why it makes much difference whether these eddies are soon converted into heat, or remain in a state of subdivision which is as nearly that of molecular vortices as any finite motion can be' (ch.12).[57]

The merger in 1860 of Aberdeen's Marischal College and King's College deprived Maxwell of his chair of natural philosophy. Failing in his bid against Tait for Edinburgh (ch.9), Maxwell was appointed to the vacant chair of natural philosophy at King's College, London, a post which he held until Easter of 1865.[58] In his inaugural lecture (October 1860) he told his academic audience that 'there are certain general laws, regulating the amount of Energy arising from given conditions, and determining the total effect of the forces called into play, which are among the most important conclusions of physical science'. Invoking Rankine's term, Maxwell stated that the 'science founded on these laws is called Energetics'.

Moreover, the 'application of these principles to natural phenomena is the special research which the present state of science points out as that from which the greatest results are to be expected in the coming age'.[59]

When Maxwell began to write 'On Physical Lines of Force', he chose to address the wider readership of the *Phil. Mag.* ('which publishes a good deal about the dynamical theories of matter and heat'[60]) rather than to continue an association with the more restricted academic audience of the Cambridge Philosophical Society. Published in four instalments (1861–2), 'On Physical Lines' aimed 'to clear the way for speculation' in the direction of understanding the *physical* nature (rather than simply the geometrical form) of lines of magnetic force. In particular Maxwell pronounced himself 'dissatisfied with the explanation founded on the hypothesis of attractive and repellent forces directed towards the magnetic poles, even though we may have satisfied ourselves that the phenomenon is in strict accordance with that hypothesis, and we cannot help thinking that in every place where we find these lines of force, some physical state or action must exist in sufficient energy to produce the actual phenomena'. His investigation focused on 'the mechanical results of certain states of tension and motion in a medium, and comparing these with the observed phenomena of magnetism and electricity'.[61]

Reviewing 'Physical Lines' a couple of years later in 'Historical Sketch of the Science of Energy', Tait explained that Maxwell had given a 'general vortex theory of magnetism, with special assumptions regarding the nature of electric currents, etc., and founded on the conservation of energy'.[62] Although in part I Maxwell employed the language of 'mechanical effect' and 'work done' rather than energy, it was in part II that he began introducing Rankine's 'actual' and 'potential energy'. Emphasizing that he had there attempted to imitate electromagnetic phenomena 'by an imaginary system of molecular vortices', he issued a subtle challenge to his opponents: 'Those who look in a different direction for the explanation of the facts, may be able to compare this theory with that of the existence of currents flowing freely through bodies, and with that which supposes electricity to act at a distance with a force depending on its velocity, and therefore not subject to the law of conservation of energy'. Weber especially would have to answer for his theory's seeming violation of energy conservation.[63]

Part I, 'The Theory of Molecular Vortices applied to Magnetic Phenomena', postulated a fluid medium which in a magnetic field would be filled with many small vortex tubes or filaments 'having their axes in the direction of the lines of force, and having their direction of rotation determined by that of the lines of force'. The intensity of the magnetic field related to the angular velocity of the molecular vortices. The rotations produced centrifugal forces which caused the vortex filaments simultaneously to expand laterally and contract longitudinally. This action was consistent with Faraday's account of the tendency of magnetic field lines to repel each other and to shorten, producing a force of attraction between unlike poles.[64]

Part II, 'The Theory of Molecular Vortices Applied to Electric Currents', extended the model to electricity. Motivated by an engineering perspective inspired by 'energetic' friends such as Rankine and the Thomson brothers,

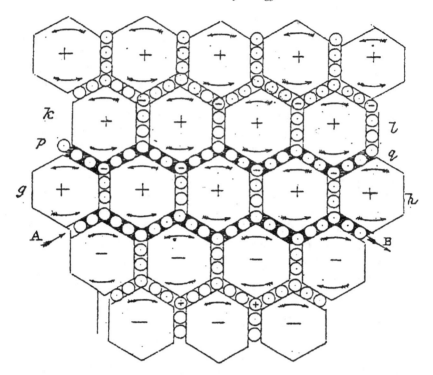

Figure 11.1 Maxwell's original 1861 representation of his theory of molecular vortices applied to electric currents. The layer of small spherical particles acted as 'idle wheels' whose transference in one direction (as distinct from their rotation) came to represent electric current (Maxwell 1861–2).

Maxwell placed small spherical particles between the vortices: 'a layer of particles, acting as idle wheels, is interposed between each vortex and the next, so that each vortex has a tendency to make the neighbouring vortex revolve in the same direction with itself' (Figure 11.1).[65] The movement or transference of these 'idle wheels' came to represent the electric current. This revised model led Maxwell to introduce into his paper a language of energy and to focus sharply on energy conservation and 'loss':

> The [moveable] particles must be conceived to roll without sliding between the vortices which they separate, so that, as long as they remain within the same molecule, there is no loss of energy by resistance. When, however, there is a general transference of particles in one direction, they must pass from one molecule to another, and in doing so, may experience resistance, so as to waste electrical energy and generate heat.[66]

The moveable particles played the part of electricity: 'Their motion of translation constitutes an electric current, their rotation serves to transmit the motion of the vortices from one part of the field to another, and the tangential pressures thus called into play constitute electromotive force'.[67]

At the same time, however, he admitted to the idle-wheel hypothesis as 'somewhat awkward' and of 'provisional and temporary character'. While he emphasized that he had not brought it forward as 'a mode of connexion existing in nature', it was 'a mode of connexion which is mechanically conceivable, and easily investigated'. Concerned to offer a *possible* explanation in terms of a continuous mechanism in opposition to action-at-a-distance force models, he later told Thomson that the 'tendency in my rotatory theory of magnetism was towards the to me inconceivable and no doubt to the misty . . . Perhaps the eminent London scavengers Mess[rs] Cleavers and Mist might find a weapon to combat the tendency'. And he explained to Tait that the vortex theory 'is built up to shew that the phenomena are such as can be explained by mechanism. The nature of this mechanism is to the true mechanism what an orrery is to the Solar System.'[68]

In Part III, 'The Theory of Molecular Vortices Applied to Statical Electricity', Maxwell attempted to resolve some of the problems of transmission of motion by endowing the material with the properties of an elastic solid, properties which in turn came to represent the electrostatic field. Given the admission of elasticity to account for transverse vibrations in the luminiferous medium (according to the undulatory theory of light), such a representation, he noted, would not be surprising. Comparing a conducting body 'to a porous membrane which opposes more or less resistance to the passage of a fluid', he suggested that a dielectric 'is like an elastic membrane which may be impervious to the fluid, but transmits the pressure of the fluid on one side to that of the other'.[69]

When acted upon by an electromotive force, a dielectric experienced a state of polarization of its parts which, like magnetic polarization, could be described 'as a state in which every particle has its poles in opposite directions [as in a bar magnet]'. Maxwell therefore argued that the 'effect of this action on the whole dielectric mass is to produce a general displacement of the electricity in a certain direction'. Although this displacement, when attained, did not amount to a current, its *variations* 'constitute currents in the positive or negative direction, according as the displacement is increasing or diminishing'. Furthermore, while the relation of displacement to the electromotive force and to the nature of the body were 'independent of any theory about the internal mechanism of dielectrics', he explained that 'we cannot help regarding the phenomena as those of an elastic body, yielding to a pressure, and recovering its form when the pressure is removed'.[70]

The molecular vortex theory now admitted of yet more modification. As before 'the magnetic medium is divided into cells, separated by partitions formed by a stratum of particles which play the part of electricity'. But when the particles were urged in any direction, their tangential action on the cells would 'distort each cell, and call into play an equal and opposite force arising from the elasticity of the cells'. Conversely, when 'the force is removed, the cells will recover their form, and the electricity return to its former position'. Employing energy methods, Maxwell considered the relation between the displacement and the force producing it, assuming the cells to be spherical. Using electromagnetic units throughout, he derived expressions for 'the energy of the medium arising from the electric displacement' involving restoring forces and displacements.[71]

From these energy expressions Maxwell deduced, as an inverse square law, the magnitude of the forces acting between two charged bodies. Comparing this 'force law' with its familiar counterpart in electrostatic measure (Coulomb's law) enabled a direct relation to be established between 'the statical and dynamical measures of electricity'. In particular he found that 'the quantity E [a constant] ... is the number by which the electrodynamic measure of any quantity of electricity must be multiplied to obtain its electrostatic measure'.[72]

Maxwell then made the dramatic assertion that he had shown, 'by a comparison of the electro-magnetic experiments of MM. Kohlrausch and Weber with the velocity of light as found by M. Fizeau, that the elasticity of the magnetic medium in air is the same as that of the luminiferous medium, if these two coexistent, coextensive, and equally elastic media are not rather one medium'. In other words he had calculated, on the basis of his extended molecular vortex model and data obtained from Weber and Kohlrausch, the theoretical velocity of transverse undulations in the 'magnetic medium'. This velocity, he reiterated, 'agrees so exactly with the [experimentally measured] velocity of light ... that we can scarcely avoid the inference that *light consists in the transverse undulations of the same medium which is the cause of electric and magnetic phenomena*'. Such a radical claim would form the core of Maxwell's 'electromagnetic theory of light'.[73]

Correspondence with his contemporaries, however, suggests that Maxwell was not satisfied with the persuasive power of his theory of molecular vortices. Even before completion of the four instalments, Monro expressed one such credibility gap:

> The coincidence between the observed velocity of light and your calculated velocity of a transverse vibration in your medium seems a brilliant result. But I must say I think a few such results are wanted before you can get people to think that, every time an electric current is produced, a little file of particles is squeezed along between rows of wheels. ... [However] I admit that the possibility of convincing the public is not the question.[74]

Convincing his scientific peers, however, *was* the question. Already in 1861 the pages of the *Phil. Mag.* were beginning to fill with a potentially damaging dispute between Maxwell and the Cambridge professor of astronomy, James Challis (1803–82), who vigorously defended his own hydrodynamical approach to electromagnetism against the perceived attack from the London professor.[75]

A 'dynamical' theory of the electromagnetic 'field'

By December 1861 Maxwell was writing privately to an old friend, H.R. Droop, that he was 'trying to form an exact mathematical expression for all that is known about electro-magnetism without the aid of hypothesis'.[76] Over the next three

years Maxwell's primary concern with accurate experimental measurements, especially those conducted at King's College on behalf of the BAAS Committee
(ch.13), reinforced his ambition to place his electromagnetic theorizing upon a
more credible foundation. Thus he wrote to his co-worker in electrical measurement, Charles Hockin, in September 1864 that he had worked out a theory of
residual charge in dielectrics which very much wanted 'determinations of the
specific induction, electric resistance, and absorption of good dielectrics, such as
glass, shell-lac, gutta-percha [the material used to insulate underwater cables]' and
that he had 'also cleared the electromagnetic theory of light from all unwarrantable
assumption, so that we may safely determine the velocity of light by measuring the
attraction between bodies kept at a given difference of potential, the value of which
is known in electromagnetic measure'.[77]

Concern about credibility, then, formed a key motivation for 'A Dynamical
Theory of the Electromagnetic Field', published in the Royal Society's *Phil. Trans.*
(1865) as Maxwell's third substantial paper on electricity and magnetism. Once
again audience was fundamental to the shaping of the memoir. As he told Tait in
1867, the paper departed from the style of 'Physical Lines': it was 'built on
Lagrange's Dynamical Equation [sic] and is not wise about vortices'.[78] Seeking
once again to go beyond a kinematical, geometrical description of electromagnetic
phenomena, Maxwell turned to a distinctive style of 'dynamical' theory which had
found recent exposition in the optical and hydrodynamical investigations of the
Lucasian professor at Cambridge, Stokes.

It was therefore no coincidence that Maxwell wrote at length to Stokes, then
Secretary to the Royal Society, in October 1864. The principal subject of the letter
was the reflection and refraction of light, particularly with respect to the conditions at bounding surfaces: 'I am not yet able to satisfy myself about the conditions
to be fulfilled at the surface except of course the condition of the conservation of
energy', Maxwell noted early in the letter. Stokes's celebrated investigations of the
undulatory theory of light from the point of view of 'dynamical theory' had since
become a standard feature of Cambridge mathematical science, incorporating a
macroscopic, non-hypothetical approach to physical theory which avoided reference to invisible, microscopic entities such as point atoms. Maxwell reported to
Stokes that he had 'now got materials for calculating the velocity of transmission of
a magnetic disturbance through air founded on experimental evidence, without
any hypothesis about the structure of the medium or any mechanical explanation
of electricity or magnetism'. Here, then, was a strong hint that the approach of the
new paper would be very different from the elaborate mechanical hypotheses of
'Physical Lines'. Even the title paid deference to the characteristic style of the
Lucasian professor.[79]

Having announced to Stokes that he had written out some of the theory and
would send it to the Royal Society in a week's time, Maxwell reported that
Thomson, the telegraph engineer Fleeming Jenkin (whom Lewis Gordon had
introduced to Thomson early in 1859) and Maxwell himself were 'devising
methods to determine this velocity = electromagnetic/electrostatic unit of electricity'. Thomson was 'going to weigh an electromotive force', Jenkin and Maxwell

were 'going to measure the capacity of a conductor both ways', and Maxwell had 'a plan of direct equilibrium between an electromagnetic repulsion and electrostatic attraction' (ch.13).[80]

In the short version read to the Royal Society in December 1865, Maxwell opened with the statement that the 'proposed [Dynamical] Theory seeks for the origin of electromagnetic effects in the medium surrounding the electric or magnetic bodies, and assumes that they act on each other not immediately at a distance, but through the intervention of this medium'.[81] The *Phil. Trans.* version explained further that the theory was called 'a theory of the *Electromagnetic Field* because it has to do with the space in the neighbourhood of the electric or magnetic bodies', the electromagnetic field being defined as 'that part of space which contains and surrounds bodies in electric or magnetic conditions'. Moreover, 'it may be called a *Dynamical* Theory, because it assumes that in that space there is matter in motion, by which the observed electromagnetic phenomena are produced'.[82] In this sense 'dynamical theory' for Maxwell followed the by now well-established course set by Stokes, Thomson and others from the late 1840s whereby specific mechanisms (such as point atoms or molecular vortices) yielded to very general assumptions of matter in motion.

Maxwell also continued his strategy of introducing the theory in the context of rival theories which he could represent as far less credible. In general these theories assumed 'the existence of substances the particles of which have the property of acting on one another at a distance by attraction or repulsion'. Above all the 'exceedingly ingenious and wonderfully comprehensive' theory of Weber 'comes to us with the more authority, as it has served to guide the speculations of one who has made so great an advance in the practical part of electric science, both by introducing a consistent system of units in electrical measurement, and by actually determining electrical quantities with an accuracy hitherto unknown'. But the assumption of particles acting at a distance with forces dependent on their velocities raised such mechanical difficulties in Maxwell's mind that he could not consider Weber's theory 'an ultimate one' even though it might serve a useful role in the coordination of phenomena.[83] His audience had thus been prepared for a 'dynamical theory' which would avoid such difficulties.

In his presentation to the Royal Society, Maxwell initially justified the probable existence of the medium on the grounds that 'the investigations of Optics have led philosophers to believe that in such a medium the propagation of light takes place'. In order to explain the propagation of light, two general properties had been attributed to the medium: first, that 'the motion of one part communicates motion to the parts in its neighbourhood'; and second, that 'this communication is not instantaneous but progressive, and depends on the elasticity of the medium as compared with its density'. Furthermore, the 'kind of motion attributed to the medium when transmitting light is that called transverse vibration'.[84] Maxwell's use of the term 'progressive' in the context of a universal medium recalls Thomson's reference to 'an endless progress, through an endless space' three years previously (ch.7) as well as his much earlier private commitment to the doctrine that 'Everything in the material world is progressive' (ch.6). As we shall

see, both natural philosophers tended to the view that 'progression', rightly interpreted, made for a perfect creation while the concomitant 'dissipation' characterized the imperfections of human beings rather than nature (ch.12).

Expanding his remarks in the *Phil. Trans.*, Maxwell admitted that the electromagnetic field might be filled with any kind of matter. Nevertheless there
was a fundamental difference between light waves and sound waves. The
removal of matter such as air was well known to produce a diminution in the
transmission of sound which would ultimately cease in a vacuum. Yet, regard-
less of however much we remove ordinary gross matter, there always remained
enough matter to receive and transmit the undulations of light and heat. Conversely, such transmission was not greatly altered when we substituted transparent bodies of gross matter for a vacuous space. Thus 'we are obliged to admit
that the undulations are those of an aethereal substance, and not of the gross
matter, the presence of which merely modifies in some way the motion of the
aether'. Filling space and permeating bodies, this etherial medium was 'capable
of being set in motion and of transmitting that motion from one part to
another, and of communicating that motion to gross matter so as to heat it and
affect it in various ways'.[85]

The etherial medium was therefore the means by which energy was transmitted
between gross bodies. In the case of radiant heat, for example, 'the undulations had
left the source of heat some time before they reached the body, and during that
time the energy must have been half in the form of motion of the medium [actual
energy] and half in the form of elastic resilience [potential energy]'. Drawing upon
Thomson's earlier considerations of the sun's energy (ch.7), Maxwell explained
that from such 'considerations Professor W. Thomson has argued that the medium
must have a density capable of comparison with that of gross matter, and has even
assigned an inferior limit to that density'.[86]

The tactic here was to embed what was potentially an exceedingly hypothetical
assumption, that of the existence of an ethereal medium, in the frameworks of
contemporary science ranging from the (by now) highly credible undulatory
theory of light to the much more recent energy cosmology. With respect to light he
also invoked the methodological views of Newton and Herschel on the doctrine of
verae causae ('true causes' in nature), and claimed that the medium to be employed
in electromagnetism had been obtained as a known entity from a hitherto
independent branch of science (light) and had thus not been introduced as an
arbitrary hypothesis serving only one branch of physics.[87]

Maxwell proceeded to argue that such a medium was demonstrably capable of
other kinds of motion and displacement than those associated with the phenomena
of light and radiant heat. These effects included the rotation of the plane of
polarized light by a magnetic field (the 'Faraday effect') as well as the effects of
electromotive forces in conductors and dielectrics. This ethereal medium formed
the basis of 'a complicated mechanism capable of a vast variety of motion, but at
the same time so connected that the motion of one part depends, according to
definite relations, on the motion of other parts, these motions being communicated
by forces arising from the relative displacement of the connected parts, in virtue

of their elasticity'. Any such mechanism, then, 'must be subject to the general laws of Dynamics'.[88]

Exploring the electromagnetic field through the phenomena of induction and attraction of currents, and mapping the distribution of magnetic fields, Maxwell sought to express the results in the form of 'the General Equations of the Electromagnetic Field', requiring at this stage some twenty equations in total involving twenty variable quantities: electric currents by conduction, electric displacements, total currents, magnetic forces, electromotive forces, electromagnetic momenta (each with three components), free electricity and electric potential.[89]

Maxwell attempted to express in terms of these quantities what he now named 'the intrinsic energy of the Electromagnetic Field as depending partly on its magnetic and partly on its electric polarization at every point'. He also made clear that he wanted his readers to view 'energy' as a literal, real entity and not simply a concept for dynamical illustration:

> In speaking of the Energy of the field. . . . I wish to be understood literally. All energy is the same as mechanical energy, whether it exists in the form of motion or in that of elasticity, or in any other form. The energy in electromagnetic phenomena is mechanical energy. The only question is, Where does it reside? On the old theories it resides in the electrified bodies, conducting circuits, and magnets, in the form of an unknown quality called potential energy, or the power of producing certain effects at a distance. On our theory it resides in the electromagnetic field, in the space surrounding the electrified and magnetic bodies, as well as in those bodies themselves, and is in two different forms, which may be described without hypothesis as magnetic polarization and electric polarization, or, according to a very probable hypothesis, as the motion and the strain of one and the same medium.[90]

Writing the Treatise on Electricity and Magnetism

'Maxwell has given a general dynamical theory of the electromagnetic field, according to which he shows the mutual interdependence of the various branches of the science, and lays down equations sufficient for the theoretical solution of any electrical problem'. Thus opened a lead paper in the *Phil. Mag.* (1869). The author was the Hon. J.W. Strutt, then a Fellow of Trinity College, Cambridge, later Lord Rayleigh, and ultimately Maxwell's successor as Cavendish professor of experimental physics. Aware that Maxwell was already well advanced with a major *Treatise*, Strutt also highlighted a contemporary demand for something more than 'scattered papers' to serve the needs of the Cambridge Mathematical Tripos. From the point of view of dynamics, though not for the whole range of natural philosophy, that need had been largely met by Thomson and Tait. It was therefore late in 1867, the year of the first and only volume of Thomson and Tait's *Treatise*, that we see the task, for electricity and magnetism, passing to Maxwell. As Tait expressed his feelings in November: 'I am delighted to hear you are going to do a Senate-House Treatise on Electricity. The sooner the better'.[91]

Maxwell's recent departure from King's College for the more leisured life at Glenlair had left him the requisite freedom to prepare a massive *Treatise on Electricity and Magnetism*. In December 1867 Tait, evidently unimpressed by Maxwell's preference for the remote country of Galloway over the polite culture of Scotland's capital city, wrote to remind him that he was a Fellow of the Royal Society of Edinburgh and to exhort him to 'Ponder this proposition' that 'A man of your *originality*, and *fertility*, and *leisure* is undoubtedly [duty] bound to furnish to the chief Society of his native land, numerous papers, however short'. But Tait recognized that his friend had a more important project in hand: 'Nevertheless I hope your Treatise on Electricity will go on soon'. By way of encouragement, Tait reported that, after only three months on the market, '*Half* the edition of our first vol. of Nat. Phil. is already sold'. Five days later Maxwell signalled his approval: 'I am glad people are buying T&T'. May it sink into their bones'.[92]

Challenged by the immense task of producing single-handedly a treatise on a par with that of Thomson and Tait, the 36-year-old laird of Glenlair prepared to transform his own 'scattered papers' into a marketable text, to incorporate the immense range of recent electrical research, and to differentiate persuasively his field approach from that of some very powerful advocates of action-at-a-distance or charge–interaction theories. Early in 1868, therefore, Maxwell told Tait of his respect for, and regret at the recent death of, Bernhard Riemann (1826–66), Gauss's successor at Göttingen:

> I now have him [Riemann] more distinct. Weber says that electrical force depends on the distance and its 1st & 2nd derivatives with respect to t [time]. Riemann says that this is due to the fact that the potential at a point is due to the distribution of electricity elsewhere not at that instant but at times before depending on the distances. In other words potential is propagated through space at a certain rate and he actually expresses this by a partial diff[erential] eq^n appropriate to propagation.[93]

With respect to the credibility of Riemann's theory compared to his own, Maxwell offered two private comments. First, 'space contains a medium capable of dynamical actions which go on during transmission independently of the causes which excited them'. This interpretation 'is no more or less than my theory divested of particular assumptions'. Second, he considered Riemann's theory as 'a fact without any etherial substratum'. Taking two points X and Y at a distance *a* moving in the same direction, he then showed on Riemann's theory that the force of X on Y and that of Y on X would be such that 'if X & Y are connected by a rigid rod, X will be pulled forward more than Y is pulled back and the system will be a locomotive engine fit to carry you through space with continually increasing velocity. See Gulliver's Travels in Laputa'. The 'failure' of Newton's third law would thus lead to a perpetual motion machine producing unlimited motive power.[94]

This evaluation enabled Maxwell to offer Tait a compelling synopsis of the comparative credibilities of Weber's, Riemann's and Maxwell's theories of electromagnetism. For Weber 'action and reaction are equal but his energy is unreclaimable'. Thus Newton's third law was upheld, but conservation of energy failed. For Riemann 'action & reaction between the gross bodies are unequal and his energy is nowhere unless he admits a medium which he does not do explicitly'. Newton's third law failed (as in the above illustration) and energy apparently had no material location. But for Maxwell himself 'action & reaction are equal only between things in contact not between the gross bodies till they have been in position for a sensible time and my energy is and remains in the medium including the gross bodies which are among it'.[95] Here again we see the very profound significance that Maxwell himself attached to the ethereal medium for the material embodiment of the energy.

By August 1868 he reported to Tait that he believed the Clarendon Press in Oxford had accepted his projected *Treatise*, thus reinforcing the continuity with Thomson and Tait's volume. He also explained that he was in the process of 'writing out the Kinematic part (Ohm's law and the theory of conduction)'. Between August 1868 and November 1869 Maxwell conducted an intensive correspondence with William Thomson with respect to almost every technical aspect of the complex *Treatise*. Around October 1869 he had mapped out the contents of the work. As written and published, the *Treatise* would consist of four major parts: electrostatics, electrokinematics (electricity in motion), magnetism, and electromagnetism (originally electrodynamics). It was a mammoth undertaking by any standards. Each part displayed encyclopaedic knowledge, demanding an unsurpassed mastery of every experimental and theoretical feature of a vast scientific territory. Yet as early as December 1869 Maxwell indicated that he was 'at the 4th of the 4 parts of my book namely Electrodynamics'.[96]

Printing out of proofs and their circulation among close colleagues such as Tait and Thomson, less for 'proof-reading' than for comments upon the mathematical and scientific 'proofs' contained therein, was under way in the spring of 1871. Staying near Dunoon on the Clyde (his father-in-law's vacation residence) in October 1871, Maxwell informed Tait of some of the tribulations that had beset the *Treatise* author. William and James Thomson, together with Helmholtz and presumably some of Maxwell's 'proofs', had been becalmed in the Scottish Hebridean Islands aboard Thomson's yacht, the schooner *Lalla Rookh*: 'I got a post card from Gair Loch, Ross-shire which enticed me to Greenock [across the Firth of Clyde from Dunoon] at 4 in the morning on Friday. I enjoyed my own society then till 4 in the afternoon of Saturday. I thought I saw L[alla].R[ookh]. creeping up by Greenock last night (Monday)'. Equally facetiously, Maxwell would urge Tait in Biblical tones to 'Proove me my prooves and again I say reproove me my reprooves'.[97]

The very extent of its subject matter gave to the *Treatise* a disorderly appearance which masked Maxwell's concern to provide in the first three parts an accurate mapping of the disparate phenomena of electrostatics, electrokinematics (which included current electricity, thermoelectricity and electrolysis) and magnetism. This mapping would then provide the foundation for his dynamical theory of the electromagnetic field constructed in the later chapters of the fourth part. As Maxwell emphasized in the preface, he had 'taken the part of an advocate rather than that of a judge', choosing to exemplify the field approach rather than to offer an impartial description of both the field and action-at-a-distance theories. As always his great icon was Faraday who, unlike the Germans (Gauss, Weber, Riemann and others), 'saw a medium where they saw nothing but distance: Faraday sought the seat of the phenomena in real actions going on in the medium'.[98]

Coinciding with the final stages of the launch of Maxwell's *Treatise*, Sir William Thomson's Presidential Address at the Edinburgh meeting of the British Association (1871) sounded a note of caution with regard to Maxwell's electromagnetic theory of light and the claim for exact equality between v (the ratio of electrostatic to electromagnetic units) and the speed of light which had seemingly provided Maxwell with empirical credibility for that theory in 1861–2. Sir William acknowledged that Weber's measurement 'verifies approximately that equality' and reminded the Association that its Committee on Electric Measurement was engaged on the 'most accurate possible determination' of v. It was nonetheless 'premature to speculate as to the closeness of the agreement' between v and the velocity of light. Significantly, the *Treatise* referred to the fact that they were 'of the same order of magnitude' and claimed that the theory, which assigned a physical reason for the equality, 'is certainly not contradicted by the comparison of the [experimental] results such as they are'. Fed by a commitment to telegraphic theory and to an elastic solid model of ether, Thomson's caution would burgeon into a fully fledged scepticism after Maxwell's death, serving to create a misleading retrospective image of Thomson and Maxwell as disparate rivals in the history of nineteenth century physical theory.[99]

In the early 1870s heated debates over Weber's electrodynamic theory had been stoked by fresh contributions from leading participants, most notably Helmholtz, Maxwell and Weber himself. Published in Germany early in 1871 and translated by G.C. Foster for the *Phil. Mag.* (1872), Weber's sixth memoir under the same general title, 'Electrodynamic Measurements', related specifically to the principle of the conservation of energy. From the outset Weber made clear that, in the face of claims that his law of electrical action contradicted the energy conservation principle, he would now show that in fact his law 'enables us to make an addition to the Principle of Conservation of Energy, and to alter it so that its application to each pair of particles is no longer limited solely to the time during which the pair does not undergo either increase or diminution of *vis viva* through the action of other bodies, but always holds good independently of the manifold relations to other bodies into which the two particles can enter'.[100] But in November 1871 Maxwell told Tait that Helmholtz's latest critique of Weber had rendered Weber's theory incredible once more, being productive of 'astonishing consequences':

> [Weber's] force has a potential which involves the square of the relative velocity. Hence in any cyclic operation no work is spent or gained. So Conservation is conserved [that is, conservation of energy holds]. But Helmholtz has shown . . . that it is possible (by Weber's Law) to produce in a material

particle carrying electricity an infinite velocity in a finite space and finite time and it appears from the formula that forthwith it is hurled with this ∞ velocity into a region where by the formula the velocity is $\sqrt{-1}$.[101]

With such 'incredible' consequences deduced from rival theories, the way ahead now seemed clear for Maxwell's own formulation of an energy-based electromagnetic theory.

In the preliminary section and first three parts of the *Treatise*, however, 'energy' considerations were much less conspicuous than in Part IV. Part I, for example, considered the energy of an electrified system; Part II discussed both Joule's law of the heating effects of an electric current and the laws of electrolysis in relation to the principle of energy conservation; and Part III spoke of the potential energy of a magnet and of a magnetic shell placed in a field of force.[102] Throughout the *Treatise*, the theory of the 'potential' received much attention. On one reading, that of orthodox action-at-a-distance theorists, the potential meant 'the result of a summation of the electrified particles divided each by its distance from a given point'. But for Maxwell the potential had to be reinterpreted in terms of a quantity satisfying a certain partial differential equation and connoting the potential energy contained in the field.[103]

By the fourth chapter of Part IV, however, 'energy' began to emerge as the keystone of the whole edifice. Having discussed in the previous chapter Helmholtz and Thomson's use of 'energy conservation' (1847–51) to link electromagnetic induction to mechanical effect (chs 5 and 7), Maxwell now considered the case of phenomena 'exhibited by the current in a wire which forms the coil of an electromagnet', namely self-induction. At first sight electricity appeared to circulate in the wire with something like momentum, as water flowing in a pipe. For instance, a sudden break in the circuit manifested a great electromotive force in the form of shocks and sparks just as a break in a pipe exhibited a sudden jet of water issuing with a velocity much greater than that due simply to the head of water. But closer analysis showed that, unlike a pipe, the wire exhibited effects which differed not only according to its form, but which were altered by the presence of other bodies such as a second wire. Thus, Maxwell insisted, 'the momentum is certainly not that of electricity in the wire'. Nevertheless:

> It is difficult . . . for the mind . . . to abandon altogether the help of this analogy. . . . The fundamental dynamical idea of matter, as capable by its motion of becoming the recipient of momentum and of energy, is so interwoven with our forms of thought that, whenever we catch a glimpse of it in any part of nature, we feel that a path is before us leading, sooner or later, to the complete understanding of the subject.[104]

Shifting attention away from the problematic 'momentum', Maxwell now argued that 'a conducting circuit in which a current has been set up has the power of doing work in virtue of this current, and this power cannot be said to be something very like energy, for it is really and truly energy'. He therefore inferred that 'a system containing an electric current is a seat of energy of some kind; and since we can form no conception of an electric current except as a kinetic phenom-enon, its energy must be kinetic energy'. Once again, however, he insisted that 'the electricity in the wire cannot be considered as the moving body in which we are to find this energy, for the energy of a moving body does not depend on anything external to itself, whereas the presence of other bodies near the current alters its energy'. Instead, we were led 'to enquire whether there may not be some motion going on in the space outside the wire, which is not occupied by the electric current, but in which the electromagnetic effects of the current are manifested'.

Avoiding the issue of the specific location and nature of these motions, Maxwell proposed to follow and enhance his 1865 'Dynamical Theory' by examining 'the consequences of the assumption that the phenomena of the electric current are those of a moving system, the motion being communicated from one part of the system to another by forces, the nature and laws of which we do not yet even attempt to define, because we can eliminate these forces from the equations of motion by the method given by Lagrange for any connected system'. He therefore devoted the following chapter to a full exposition of the equations of motion of a connected system. But Maxwell's approach was that of his fellow scientists of energy rather than Lagrange and Hamilton:

> The aim of Lagrange was to bring dynamics under the power of the calculus ... Our aim, on the other hand, is to cultivate our dynamical ideas. We therefore avail ourselves of the labours of the mathematicians, and retranslate their results from the language of the calculus into the language of dynamics, so that our words may call up the mental image, not of some algebraical process, but of some property of moving bodies.[105]

Just prior to publication of his *Treatise* the author despatched an urgent post-card to Tait inquiring as to the whereabouts of his proofs and informing the Edinburgh professor that he would send him 'shortly some remarks on T&T' for [the] next edition'. He then explained that he had been 'overhauling the Equations of Motion ... beginning with impulsive force'. The changes, he believed, 'constitutes an improvement in my book and a preparation for Electro-kinetics and Magnetic action on light'. In the published *Treatise* Maxwell acknow-ledged with respect to his exposition of the equations of motion of a connected system that much of it was 'suggested by the investigation in Thomson and Tait's *Natural Philosophy*, especially the method of beginning with the theory of impulsive forces'. Illustrative of the very close interactions between Maxwell and Thomson and Tait, the method had been inspired by T&T', developed as the dynamical foundation for the theory of the electromagnetic field in Maxwell's *Treatise* (1873) and subsequently given far greater exposition in the second edition of Thomson and Tait (1879).[106]

Expressed most vividly in Maxwell's later bell-ringing analogy (ch.10), the method was essentially that of the 1865 'Dynamical Theory'. Based upon a general physical analogy between electromagnetic and dynamical variables (velocity, momentum, angular momentum, and potential and kinetic energy), it enabled Maxwell to construct the 'general equations of the electromagnetic field'. As stated in chapter nine of his *Treatise*, these equations included those of magnetic induction, electromotive force, electromagnetic force, magnetization, electric currents, electric displacement, and conductivity: some thirteen principal equations in total. Only after Maxwell's death would the list be transformed into the set of four vector equations known as 'Maxwell's equations'(ch.14).[107]

Treatise readers would compete over a complex legacy from which no single and unambiguous meaning could be extracted. Reactions varied from William Thomson's complaints that he could not get a satisfying mechanical model for electromagnetism to entirely new views in which electromagnetism and 'energetics' replaced mechanics as the fundamental physical reality (ch.14). But, as Tait had pointed out in his contemporary review, the *Treatise* aimed primarily to upset the notions of action-at-a-distance. In the attainment of this goal, energy considerations had come to replace those of force. As Maxwell himself concluded in his *magnum opus*, energy meant that the 'idea of a medium cannot be got rid of':

> But in all of these [action-at-a-distance] theories the question naturally occurs:- If something is transmitted from one particle to another at a distance, what is its condition after it has left the one particle and before it has reached the other? If this something is the potential energy of the two particles . . . , how are we to conceive this energy as existing in a point of space, coinciding neither with the one particle nor with the other? In fact, whenever energy is transmitted from one body to another in time, there must be a medium or substance in which the energy exists after it leaves one body and before it reaches the other, for energy, as Torricelli remarked, 'is a quintessence of so subtile a nature that it cannot be contained in any vessel except the inmost substance of material things'.[108]

CHAPTER 12

Demons *versus* Dissipation

Now one thing in which the materialist (fortified with dynamical knowledge) believes is that if every motion great & small were accurately reversed, and the world left to itself again, everything would happen backwards: the fresh water would collect out of the sea and run up the rivers and finally fly up to the clouds in drops which would extract heat from the air and evaporate and afterwards in condensing would shoot out rays of light to the sun and so on. Of course all living things would regrede from the grave to the cradle and we should have a memory of the future but not of the past.

The reason why we do not expect anything of this kind to take place at any time is our experience of irreversible processes, all of one kind, and this leads to the doctrine of a beginning & an end instead of cyclical progression for ever.

James Clerk Maxwell explains the untenability of strict materialist dogma to
Mark Pattison, 1868[1]

Such fantastic visions of a reversible course of nature were to become a common refrain among the scientists of energy, most notably Maxwell and Thomson, in their offensive against anyone who might seek to further the cause of materialism by appropriating conservation of energy without also acknowledging its dissipation. The contrasting appeal to experience was an exhortation to students to grasp for themselves the great truths of nature which would stand in harmony with the great truths of Christianity and therefore to appreciate the great falsehoods of atheistic philosophies and disembodied systems of mere words and symbols. As Maxwell told the Rector of Lincoln College Oxford in the same letter, 'I happen to be interested in speculations standing on experimental & mathematical data and reaching beyond the sphere of the senses without passing into words and nothing more'.[2]

Maxwell and his allies placed the doctrine of dissipation, embodied solidly in scientific and engineering practice, centre stage in the science of energy. It was dissipation of energy which gave fresh meaning to the Christian perspective on beginnings and endings to the visible cosmos. By pointing to the necessity of a beginning, the original divine creation of matter and of energy, the doctrine purported to undermine all those creeds which espoused a self-regulating, self-sustaining and eternal material world (chs 2, 6 and 7). But dissipation of energy also linked together the natural and moral orders. For Thomson dissipation was a waste of energy to man rather than a waste in nature (ch.6). For Maxwell too, human beings, imperfect physically as finite beings and imperfect spiritually as fallen creatures, needed to recognize those limitations and failings as a step towards a proper relationship to God and to His creation. In Maxwell's personal formulation the complementary tasks of the natural philosophy professor and the minister of Christ were to facilitate that recognition (ch.11).

With respect to the physical world, then, it was the primary duty of the natural philosopher to guide students to see for themselves the truth that, since the laws of nature were unalterable by human beings, man had to work according to those laws if he were to avoid failure (ch.11). Man therefore needed to acknowledge in humility his limitations. Unable to create, destroy or restore energy, man could nevertheless work with the laws to direct available energy by means of his will and for his use. As Maxwell later summed up for the *Encyclopaedia Britannica* (1878):

> Available energy is energy which we can direct into any desired channel. Dissipated energy is energy which we cannot lay hold of and direct at pleasure, such as the energy of the confused agitation of molecules which we call heat. . . . [The] notion of dissipated energy could not occur to a being who could not turn any of the energies of nature to his own account, or to one who could trace the motion of every molecule and seize it at the right moment. It is only a being in the intermediate stage, who can lay hold of some forms of energy while others elude his grasp, that energy appears to be passing inevitably from the available to the dissipated state.[3]

Limited in the physical world, man was likewise subject to moral imperfections and failings. With respect to the moral world, it was thus the duty of the Christian minister to lead his flock to acknowledge the imperfect nature of man and to recognize that the nature of his actions depended on the state of his will (ch.11). For the scientists of energy in general free will was a basic assumption founded on a direct personal perception of a gift bestowed on man by God. With their Calvinist roots these Scottish presbyterians regarded man, made in the image of God, as a creature of will rather than of reason. But while Maxwell could no longer regard man as possessed of a depraved will, he held that human will had to be developed and perfected through acceptance of another divine gift, that of grace, in order to achieve union with Christ. This creed, uniting both the physical and spiritual worlds, was summed up in a single verse from the Epistle of James, quoted in Greek by the dying Maxwell: 'Every good and every perfect gift is from above, and cometh down from the Father of lights, with whom is no variableness, neither shadow of turning'. Rich in metaphor yet simple in its message, the text spoke to

Christians of the unchangeableness and constancy of God, source of all natural and spiritual perfection freely offered to human beings.[4]

Standing at the core of this presbyterian theology, then, was the acceptance and deployment by man of certain free gifts from God: grace, free will, the energies of nature, and those gifts given to individuals (as teachers, ministers, natural philosophers, or mathematicians, for instance) essential to the spiritual division of labour. Only when the limitations and failings of man as a natural and spiritual being were acknowledged would the full benefits of these gifts be realized. There was to be no justification by man's own efforts for justification came through acceptance and acknowledgment of the divine gifts. Only then could man deploy his liberated will to direct those gifts, whether material, intellectual or spiritual, for beneficial ends. It was a democratic creed which recognized not the equality of all human beings but the endowment of each and every individual with talents which needed to be used to maximum effect.

The Scottish context of Maxwell and his friends, with its 'democratic' traditions, placed a high premium on the pedagogical value of scientific texts. Symbolism, mystification and complex displays of mathematical skill echoed the evils of Roman Catholicism, whereas practicality, direct experience and simplicity of problem-solving technique reflected the traditions of presbyterian Scotland. Just as ministers of the reformed Kirk presented themselves not as guardians of esoteric knowledge but as teachers of the people (often by analogy and parable), so the scientists of energy advocated a physics with strong visualizable foundations accessible to everyone in contrast to the far more elite symbolic formulations of their continental competitors.

Constructing molecular physics: Clausius versus Maxwell

Appointed professor of mathematical physics at the new Zurich Polytechnic in Switzerland from 1855, Clausius soon began a series of investigations into molecular physics which focused specifically on the nature of heat and which would eventually earn for him the credit of being one of the principal founders of what Maxwell later named 'the kinetic theory of gases'. 'On the Nature of the Motion Which We Call Heat' (1857) opened with the claim that even prior to Clausius's 1850 memoir on the mechanical theory of heat he had 'a distinct conception of this motion' but that he had deliberately 'avoided mentioning this conception, because I wished to separate the conclusions which are deducible from certain general principles from those which presuppose a particular kind of motion'.[5]

The German physicist August Krönig (1822–79), a founder member of the Berlin Physical Society, had recently published a paper on gas theory in which he assumed that 'the molecules of gas do not oscillate about definite positions of equilibrium, but that they move with constant velocity in right lines until they strike against other molecules, or against some surface which is to them impermeable'. Acknowledging Krönig's priority in the publication of views which Clausius himself shared, the Zurich professor nonetheless made clear that he believed that rotatory and vibratory, as well as translatory (straight-line), motions were involved. Indeed, the mathematical part of Clausius' paper claimed to show decisively that 'the *vis viva* of the translatory motion does not alone represent the whole quantity of heat in the gas, and that the difference is greater the greater the number of atoms of which the several molecules of the combination consist'. Thus he insisted that 'besides the translatory motion of the molecules as such, the constituents of these molecules perform other motions, whose *vis viva* also forms a part of the contained quantity of heat.[6]

In the second half of his 1857 paper Clausius deduced an expression 'which shows in what manner the pressure of the gas on the sides of the vessel depends upon the motion of its molecules', that is, the ideal gas law:

$$p = mnu^2/3v$$

where p is gas pressure, v the volume, m the mass of a molecule, n the total number of molecules, and u the mean velocity.

Clausius's model contained the permanent gas in a vessel bounded by two large parallel planes and two narrow strips. In accordance with 'the laws of probability', he assumed that 'there are as many molecules whose angles of reflexion fall within a certain interval ... as there are molecules whose angles of incidence have the same limits, and that, on the whole, the velocities of the molecules are not changed by the side'. In other words an individual molecule might not actually obey the laws governing elastic spheres colliding with an elastic plane, but there would be no difference in the final result by assuming that 'for each molecule the angle and velocity of reflexion are equal to those of incidence'. He therefore envisaged each molecule moving to and fro between the large sides 'in the same directions as those chosen by a ray of light between two plane mirrors, until at length it would come in contact with one of the small sides; from this it would be reflected, and then commence a similar series of journeys to and fro'.[7]

Admitting that in reality 'the greatest possible variety exists amongst the velocities of the several molecules', Clausius introduced a second major 'probabilistic' assumption, namely, that 'we may ascribe a certain mean velocity to all molecules'. This mean velocity had to be chosen such that 'the total *vis viva* of all the molecules may be the same as that corresponding to their actual velocities'. Although unable to assign values to m and n, knowledge of the total mass of the gas (mn) enabled him to calculate the mean velocities of the molecules of oxygen, nitrogen and hydrogen. Again he stressed, however, that 'it is possible that the actual velocities differ materially from their mean value'. Finally, he showed that the ratio of the *vis viva* of translation to the total *vis viva* was a function of specific heats at constant pressure and constant volume.[8]

Criticism of Clausius's views on the nature of heat (together with the similar views of Krönig and Joule) appeared soon afterwards in Poggendorff's *Annalen*. There the Dutch physicist and meteorologist C.H.D. Buys-Ballot (1817–90) objected that 'if the molecules moved in straight lines, volumes of gases in contact would necessarily mix with one another', a result inconsistent with, for example, observations of clouds of smoke which often remained extended in immovable layers for a considerable length of time. Similarly, a pungent gas evolved in one part of a room was not smelt immediately in another part. Yet the theory assumed that 'the particles of gas must have had to traverse the room hundreds of times a second'. In order to counter this threat to credibility, Clausius introduced what would become his most celebrated innovation in gas theory, the 'mean length of the paths described by the separate molecules of gaseous bodies on the occurrence of molecular motion' (subsequently known as the 'mean free path').[9]

Building on his earlier (1857) notion that the portions of the path of a molecule influenced by molecular forces (the spheres of action of other molecules) must be negligible in a gas compared to those portions uninfluenced by such forces, Clausius argued that the latter portion could still be very small compared to the dimensions of a container. He then proposed the idea of 'a sphere of action' as one around the centre of gravity of a molecule and beyond whose boundary molecular effects were insensible. Imagining a given space with 'a great number of molecules moving irregularly about amongst one another', he invited his readers to 'select one of them to watch'. They would then see that 'such a one would ever and anon impinge upon one of the other molecules, and bound off from it'. The basic question was therefore to determine 'how great is the mean length of the path between two such impacts', or, more precisely: '*how far on an average can the molecule move, before its centre of gravity comes into the sphere of action of another molecule*'.[10]

Probabilistic considerations, applied to a model in which free molecules approached successive layers of gas molecules considered as fixed, led to the conclusion that a molecule was unlikely to travel beyond its mean path. Given the likely small length of this path compared to a room, for example, Clausius claimed that his theory did not lead to Buys-Ballot's conclusion 'that two quantities of a gas bounding one another must mix with one another quickly and violently'. Instead, 'only a comparatively small number of atoms can arrive quickly at a great distance, while the chief quantities only gradually mix at the surface of their contact'.[11]

Publication of a translation of Clausius's paper for the *Phil. Mag.* (1859) stimulated Maxwell into a major investigation of gas theory. He perceived that Clausius had 'determined the mean length of path in terms of the average distance of the particles, and the distance between the centres of two particles when collision takes place', but had offered 'no means of ascertaining either of these distances'.[12]

Writing to Stokes at the end of May 1859, however, Maxwell explained that his recent reading of Clausius had made him think that 'it might be worth while examining the hypothesis of free particles acting by impact and comparing it with phenomena which seem to depend on this "mean path"'. He had therefore 'begun at the beginning and drawn up the theory of the motions and collisions of free particles acting only by impact, applying it to internal friction of gases, diffusion of gases, and conduction of heat through a gas (without radiation)'. From a theory of gaseous friction he arrived at numerical figures for the 'mean path' (1/447,000 of an inch) and the number of collisions per second (8,077,000,000). Experiments by Thomas Graham on gas diffusion, on the other hand, seemed to yield a value for the 'mean path' much larger than that derived from friction.[13]

Far more sceptical in tone than Clausius, Maxwell told Stokes that he simply did not know 'how far such speculations may be found to agree with facts' and that 'even if they do not it is well to know that Clausius' (or rather Herapath's) theory is wrong'. In the same letter he also admitted:

> as I have found myself able and willing to deduce the laws of motion of systems of particles acting on each other only by impact, I have done so as an exercise in mechanics. Now do you think there is any so complete a refutation of this theory of gases as would make it absurd to investigate it further so as to found arguments upon measurements of strictly 'molecular' quantities before we know whether there be any molecules [?].[14]

Two features emerge from these rather enigmatical remarks. First, it seems that this 'exercise in mechanics' involved the construction of a purely mechanical system of particles which might – or might not – turn out to be a good representation of gaseous phenomena. Maxwell made clearer his aims to Stokes during October 1859: 'I intend to arrange my propositions about the motions of elastic spheres in a manner independent of the speculations about gases, and I shall probably send them to the Phil. Magazine, which publishes a good deal about the dynamical theories of matter & heat'.[15]

Maxwell's tactic of distancing the mechanical representation from the physical phenomenon was graphically expressed in the title of his published paper, 'Illustrations of the Dynamical Theory of Gases' (1860), and further elaborated in the introduction: 'If the properties of such a [mechanical] system of bodies are found to correspond to those of gases, an important physical analogy will be established, which may lead to more accurate knowledge of the properties of matter'. On the other hand 'If experiments on gases are inconsistent with the hypothesis of these propositions, then our theory, though consistent with itself, is proved to be incapable of explaining the phænomena of gases'.[16] By his methodological style here, Maxwell identified himself with that of his mentor, William Thomson, whose 'mechanical representations' and 'dynamical illustrations' of electrical and magnetic phenomena had formed a principal inspiration for the young Maxwell's electrical researches (ch.11).[17]

Second, Maxwell was confiding to Stokes a deeply held conviction that lack of knowledge concerning the very existence of molecules called into question the whole theory of gases. Such an apparently sceptical line certainly reflects parallel strategies in his contemporaneous construction of credible electromagnetic theories (ch.11). Above all, his caution relates to a British preference for a *continuum* theory of matter in which gross matter was ultimately reducible to the motions of an all-pervading etherial medium. Any particulate, atomistic theory of gases or other states of matter could therefore be at best a mere approximation.[18]

From the outset Maxwell did not pretend to claim privileged knowledge of the particles of matter. 'Instead of saying that the particles are hard, spherical, and elastic, we may if we please say that the particles are centres of force, of which the action is insensible except at a small distance, when it suddenly appears as a repulsive force of very great intensity', he explained in the introduction to his 1860 paper. 'It is evident', he asserted without elaboration, 'that either assumption will lead to the same results'.[19] In his 'Illustrations', however, Maxwell opted for convenience to employ the language of perfectly elastic spherical bodies:

> If a great many equal spherical particles were in motion in a perfectly elastic vessel, collisions would take place among the particles, and their velocities would be altered at every collision; so that after a certain time the *vis viva* will be divided among the particles according to some regular law, the average number of particles whose velocity lies between certain limits being ascertainable, though the velocity of each particle changes at every collision.[20]

Like Clausius, Maxwell recognized that the 'particles have not all the same velocity'. But, unlike Clausius, he did not simply ascribe 'a certain mean velocity' to all molecules. Rather, 'the velocities are distributed according to the same formula as the errors are distributed in the theory of least squares'. This innovation led to a formula for the 'mean path' with a different coefficient to that of Clausius.[21]

Within a few years, however, Maxwell was openly confessing the inadequacy of his 'Illustrations': 'I . . . gave a theory of diffusion of gases, which I now know to be erroneous; and there were several errors in my theory of the conduction of gases, which M. Clausius has pointed out in an elaborate memoir on that subject'. As a result he presented a new memoir, 'On the Dynamical Theory of Gases', to the Royal Society in 1866. Published both in the *Phil. Trans.* (1867) and the *Phil. Mag.* (1868), this memoir did much to enhance Maxwell's credibility as a leading British natural philosopher. Coinciding approximately with the appearance of Thomson and Tait's *Treatise* (ch.10) and with his own 'Dynamical Theory of the Electromagnetic Field' (ch.11), it also predated by a very short time Thomson's attempts to provide a continuum theory of matter for the 'science of energy' through the vortex atom.[22]

'Theories of the constitution of bodies suppose them either to be continuous and homogeneous, or to be composed of a finite number of distinct particles or molecules', Maxwell began. While 'the properties of a body supposed to be a uniform *plenum* may be affirmed dogmatically', he stated that such homogeneous properties 'cannot be explained mathematically' and that he was unaware that any theory of this kind had yet been proposed. Molecular theories also presented mathematical difficulties and 'till they are surmounted we cannot fully decide on the applicability of the theory'. But, he emphasized, we are able 'to explain a great variety of phenomena by the dynamical [molecular] theory which have not been hitherto explained otherwise':[23]

> I propose in this paper to apply this theory to the explanation of various properties of gases, and to show that, besides accounting for the relations of pressure, density, and temperature in a single gas, it affords a mechanical explanation of the known chemical relation between the density of a gas and its equivalent weight, commonly called the Law of Equivalent Volumes. It

also explains the diffusion of one gas through another, the internal friction of a gas, and the conduction of heat through gases.

Maxwell, however, now shifted from previous elastic sphere models to one of 'small bodies or groups of smaller molecules repelling one another with a force whose direction always passes very nearly through the centres of gravity of the molecules, and whose magnitude is represented very nearly by some function of the distance of the centres of gravity'. Guided by the results of his own (and his wife Katharine's) experiments on the viscosity of air at different temperatures, he inferred that 'the repulsion is inversely as the *fifth* power of the distance'.[24]

In contrast to confidence in a specific 'dynamical theory of gases', Maxwell's 1866 version aimed at as general a formulation as possible, with minimum construction of hypotheses concerning the invisible inner character of gases. Molecules in particular were now simply defined as those portions of a gas 'which move about as a single body'. As for their specific, hypothetical properties, Maxwell emphasized the options:

> These molecules may be mere points, or pure centres of force endowed with inertia, or the capacity of performing work while losing velocity. They may be systems of several such centres of force bound together by their mutual actions; and in this case the different centres may either be separated, so as to form a group of points, or they may be actually coincident, so as to form one point.
>
> Finally, if necessary, we may suppose them to be small solid bodies of determinate form; but in this case we must assume a new set of forces binding the parts of these small bodies together, and so introduce a molecular theory of the second order.[25]

From such remarks it is evident that Maxwell was by no means opposed to considerations of the nature of molecules. By showing the range of possibilities, he was instead emphasizing a very limited knowledge of this invisible world. A commitment to any particular hypothesis could not be justified in the present state of human knowledge.[26]

During the 1860s and early 1870s, Maxwell was reworking the meaning of 'dynamical theory', not in the sense of definite hypotheses on the detailed nature of molecular mechanisms, but within a British tradition of 'dynamical' natural philosophy. 'Dynamical theories', in opposition to molecular hypotheses, had found their most enthusiastic exponents in Stokes (hydrodynamics and light) and Thomson (heat, electricity and magnetism). Now Maxwell had begun to develop parallel perspectives on gas theory and electromagnetic theory. Revealingly, Maxwell's generalized definition of 'molecules' coincided with his introduction of the language of energy into his 'Dynamical Theory of Gases'. Thomson and Tait's *Treatise* (1867) would provide some of the most powerful tools for the new generalized 'dynamical' approaches to physical phenomena, most obviously with respect to electromagnetism (ch.11) but also in relation to gas theory.[27] Henceforth, he would debate the issues of molecular physics within the framework of the 'science of energy'.

Interpreting the Second Law of Thermodynamics

Writing to Thomson in 1854, Stokes commented: 'I think it possible that something interesting may turn up [in the course of pendulum experiments] from the value of the index of friction, which is as it were the door through which vis viva passes from the mechanical state (observable motion) to the molecular state (heat)'.[28] The conversion of work into heat during fluid friction, consistent with a dynamical theory, was by now a very familiar theme. But the *irrecoverable* nature of the energy thus transformed had not been explained in terms of dynamical principles. Thomson in particular had founded the second fundamental proposition (Carnot and Clausius) of the dynamical theory of heat on a universal, cosmological axiom (ch.6), but it remained peculiarly inaccessible to dynamical explanation. Thus in an apparently casual comment of May 1855 Maxwell asked Thomson an unsettling question: 'By the way do you profess to account for what becomes of the vis viva of heat when it passes thro' a conductor from hot to cold? You must either modify Fourier's laws or give up your theory, at least so it seems to me'.[29]

Throughout 1857 the theme of fluid friction again preoccupied Thomson, both in his private notebook and in correspondence with Stokes and Maxwell. For some time he had been engaged in an ambitious programme of dynamical theory in which all the branches of natural philosophy would be comprehended not only in relation to one another but above all in terms of matter and motion (energy). Thomson increasingly looked to the science of hydrodynamics, and to Stokes as master of the science, for answers. Much vexed by the intricate problems, he confessed to Stokes in May:

> I have been [in] great difficulties about hydrodynamics for some time. I thought I had made out that a mote [or solid particle] in a perfect (non frictional) liquid, would if set into any state of motion, end with its centre of gravity at rest and all the energy of the given motion transformed into energy of rotation of the mote & of corresponding motion of the surrounding liquid. I am however shaken from this comfortable theory, and I fear it comes to nothing.[30]

He then drew Stokes's attention to 'a curious illustration' which Clerk Maxwell had shown him 'a long time ago'. A rectangular slip of paper, falling from a height with its length almost horizontal, acquired rotatory motion about its longitudinal axis which in Thomson's opinion was 'inexplicable without taking into account the viscosity of air'. He concluded that were his theory to have held, 'the same kind of phenomenon would take place in a perfect liquid'.

Maxwell discussed the illustration with Thomson in November 1857. He explained the conditions for the phenomenon, notably 'An excess of pressure on the first [lower] surface of the strip over that behind', with the excess 'greatest at the foremost edge of the strip and near it'. These two effects would increase 'with the relative velocity of the strip and the medium'. The results included 'A permanent rotation of the strip indifferently either way round'. He then focused on the problematic question of reversibility by reminding Thomson that 'In May you thought that these effects would take place in an incompressible fluid without friction, and now you think that opinion a delusion, because if all motions at any instant were reversed all would go back as it came'.[31]

Challenging Thomson to provide a rationale of why such a reversal should be impossible, Maxwell continued: 'Now I cannot see why, if you could gather up all the scattered motions in the fluid, and reverse them *accurately*, the strip should not fly up again. All that you need is to catch all the eddies, and reverse them not approximately, but accurately'. He elaborated a 'thought experiment':

> If you pour a perfect fluid from any height into a perfectly hard or perfectly elastic basin its motion will break up into eddies innumerable forming on the whole one large eddy in the basin depending on the total moments of momenta for the mass.
>
> If after a given time say 1 hour you reverse every motion of every particle, the eddies will all unwind themselves till at the end of another hour there is a great commotion in the basin, and the water flies up in a fountain to the vessel above. But all this depends on the *exact* reversal for the motions are *unstable* and an approximate reversal would only produce *a new set of eddies multiplying by division*.

Searching for a rationale for the impossibility of such a spectacular reversal of the course of nature, Maxwell seemed to find it in a tendency in nature from unstable to stable motion. He explained that he did not 'see why the unstable motion of a perfect fluid should not produce eddies which can never be gathered up again except by miracle'. Given that 'the diminution of pressure at the back of a body depends on the formation and dissipation of eddies by unstable motion', he suggested that it arguably made little 'difference whether these eddies are soon converted into heat, or remain in the fluid in a state of subdivision which is as nearly that of molecular vortices as any finite motion can be'. Maxwell therefore seemed to be speculating that such molecular vortices represented an ultimate and stable form of motion, unalterable save by divine intervention as in a miracle and thus not otherwise reversible into those unstable states from whence the molecular vortices had derived.

This radical speculation would probably have astonished even Thomson. But Thomson became used to Maxwell's fertile imagination over the next decade, especially with Maxwell's employment of a molecular vortex model to construct an electromagnetic theory of light in 1861 (ch.11). Thomson, too, articulated a vortex theory of matter by 1867.[32] Maxwell's 1857 letter further shows just how seriously he took Rankine's molecular vortex hypothesis at this stage, specifically in relation to heat interpreted as eddies in a fluid and in relation to the conversion of work into heat (fluid friction and the dissipation of energy). Only later would Maxwell express his reservations about such a specific hypothesis and argue instead for a more general 'dynamical theory'.

Since his formulation of the doctrine of universal dissipation of energy in the early 1850s, Thomson had held to the view that the will of living creatures could direct, but not reverse, the natural tendency for energy towards diffusion in the form of heat. During the early 1860s, the question of the free will of animated creatures (especially humans) was much discussed among the Thomson brothers and Maxwell. Above all, they shared a common dislike of the mechanistic and deterministic views of life which their friend Helmholtz appeared to espouse and which their rival Tyndall would certainly seek to exploit. Helmholtz's 1861 Royal Institution lecture 'On the Application of the Law of Conservation of Force to Organic Nature' reiterated his earlier commitment to the extermination of 'vital forces' and to the subjection of biological phenomena to conservation of force (ch.7). He thus asserted with respect to forces in living creatures that 'there cannot exist any arbitrary choice in the direction of their actions'.[33]

In a letter of April 1862 to Lewis Campbell, Maxwell articulated his opposition to such determinism. He agreed with Helmholtz that 'the soul is not the direct moving force of the body' because if it were 'it would only last till it had done a certain amount of work, like the spring of a watch'. But, he argued, the soul does not, like food, perish 'in the using'. The soul *directed* the action of matter and energy: 'when a man pulls a trigger it is the gunpowder that projects the bullet, or when a pointsman shunts a train it is the rails that bear the thrust'. Free will thus appeared to take advantage of instabilities and discontinuities in mechanical systems. Content to suggest by analogy the possibility of free will, Maxwell did not of course claim to provide an account of its nature. But the strength of his conviction as to the existence of human free will derived from his debt and commitment to a Christian faith with deep presbyterian roots (ch.11). And such discussions of 'directing power' provided a major context for Maxwell's 'statistical' interpretation of the second law of thermodynamics.[34]

In December 1867 Tait wrote to Maxwell to ask him if he were 'sufficiently up to the history of Thermodynamics to critically examine & put right a little treatise I am about to print – and will you kindly apply your critical powers to it'. A few days later Maxwell replied in gentlemanly fashion with a denial that he knew 'in a controversial manner the history of thermodynamics, that is I could make no assertions about the priority of authors without referring to their actual works'. But he offered to assist in enhancing the credibility of Tait's text: 'If I can help you in any way with your book I shall be glad as any contributions to that study are in the way of altering the point of view here & there for clearness or variety and picking holes here & there to ensure strength & stability'. He urged Tait to 'make something of the theory of the absolute scale of temperature by reasoning pretty loud about it and paying it due honour at its entrance'. But it was with respect to picking a hole that Maxwell developed a new interpretation of the second law of thermodynamics 'that if two things are in contact the hotter cannot take heat from the colder without external agency' (ch.5).[35]

In his original formulation of what became known as 'Maxwell's demon', he

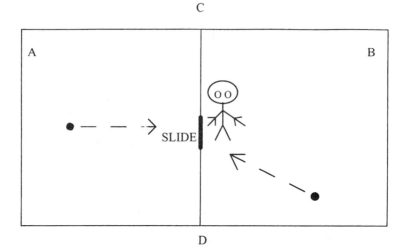

Figure 12.1 In this thought experiment, subsequently known as 'Maxwell's demon', the two vessels of equal volume but unequal temperature (A being hotter than B) are separated by a fixed diaphragm CD. A 'finite being' can open and close a hole in CD by means of a sliding door without mass. If the 'very observant and neat fingered being' knows the paths and velocities of all the gas molecules in A and in B, he may allow a molecule in A heading for the hole to pass into B, and a molecule in B heading for the hole to pass into A. By selecting molecules with less than the mean square velocity of B to pass from A into B, and molecules with more than the mean square velocity of A to pass from B into A, the number of molecules in A and B will remain unchanged, but the temperature of A will increase and that of B will decrease without any work being done. Such a skilful being, if it existed, could therefore reverse the 'natural' tendency of heat to pass from hotter to colder regions. But for human beings without such skills, the second law remained inviolable.

supposed two vessels A and B divided by a diaphragm CD. Both vessels contained an equal number of 'elastic molecules in a state of agitation which strike each other and the sides' (Figure 12.1). But the particles in A were allowed 'the greatest energy of motion'. He then explained that even if all the molecules in A have equal velocities, 'if oblique collisions occur between them their velocities will become unequal & I have shown that there will be velocities of all magnitudes in A and the same in B only the sum of the squares of the velocities is greater in A than in B'. Furthermore, when a molecule was reflected from the fixed diaphragm 'no work is lost or gained'. Similarly, if a molecule were allowed to go through a hole in the diaphragm instead of being reflected no work would be lost or gained but 'its energy would be transferred from the one vessel to the other'.[36]

Having thus set up the thought experiment in simple physical terms, Maxwell now introduced in dramatic fashion a living (microscopic) actor possessed only of the capacity to observe individual molecules, to know their paths and velocities, and to open and close a frictionless sliding door without mass positioned in the diaphragm:

> Now conceive a finite being who knows the paths and velocities of all the molecules by simple inspection but who can do no work except to open and close a hole in the diaphragm by means of a slide without mass.
>
> Let him first observe the molecules in A and when he sees one coming the square of whose velocity is less than the mean sq[uare] vel[ocity] of the molecules in B let him open the hole & let it go into B. Next let him watch for a molecule in B the square of whose velocity is greater than the mean sq. vel. in A and when it comes to the whole [sic] let him draw the slide & let it go into A keeping the slide shut for all other molecules.
>
> Then the number of molecules in A & B are the same as at first but the energy in A is increased and that in B diminished, that is the hot system has got hotter and the cold colder & yet no work has been done, only the intelligence of a very observant and neat fingered being has been employed.

The message to be extracted from this scientific parable, then, was simply that 'if heat is the motion of finite portions of matter and if we can apply tools to such portions of matter so as to deal with them separately then we can take advantage of the different motions of different portions to restore a uniformly hot system to unequal temperatures or to motions of large masses'. But, concluded Maxwell, human beings cannot, 'not being clever enough'.

Recognizing the potency of Maxwell's parable, Tait forwarded the letter to Thomson with pencilled postscript urging him to 'say what occurs to you about the enclosed *which return* speedily'. Thomson's brief response endorsed the message with the terse remark: 'Very good. Another way is to reverse the motion of every particle of the Universe, and preside over the unstable motion thus produced'. As we have seen, Maxwell developed this fantasy in his correspondence with the reforming Rector of Lincoln College in 1868. A couple of years later he reiterated the vision to J.W. Strutt (later Lord Rayleigh):

> If this world is a purely dynamical system and if you accurately reverse the motion of every particle of it at the same instant then all things will happen backwards till the beginning of things the rain drops will collect themselves from the ground and fly up to the clouds, &c &c and men will see all their friends passing from the grave to the cradle till we ourselves become the reverse of born, whatever that is.[37]

Around the same period Maxwell drew up a neat Catechism, mirroring the Church of Scotland's 'Shorter Catechism' (1648) which began with the question 'What is the chief end of man'?:

Concerning Demons.

1. Who gave them this name? Thomson.

2. What were they by nature? Very small BUT lively beings incapable of doing work but able to open and shut valves which move without friction or inertia.

3. What was their chief end? To show that the 2nd Law of Thermodynamics has only a statistical certainty.

4. Is the production of an inequality of temperature their only occupation? No, for less intelligent demons can produce a difference in pressure as well as temperature by merely allowing all particles going in one direction while stopping all those going the other way. As such value him. Call him no more a demon but a valve like that of the hydraulic ram, suppose.

Maxwell was not of course here objecting to Thomson's use of the term demon, but simply considering a being of less skill who merely played the role of a valve or switch, without exerting forces on molecules.[38] For his part Thomson publicly presented the Maxwellian demon to the Royal Society of Edinburgh in 1874 as 'an intelligent being endowed with free will and fine enough tactile and perceptive organization to give him the faculty of observing and influencing individual molecules of matter'. This finite being possessed the same gifts, free will and intelligence, as Thomson and Maxwell ascribed to man who could direct though probably not reverse the energies of nature. But the demon's capacity to direct at the molecular level gave him the ability to sort the motions of individual molecules.[39]

Thomson's RSE presentation in fact led him to consider types of demon endowed with superior skills which enabled them to transfer energy from one molecule to another. He here invoked a distinction between 'abstract' and 'physical dynamics' introduced in the *Treatise* (ch.10). Abstract dynamics applied to an idealized world of an isolated, finite system of discrete particles and conservative forces. Total reversal was conceivable in such a world. Physical dynamics, on the other hand, applied to the infinite real world of imperfectly isolated and non-conservative systems which involved friction and diffusion. Here reversibility was inconceivable. To develop the point, he showed how an army of demons with molecular cricket bats could reverse the motions of individual molecules and thus reverse dissipation of energy in an isolated vessel or bar containing a finite number of molecules. But real physical systems, which could never be isolated in practice and which therefore involved an infinite number of molecules, would require an infinite army of demons: 'Do away with this impossible ideal [of isolation], and believe the number of molecules in the universe to be infinite; then we may say one-half of the bar will never become warmer than the other'.[40]

Neither Thomson nor Maxwell, however, were enthusiasts for a universe of discrete particles. Particle dynamics provided a valuable approach to the real world, but it did not necessarily reflect the true complexity of even inanimate nature. For Thomson in particular an irreducible and undifferentiated continuum or universal ether seems to have provided the right kind of medium for an artificer of infinite directing power, rendering the second law wholly inviolable to any human artifice. Indeed, such issues of the infinite seemed to remove any ultimate understanding of either the nature of free will or the second law of thermo-dynamics beyond the scope of human minds.[41]

The North British interpretation of the Second Law of Thermodynamics found wide public expression in Balfour Stewart and P.G. Tait's *Unseen Universe or Physical Speculations on a Future State* (1875). A former pupil (1845–6) and later assistant (1856–9) in Forbes's natural philosophy class, the Edinburgh-born Stewart became Director of Kew Observatory in 1859 and natural philosophy professor at Owens College, Manchester, in 1870. Through the BAAS and his Edinburgh connections, Stewart was an able popularizer of the science of energy, especially through his textbook *The Conservation of Energy* (1874). But during the writing of the *Treatise*, Tait had told Thomson that 'Stewart and I have simul-taneously struck on an idea of which "Conservation of Energy" is a particular case – and we intend to develop from it some tremendous consequences . . . It is a gushing, gasping idea . . . As it will at once obliterate everybody who pretends to science but is not acquainted with it, we propose shortly to initiate *you*, in order that you may not be lost in the stramash [excitement] that will follow its publi-cation'.[42] While we cannot be certain that this 'gasping idea' was that which even-tually emerged as *The Unseen Universe*, Tait was here announcing his and Stewart's intention of policing natural philosophy against the threat of scientific pretenders of every variety.

John Tyndall's notorious BAAS presidential address at the Belfast meeting in 1874 seemed, at least to the North British, to misuse energy conservation to promote an anti-Christian vision of materialism and determinism. Tyndall claimed that energy conservation ' "binds nature fast in fate" to an extent not hitherto recognized, exacting from every antecedent its equivalent consequent, from every consequent its equivalent antecedent, and bringing vital as well as well as physical phenomena under that law of causal connexion which, so far as the human under-standing has yet pierced, asserts itself everywhere in nature'. Moreover, Tyndall enthusiastically deployed the modern doctrines of natural selection (Darwin) and evolutionary psychology (Spencer), together with a survey of more ancient atomic doctrines, to promote a materialistic, naturalistic perspective whereby we may discern *within* matter itself 'the promise and potency of all terrestrial Life'. In fact, Tyndall denied a commitment to 'rank materialism', professing instead an almost pantheistic belief that 'the whole process of evolution is the manifestation of a Power absolutely inscrutable to the intellect of man'. But whatever the ultimate nature of that 'Power', it was not the power of the Christian Artificer 'fashioned after the human model'.[43]

Written in the context of such threats to Christian orthodoxy, the (originally anonymous) authors of *The Unseen Universe* announced that their general aim was to try 'to show that the presumed incompatibility of Science and Religion does not exist' and that their specific task was to show 'that immortality is strictly in accord-ance with the principle of Continuity (rightly viewed); that principle which has been the guide of all modern scientific advance'. To begin with, the authors' assumptions were voluntaristic, with an orthodox insistence upon divine will and governance: 'God the Father appears to be regarded [by most of the Christian

Churches] as the Being or Essence in virtue of whom the Universe exists. Thus in reciting the Apostles' Creed the Christian disciple says: "I believe in God, the Father Almighty, maker of Heaven and Earth", and the laws of the Universe are regarded by Christian theologians as being expressions of the will acting in conformity with the character of this Being'. They then drew the conclusion that 'the will of man is accomplished in conformity with the laws of the universe, while on the other hand the will of God . . . constitutes in itself the laws of the universe'.[44]

As with the other North British scientists of energy, the visible universe was transitory and perishable. Indeed, Stewart and Tait quoted the very scriptural verses cited by Scottish natural philosophers and theologians from Chalmers onwards: *Psalm* 102 on the waxing old of the heavens and the earth; second *Corinthians* on the temporal nature of that which is seen and on the eternal nature of that which is unseen; and second *Peter* on the end of the universe and the vision of a new heaven and earth (chs 2,6 and 7). For Stewart and Tait, then, 'the visible universe must, *certainly in transformable energy, and probably in matter*, come to an end. We cannot escape this conclusion'. Immortality was thus 'impossible in such a universe'.[45]

By making 'Continuity' fundamental, however, Stewart and Tait departed from orthodoxy. All absolute discontinuities, such as beginnings and endings, were to be abolished. They attempted therefore to forge continuity between the visible world and the invisible realm by supposing that the energy of visible matter had originally derived from, and would eventually return to, the energy of invisible ether. The indestructible energy of the 'Great Whole' was infinite in space and time but the energy of the visible universe, part only of the Whole, was constantly being dissipated according to the Second Law.[46]

Matter too, if it were to consist of vortex atoms, might embody such characteristics. But Thomson's recent idea that 'what we call matter may consist of the rotating portions of a perfect fluid which continuously fills space' breached continuity because such a frictionless fluid rotation could neither be created nor destroyed but appeared to require a creative act. Stewart and Tait supposed instead that the material universe was 'composed of a series of vortex-rings developed from an invisible universe which is not a perfect fluid'. The matter of the present universe, like its available energy, was therefore also finite, passing from the invisible to the visible and ultimately back to the invisible universe whose etherial and energetic gradations allowed for the existence of all manner of 'superior intelligences' and for the immortality of man.[47]

The authors furthermore speculated that the whole universe need not necessarily be subject to dissipation of energy:

> the laws of energy are rather generalisations derived from our experience than scientific principles, like that which we call the *Principle of Continuity*. There would be no confusion of thought introduced if these laws should be found not to hold, or to hold in a different way in the unseen universe. Nor can we regard the law of dissipation as equally fundamental with that of the Conservation of Energy. What is to prove it in the unseen? We have been

shown . . . how Clerk-Maxwell's demons (though essentially finite intelligences) could be made to restore energy in the present universe without spending work. Much more may be expected in a universe free from gross matter.[48]

Claims such as these help us to understand why William Thomson and Maxwell did not like *The Unseen Universe*. Not only did it appear to speculate far beyond the boundaries of the visible world, but the weakening of the scientific doctrines of conservation and dissipation of energy in favour of a principle of continuity undercut the strong commitment of both Maxwell and Thomson to belief in God's absolute creation of matter and energy *ex nihilo* and thus above all to a belief in divine omnipotence.[49]

Securing the Science of Thermodynamics

'The science of Thermodynamics is now securely founded upon bases almost as simply enuntiated [sic], and quite as impregnable, as Newton's *Laws of Motion*; and the opposition which it even yet occasionally meets with is therefore quite as absurd as would be a denial of the main conclusions of the *Principia*'. Thus wrote Tait in his prefatory remarks to the second edition of his textbook on thermodynamics (1877). Reviewing the same edition for *Nature* in 1878, Maxwell similarly claimed that 'within a short time and by a small number of men' thermodynamics had been transformed 'from the condition of a vague anticipation of nature to that of a science with secure foundations, clear definitions, and distinct boundaries'.[50] In his retrospective review of Clausius's scientific career, written for the American Academy of Arts and Sciences at Boston in 1889, Josiah Willard Gibbs (1839–1903) paraphrased Maxwell's remarks on the science of thermodynamics, claiming that since Clausius's 1850 paper (ch.5) 'the foundations of the science were secure, its definitions clear, and its boundaries distinct'.[51]

Written from the vantage point of the 1870s and 1880s, such remarks mask the very intense series of debates which characterized the science of thermodynamics in the making. In this section I argue that disagreements over credible foundations and formulations from the early 1850s to the early 1870s dramatically illustrate the historical contingency of thermodynamics. In particular, conceptual controversies over the interpretation of the second law reveal major differences in scientific culture between North British scientists of energy and German physicists.

Rankine's *Manual of the Steam Engine* (1859) claimed 'to explain the scientific principles of the action of "PRIME MOVERS", or machines for obtaining motive power, and to show how those principles are to be applied to practical questions'. With its lengthy chapter devoted to the 'principles of thermodynamics', Maxwell had good reason to describe Rankine's textbook as the first treatise on thermodynamics (ch.8). Clausius's *Mechanical Theory of Heat* (1867), on the other hand, was largely a reprint of the author's nine memoirs on the subject, centred on the doctrine of heat as *vis viva* but omitting more speculative papers on gas theory. Only with Tait's *Thermodynamics* (1868), however, was the new science displayed in the title and embedded in the content of a textbook that treated thermo-dynamics as a major branch of natural philosophy.[52]

According to Tait, *Thermodynamics* was prompted by two principal considera-tions. First, 'Representations were made by various men of science, especially engineers, to the publishers of the *North British Review*' urging a reprint in sepa-rate form of Tait's two historical articles of 1864 (ch.9). Second, Tait needed a short and elementary textbook on heat and energy for use in his Edinburgh natural philosophy class 'until the publication of the volume of Thomson and Tait's *Natural Philosophy* in which they will be treated'. Tait's *Thermodynamics* should thus be seen both as part of the grand Thomson and Tait *Treatise* project and as the product of those men of science most involved with promotion of the science of energy. In particular, Tait enhanced the credibility of his text by acknowledging the aid of Professors Jenkin (ch.13), Maxwell, and Rankine, 'to whose kind assis-tance the volume is largely indebted for accuracy and completeness'. He also explained that he had added more advanced material 'mainly taken from the scat-tered papers of Sir W. Thomson'.[53]

On the other hand, recent texts on heat by Tyndall and by Stewart failed to cover the ground: 'Tyndall's is designed specially for the non-scientific public, and contains, even in its latest edition, but slight references to the second law of Thermodynamics'. Tait's *Thermodynamics* was to be of a different order. Just prior to publication of the first edition, Maxwell wrote enthusiastically to the author: 'I, personally, am satisfied with the book as a development of T' [Tait] and as an account of a subject where the ideas are new and as I well know almost *unknown* to the most eminent scientific men'.[54] A decade later, in his review of the second and much revised edition, Maxwell publicly extolled the virtues of Tait's work:

> it is impossible to compare this book either with so-called popular treatises or with those of a more technical kind. . . . [Professor Tait] serves up his strong meat for grown men at the beginning of the book, without thinking it necessary to employ either the language of the nursery or of the school; while for younger students he has carefully boiled down the mathematical elements into the most concentrated form, and has placed the result at the end as a *bonne bouche* [titbit] . . . [The author] is at no pains to conceal his own idiosyncrasies, or to smooth down the obtrusive antimonies of a vigor-ous mind into the featureless consistency of a conventional philosopher. . . . This sort of writing, however unlike what we might expect from the con-ventional man of science, is the very thing to rouse the placid reader, and startle his thinking powers into action.[55]

But Tait's style startled Clausius into action. While consulting Maxwell on the draft text in 1867, Tait had already admitted that Clausius and others 'have cut up very rough about bits referring to them'.[56] The principal bone of contention was over Tait's appropriation of Clausius's new term 'entropy'. In Britain, readers of the *Phil. Mag.* had been introduced to 'entropy' through translations of two of Clausius's papers.[57] But Tait seized the opportunity afforded by the still malleable and relatively unfamiliar nature of the term to announce in his textbook:

It is very desirable to have a word to express the *Availability* for work of the heat in a given magazine; a term for that possession, the waste of which is called *Dissipation*. Unfortunately the excellent word *Entropy*, which Clausius has introduced in this connexion, is applied by him to the negative of the idea we most naturally wish to express. It would only confuse the student if we were to endeavour to invent another term for our purpose. But the necessity for some such term will be obvious from the beautiful examples which follow. And we take the liberty of using the term Entropy in this altered sense . . . The entropy of the universe tends continually to zero.[58]

Maxwell, writing to Tait several years later and in the aftermath of many critical exchanges between the protagonists, remarked with polite understatement that 'When you wrote the Sketch [*of Thermodynamics*] your knowledge of Clausius was somewhat defective. Mine is still, though I have spent much labour upon him and have been occasionally rewarded'.[59] Whether or not his knowledge of Clausius was 'defective', Tait unhesitatingly annexed the term 'entropy' as a convenient name for the 'available energy', a concept of special significance for the 'engineering' perspective of the science of energy in Britain.

Published just three years after Tait's *Thermodynamics*, Maxwell's own textbook, *Theory of Heat* (1871), reached its fourth edition in as many years. It was also published in New York in 1872. Not limited to thermodynamics, its contents ranged over the whole experimental and theoretical subject matter of heat as a branch of physics. Most strikingly, it aimed to present state-of-the-art material, including, for the first time in public, the 'demon' in order to illustrate the 'limitation of the second law of thermodynamics'.[60]

Throughout Maxwell's textbook the 'science of energy' was everywhere manifest. Stressing the importance of 'elementary dynamical principles' in an early chapter, for example, Maxwell adopted Thomson and Tait's new terminology of 'kinetic' and 'potential' energy and explained that 'the recent advances in the science of heat have been chiefly due' to the 'Principle of the Conservation of Energy'. In a much later chapter 'On the Molecular Theory of the Constitution of Bodies' he replaced his previous language of 'dynamical theory' with the new phrase 'kinetic theory of gases'. The change directly reflected Thomson and Tait's usage. 'Kinetic' now expressed energy of motion (such as molecular motion), while 'dynamical' referred to those fundamental general principles (such as conservation of energy) which regulated all the phenomena of nature.[61]

Soon after publication of the first edition, Clausius protested vehemently to the editors of the *Phil. Mag*:

[In Maxwell's text] my writings are left quite unmentioned; and my name occurs only once, when it is said I introduced the word *entropy*; but it is added that the theory of entropy had already been given by W. Thomson. Hence any one who derives his knowledge of the matter solely from this book must conclude that I have contributed nothing to the development of the mechanical theory of heat.[62]

Maxwell privately admitted to Tait that he had 'imbibed' his thermodynamics from sources other than Clausius and had in his *Heat* 'been unconsciously acted on by the motive not to speak about what I don't know'. He now aimed to remedy that deficiency: 'In my spare moments, I mean to take such draughts of Clausiustical Ergon [work] as to place me in that state of disgregation in which one becomes conscious of the increase of the general sum of Entropy'.[63] In his subtle way Maxwell thus appeared not to go along with Tait's redefinition of entropy.

Clausius's protest drew a much sharper public response from Tait. Invoking the authority of Helmholtz's popular lectures, Tait fanned the flames of controversy: 'When Professor Clausius succeeds in making his own countrymen regard him as the discoverer of the Dissipation of Energy . . . it will be time enough to complain that foreigners do not give him that credit'. Tait, however, also used the occasion to discredit Clausius's physics by attacking his demonstration of Carnot's theorem:

> As regards the question to whom is due the credit of first correctly adapting Carnot's magnificently original methods to the true theory of heat, it is only necessary to compare the *Axiom* of Professor Clausius's first paper (the only one which has a chance of priority over Thomson) with the behaviour of a thermo–electric circuit in which the hot junction is at a temperature higher than the neutral point, and where therefore heat *does, of itself, pass from a colder to a hotter body*. A thermo–electric battery, worked with ice and boiling water, is capable of raising to incandescence a fine wire, giving another excellent instance of the fallacy of the so–called axiom.[64]

Allowing Clausius credit for rendering 'many services to science, especially in the Kinetic Theory of Gases', Tait unreservedly condemned his contributions to thermodynamics. Indeed, Clausius 'has done, and seems to take credit in doing, uncompensated [that is, 'irreversible'] mischief by his introduction of what he calls *innere Arbeit* [internal work] and *Disgregation*'. Given our present ignorance of the nature of matter, these ideas were speculative, hypothetical and therefore harmful to science. 'No one', Tait acidly concluded, 'will dispute his full claim to originality as regards *them*'.

Responding, Clausius professed to 'take no notice of the somewhat irritated tone' of Tait's reply. He claimed that Tait had not contested his priority but had instead challenged the validity of his scientific investigations. He first reminded readers of the *Phil. Mag.* that the very axiom declared fallacious by Tait had been 'immediately acknowledged as correct by Sir W. Thomson, and since then has likewise been made use of by many other authors for the same demonstration'. Clausius thus suggested that Tait alone could not destroy Clausius's scientific credibility.[65]

He then engaged directly with the case of a thermoelectric battery cited by Tait as evidence against his axiom. Comparing the thermoelectric case to a steam engine, he argued that the passage of heat from a lower to a higher temperature, raising a wire to incandescence, did not in fact take place 'of itself' but occurred simultaneously with the passage of a quantity of heat from higher to lower temperature just as the work done by a steam engine could be employed to raise the temperature of a body simultaneously with the passage of heat from boiler to condenser. Thus the phrase 'of itself' meant '*without the simultaneous occurrence of another change serving for compensation*'. He further dismissed the evidence of the thermoelectric circuit on the grounds that the circumstances of the phenomenon were not as yet fully known.

Clausius finally defended his division of 'internal' and 'external work' in relation to the work accomplished by heat in the change of state of a body as one employed by all authors on the subject of the mechanical theory of heat. 'Disgregation' too he vigorously promoted: 'investigations by Boltzmann and myself have just been published by which it acquires a universal mechanical significance; and although the investigations relative to this subject are not yet concluded, I believe they already make manifest that the introduction of this idea was dictated by the nature of the subject'.

In his reply Tait insisted that the question was 'Who first *correctly* adapted Carnot's magnificently original methods to the true Theory of Heat?' He accused Clausius of rushing into print 'with a proof which has afterwards to be explained and patched up' rather than waiting 'till one hits on a complete and irrefragable demonstration'. Indeed, he claimed that Clausius's first paper nowhere contained an explicit statement of the axiom, nor did it contain the phrase '*von selbst* [of itself]'. By implication, Thomson had been the first to offer a 'correct' demonstration, but Thomson had been too generous by far to Clausius.[66]

These disputes, ostensibly conducted over issues of scientific honour, thinly concealed deep divergences of scientific style between promoters of 'the science of energy' and its continental rivals. Thus, almost a year before the public controversy between Tait and Clausius, Maxwell asked Tait if he had been introduced to Clausius's concepts of 'Virial' and 'Ergal'. Ergal, he explained, 'is an old friend, being equal to the potential of a system on itself'. But Clausius had found new tasks for his latest creations:

> Clausius is now working along with those eminent artistes [Ergal and Virial] at the 2nd Law of $\Theta\Delta^{cs}$ but as far as I can see they have not yet furnished him with the dynamical condition of the equilibrium of temperature. This is got by the celebrated principle of Assumption and Resumption.[67]

Maxwell's wry comments on Clausius's scientific style reflected a profound distaste for certain kinds of 'theoretical' physics. That distaste, derived from the conviction that fundamental physical concepts (such as energy and entropy) should be capable of direct measurement, was one shared by his immediate circle and promoted most rigorously by Thomson who was later to point to similar evils within Maxwell's own electromagnetic theory (chs 11 and 14). In the present context the sin manifested itself in the very different approaches to thermodynamics advocated by German academic physicists.

Evidence of these sharply differing scientific cultures permeated the literature of the protagonists themselves. In his 1878 review of Tait's *Thermodynamics*, for example, Maxwell praised Clausius for his handling of entropy, and particularly for his establishment of the 'most important theorem in the whole science' that 'when heat passes from one body to another at a lower temperature, there is always an increase of the sum of the entropy of the two bodies, from which it follows that the entropy of the universe must always be increasing'. But he then reprimanded Clausius for his 'theoretical' approach to thermodynamics:

> Clausius, having begun by breaking up the energy of the body into its thermal and ergonal content, has gone on to break up its entropy into the transformational value of its thermal content and the disgregation.
>
> Thus both the energy and the entropy, *two quantities capable of direct measurement*, are broken up into four quantities, all of them quite beyond the reach of experiment, and all this owing to the actual heat which Clausius . . . suffered to remain in the body.[68]

By December 1873, however, Maxwell, 'under the conduct of Professor Willard Gibbs', had been 'led to recant an error I had imbibed from your [Tait's] $\Theta\Delta^{cs}$ namely that the entropy of Clausius is *unavailable energy* while that of T[homson] is available energy'. Rather, 'The entropy of Clausius is neither the one nor the other it is only Rankine's Thermodynamic function'. Gibbs's first scientific paper, 'Graphical Methods in the Thermodynamics of Fluids', had recently been presented to the Connecticut Academy in the United States but the young professor of physics at Yale University had taken steps to build up his personal scientific credibility through the distribution of reprints of his papers to everyone who mattered in the European field of thermodynamics. Maxwell quickly became an enthusiast for Gibbs's work and was soon making models of Gibbs's thermodynamic surfaces.[69]

Son of a Yale College professor, Josiah Willard Gibbs (1839–1903) had graduated from Yale in 1858 and enrolled as a student of engineering in the new graduate school. Like the first generation of British thermodynamicists (notably Rankine and James Thomson), this experience of academic engineering, with its geometrical and practical emphasis, guided his subsequent work. Receiving his Ph.D in 1863, he taught at Yale for three years before spending one year at each of the universities in Paris, Berlin and Heidelberg. Appointed professor of mathematical physics at his old university in 1871, his first paper took up Clausius's little-known concept of entropy given in differential form by the equation:

$$dS = dQ/T \text{ (for reversible processes)}$$

where S represented the entropy of the system, dQ the heat added and T the absolute temperature. The first law could be written as:

$$dU = dQ + dW$$

or

$$dU = dQ - PdV$$

where U represented the internal energy of the system, dW the external work done on the system, and P and V the pressure and volume respectively.

The differentials for heat and work were not differentials of a state function and were inexact or nonintegrable. In order to eliminate them, Gibbs combined these versions of the first and second laws of thermodynamics into the equation:

$$dU = TdS - PdV$$

The equation now contained only the state variables of the system, heat and work having been eliminated.[70]

Just as pressure–volume diagrams (Watt's 'indicator diagram' and Rankine's 'diagram of energy') had long formed part of the science of thermodynamics especially in its relation to practical engineering, so Gibbs introduced other diagrams such as temperature–entropy and volume–entropy. Here Gibbs had transferred his early enthusiasm for engineering science to one of the newest branches of physics, giving to his formulation of the science of thermodynamics that geometrical style which readily appealed to Maxwell but which initially found little favour among Continental physicists.

Gibbs's second paper (also published in 1873) extended his treatment to three dimensions. The geometrical properties of a surface came to represent the basic thermodynamic equation of a pure substance in its various phases of solid, liquid and gas, the three axes corresponding to energy, entropy and volume. The surface specifically allowed for a representation of the 'critical point', recently investigated by the Belfast chemist Thomas Andrews, at which the states of liquid and gas are identical.[71]

Extending his earlier work from pure substances to heterogeneous mixtures, Gibbs set out his distinctive approach to thermodynamic equilibrium in a massive paper 'On the Equilibrium of Heterogeneous Substances' (1875–8): 'It is an inference naturally suggested by the general increase of entropy which accompanies the changes occurring in any isolated material system that when the entropy of the system has reached a maximum, the system will be in a state of equilibrium'. His necessary and sufficient condition for the equilibrium of any isolated system was that 'in all possible variations of the state of the system which do not alter its energy, the variation of its entropy shall either vanish or be negative', that is, thermodynamic equilibrium entails maximum entropy of the system. Alternatively, the condition could be stated as one in which 'in all possible variations of the state of the system which do not alter its entropy, the variation of its energy shall either vanish or be positive', that is, as with the far more limited cases of mechanical equilibrium, thermodynamic equilibrium involves minimum energy.[72]

For cases of chemical equilibrium Gibbs extended his equation to take account of changes in internal energy brought about by changes in the mass of the chemical 'components'. These 'components' were chemical species whose quantities could be varied independently from one another. The additional term involved summing the products of the 'chemical potential' and the mass of each component. By the 'chemical potential' of a component Gibbs meant a rate of change of energy with respect to the mass of that constituent. To fulfil the equilibrium condition in a heterogeneous system consisting of several homogeneous phases (liquids, for example), Gibbs argued that the temperature, pressure and chemical potential of each independent chemical component must have the same values throughout the system. He then specified, as his 'phase rule', that the number of independent variations in a system of r coexistent phases having n independent chemical components was $(n + 2 - r)$.[73]

Gibbs's account of the conditions for chemical equilibrium may be readily illustrated by a simple chemical equation such as:

$$2H_2 + O_2 = 2H_2O$$

Equilibrium can be expressed by the equation in the form:

$$2H_2 + O_2 - 2H_2O = 0$$

Gibbs's equilibrium condition then required that the sum of the product of the integer preceeding each chemical component and the chemical potential of that component be zero.[74]

In a conference address at South Kensington in May 1876, Maxwell offered a contemporary interpretation of Gibbs's long and complex memoir. He began by distinguishing 'two lines along which dynamical science is working its way to undermine at least the outworks of Chemistry': molecular physics and thermodynamics. Indeed, he noted that 'the chemists of the present day, instead of upholding the mystery of their craft, are doing all they can to open the gates to the enemy'. Following the second line of attack, Gibbs, in Maxwell's opinion, seemed to 'enable us, without any lengthy calculations, to comprehend the relations between the different physical and chemical states of bodies'.[75]

Maxwell then pointed out that in studying the properties of a homogeneous mass of fluid consisting of n component substances, Gibbs took as his principal function the energy of the fluid. This function depended on the volume and the entropy, together with the masses of its n components, making $n+2$ independent variables. This class of variables constituted the *magnitudes* of the system.

Differentiating the energy with respect to each of the above variables yielded $n+2$ other variables. By differentiating with respect to volume, we obtained the pressure of the fluid (sign reversed); differentiating with respect to entropy yielded temperature on the thermodynamic scale; and differentiating with respect to the mass of any one of the component substances gave 'potential' (Gibbs's term). This new class of variables constituted the *intensities* of the system:

> These $n + 2$ new quantities, the pressure, the temperature, and the n potentials of the component substances, form a class differing from the first set of variables. They are not quantities capable of combination by addition, but denote the intensity of certain physical properties of the substance. Thus the pressure is the intensity of the tendency of the body to expand, the temperature is the intensity of its tendency to part with heat; and the potential of any component substance is the intensity with which it tends to expel that substance from its mass.[76]

Less committed to a programme of mechanical reduction and molecular physics than many of his continental contemporaries, Maxwell had perceived the power of

Gibbs's approach for bringing the territory of chemistry firmly and irreversibly into the empire of energy physics. But his account remained buried in conference reports largely unknown to late nineteenth-century chemists.

Although by 1873 Maxwell also expressed 'great respect for the elder of those celebrated acrobats, Virial and Ergal, the Bounding Brothers of Bonn' which he had earlier hoped might 'from their throncs be cast' to 'end their strife with suicidal yell', he nevertheless retained severe reservations about German approaches to thermodynamics. He had recently been studying not only Clausius but also Ludwig Boltzmann (1844–1906) whom he at first only knew 'as a student of the ultimate distribution of vis viva in a swarm of molecules'.[77] But, as Maxwell told Tait in August 1873, he and Boltzmann differed radically in their respective scientific styles:

> By the study of Boltzmann I have become unable to understand him. He could not understand me on account of my shortness and his length was and is an equal stumbling block to me. Hence I am very much inclined to join the glorious company of supplanters and to put the whole business in about six lines. Boltzmann's aim is to settle the equilibrium of kinetic energy among a finite number of bodies.[78]

In contrast to his reservations about German thermodynamics, Maxwell became increasingly impressed by the work of Gibbs. Thus in 1874 Maxwell urged Tait with respect to preparations for the second edition of *Thermodynamics* to 'read Prof. J. Willard Gibbs on the surface whose coordinates are Volume, Entropy and Energy ... He has more sense than any German'. Maxwell particularly drew Tait's attention to Gibbs's paper on the equilibrium of heterogeneous substances which he remarked was 'Refreshing after H[erbert] Spencer on the Instability of the Homogeneous'.[79]

Maxwell's disquiet about German thermodynamics, however, centred on the fundamental question of the nature of the second law. German physicists in general, he suggested, were engaged in a futile quest to reduce the second law to mechanics:

> it is rare sport to see those learned Germans contending for the priority of the discovery that the 2nd law of $\Theta\Delta^{cs}$ is the Hamiltonische Princip [Hamilton's principle of least action in dynamics] when all the while they *assume* that the temperature of a body is but another name for the vis viva of one of the molecules a thing which was suggested by the labours of Gay Lussac, Dulong &c but first deduced from dynamical statistical consider- ations by dp/dt [Maxwell]. The Hamiltonische Princip, the while, soars along in a region unvexed by statistical considerations while the German Icari flap their waxen wings in nephelo coccygia [cloud cuckooland] amid those cloudy forms which the ignorance and finitude of human science have invested with the incommunicable attributes of the invisible Queen of heaven.[80]

Writing to Tait in October 1876, Maxwell again accused Clausius and others of trying 'to degrade [Hamilton's dynamical principle of least action] into the 2nd law of $\Theta\Delta^{cs}$ as if any pure dynamical statement would submit to such an indignity'.[81]

At the end of 1877 *Nature* had invited Maxwell to review the second edition of Tait's *Thermodynamics*. The Cavendish professor seized the opportunity to confer credibility on those features of North British thermodynamics which distinguished the science from that of its continental practitioners. Unlike his impetuous friend Tait, however, he looked favourably on some, though only some, of Clausius's contributions. He focused on the second law in relation to the formulations of the 'three great founders' of the science, Rankine, Clausius and Thomson. At the time of Maxwell's review, preparations were under way for a collection of the late Professor Rankine's scientific papers, many on thermodynamic topics. Asked by Tait to assist in the weeding out of 'those papers of Rankine's which it would not do honour to his memory to print', Maxwell replied that he would 'not do the Ranking of rankine . . . Let the irreprintable be selected by editors who can appreciate what they are rejecting'.[82]

Rankine had long been one of the inner circle of the 'scientists of energy', a close and valued colleague of Thomson at Glasgow University from 1855 until his death in 1872, and one whose name appeared often in the correspondence between Maxwell and Tait. In 1876, for example, Maxwell told Tait that he thought 'Rankine by introducing his thermodynamic function ϕ which is $\int dQ/T$ made a great hit because ϕ is a real quantity whereas Q is not'. He also expressed the opinion that 'Rankine's statements are *identical* with C[lausius] but T° [Thomson] are only equivalent'. As he put the issue in his 1878 review, 'Both Rankine and Clausius have pointed out the importance of a certain function, the increase or diminution of which indicates whether heat is entering or leaving the body. Rankine calls it the thermodynamic function, and Clausius the entropy'.[83]

On the other hand, we have noted Maxwell's decreasing enthusiasm for molecular vortex theories (ch.11). He communicated to Tait his continuing scepticism regarding human knowledge of molecular mechanisms: 'With respect to our knowledge of the condition of energy inside a body, both Rankine & Clausius pretend to know something about it'. For Maxwell, all that human beings know is 'how much goes in and comes out and we know whether at entrance or exit it is in the form of heat or work but what disguise it assumes when in the privacy of bodies . . . is known only to R[ankine], C[lausius], and Co'.[84]

For their part, Clausius and Thomson had introduced the second law 'as an axiom on which to found Carnot's theorem that the efficiency of a reversible engine is at least as great as that of any other engine working between the same limits of temperature' (chs 5 and 6). Clausius's 1850 formulation had referred to 'the general deportment of heat, which everywhere exhibits the tendency to equalise differences of temperature, and therefore to pass from the warmer to the colder body'. While the 'axiom' was 'irrefragable' in its 'obvious and strict sense', Maxwell drew attention to a flaw in the 'demonstration' of Carnot's theorem. 'Proof' rested on showing that a violation of Carnot's theorem could lead to the communication of heat from a colder to a hotter body, that is, to a contradiction of the 'axiom'. But, Maxwell argued, even in this hypothetical process:

every communication of heat is from a warmer to a colder body. When the heat is taken from the cold body it flows into the working substance which is at that time still colder. The working substance afterwards becomes hot, not by communication of heat to it, but by change of volume, and when it communicates heat to the hot body it is itself still hotter.

It is therefore hardly correct to assert that heat has been transmitted from the colder to the hotter body. There is undoubtedly a transfer of energy, but in what form this energy existed during its middle passage is a question for molecular science, not for pure thermodynamics.[85]

Maxwell also evaluated Clausius's 1864 modification of the 'axiom' into the form 'that heat cannot of itself pass from a colder to a warmer body'. Explaining that the phrase 'of itself' was not aimed at the exclusion of self-acting machinery such as an engine, driven by the passage of heat between 200° and 100° and driving 'a freezing machine so as to take heat from water at 0°, and so freeze it' or the same engine driving a friction brake 'so as to generate heat in a body at 500°'. In order to rule out such exceptions, Maxwell thus claimed that we would need to exclude 'all bodies except the hot body, the cold body, and the working substance'. In this claim we may note the similarity to Thomson's 1851 formulation of Clausius's 'axiom': 'It is impossible for a self-acting machine, unaided by any external agency, to convey heat from one body to another at a higher temperature'.

Clausius's final modification of his axiom appeared in the second edition of his *Heat* (1876) when the phrase 'without compensation' replaced 'of itself'. In Maxwell's view, this phrase, properly interpreted, yielded 'a complete and exact' statement of the axiom, but 'in order to understand it we must have a previous knowledge of the theory of transformation-equivalents, or in other words of entropy, and it is to be feared that we shall have to be taught thermodynamics for several generations before we can expect beginners to receive as axiomatic the theory of entropy'.

Thomson's 1851 axiom, that 'It is impossible, by means of inanimate material agency, to derive mechanical effect from any portion of matter by cooling it below the temperature of the surrounding objects' (ch.6), likewise came under scrutiny in Maxwell's review. He argued that if we allowed air to expand, 'we may derive mechanical effect from it by cooling it below the temperature of the coldest of surrounding objects'. Thus some further restriction, such as the condition that 'the material agency is to be left in the same state at the end of the process as it was at first, and also that the mechanical effect is not to be derived from the pressure of the hot or of the cold body', was needed to render the axiom strictly true.[86]

The overall aim of Maxwell's critique of the foundations of thermodynamics was to provide, once more, a credible understanding of the peculiar nature of the second law: 'It is probably impossible to reduce the second law of thermodynamics to a form as axiomatic as that of the first law, for we have reason to believe that though true, its truth is not of the same order as that of the first law', that is:

The second law relates to that kind of communication of energy which we call the transfer of heat as distinguished from another kind of communication of energy we call work. According to the molecular theory the only difference between these two kinds of communication of energy is that the motions and displacements which are concerned in the communication of heat are those of molecules, and are so numerous, so small individually, and so irregular in their distribution, that they quite escape all our methods of observation; whereas when the motions and displacements are those of visible bodies consisting of great numbers of molecules moving all together, the communication of energy is called work.[87]

If we then supposed 'our senses sharpened to such a degree that we could trace the motions of molecules as easily as we now trace those of large bodies . . . the distinction between work and heat would vanish' as the communication of heat would be seen as a communication of energy in the form of work.

Foundations for the second law could thus be laid from two directions. First, we construct the foundation from 'our actual experience in dealing with real bodies of sensible magnitude', that is, 'in thermodynamics proper, in which we deal only with sensible masses and their sensible motions'. And second, we deduce the second law from molecular science, that is, 'on the hypothesis that the behaviour of [gross] bodies consisting of millions of molecules may be deduced from the theory of the encounters of pairs of molecules, by supposing the relative frequency of different kinds of encounters to be distributed according to the laws of probability'. Unlike the first law which, as the principle of energy conservation, could 'be proved mathematically true if real bodies consist of matter . . . acted on by forces having potentials', the truth of the second law was statistical and not mathematical, depending as it did 'on the fact that the bodies we deal with consist of millions of molecules, and that we can never get hold of single molecules'.

Referring to Thomson's 1874 paper 'On the Kinetic Theory of the Dissipation of Energy', Maxwell explained that Thomson had 'shown how to calculate the probability of the occurrence within a given time of a given amount of deviation from the most probable distribution of a finite number of molecules of two different kinds in a vessel', with specific illustration for the case of gas diffusion. He suggested that Thomson's method might be extended to other cases of energy dissipation resulting from molecular action such as heat conduction and internal friction. In all these cases, he claimed, there was a tendency towards a state of 'equilibrium of exchanges of the molecules and their momenta and energies between the different parts of the system'.[88]

At the same time, individual molecules, or a finite number of molecules, deviated 'very considerably from the theoretical mean of the whole system, because the molecules which form the group do not submit their procedure as individuals to the laws which prescribe the behaviour of the average or mean molecule'. This conclusion had major consequences for the second law which:

> is continually being violated, and that to a considerable extent, in any sufficiently small group of molecules belonging to a real body. As the number of molecules in the group is increased, the deviations from the mean of the whole become smaller and less frequent; and when the number is increased

till the group includes a sensible portion of the body, the probability of a measurable variation from the mean occurring in a finite number of years becomes so small that it may be regarded as practically an impossibility.

Thus we had reason for believing 'the truth of the second law to be of the nature of a strong probability, which, though it falls short of certainty by less than any assignable quantity, is not an absolute certainty'. It followed that attempts to deduce the law from purely dynamical principles without introducing any elements of probability, could not adequately explain the law. Maxwell therefore concluded his review by indicating that those attempts by Clausius, Szily von Nagy-Szigeth (1838–1924, professor of mathematical physics in Budapest) and J.H.J. Müller (1809–1875, professor of physics at the University of Freiburg) had 'already indicated their unsoundness by leading to determinations of physical quantities which have no existence, such as the periodic time of the alternations of the volume of particular gases'.[89]

Long after Maxwell had gone, Thomson and Tait maintained this powerful commitment against disembodied mathematical theory and in favour of visualizable processes and experimentally-grounded concepts. Thus Tait's 1892 review of Poincaré's *Thermodynamique* denounced the text as one which 'exhibits a lavish, almost a reckless, use' of mathematical power, rendering it of little value to the student. Indeed, in choosing 'to play almost exclusively the part of the pure technical analyst' the author 'looks with indifference, if not with absolute contempt, on the work of the lowly experimenter'. In content, as well as style, Tait condemned the Frenchman's text above all for 'the entire ignoration of the true (i.e. the statistical) basis of the second Law of Thermodynamics' and for giving 'an altogether imperfect notion of the true foundation for the reckoning of absolute temperature'. Poincaré had been guilty in the eyes of the British school of 'energy physics' of introducing theoretical terms, like Clausius's 'disgregation', 'as mere empty names employed to conceal our present ignorance'. Endorsing fully Tait's indictment, Lord Kelvin (as Thomson had recently become) noted tersely but tellingly on Tait's proof sheet: 'OT OK K'!![90]

CHAPTER 13

Energy and Electricity: 'the Apparatus of the Market Place'

> In order to compare things measured by different persons it is necessary to assume a standard of measure. Some one must indicate a particular specimen of the thing to be measured as the standard of that quantity and every one else must agree to measure by that standard or by copies of it. This method introduces experiment into mathematics, and greatly disturbs the easy elegance of the adherents of proportions [that is, mathematicians] who are not accustomed to the apparatus of the market place . . .
>
> *James Clerk Maxwell tries his hand at a draft exposition of mechanical principles*
> *(c.1857)*[1]

Constructed by the scientists of energy during the 1860s, a work-based absolute system of measurement (based on fundamental units of mass, length and time) was simultaneously designed as a scientific and commercial system of measurement for imperial and international ends. As Thomson advised Maxwell in 1872: 'when electrotyping, electric light, &c become commercial we may perhaps buy a microfarad or megafarad of electricity. . . . I am glad you agree with me that if there is to be a name given it had better be given to a real purchaseable tangible object, than to a quantity of electricity'. Maxwell himself had long been convinced of the importance of universal units and standards for scientific and commercial exchange: 'The man of business requires these standards for the sake of justice, the man of science requires them for the sake of truth, and it is the business of the state to see that our . . . measures are maintained uniform'.[2]

This chapter highlights the historical contingency of a work-based absolute system of measurement. Scientists of energy promoted a system which *appeared* at first sight to follow that of Carl Friedrich Gauss (1777–1855) and Wilhelm Weber (1804–91) but which was in fact grounded on 'work' rather than 'force' as the great connecting link between all physical measurements. As presented by Gauss and Weber, an absolute system of measurement might have remained the property of elite physicists. In the hands of the scientists of energy, the absolute system was transformed into a package embodying a harmony of theory and practice, designed to appeal at once to natural philosophers, engineers, and commercial electricians. Exploiting their dominent role in key BAAS Committees, the scientists of energy supplied telegraph engineers with the accurate standards of resistance demanded, whether for the economical location of faults or for the increasingly expensive contracts with cable manufacturers.[3]

From the mid-1850s submarine telegraphy had become the most public context for the practice of British electrical (especially electromagnetic) science. British telegraph engineers and construction companies increasingly dominated the world market for underwater cables which served above all to unify the British Empire politically, economically and militarily. At the same time Faraday's investigation of the role of dielectrics – materials which allowed transmission of electricity yet were non-conductors – came to be seen as fundamental to an understanding of underwater cables consisting of a copper conducting core surrounded by a gutta percha insulator which functioned as the dielectric.[4] With emphasis upon the intervening medium and the energy that it contained, British field theorists constructed a major rival to continental theories of electromagnetism (chs 11 and 14).

Recognizing the demand for accurate electrical standards, the scientists of energy established a British Association Committee whose chief function was the determination of a standard of electrical resistance by various experimental methods. Conducted principally through the BAAS, their campaign to promote an absolute system resulted in a distinctive role for the physicist in electrical engineering, arousing no small amount of opposition from established practitioners of the art, particularly the celebrated German electrician and founder of the great electrical manufacturing company, Werner Siemens (1823–83). Maxwell, however, pointed to wider implications in the preface to his *Treatise*:

> The important applications of electromagnetism to telegraphy have also reacted on pure science by giving a commercial value to accurate electrical measurements, and by affording to electricians the use of apparatus on a scale which greatly transcends that of any ordinary laboratory. The consequences of this demand for electrical knowledge, and of these experimental opportunities for acquiring it, have been already very great, both in stimulating the energies of advanced electricians, and in diffusing among practical men a degree of accurate knowledge which is likely to conduce to the general scientific progress of the whole engineering profession.[5]

'Accurate measurement' characterized much of the theory and practice of electrical research in the second half of the nineteenth century. The opening pages of Maxwell's *Treatise*, for instance, were laden with remarks concerning measurement: 'The most important aspect of any phenomenon from a mathematical point of view is that of a measurable quantity. I shall therefore consider electrical phenomena chiefly with a view to their measurement, describing the methods of measurement, and defining the standards on which they depend'.[6] It was also no coincidence that, beginning with Thomson's Glasgow physical laboratory in the

1850s, this period marked the establishment of British university physics laboratories for teaching and research: for example, Tait's at Edinburgh (1868), the Clarendon at Oxford (1870), and the Cavendish at Cambridge (1874).[7]

Heavily involved in the production of instruments, data, units and standards, the scientists of energy rapidly facilitated and created demand from electricians, telegraph projectors and the state. Correspondingly, their credibility rose to its highest level in the imperial and international markets. The deployment of an absolute system in networks of imperial telegraph cables represented an explicit embodiment of energy physics in space: the cultural and physical spaces of imperial power that exerted discipline, control and authority over a quarter of the earth's peoples. As Henri Lefebvre's perceptive analysis suggests, the ideology of a successful cultural elite must generate space, must become inscribed in space if it is to avoid disappearing into disembodied and impotent realms of mere signs, abstract descriptions and fantasies. This inscription is precisely what the scientists of energy achieved in the 1860s.[8]

'Absolute' measurements in magnetism and electricity: Gauss and Weber

German astronomers had long been entrusted by their state patrons with tasks of mapping not merely the heavens but also the German lands. Accurate surveying required precision optical and other measuring instruments which by the early nineteenth century were being designed and constructed by German makers to the highest available standards. To accommodate these instruments and their skilled users new observatories in Berlin, Munich and Göttingen had opened by 1830, providing Gauss especially with the resources for a new programme of physical research.[9]

Alexander von Humboldt (1769–1859) had already set up in 1828 a small observatory for earth magnetism in Berlin, from which he attempted to coordinate measurements with observers elsewhere. At the University of Göttingen, however, Gauss was characteristically contemptuous of his predecessors for 'so much that is vague, meaningless, illogical' both in their theories and their instrumentation. He thus constructed new foundations for the theory and practice of earth-magnetic science. Those foundations began with the 'most precise observations [which] can be expected only of those mathematicians who are familiar with the finest means of observation, namely, the practical astronomers'.[10]

Initially concerned with terrestrial magnetism, Gauss devised precision instruments to measure parameters such as magnetic declination (the angular deviation of the compass needle to the east or west of true north) and magnetic intensity. An informal 'Magnetic Union', largely controlled by Gauss and Weber from Göttingen's new magnetic observatory, promoted world-wide collaboration in the quest for a greater understanding of the nature of the earth's magnetism.[11]

Inspired in part by Gauss's work, British gentlemen of science persuaded their own government of the prestige to be gained by the launch of a 'Magnetic Crusade' in the late 1830s to map the earth's magnetism on an Empire-wide scale, including the little-known Southern polar regions. Mapping was above all about power. Blank spaces on maps, whether geographical or magnetic, implied a power vacuum. Through a co-ordinated campaign of navy, army and men of science Britain eliminated many of those blank spaces and, by means of her Admiralty charts, asserted world-wide maritime power.[12]

The 'Magnetic Crusade' paralleled other BAAS concerns such as meteorology (the science of the earth's atmosphere) and tidology (the science of the tides). Systematic and accurate *field* measurements, often expressed in visual, geometrical form, provided the basis for the establishment of quantitative laws and, ultimately, for a unifying mathematical theory. Whether in the field, in the observatory, or in the laboratory, accurate measurement quickly became the hallmark of nineteenth-century physical science.[13] In the preface to his *Treatise* Maxwell would acknowledge the specific debt perceived by his own generation to Gauss:

> [Gauss] brought his powerful intellect to bear on the theory of magnetism, and on the methods of observing it, and he not only added greatly to our knowledge of the theory of attractions, but reconstructed the whole of magnetic science as regards the instruments used, the methods of observation, and the calculation of the results, so that his memoirs on Terrestrial Magnetism may be taken as models of physical research by all those who are engaged in the measurement of any of the forces of nature.[14]

The appointment of the young Wilhelm Weber to the vacant chair of physics at Göttingen in 1831 initiated a collaboration on earth magnetism with Gauss which lasted well beyond Weber's dismissal from the chair in 1837 following his refusal to take the oath of loyalty to the new Hannoverian king. By about 1832 Weber was carrying out experiments according to Gauss's specification, while Gauss himself sought to construct instruments adapted to the measurement of magnetic quantities. At the end of the year Gauss presented a paper concerned with the determination of the intensity of earth magnetism not in relative measure but in absolute units, that is, in mechanical measures. As Weber extended his own researches from magnetism to electricity and electromagnetism during the years that followed, absolute measurement continued to serve as the foundation for his experimental and theoretical investigations.[15]

With its international readership, Poggendorff's *Annalen* provided an appropriate 'universalizing' medium for Weber's discourse 'On the Measurement of Electric Resistance According to an Absolute Standard' (1851). 'If there are measures for time and space', Weber began, 'a special *fundamental measure* for *velocity* is not necessary; and in like manner no special *fundamental measure for electric resistance* is needed if there are measures for electromotive force and for intensity of the current. . . . Upon this depends the reduction of the measurements of electric resistance to an absolute standard'.[16]

Arguing strongly for the superiority of such an absolute system over a 'specific' or relative one, Weber claimed that if copper were employed as the standard '*the specific resistance of copper must be given as unit* for the specific resistance of conducting substances', that is, 'a special *fundamental measure for specific resistances* would be necessary'. In this case, then, there would be no saving in the number of the fundamental measures. Furthermore, on account of differences in resistance of even the most chemically pure of metals, 'if one physicist referred his rheostat and multiplicator to copper wire a metre in length and 1 millimetre thick, other physicists could not be sure that his copper wire and theirs had the same *coefficient of resistance*, that is, whether the *specific* resistance of all these wires was the same'. In other words, the superiority of an absolute system lay in its advantages as a universal, rather than a merely local, system of measurement.[17]

Weber's system was founded on absolute measures of electromotive forces and intensities. Electromotive force was to be measured in terms of the force exerted by the earth's magnetism upon a closed conductor. Electromotive intensity was defined as 'the intensity of that current which, when it circulates through a plane of the magnitude of the unit of measure, exercises, according to electro-magnetic laws, the same action at a distance as a bar magnet which contains the unit of measure of bar magnetism'. Underlying these definitions was the assumption that Gauss had provided absolute measures of bar and earth magnetism in his 1833 memoir. Weber's absolute system for electric resistance was thus largely an extension of Gauss's absolute system for magnetic measurement, resting on *force* laws, the laws of magnetic forces acting at a distance. Weber showed that in fact his unit of resistance depended merely on the measure of a velocity. But the extreme smallness of this unit (less than 10^{-8} times the resistance of a mile of copper wire), together with differences among his own experimental determinations, rendered his system distinctly unattractive to telegraph engineers.[18]

The remainder of the paper was devoted to detailed analysis of the practical methods of measuring electric resistance according to an absolute standard and to comparisons of the results with a standard of resistance established by M.H. Jacobi. A final section considered the determination of the constants involved in the laws of electrical action, particularly his own 'general fundamental law' which represented electric force as an expression of the electric masses, their distance, relative velocity and acceleration.[19] But it was Weber's absolute system of measurement, rather than his electromagnetic theory, which would prompt the North British scientists of energy to rebuild the system into one based not upon an abstract concept of force but upon the practical measure of work.

'Absolute' measurements and mechanical effect: Thomson's perspective

By 1845, the year of his graduation from Cambridge as Second Wrangler and First Smith's Prizeman, 21-year-old William Thomson had already published in the pages of the *Cambridge Mathematical Journal* a series of papers on the mathematical theory of electricity. While composing these early papers, William was very conscious of living in the shadow of the mighty Gauss. In 1842, for instance, the undergraduate wrote home to admit that 'as I feared, he [Gauss] has proved exactly the same theorems as I had done in my previous paper, and in nearly the same way'. In 1843 his brother John, attending the Glasgow course of natural philosophy, explained that Professor Nichol 'had given me some of (your rival!!) Gauss's observations regarding magnetism to translate for him'.[20]

William's initial discussions with his brother James concerning the production of mechanical effect in heat engines (with particular reference to the waterfall analogy) had taken place in August 1844 (ch.3). Shortly afterwards, he began thinking over the physical meaning of a general minimum condition of Gauss for proving the existence and uniqueness of electrical distributions on conducting surfaces. His answer to something which had 'long appeared very mysterious' to him lay in regarding Gauss's minimized function as an expression of the *mechanical effect* or work contained in the system. The electrical intensity would distribute itself so as to reach the lowest level of total mechanical effect just like the working substances in waterwheels and heat engines. But it was during the spring of 1845, while voluntarily assisting Regnault in his physical laboratory, that he developed more fully his solution.[21]

Regnault's accurate measurements on the properties of steam and other gases were being funded by the French government with a view to improving the efficiency of heat engines. Thomson sought to acquire from Regnault experimental skills in order to enhance his prospects as a candidate for the Glasgow chair of natural philosophy. Regnault, however, also inspired Thomson's life-long enthusiasm for accurate measurement in laboratory research. He much later acknowledged his principal debt to the French *physicien* as 'a faultless technique, a love of precision in all things, and the highest virtue of the experimenter – patience'.[22]

These interests coincided with Thomson's continuing concerns with the mathematical theory of electricity. The editor of the French *Journal de mathématique*, Joseph Liouville (1809–82), had asked him to prepare a paper reconciling Poisson's action-at-a-distance theory of electrical forces with experiments of Faraday and others which appeared to undermine Poisson. Focusing on the problem of two electrified spherical conductors, Thomson had before him both the measurements of the experimental philosopher William Snow Harris, purporting to disprove the inverse square law for the force between two such conductors, and the failure of Poisson to obtain a general mathematical solution to the two spheres problem because his sum over the microscopic forces that he postulated between point masses of electricity had yielded exceedingly complex integrals.[23]

Thomson's 'reconciliation' employed his previous insight into the physical meaning of Gauss's minimum condition in terms of the total mechanical effect contained in an electrical system. The mechanical effect given out or taken in by an electrical system was precisely analogous to the work done or work absorbed by a waterwheel or heat engine. In the electrical case, potential was analogous to the height of a waterfall or the temperature of the boiler, while quantity of electricity was analogous to mass of water or quantity of heat. Thus the mechanical effect produced by an infinitesimal change in distance between two conducting spheres would be given by the total ponderomotive force (Thomson's term for the force tending to move masses) exerted by the spheres upon one another multiplied by that change in distance. The resulting expression for the total ponderomotive force enabled him to make detailed calculations showing that the inverse square law was not violated. Ponderomotive force was to be conceptualized, not in terms of the sum of microscopic forces acting at a distance between point masses of electricity, but in terms of total work contained in the electrical system.[24]

Thomson simultaneously recognized that his new conception would allow *mechanical* or *absolute* measurements to be made of electrical quantities: 'This [solution] has confirmed my resolution to commence experimental researches, if ever I make any, with an investigation of the absolute force of statical electricity. As yet each experimenter has only compared intensities by the dev[iatio]^{ns} of their electrometers. They must be measured by pounds on the square inch, or by "atmospheres"'.[25] Measurements of electrical phenomena and of steam were to be regarded in precisely the same way: in absolute, mechanical and, above all, practical, engineering terms. The contrast not only to Laplace and Poisson, but especially to Faraday's non-mechanical perspective, was striking.[26]

Under his new system electrical 'pressure' would be measured by a simple balance, the very 'apparatus of the market place' as Maxwell would term it. Constructed in 1853 and specifically designed to measure electrical potential, Thomson's first absolute electrometer employed this principle by balancing the total force between two spherical conductors against weights. His subsequent instruments used flat parallel discs in which potential between them was inversely proportional to the distance and total force inversely proportional to the square of that distance.[27]

In these cases Thomson interpreted the ponderomotive forces in terms of a gradient in mechanical effect or capacity for work. 'Potential' was no longer seen as a potential function for forces between point masses but as potential energy with a capacity for doing work. As he expressed the point in 1860: 'The amount of work required to move a unit of electricity from any one position to any other position, is equal to the excess of the electric potential of the second position above the electric potential of the first position'.[28] In the same way the balance principle was construed in terms of the principle of virtual velocities (ch. 10): the forces balanced when the quantities of work done during any movement cancelled one another. The system as a whole had reached minimum mechanical effect.[29]

The practicality of this approach meant that electricity could be treated without reference to any specific theory concerning its nature. Now its quantity was to be measured simply in terms of the work done by the fall of a quantity of electricity through a potential, just in the way that work was done by the fall of a mass of water through a height. In 1848 Thomson's absolute scale of temperature utilized the same notion of absolute measurement which had already been developed for the electrical case (ch.3). Also in that year Thomson examined ponderomotive forces acting on inductively magnetized objects in terms of mechanical effect, a move integral to his and Maxwell's translation of action-at-a-distance into 'field' theory.[30]

Thomson's first public expression of his commitment to a system of absolute units for electrical measurement coincided both with his reading of Weber's contribution to Poggendorff's *Annalen* and with the peak of activity on the 'Dynamical Theory of Heat' series in late 1851. Published in the *Phil. Mag.*, 'Applications of the Principle of Mechanical Effect to the Measurement of Electro-motive Forces, and of Galvanic Resistances, in Absolute Units' opened with a reminder of an earlier presentation to the BAAS (1848) which applied the principle of mechanical effect to electric currents (ch.5). Now he claimed to have extended the original application and to have 'found that the same general principle of mechanical effect is sufficient to enable us to found on a few elementary facts, a complete theory of electro-magnetic or electro-dynamic induction'. But his primary focus was on 'practical applications with reference to the measurement of electro-motive forces arising from chemical action, and to the system of measurement of "galvanic resistance in absolute units", recently introduced by Weber'.[31]

Thomson showed that 'universally, the mechanical value of the work done in a unit of time by an electro-motive force, F, on a circuit through which a current of strength γ is passing, is $F\gamma$' where F and γ are given in Weber's absolute electro-magnetic units. Taking the mechanical value of a quantity of heat H to be JH, where J is Joule's mechanical equivalent of heat, Thomson found for the heat produced by a current ($JH = F\gamma$). Substituting from Ohm's law ($k = F/\gamma$) gave an expression for the resistance ($k = JH/\gamma^2$) in absolute measure. He concluded:

> It is very much to be desired that Weber's direct process, and the indirect method founded on estimating, according to Joule's principles, the mechanical value of the thermal effects of a galvanic current, should be both put into practice to determine the absolute resistance of the same conductor, or that the resistance of two conductors to which the two methods have been separately applied, should be accurately compared. . . . it would afford a most interesting illustration of those principles by which Mr Joule has shown how to trace an equivalence between work spent and mechanical effect produced, in all physical agencies in which heat is concerned.[32]

Over the next decade Thomson's activities became increasingly centred on two closely-interwoven issues: the development of his physical laboratory in Glasgow College; and the mathematical and experimental problems of long-distance submarine telegraphy. From the 1840s the practical employment of electromagnetic effects for a vast continental network of land telegraph lines both in Europe and North America had been well under way. By 1851, the year of the first great international exhibition at the Crystal Palace in London's Hyde Park, a viable submarine telegraph cable connected England and France. Accompanying the realization of its commercial possibilities came an increasing demand for the accurate measurement of electromagnetic quantities. In particular the swift and inexpensive location of faults in underground and underwater cables required resistance coils calibrated in terms of feet of copper wire.[33]

Unforeseen retardation effects in long submarine cables, first observed by Werner Siemens, threatened the financial future of many of the ambitious telegraph projects. Faraday's qualitative diagnosis of the problem as one of treating the cable as a Leyden jar of vast capacity for electric charge provided the starting point for Thomson's mathematical analysis in 1854. His 'law of squares' showed the dependence of the retardation effect on resistance and inductive capacity, and created a new market for accurate knowledge of electrical quantities. One function of his physical laboratory was to provide that data. Employing his 1851 method of determining resistances in absolute measure, he found great variation in the resistance of different specimens of supposedly pure copper wire manufactured by different firms for use in submarine telegraphs. The effects on the commercial transmission of signals over long distances would be to reduce profitability and perhaps even render the project unworkable. Accurate measurement of resistances during manufacture would introduce quality control, and hence greater commercial stability, into the telegraph industry.[34]

By the late 1850s Thomson had begun a close association with the first trans-Atlantic project. As Maxwell noted wryly: 'I was writing great screeds of letters to Professor Thomson about those [theoretical problems of Saturn's] Rings, and lo! he was a-laying of the telegraph which was to go to America, and bringing his obtrusive science to bear upon the engineers, so that they broke the cable with not following (it appears) his advice'.[35] Although completed, the 1858 cable ceased transmitting signals after a few weeks. Other long-distance cables shared much the same fate. The result was a joint Board of Trade/Atlantic Telegraph Company Inquiry which published its findings in 1861. Given the scientific representation on the Committee of Inquiry, it was not surprising that the practical electrician W.O.W. Whitehouse was made to represent a 'trial-and-error' philosophy and thus became the principal scapegoat for telegraphic failures. The Inquiry's findings came down heavily in favour both of a laboratory-centred approach, characterized by accurate measurement, and of a system of absolute units. This Inquiry provided one stimulus for setting up a BAAS Committee to consider and agree units and standards of electrical resistance.[36]

A second, related stimulus for the BAAS telegraphers was the *Phil. Mag.* translation early in 1861 of Werner Siemens's recent 'Proposal for a new Reproducible Standard Measure of Resistance to Galvanic Currents'. Siemens claimed that his method had been suggested by the 'want of a generally received standard measure of current resistance, and the great inconveniences thence arising, especially in technical physics'. On the one hand, he rejected any standards based on metal wires, largely on account of problems with purity of the metal, the molecular state of the wire and constancy over time.[37] On the other hand, Siemens had no liking for absolute standards:

> The absolute standard measure appeared to me equally ill adapted to general use. It can only be reproduced by means of very perfect instruments, in places especially arranged for the purpose, and by those possessed of great manual dexterity; and moreover it is liable to the grave practical objection, that it does not exhibit itself in a physical form; and lastly, the numbers it involves are exceedingly inconvenient on account of their magnitude.[38]

In a follow-up account Siemens further denounced the absolute system as one fraught with too great an inaccuracy even for technical purposes: 'the most experienced and skilful physicists, even with the best instruments and most appropriate localities, are not able to determine resistance in absolute measure which do not vary several per cent'. These claims brought a direct response from Augustus Matthiessen, FRS, who was concerned to defend his own proposed metal alloy standards. Matthiessen's reply also quoted ringing endorsements by Thomson of the value and accuracy of Weber's system of absolute measurement.[39]

Siemens's preferred method was his own mercury standard based on the resistance at 0°C of a column of pure mercury one metre in length and one square millimetre in cross-section. In his view, this approach offered the 'only practicable method of establishing a standard measure which should satisfy all requirements, and especially which any one could reproduce with ease and with sufficient accuracy'. Mercury had the advantages of almost perfect purity, a consistent molecular state, a resistance more independent of temperature than other metals, and a high specific resistance which yielded small and convenient numerical comparisons founded on it as the standard.[40]

Siemens's refrain concerned the key issue of reproducibility. Although the issues were by no means simply a case of 'practical men' versus 'scientific men', his hostility to the absolute system appeared to derive from the belief that practical engineers were being excluded from control of the standards. As a practical man himself, Siemens clearly resented the prospect that physicists would control the selection and determination of standards, leaving engineers dependent upon them. His hostility to absolute systems in general can also be best understood as stemming from a reaction against the force physics of Gauss and Weber rather than from Thomson's absolute system which was rooted largely in a practical tradition of work.[41]

'Absolute' measurements and BAAS standards

At the Manchester meeting of the British Association (1861) the telegraph engineers Latimer Clark and Sir Charles Bright tabled a paper appealing for a standardized system of electrical units. Unable to attend in person on account of a serious leg injury, Thomson had already been preparing his own ground for the establishment of a BAAS Committee 'On Standards of Electrical Resistance'. What might have been an embarrassing clash between practical electricians and natural philosophers was largely avoided through the diplomatic skill of the most recent recruit to the North British scientific network, Fleeming ('the Fleemingo') Jenkin.[42]

Although born in Kent, Jenkin had been educated alongside Tait and Maxwell at Edinburgh Academy. Family circumstances took him to Frankfurt, Paris and Genoa, and it was from the University of Genoa that he graduated Master of Arts with first class honours before returning to Britain in 1851. Apprenticed to Fairbairn's mechanical engineering works in Manchester and then appointed as a draughtsman at Penn's marine engineering works in Greenwich, Jenkin subsequently worked closely with Lewis Gordon on telegraphic engineering from 1857. Gordon introduced him to Thomson two years later, and so began a long association which ranged from the marketing of joint telegraphic patents to common perspectives on cosmology, evolutionary theory, political economy, and free will. But it was his work on the BAAS Committee, most notably his advocacy of an absolute system of units, which would link him most conspicuously to 'the science of energy' in the 1860s. Jenkin's loyal campaigning on behalf of the scientists of energy was soon rewarded with an appointment to a new engineering chair at the University of Edinburgh in 1868.[43]

In Thomson's absence from the 1861 BAAS meeting, Jenkin acted on his behalf. Reporting on the reception afforded Thomson's papers on the age of the earth and sun (chs 7 and 9), Jenkin affirmed that the 'absolute unit question also progresses well' and suggested that Thomson's advance lobbying had begun to pay dividends:

> [Dr Augustus] Matthiessen got your letter – [Sir Charles] Wheatstone framed a resolution not altogether a bad one and the Committee is appointed and has (Committee Recommendation consenting) £50 for its use – You and I are to redetermine Weber's Unit; compare it with the one he will furnish – Matthiessen and Williamson are to try the permanency of wire and its reproductibility (?) (my word) – Today Sir C. Bright and Latimer Clark have come down today [sic] and want to get on the Committee which will report on the two subcommittees above named.[44]

Jenkin further reported that the General Committee would probably be made up of Thomson, Wheatstone, Matthiessen, W.H. Miller, A.W. Williamson (all FRS's) and Jenkin himself. This list, later made official, also confirmed the initial dominance of scientific over practical men, although by 1863 the Committee had been expanded to include three practical electricians and telegraphers (Bright, Cromwell Varley and C.W. Siemens) closely allied to Gordon, Thomson and Jenkin, and three additional 'energy' people (Joule, Stewart and Maxwell).[45]

In the same letter Jenkin also noted that Bright and Clark 'seem to have found the necessity of a dependence between the various electrical units and had even prepared a paper on the subject but I got hold of them and pointed out that it was all done already'. Contrary to the impression given in the published *BAAS Report*, therefore, Jenkin's account suggests that Bright and Clark played no role in the setting up of the Committee. Thus at the news that the Committee had already been fixed, 'Latimer Clark looked delighted and is eager to have it all explained – they laid their paper on the table without reading it, I understand, and it will be referred to our Committee'.[46]

The specific aim of the BAAS Committee, pursued in six principal reports published between 1862 and 1869, was to determine, first, 'what would be the most convenient *unit* of resistance' (both 'practical and purely scientific') and second, 'what would be the best form and material for the *standard* representing that unit'. The Committee emphasized the distinct and independent nature of these two tasks: the former concerned the adoption of a system of units (such as an absolute system), while the latter concerned the choice of metal (mercury or gold, for example) and construction of the standard.[47]

In its choice of a unit of resistance the Committee faced the problem of satisfying simultaneously several criteria. On the one hand, a foot of copper or a metre of mercury, like other units 'founded simply on some arbitrary length and section or weight of some given material more or less suited for the purpose', did not possess the desired qualities of definite relations to other units of the system and to the unit of work in particular. In this respect they did not appear 'very scientific standards'. Nevertheless, 'they produced a perfectly definite idea in the minds even of ignorant men, and might possibly, with certain precautions, be both permanent and reproducible'. On the other hand, Weber's unit satisfied the relationship to other units but, having 'no material existence', fell short on the criteria of permanence and reproducibility:

> for if the absolute or Weber's unit were adopted *without qualification*, the material standard by which a decimal multiple of convenient magnitude might be practically represented would require continual correction as successive determinations made with more and more skill determined the real value of the absolute unit with greater and greater accuracy. Few defects could be more prejudicial than this continual shifting of the standard. This objection would not be avoided even by a determination made with greater accuracy than is expected at present, and was considered fatal to the *unqualified* adoption of the absolute unit as the standard of resistance.[48]

The Committee therefore proposed a compromise combining the advantages of the arbitrary material standard *and* the absolute system: 'a material standard should be prepared in such form and materials as should ensure the most absolute permanency; that this standard should approximate as nearly as possible in the present state of science to ten millions of metre/seconds [the absolute value], but that, instead of being called by that name, it should be known simply as the unit of 1862, or receive some other simpler name'. Then, 'from time to time, as the advance of science renders this possible, the difference between this unit of 1862 and the true ten millions of metre/seconds should be ascertained with increased accuracy, in order that the error resulting from the use of the 1862 unit in dynamical calculations instead of the true absolute unit may be corrected by those who require these corrections, but that the material standard itself shall under no circumstances be altered in substance or definition'.[49]

In order to implement the proposal, both 'the degree of accuracy with which the material standard could at present be made to correspond with the metre/second' and 'the degree of permanency which could be ensured in the material standard when made' had to be determined. With respect to the former question, Thomson had sent resistance coils to Weber for an accurate determination of their resistance. The Committee, however, felt that 'the results of no one man could be accepted without a check'. Thus Thomson and Jenkin began to use an independent method to check on Weber's result, but pressure of work in connection with the International Exhibition delayed the experiments. The Committee recommended that if Weber's results accorded within one per cent of the new determinations then 'provisional standards shall be made of German-silver wire . . . and at once issued to all interested in the subject, without waiting for the construction of the final material standard'. That final standard, Jenkin explained, depended on the outcome of laborious, and as yet inconclusive, experiments with respect to the question of permanency, treated from a chemical point of view by Professors Williamson and Matthiessen in two supplementary reports.[50]

At its second meeting the Committee directed Jenkin to prepare a circular informing selected 'foreign men of science' of the appointment of the BAAS Committee and inviting suggestions 'in furtherance of its object'. The circular presented the BAAS as an influential institution serving an international cause and desired that 'such a standard may be selected as will give very general satisfaction; and, if approved by you, that you will kindly take an interest in procuring its general adoption'. The Committee named eleven such 'foreign men' of science of whom no less than six were located in German States (G.T. Fechner at Leipzig, Gustav Kirchhoff at Heidelberg, Franz Neumann at Königsberg, Poggendorff and Werner Siemens in Berlin, and Wilhelm Weber at Göttingen). The remainder consisted of one Swede (Edlund of Upsala), one American (Henry in Washington), one Italian (Matteucci in Turin), one Frenchman (Pouillet in Paris) and one expatriate German (M.H. Jacobi in St Petersburg).[51] Thus it was the Germans, strong in electrical science and familiar with the researches of their countrymen Gauss and Weber, who would be most likely to promote the cause of absolute units and confer maximum international credibility on the whole project.

The Committee's 1863 Report (apparently drawn up by Thomson) extolled the virtues of an absolute system without reservation. It used its expanded membership to suggest that it was no longer bound by the restricted goal, that of determining the best unit of electrical resistance, for which the 1861 Committee had been established. The Committee had previously expressed doubts, however, 'as to the degree of accuracy with which this admirable system could be or had been reduced to practice'. While there was little problem about the measurement of electromotive forces, currents and quantity, the doubts centred on the 'accuracy with which the absolute resistance of some one standard conductor could be measured', a standard from which copies 'could be multiplied at will with any desired precision'. Since the 1862 Report, however, a dramatic change had been wrought in the confidence to be placed in the enterprise:

> The Committee are happy to report that these doubts have been dispelled by the success of the experiments, made for the Committee by Professor J. Clerk Maxwell, Mr. Balfour Stewart, and Mr. Fleeming Jenkin, according to the method devised by Professor W. Thomson. These experiments have been actively prosecuted at King's College for the last five months with continually increasing success, as, one by one, successive mechanical and electrical improvements have been introduced, and the various sources of error discovered and eliminated.[52]

Even though the Committee had decided not to issue standard coils at the BAAS meeting (1863) on account of further improvements still to be implemented and to avoid inconvenient corrections, 'the results already obtained leave no room for doubt that the absolute system may be adopted, and that the final standard of resistance may be constructed without serious delay'. Yet the original doubts had been well founded: the Committee confessed that the experiments revealed that resistance coils previously constructed from absolute determinations 'do not agree one with another within 7, 8, or even 12 per cent'.[53]

Such hostages to fortune threatened to undermine the carefully orchestrated campaign to build international credibility. As Simon Schaffer has shown, Werner Siemens, armed with the authority of his own 'geometrical' mercury standard of electrical resistance widely used in continental telegraph systems, challenged the accuracy of the British Association's experimental work. The King's College results, Siemens noted in 1866, differed among themselves by up to 1.4 percent: 'by what means the Committee holds itself justified ... in concluding upon an error of only 0.1 percent, I am totally unable to imagine'.[54] Siemens had additional reasons for defending his mercury unit: 'the great majority of calculations with electrical resistances', he noted, 'belonged to the geometrical and not to the dynamical domain, and that the reproducible unit with a geometrical foundation proposed by me might just as well be called an absolute one as [that] which was proposed as a unit on the English side'.[55]

As though in anticipation of just such rival bids for 'absolute' units, the 1863 BAAS Report placed centre stage 'a full explanation of the meaning of absolute measurement, and of the principles by which absolute electrical units are determined'. The official reason offered for this exposition was that 'the only information on the subject now extant is scattered in detached papers by Weber, Thomson, Helmholtz, and others, requiring considerable labour to collect and understand'. The exposition began with an explanation of the term 'absolute': 'the measurement, instead of being a simple comparison with an arbitrary quantity of the same kind as that measured, is made by reference to certain fundamental units of another kind treated as postulates'. In the Committee's language, the term 'absolute' was 'intended to convey the idea that the natural connexion between one kind of magnitude and another has been attended to, and that all the units form part of a coherent system'. The words 'relative' and 'arbitrary' were here juxtaposed to 'absolute' and 'natural'. 'Absolute' in particular connoted a certain fixed, unchangeable quality, having already received an embodiment in the 'absolute temperature scale' (ch.3).[56]

The Committee illustrated these points in an elementary way, suitably targeted at practical engineers, by reference to the 'relative' power of an engine expressed 'as equal to the power of so many horses' without reference to units of space, mass, or time ('although all these ideas are necessarily involved in any idea of work'). On the other hand, the power of an engine 'when expressed in foot-pounds is measured in a kind of absolute measurement, i.e. not by reference to another source of power, such as a horse or a man, but by reference to the units of weight and length simply – units which have been long in general use, and may be treated as fundamental'. But the unit of force (weight) assumed here, arbitrarily chosen and dependent upon latitude, was not an absolute measure. The Committee therefore adopted Gauss's dynamical unit, following a trend set by Thomson in his early Glasgow lectures and later set forth in the *Treatise* (ch.10).

From a practical point of view the absolute system avoided 'useless coefficients in passing from one kind of measurement to another'. For example, if the power of an engine is to be 'deduced from the pressure on the piston and its speed, it is given in foot-pounds or metre kilogrammes per second by a simple multiplication; to obtain it in horse-power, the coefficients 33,000 or 550 must be used'. A factor of this sort was 'a very serious annoyance'. This commitment to practical economy, to the elimination of 'wasteful' coefficients which served no useful role and which simply absorbed time and clouded rather than advanced the processes of learning and understanding, was also typical of Thomson's commitment to simplifying techniques and the 'saving of brains' in all spheres of scientific activity.[57]

Determination of electrical units in the absolute system required practitioners simply to 'follow the natural relation existing between the various electrical quantities, and between these and the fundamental units of time, mass, and space'. The four 'electrical phenomena susceptible of measurement' (electromotive force, current, resistance and quantity) were expressed by two equations (Ohm's law and Faraday's law of quantity). The absolute system, then, followed nature's economy; other systems introduced wasteful and arbitrary coefficients which had no counterpart in the system of nature.

From Ohm's law and Faraday's relation of quantity of electricity to current and time it followed 'that only two of the electrical units could be arbitrarily chosen, even if the natural relation between electrical and mechanical measurements were disregarded'. Thus units of electromotive force (such as a Daniell's cell) and resistance (a metre length of mercury with one millimetre section, for example) might be chosen, yielding current and quantity. In the Committee's view this 'system would be coherent; and if all mechanical, chemical, and thermal effects produced by electricity could be neglected, such a system might perhaps be called absolute'. But such effects could not be simply overlooked:

> all our knowledge of electricity is derived from the mechanical, chemical, and thermal effects which it produces, and these effects cannot be ignored in a true absolute system. Chemical and thermal effects are, however, now all measured by reference to the mechanical unit of work; and therefore, in forming a coherent electrical system, the chemical and thermal effects may be neglected, and it is only necessary to attend to the connexion between electrical magnitudes and the mechanical units. What, then, are the mechanical effects observed in connexion with electricity?

The first of these mechanical effects related directly to Joule's researches (ch.4) which had shown that 'whenever a current flows through any circuit it performs work, or produces heat or chemical action equivalent to work'. This work and its equivalents were 'experimentally proved by Dr. Joule to be directly proportional to the square of the current, to the time during which it acts, and to the resistance of the circuit; and it depends on these magnitudes only'. The result was a third 'fundamental equation affecting the four electrical quantities', representing 'the most important connexion between them and the mechanical units'. Thus 'the unit current flowing for a unit of time through a circuit of unit resistance will perfom a unit of work or its equivalent'.

The second of the mechanical effects involved the inverse square law of electrical attraction (or repulsion) and yielded a fourth equation. Here 'the unit quantity [of electricity] should be that which at a unit distance repels a similar and equal quantity with unit force'. Consequently, the four equations 'are sufficient to measure all electrical phenomena by reference to time, mass, and space only, or, in other words, to determine the four electrical units by reference to mechanical units':

$$I = E/R \text{ (Ohm's law)} \tag{1}$$

$$Q = It \text{ (Faraday's relation)} \tag{2}$$

$$W = I^2Rt \text{ (Joule's law)} \tag{3}$$

$$F = Q/d^2 \text{ (Coulomb's law)} \tag{4}$$

where I = current, E = electromotive force, R = resistance, Q = quantity of electricity, t = time, d = distance, F = force, and W = work. Equation (4) at once determines the [electrostatic] unit of quantity, which, by (2), determines the unit current. The unit of resistance is then determined by (3), and the unit of electromotive force by (1). This 'electrostatical' system of absolute units had been so-called by Weber, though without reference to (3), that is, 'without reference to the idea of work, introduced into the system by Thomson and Helmholtz'.[58]

Since 'at present the chief applications of electricity are dynamic', a more useful system would retain the first three equations and replace (4) by an electromagnetic relation. Thus 'the force exerted on the pole of a magnet by a current in its neighbourhood is a purely mechanical phenomenon', such that

$$f = ILm/k^2 \tag{5}$$

where f = force, m = magnetic strength of the pole (in mechanical units), k = radius of the conductor placed in a circle round the pole, and L = length of conductor. It then followed that 'the unit length of the unit current must produce the unit force on a unit pole at the unit distance'. Equations (1), (2), (3), and (5) yielded the 'electromagnetic' system of units.

The Committee then explained how the four principal electrical measurements could be practically made in such electromagnetic units. Provided the horizontal force (H) of the earth's magnetism were known, currents were easily measured by means of a tangent galvanometer and an equation derived from (5):

$$I = Hk^2/L \tan d \tag{6}$$

where d = deflection produced by current. Quantity of electricity, on the other hand, could be measured by the swing of a galvanometer needle when an instantaneous discharge is allowed to pass through it. That left only resistance and electromotive force: '[If] we could measure resistance in absolute measure, the whole system of practical absolute measurement would be complete, since, when the current and resistance are known, equation (1) (Ohm's law) directly gives the electromotive force producing the current'.

The absolute resistance of a wire could be measured in various ways. First, from equation (3), Thomson in 1851 'had determined absolute resistance by means of Dr. Joule's experimental measurement of the heat developed in the wire by a current; and by this method he obtained a result which agrees within about 5 per cent. with our latest experiments'. The Committee judged this method, the simplest of all in mental conception, as 'probably susceptible of very considerable accuracy'.

Second, 'Indirect methods depending on the electromotive force induced in a wire moving across a magnetic field have . . . now been more accurately applied'. In the simplest (though barely practicable) case, a straight conductor with its two ends resting on two conducting rails of large section in connection with the earth would move perpendicularly to the magnetic lines of force and to its own length. The action of the magnetic force on the current developed would resist the motion. Given uniform motion, the work done by the current would be equivalent to that done in moving the conductor against the resistance.

A more practicable method involved a circular coil of known dimensions revolving with known velocity about an axis in a magnetic field of known intensity. Although Weber had employed the method many times, the laborious determination of the intensity of the magnetic field, and its inconstant value for the earth, undermined the value of the method. Not surprisingly given its constitution, the Committee instead endorsed Thomson's method which required only measurements of a velocity and a deflection.[59]

The 1863 Report concluded with a reminder of the earlier recommendation that 'the coils, when issued, will not be called absolute units . . . so that any subsequent improvement in experimental absolute measurement will not entail a change in the standard, but only a trifling correction in those calculations which involve the correlation of physical forces'. Preparing for the moment when these coils would be replicated and disseminated as the material embodiment of absolute standards of resistance, the Committee announced that the units would henceforth be known as the 'units of the British Association'.[60] The goal of imperial and international credibility was, it seemed, within reach.

Two years later that moment was at hand. The 1865 Report stated that 'the object for which they [the members of the Committee] were first appointed has now been accomplished': 'The unit of electrical resistance has been chosen and determined by fresh experiments; the standards have been prepared, and copies of these standards have been made with the same care as was employed in adjusting the standards themselves; seventeen of these copies have been given away, and sixteen have been sold'. Recipients of free copies included Directors of Public Telegraphs in nine continental states as well as India and Australia. Leading British commercial operators, who dominated world telegraphic markets, promised to use the unit exclusively. Skilled as ever in promoting its products, the Committee aimed to send the standard coils not to institutions 'where they would probably have laid [sic] on a shelf useless and unknown, but rather to distribute them widely, where they might become available to practical electricians'.[61]

A final brief notice from the Committee in 1870 recommended its replacement by three smaller committees to determine and issue, first, a condenser representing the unit of capacity; second, a gauge for showing the unit difference of potential; and third, an electrodynamometer adapted to measure the intensity of currents in a decimal multiple of the absolute measure. The Committee noted with evident satisfaction that its principal achievement had been to make the absolute measure of resistance 'a tangible and practical operation', a choice already 'ratified by men of science over a great portion of the globe'.[62]

Competition from Werner Siemens, however, proved difficult to eliminate. In 1869–70, for example, Friedrich Kohlrausch and W.A. Nippoldt at Weber's Göttingen magnetic observatory found a 2 per cent error in the British Association units and came down decisively in favour of Siemens's mercury standard as the best means of resistance testing. A few years later the American physicist Henry Rowland (1848–1901), with the establishment of a laboratory of his own at the new Johns Hopkins University in mind, travelled around British and German physics laboratories. Having seen Kohlrausch's methods at first hand and having stayed with Maxwell at his country house, Rowland soon published in the *Phil.Mag.* a critical assessment of Kohlrausch's results. Blaming the discrepancy on Kohlrausch's neglect of self-induction between the elements of his circuit, Rowland appeared to lend fresh credibility to the troubled BAAS standards.[63]

Although absolute measurement of resistance remained the central goal of the BAAS Committee throughout the 1860s, a closely related task was the accurate determination of v, the ratio between electrostatic and electromagnetic units (ch.11). Since at least 1854 Thomson had believed that the value of this ratio was integral to determining the commercially important issue of the speed of telegraph signalling. In correspondence with Stokes and Airy he derived rough estimates of its value. Soon afterwards Wilhelm Weber and Rudolf Kohlrausch obtained an experimental value for the ratio which related not to commercial telegraphy but to Weber's electromagnetic theory. Throughout the 1860s Thomson, Jenkin and other telegraphers continued to push for accurate redeterminations of the ratio. So too did their friend James Clerk Maxwell to whom fell the task of leading the experimental work on v at King's College.[64]

For Maxwell there was much personal credibility at issue. Having constructed in 1861–2 an electromagnetic theory to rival that of Weber, Maxwell had concluded that light consisted of transverse electromagnetic waves in a mechanical ether (ch.11). He rested his case to a large extent on the experimental coincidence between v and the speed of light. Drawing upon Göttingen's most recent values of v and measurements of the velocity of light conducted by Hippolyte Fizeau in Paris from 1849, Maxwell could claim a 'coincidence' with a difference of over 1 percent. That difference, however, soon widened to more than 4 percent due to Léon Foucault's redeterminations in mid-1862 and their endorsement by Sir John Herschel and the Royal Astronomical Society early in 1864. But it was Thomson's enduring scepticism concerning Maxwell's claims, combined with his continuing recognition of the value of v for commercial telegraphy, which helped drive the redetermination of v as part of the British Association Committee's electrical standards project.[65]

Maxwell's early results for v placed its value not only well below the Göttingen values which he had initially used, but well below Foucault's diminished velocity of light. Since 'the only measurement which must be referred to a material standard is that of [Willoughby Smith's] great resistance', Maxwell increasingly found fault with the British Association resistance standard. Coincidence of some of his results for v with those in Glasgow and elsewhere in London tended to confirm his mistrust of the BAAS standard. This mistrust, together with problems regarding inadequate material and human resources, was eventually countered by his move to Cambridge and the new Cavendish Laboratory in the early 1870s. There he launched a programme, employing new experimental methods, for the redetermination of the resistance standard, effectively a prerequisite for v. The other scientists of energy, notably Thomson, Stewart and Joule, backed his efforts. As Joule told Maxwell in 1876: 'We agreed that it would be most important if you would undertake fresh experiments for redetermining the unit of resistance. I earnestly wish you would do this and then my results [on the mechanical equivalent of heat determined by electrical method] will sink or swim'.[66]

Even with the greater resources at Maxwell's disposal, and his experienced management of them, his authority was by no means total. In 1877–8, for example, his American friend Rowland obtained values of resistance in Baltimore which caused yet more deviations from the value of v desired by Maxwell's theory. Cambridge and Baltimore continued to disagree over resistance measurements into the 1880s. But it was telegraph engineering, rather than physics, which first secured the value of v. In mid-1878, little more than a year before Maxwell's death, two former students of William Thomson, W.E. Ayrton and John Perry, presented the results of experiments to determine v which they had conducted in Japan. Charles Hockin, who had worked closely with Maxwell at King's College, confirmed the value in 1879 as did J.J. Thomson (1856–1940) in Cambridge four years later. With these results from diverse laboratory sites, the match between Foucault's velocity of light and v now appeared credible.[67]

As Maxwell's successor in the Cavendish Chair of Experimental Physics, Lord Rayleigh and his team relaunched the electrical standards programme. In particular the widespread distribution of standard resistance boxes established the authority of Cambridge as the leading site for accurate and reliable electrical measurement. Even William Thomson began to send resistances to Cambridge for calibration. It was furthermore Rayleigh's circle that would lead the campaign for the National Physical Laboratory around the turn of the century.[68]

Meanwhile the first International Congress on electrical standards had been held at Paris in September 1881. Helmholtz and Thomson were foreign vice-presidents, and publicly disputed whether or not to accept Werner Siemens's unit of resistance or the BA unit. The initial outcome was the acceptance of the British Association ohm represented by a mercury column of appropriate length to be decided after further research. This decision represented a victory for the supporters of an absolute system. Persuaded to comply, Siemens was reported to have said 'with a sigh' that 'it was surrendering an important part of his life's work'.[69] The Congress also agreed to the names of the units: ohm, volt, farad, coulomb, and ampere (rather than weber). In 1884 the fourth meeting of the Congress agreed to a standard of resistance to be called the 'legal ohm' (to avoid confusion with both Siemens's standard and the BA ohm) represented by the resistance of a column of mercury 106 centimetres long and one square millimetre in section.[70]

Concluding a survey of the recent history of electrical units of measurement, William Siemens affirmed in his presidential address to the British Association in 1882 that upon the accurate determination of the ohm 'will depend the whole future system of measurement of energy of whatever form'.[71] In this chapter I have shown how the scientists of energy promoted that absolute, energy-based system of measurement. Although not all the scientists of energy (most notably James Thomson, Rankine and Tait) participated directly in the British Association Committee on standards of electrical resistance, the Committee represented a consolidation of the new science of energy into a formal and powerful grouping. William Thomson, Jenkin, Maxwell, Joule and Stewart together ensured that the system of measurement adopted by the BAAS represented that science of energy and thereby remained under the firm control of elite physicists and scientific engineers.

As we have seen, however, enhancement of credibility on an international stage was not an automatic progression. Only by actively exploiting those practical and cultural resources at their disposal was that credibility constructed. Thus the resources of submarine telegraphy, contacts with laboratory sites across the Western world and in Japan, and the capacity to distribute material standards embodying the new science were all deployed to that end. In his obituary notice of Lord Kelvin (1908), Sir Joseph Larmor might have been writing of the whole group when he vividly encapsulated the intimate relationship between units of measurement and the science of energy:

> If one had to specify a single department of activity to justify Lord Kelvin's fame, it would probably be his work in connexion with the establishment of the science of Energy, in the widest sense in which it is the most far-reaching construction of the last century in physical science. This doctrine has . . . furnished a standard of industrial values which has enabled mechanical power in all its ramifications, however recondite its sources may be, to be measured with scientific precision as a commercial asset.[72]

CHAPTER 14

Sequel: Transforming Energy in the Late Nineteenth Century

[Poynting's idea of continuity in the existence of energy is] an extension of the principle of the conservation of energy. The conservation of energy was satisfied by the *total quantity* remaining unaltered; there was no individuality about it: one form might die out, provided another form simultaneously appeared elsewhere in equal quantity. On the new plan we may label a bit of energy and trace its motion and change of form, just as we may ticket a piece of matter so as to identify it in other places under other conditions; and the route of the energy may be discussed with the same certainty that its existence was continuous as would be felt in discussing the route of some lost luggage which has turned up at a distant station in however battered and transformed a condition.

Oliver Lodge publicly endorses J.H. Poynting's
interpretation of energy transfer
in an electromagnetic field[1]

From almost any subsequent vantage point Thomson and Tait's *Treatise on Natural Philosophy* (1867) and Clerk Maxwell's *Treatise on Electricity and Magnetism* (1873) appear as the most enduring embodiments of the science of energy. But these treatises, seemingly definitive, were in historical terms to prove highly malleable. This final chapter seeks to chart some of that malleability in the years following the death of James Clerk Maxwell. Indeed the rapid reinvention of Maxwell was symptomatic that the scientists of energy, overtaken by age and death, were no longer on an upward spiral of scientific credibility. As new groups vied with one another for scientific authority, the science of energy was to be reshaped beyond recognition.

By the 1880s the science of energy was fast slipping from the control of its original British promoters. Rankine and Maxwell had already gone from the scene. During the coming decade death would exact a further toll with the passing of Jenkin, Stewart and Joule. Thomson and Tait alone would continue to assert their authority over physics in Britain. But against the new generations of physical scientists – theoretical and experimental physicists as well as physical chemists – Thomson especially began to look increasingly conservative, a survivor from a past era of natural philosophy.

In contrast the rising generations began to recast the energy doctrines for their own purposes. A self-styled British group of 'Maxwellians', comprising G.F. FitzGerald (1851–1901), Oliver Heaviside (1850–1925) and Oliver Lodge (1851–1940), reinterpreted Maxwell's *Treatise* for their own ends and in accordance with energy principles. But for them 'Maxwell was only half a Maxwellian', as Heaviside noted wryly in 1895 after he and his associates had wrought a transformation in Maxwell's original perspective.[2] Later 'Maxwellians' increasingly located energy in the field around an electrical conductor, tended to carry mechanical model building to extremes, and began to reify energy rather than regard it as mechanical energy or the capacity to do work. It was, above all, this fundamental link between matter and energy, whereby all energy was ultimately regarded as mechanical energy measured in terms of work done, that had characterized the scientists of energy.

William Thomson (by then Lord Kelvin), on the other hand, generally treated the newer electromagnetic views with unreserved contempt: 'It is mere nihilism, having no part or lot in Natural Philosophy, to be contented with two formulas for energy, electrostatic and electromagnetic, and to be happy with a vector and delighted with a page of symmetrical formulas'.[3] There was for him no substitute for dynamical theory, properly understood in conceivable and measurable terms. Although initially cautious about Maxwell's electromagnetic theory of light and always antagonistic to Maxwell's and Tait's seduction by vector notation, his personal, academic and cultural links to his protege strikingly contrast to his disagreements with the self-styled heirs to Maxwell's electromagnetic legacy.

These disagreements, however, appear relatively mild when set against the rise of the so-called 'Energeticist' school in Germany. This school marked a far more radical departure from the 'science of energy'. Led by the physical chemist Wilhelm Ostwald (1853–1922), the Energeticists rejected atomistic and other matter theories in favour of a universe of 'energy' extending from physics to society. Any remaining link between matter and energy had been decisively severed. The aged Lord Kelvin responded to this trend by writing disparagingly to Stokes's successor in the Lucasian chair of mathematics at Cambridge, Joseph Larmor, in 1906:

> Young persons who have grown up in scientific work within the last fifteen or twenty years seem to have forgotten that energy is not an absolute existence. Even the Germans laugh at the 'Energetikers'. I do not know if even Ostwald knows that energy is a capacity for doing work; and that work done implies mutual force between different parts of one body relatively movable, or between two bodies or two pieces of matter, or between two atoms of matter, or between an atom of matter and an electron, or, at the very least, between two electrons.[4]

Transforming Maxwell: the energy of the electromagnetic field

A striking characteristic of the 'Maxwellians' (FitzGerald, Lodge and Heaviside) was their distance from the British mathematical establishment represented in the University of Cambridge, most notably by the famous Mathematical Tripos. Cambridge regarded Maxwell's *Treatise* as a resource for abstruse mathematical problems, especially through the application of Lagrangian methods, rather than as a key to physical nature. The Maxwellians, however, directed their attention not to the applications of Lagrangian methods but to the distibution and flow of electromagnetic energy.[5]

FitzGerald's academic career centred on Trinity College Dublin, already celebrated for its association with the distinguished nineteenth-century Irish mathematicians and natural philosophers Sir William Rowan Hamilton (1805–65), Humphrey Lloyd (1800–81) and James MacCullagh (1809–47). Following his undergraduate career (1867–71), six further years of study was rewarded with a coveted fellowship. In 1881 FitzGerald became Erasmus Smith professor of Natural and Experimental Philosophy, a post which he retained until his death three decades later.[6]

During the arduous preparations for the Trinity fellowship examinations, FitzGerald began to focus his attention on the physical optics of MacCullagh. In 1839 MacCullagh had employed a form of optical ether possessing rotational elasticity to account for the standard phenomena of reflection, refraction and polarization. In contrast, Green, Stokes and other Cambridge mathematicians promoted an elastic solid ether while criticizing MacCullagh's ether for its physical implausibility.[7]

Working outside Cambridge orthodoxy, FitzGerald published in 1879 a paper (refereed by Maxwell at the very end of his life) which both revived MacCullagh's theory and, using his countryman's techniques, extended Maxwell's theory of electromagnetism to reflection and refraction of light. FitzGerald linked Maxwell's expressions for the kinetic and potential energy of the ether to equivalent terms in MacCullagh's theory. Subject to the least action principle, these energy expressions accounted for the complete behaviour of a dynamical system. Despite formidable difficulties with this attempt to reconcile MacCullagh and Maxwell, FitzGerald was convinced of the need to break with the elastic solid orthodoxy which had long kept MacCullagh's theory on the margins of respectability and which now threatened the very survival of Maxwell's electromagnetic theory. Strengthened by FitzGerald's friendship with Lodge and later with Heaviside, that conviction shaped the development during the 1880s of a new and credible body of 'Maxwellian' doctrine.[8]

Lodge spent his early years in his father's business selling clay to Staffordshire potters. Inspired by Tyndall's lectures at the Royal Institution, Lodge took advantage of the diverse opportunities offered in Victorian London for a scientific education. The ambitious Oliver thus began a career in physics in the early 1870s as assistant to G.C. Foster, professor of physics at University College and translator of many German papers on physics for the *Phil. Mag.* During the same period,

Lodge became acquainted with, and published two papers on, Maxwell's electromagnetic theory. By 1881 he had been appointed to the new chair of physics at Liverpool University. Like FitzGerald, whom he first met at the Dublin meeting of the BAAS (1878), he possessed from the outset a fanatical commitment to mechanical model-building but his skills at popular exposition and experimentation, combined with a lack of mathematical training, complemented rather than duplicated those of the Dubliner.[9]

Particularly through his textbook, *Modern Views of Electricity* (1889), Lodge played a vital role in promoting the new 'Maxwellian' perspectives on electromagnetic theory at pedagogical and popular levels. Serialized in *Nature*, *Modern Views* went through three editions in Britain and the United States (1889, 1892 and 1907) as well as translations into Russian (1889), French (1891) and German (1896). Its popular success brought criticism as well as acclaim. The French physicist and philosopher of science Pierre Duhem (1861–1916) complained of what he saw as its tendency, through its prolific use of mechanical models, to degrade physics to the level of industry: 'We thought we were entering the tranquil and neatly ordered abode of reason, but we find ourselves in a factory'.[10]

During the decade prior to *Modern Views*, Lodge had become adept at promoting physics in general and energy physics and field theory in particular. In October 1879 he published 'An Attempt at a Systematic Classification of the Various Forms of Energy' in the *Phil. Mag.* (Figure 14.1). His accompanying letter to the editors (including Sir William Thomson) explained that while he had been 'writing a little elementary manual of mechanics lately for [the Edinburgh publishers] Messrs. Chambers his attention had been directed to a certain amount of vagueness and loose language which appears to be current in modern statements concerning energy'.[11]

Presenting his 'Systematic Classification' in twenty simple steps, Lodge rejected as inadequate the more familiar definition of energy as 'the power of doing work'. To illustrate the point he resorted to elementary principles of political economy. A sovereign, for example, had 'an infinite power of buying goods . . . , twenty-shillings worth being bought whenever it is transferred from one man to another'. Lodge explained that the 'proper statement is that a sovereign usually *confers upon the man that possesses it* a certain buying power, which power he loses when he has transferred it; and in this sense money is a power of buying goods'. Hence energy 'is power of doing work in precisely the same sense as capital is the power of buying goods'.[12]

But just as money does not necessarily confer upon its owner any buying power 'because there may not be any accessible person to buy from; and if there be, he may have nothing to sell', so energy usually, but not necessarily, 'confers upon the body possessing it a certain power of doing work, which power it loses when it has transferred it'. He therefore set out a pedagogical analogy:

Energy corresponds to capital.
Doing work [upon a body] corresponds to buying.
Doing antiwork [i.e. work done by a body] corresponds to selling.

BODIES.	ENERGY OF MOTION, OR KINETIC ENERGY.		ENERGY ALTERNATELY KINETIC AND POTENTIAL.	ENERGY OF STRESS, OR POTENTIAL ENERGY.
	Rotation.	*Translation.*	*Vibration.*	*Strain &c.*
Planetary masses	1. E. g. Earth's diurnal motion.	2. E. g. Earth's annual motion.	3. E. g. The moon's libration. Tides. Pendulums.	4. Energy of gravitation. E. g. A bead of water. A raised weight.
Ordinary masses	5. E. g. Fly-wheel.	6. E. g. Cannon-ball. Rivers.	7. Sound-vibrations. E. g. Tuning-fork.	8. Energy of strained elastic bodies. E. g. Watch-springs.
Particles or molecules ...	9. Part of the heat-energy of fluids.	10. Most of the heat-energy of gases.	11. Heat-energy of solids.	12. Energy of molecular stresses. E. g. "Internal work."
Atoms	13. 14. Unknown motions which take place during the act of chemical combination and during dissociation. The translation of atoms is observed in electrolysis.		15. The period of atomic vibration is observed by the spectroscope.	16. Energy of chemical affinity.
Something	17. Magnetism.	18. Electric currents.	19. (1) Discharge of accumulators. (2) Radiation.	20. (1) Electrostatic stress. (2) Electromagnetic stress.

Figure 14.1 Lodge's classification of energy in relation to different classes of bodies. He divided energy into three classes (kinetic, potential and alternatively kinetic and potential) and bodies into five classes (from planets down to aether) (Lodge 1879: 282).

The transfer of capital is accompanied by two equal opposite acts, buying and selling; and it is impossible for one to go on without the other. Hence the algebraic sum of all the buying in the world is always zero: this is the law of conservation of capital.

Lodge thus claimed that the 'power of doing work conferred upon a body by the possession of energy does not depend upon the absolute quantity of that energy only, but on its transferability'. He then he distinguished between 'a high or available form of energy' (capable of being 'guided, and all, or nearly all, transferred to any body at pleasure' and hence of doing 'useful' work) from 'a low or unavailable form of energy . . . nearly incapable of being guided, and which transfers itself in directions not required', yielding 'useless' work.[13]

Following the teachings of the masters of 'the science of energy' (especially Maxwell), Lodge emphasized that the 'distinction between high and low forms of energy is a relative one, and depends on our present power of dealing with matter'. Masses of matter 'comparable to our own bodies in size can be handled and dealt with singly; and so they can in general be caused to do work upon, and therefore transfer their energy at pleasure to, any of the numerous accessible bodies which are competent to receive it'. But planets, which could also be treated singly, offered little opportunity for the transfer of energy (apart from tidal energy 'now available to us' for useful work). For the most part, then, 'the kinetic energy of the earth is of no more use to us than a bank-note to Robinson Crusoe'.[14]

Relatively unavailable too was the energy of moving molecules 'because we can only deal with them statistically and not individually'. In a witty footnote Lodge again resorted to the language of elementary political economy for pedagogical purposes:

An analogy may be drawn between the molecular energy of a body and the money of a bank; of which a reserve fund is kept for internal transfer and transactions between customers, while the excess gets invested in external concerns which have a deficiency, and so becomes available for doing useful work. To make the analogy more complete, the clerks should be uniformly dishonest, or the coffers insecure, so that stored money should dribble away.[15]

Within this framework of energy transfer Lodge began to consider in the same year (1879) the possible production of electromagnetic waves, something which Maxwell himself apparently did not contemplate. FitzGerald, indeed, initially read Maxwell's claim for the mathematical identity of his results with the German action-at-a-distance theories (which took no account of a non-conductor) as an argument against the possibility of energy being transferred in the form of electromagnetic waves from an electrical system (such as a closed system of perfect conductors) into non-conducting space. Instead, FitzGerald took the view, similar to that of Maxwell himself, that light was not produced electrically, but probably as the result of an atomic interaction between matter and aether. FitzGerald's scepticism, made public before the Swansea meeting of the British Association (1880), though not published in the *Report*, almost certainly served to direct Lodge's attention away from the subject until 1888, the year in which Heinrich Hertz (1857–94) announced the detection of electromagnetic waves.[16]

By 1882, however, FitzGerald had effectively reversed his stance against the production of electromagnetic waves. His reading of a key passage in Lord Rayleigh's *Theory of Sound* (1877) led him to the conclusion that for an oscillating current 'all the energy is gradually transferred to the medium', contrary to action-at-a-distance theories. As he wrote to the physicist Henry Rowland at Johns Hopkins University in Baltimore in 1882:

> it is most likely that all periodic electric and magnetic [oscillations] are accompanied by a loss of energy just like, and in fact the same thing as, the radiation of heat. In fact it extends the dissipation of energy by radiation to the case of periodically working electric and electro-magnetic machines.[17]

At the Southport meeting of the BAAS (1883) FitzGerald read two brief papers: 'On the Energy Lost by Radiation from Alternating Currents' and 'On a Method of Producing Electromagnetic Disturbances of Comparatively Short Wavelengths'. Electromagnetic waves of 10 metres wavelength or less, he announced, would be possible. Yet although FitzGerald had already worked out the equations for such energy radiation, electromagnetic waves remained experimentally undetected for five more years.[18]

The third leading member of the 'Maxwellian' group, Heaviside, excelled the others in eccentricity and remained throughout his life on the fringes of English scientific society. Heaviside grew up in comparative poverty and ill-health in the seedier parts of London's Camden Town. His early scientific education seems to have been largely of his own making. But thanks to the influential Charles Wheatstone (1802–75), an uncle by marriage, the young Heaviside began in 1867 a practical career in electric telegraphy.[19]

Following publication of 'On the Best Arrangement of Wheatstone's Bridge for Measuring a Given Resistance with a Given Galvanometer and Battery' in the *Phil. Mag.* (1873), Heaviside's mathematical skill (the more noteworthy because he lacked a Cambridge mathematical education) attracted the attention of Sir William Thomson. Backed by Thomson and Sir William Siemens, Heaviside was admitted in 1874 to the new Society of Telegraph Engineers in the face of opposition from Post Office engineers and officials who apparently regarded him as a mere telegraph clerk. Illness in the same year prompted him to abandon practical telegraphy in favour of private, almost reclusive, electrical research.[20]

From the mid-1850s Thomson had been offering a mathematical analysis (based on Fourier's methods for treating problems of heat flow) of the distortion and retardation of signals which occurred in undersea telegraphy at a time when the first long-distance cables (notably the trans-Atlantic) were being planned. Experiencing such retardation effects for himself on the Anglo-Danish cable at Newcastle in the early 1870s, Heaviside extended Thomson's theory to take into account the phenomenon of self-induction (first described by Faraday and manifested as a tendency of electric currents to oppose any changes in their strength, rather as a machine flywheel). These concerns focused Heaviside's attention less on the electric conductors themselves and more on the space around them.[21]

Maxwell's *Treatise* had offered its early readers a somewhat ambivalent perspective. On the one hand, the energy of the current in a wire and the electrostatic energy could be expressed in terms of vector and scalar potentials. If these potentials were read as fundamental, the *Treatise* could be construed as retaining an action-at-a-distance approach which located the energy of a steady current *within* the wire. On the other hand, the energy of the current could also be expressed in terms of magnetic force and the electrostatic energy in terms of electric force, both expressions placing emphasis on the surrounding medium (ch.11). Five years after Maxwell's death, J.H. Poynting (1852–1914) published an extended treatment in terms of the energy of the electromagnetic field, marking a decisive shift away from potentials to a location of the energy in the field.[22]

Poynting's 'On the Transfer of Energy in the Electromagnetic Field' appeared in the *Phil. Trans.*(1884). Poynting had left Owens College in Manchester for Cambridge in 1872 where he graduated third wrangler four years later. For a brief period before Maxwell's death Poynting worked at the Cavendish Laboratory under the direction of the professor. From 1880 he served as physics professor at Birmingham's new Mason College of Science. His thoroughgoing study of Maxwell's theory inspired the Royal Society paper that Lodge publicly described in 1885 as 'a most admirable and important paper . . . which cannot but exert a distinct influence on all future writings treating of electric currents'.[23]

The fundamental feature of Poynting's investigation was the notion of continuity in the existence of energy. His consequent search for energy in the intervening space and his study of the paths by which it travelled led him to a new theorem about the electromagnetic field: the rate at which energy entered a region of space, where it was stored in the field or dissipated as heat, depended only on the values of the electric and magnetic forces at its boundaries.[24]

A simple but striking application of his 'flux theorem' produced the result that electromagnetic energy did not flow inside a current-carrying wire but flowed through the surrounding space (dielectric) to enter the wire through its sides. Similarly, an application of the theorem to the discharge of a condenser showed that the electromagnetic energy flowed out from between the condenser plates and radially into the current-carrying wire. Poynting summed up his reading of Maxwell's theory as one in which 'currents consist essentially in a certain distribution of energy in and around a conductor, accompanied by the transformation and consequent movement of energy through the field'.[25] The field was now the location of electromagnetic energy.

Lodge's enthusiastic endorsement of Poynting's paper was based on what he saw as 'this new form of the doctrine of the conservation of energy' which was 'really much simpler and more satisfactory than in its old form'. He then claimed that the new version could be demonstrated from two premisses: Newton's 'law of motion' and the denial of action-at-a-distance. With respect to Newton's laws, he believed them to be 'really three very important aspects of one law' which could be expressed in a number of ways including the equality of action and reaction. His preferred expression was that 'Force is always one component of a stress'. Denial of action-at-a-distance required simply the premiss: 'If a stress exist between two bodies they must be in contact'.[26]

Lodge then argued that conservation of energy followed from these two basic premisses:

> not only conservation, but conservation in the new form, viz. the *identification* of energy; thus: If [body] A does work on [body] B it exerts force on it through a certain distance; but (Newton's law) B exerts an equal opposite force, and (being in contact) through exactly the same distance; hence B does an equal opposite amount of work, or gains the energy which A loses. The stress between A and B is the means of transferring energy from A to B, directly motion takes place in the sense AB. And the energy cannot *jump* from A to B, it is transferred across their point of contact, and by hypothesis their 'contact' is absolute: there is no intervening gap, microscopic, molecular, or otherwise. The energy may be watched at every instant. Its existence is continuous: it possesses identity.[27]

Taking seriously the possible objection that two pieces of matter (for example, two molecules A and M) were never in contact, Lodge countered that the 'energy may be transferred from A to M, but not directly'. Thus 'A can act on B, transferring its energy to B, B can act on C . . . handing on the energy to L, which is in contact with, and can act on, M, doing work on it and giving up to it the energy lost by A'. Although he took care to avoid a commitment to the specific nature of the intervening steps, he concluded that 'of course one supposes them to be successive portions of the perfectly continuous space-filling medium Aether'.[28]

As part of his overall goal of constructing and promoting a *reformed* language, Lodge then subjected 'potential energy' to a severe critique. He claimed that the 'usual ideas and language current about potential energy are proper to notions of action at a distance'. While it 'was easy enough to take account of it in the formulae, . . . it was not easy or possible always to form a clear and consistent mental image of what was physically meant by it'. For example, a raised weight such as a stone gained potential energy: but how could 'an inert and quiet stone' be said to possess energy? To reply that ' "the system of earth and stone possesses energy in virtue of its configuration" ' was true, but also 'foggy'. Admitting universal contact action dispelled the fog: 'the energy is seen to be possessed, not by stone or earth or by both of them, but by the medium which surrounds and presses them together'.[29]

At the same time, however, Lodge confessed that he did not intend to claim that 'the natures of gravitation, elasticity, cohesion, &c. become clear'. To account for the stress in the medium was 'a much higher and more difficult problem'. He only wanted his readers to recognize the '*seat* of the energy' in the surrounding medium.[30] Yet the demand for an explanation, or at least illustration, of how the medium might operate brought forth from Lodge and his contemporaries that immense proliferation of mechanical models for which Duhem was to indict most Victorian physicists.

This new emphasis on energy transfer in physical theory in general and electromagnetic theory in particular helps explain why the British liking for mechanical models reached such a peak in the 1880s. Philosophers of science have often cited Sir William Thomson's remark in his 'Baltimore Lectures' (1884) concerning electromagnetism and the wave theory of light: 'I never satisfy myself until I can make a mechanical model of a thing. If I can make a mechanical model I can understand it. As long as I cannot make a mechanical model all the way through I cannot understand; and that is why I cannot get the electro-magnetic theory'.[31] If Thomson appeared in retrospect as the most uncompromising exponent of mechanical models in Victorian physical science, his was by no means a lone voice. Just as his contemporaries (especially Rankine and Maxwell) had once placed great emphasis on mechanical hypotheses, so too did members of the younger generation of British physicists.

Mechanical models served a variety of functions: illustrative and pedagogical; directing enquiry among otherwise recondite phenomena; demonstrating the possibility that those phenomena could in principle admit of some kind of mechanical explanation; and even offering the prospect that the model itself reflected more or less accurately the 'true mechanism' of nature. Above all, mechanisms provided an exemplar of continuity in energy transfer: in proper mechanisms there could be no action-at-a-distance.[32]

FitzGerald's ingenious 'wheel and band' model (1884–5) offered one such mechanical model to illustrate the nature of energy transfer involved in Poynting's new theorem. As Fitzgerald told Lodge: 'I have been constructing a model ether . . . as I want it to illustrate Poynting's great discovery that the energy of an electric current must come in at its sides and is not carried along with the current'. Fitzgerald subsequently developed this illustrative model into a wide-ranging mechanical representation in two dimensions of Maxwell's electromagnetic equations.[33]

Not content with illustrative models, FitzGerald simultaneously sought a 'realistic' model of the ether in the form of a 'vortex sponge'. In its conception the vortex sponge model owed much to William Thomson's earlier attempts to construct a vortex theory of matter. Ironically, it was in part to counter Thomson's commitment to an elastic-solid ether in his 'Baltimore Lectures' (1884) that FitzGerald put his faith in a hydrodynamical model which aimed to explain all the phenomena of nature in terms of matter in motion, or in terms of kinetic energy. Thomson too became enthusiastic for a time. But by 1890, weighed down by its many conceptual difficulties, the vortex sponge ether lost much of its appeal.[34]

The spectacular proliferation and conspicuous collapse of such grand speculative theories prompted some British physicists to argue in favour of mathematical laws rather than mechanical hypotheses. This approach had a long pedigree. Thomson and Maxwell, for example, had appeared to oscillate between a commitment to laws ('mathematical theory' in Thomson's phrase) on the one hand and a belief in the ultimate mechanical nature of the physical world while reserving their scepticism for specific mechanical hypotheses. A similar attitude characterized some of the later 'Maxwellians'. In the early 1890s Poynting criticized Lodge's seeming belief in the truth of his mechanical models. Emphasizing instead the illustrative role of mechanical models and the formulation of electromagnetic laws, he urged physicists to 'leave the ether out of account'.[35]

With rather less scepticism towards mechanical explanation, Heaviside too had far less enthusiasm for mechanical models than his friends Lodge and FitzGerald. From 1883 he began publishing in the *Electrician* a series of articles on 'The Energy of the Electric Current'. At first he followed the conventional perspective of regarding the energy of the electric current as analogous to the flow of a fluid in a pipe. A relatively small amount of energy passed from the wire into the surrounding medium, there to be stored by the rotation of 'flywheels' in the ether. Published in mid-1884 (and apparently without knowledge of Poynting's work), Heaviside's 'Transmission of Energy into a Conducting Core' shifted the emphasis away from the flow of energy in the wire to considerations of the flow of energy in the field. Basing the analysis on the conviction that one needed to follow the paths of electromagnetic energy, Heaviside independently arrived at the energy flux theorem six months after Poynting.[36]

Heaviside was not bound by academic orthodoxy. While Cambridge mathematicians used Thomson and Tait's *Treatise* as a textbook resource for Lagrangian and Hamiltonian dynamics, Heaviside seized upon the 'principle of activity' (ch.10) which expressed Newton's third law in terms of force times velocity or rate of working (action) equivalent to rate of increase of kinetic energy (reaction). Realizing that the 'principle of activity' could not be formed directly from Maxwell's equations (by finding the rate at which the electric and magnetic forces act on the 'velocities' or currents), Heaviside set about reworking Maxwell's 'General Equations of the Electromagnetic Field'. Vector and scalar potentials permeated Maxwell's original list of thirteen principal equations. As Bruce Hunt has shown, from eight of these equations Heaviside obtained three equations which contained no potentials. He finally constructed a fourth equation expressing Faraday's law of electromagnetic induction relating rate of change of magnetic force to electric force, again without potentials. Rate of working and energy flow could then be derived directly from these four concise and symmetrical equations.[37]

First published in the *Electrician* early in 1885, these four equations became known as 'Maxwell's Equations'. Heaviside's formulation had eliminated all remnants of action-at-a-distance embodied in the potentials employed by Maxwell. The theoretical issue of the electric potential in particular was debated publicly (though characteristically in the absence of the retiring Heaviside) at the Bath meeting of the BAAS (1888) when general agreement was reached as to the desirability of Heaviside's elimination of potentials. The full emphasis was now on the energy of the field.[38]

In the summer of 1888 came news from Germany of Hertz's experimental generation and detection of electromagnetic waves in free space. Working with Helmholtz in Berlin for several years, Hertz was well acquainted with his master's generalized electrodynamics which rendered the rival theories of Weber, Neumann and Maxwell special cases of his own. While in Britain action-at-a-distance theories had long since received short shrift, in Germany it was Maxwell's field theory which looked out of place among action-at-a-distance orthodoxy. All the theories tended to account equally well for standard experimental results, but Maxwell's theory differed in its prediction of electromagnetic effects from dielectric displacement currents. As early as 1879 Helmholtz had urged Hertz to try to detect such effects. Given formidable experimental difficulties, it was not until late 1887 that he succeeeded to his own satisfaction. Recognizing that the key feature of field theory lay in electromagnetic action in space (rather than material dielectrics), Hertz quickly moved to produce and detect electromagnetic waves in free space, that is in a vacuum.[39]

The initial credibility conferred upon Hertz's experimental work was largely contingent upon the forum of the BAAS at Bath. A formidable array of men of science with electrical expertise assembled in the ancient spa town: Sir William Thomson, Lord Rayleigh, FitzGerald, Rowland and Lodge. As president of Section A it was FitzGerald who made the formal announcement. Predictably, FitzGerald, with his deep commitment to Maxwellian field theory, chose to represent Hertz's achievement as confirmation of that field theory against action-at-a-distance: 'Henceforth I hope no learner will fail to be impressed with the theory – hypothesis no longer – that electromagnetic actions are due to a medium pervading all space, and that it is the same medium by which light is propagated; that non-conductors can, and probably always do, as Professor Poynting has taught us, transmit electromagnetic energy'.[40]

Whereas in Germany Hertz's discovery might initially have gone unnoticed, the BAAS meeting provided Hertz with rapid publicity. From the leading authorities in mathematical physics to the principal newspapers came enthusiastic responses. In the next few years translations of Hertz's papers appeared in the most widely read scientific journals (*Phil. Mag.* and *Nature*), culminating in a collection of Hertz's papers (1892) with a preface by no less an authority than Sir William Thomson.[41]

Although in the few remaining years of his short life Hertz interacted closely with the British 'Maxwellians', the German physicist differed markedly in his interpretation of Maxwellian doctrine. Defining Maxwell's theory as 'Maxwell's system of equations' in his *Electric Waves* (1892), Hertz appeared to separate the mathematical content of Maxwell's theory from its physical foundations of the electromagnetic field. In large part Hertz's move reflected his desire to avoid the deep conflict over physical theory which divided Continental traditions (now exemplified by Helmholtz's electrodynamics) from the British conceptions (embodied above all in Heaviside's Maxwellian exposition).[42]

The British 'Maxwellians' of course would have none of this 'nihilism'. A review of Hertz's *Electric Waves* for *Nature* (1893), probably by FitzGerald, noted that Hertz 'seems content to look upon Maxwell's theory as the series of Maxwell's equations'. But, he remonstrated, 'Any exposition of Maxwell's theory which does not clearly put before the reader that energy is stored in the ether by stresses working on strains, is a very incomplete representation of Maxwell's theory'. Crucial to the 'purified' Maxwellian doctrine of FitzGerald, Lodge and Heaviside, then, was the physical conception of energy flux in a dynamical ether.[43]

From the point of view of the 'Maxwellians', just as corrupt an interpretation of the sacred *Treatise* lay with the Cambridge mathematicians. The post-Maxwell generation of J.J. Thomson and Larmor, among others, were guilty of following the Lagrangian letter and not the aetherial spirit of the *Treatise*. Beginning in the 1890s and culminating in his *Aether and Matter* (1900), Joseph Larmor made the principle of least action central to his electrodynamics. For Heaviside, least action was 'a golden or brazen idol' for Cambridge men to worship. For FitzGerald, it produced an 'analytical juggle'. But for Larmor least action 'comprehends in itself the whole of mechanical science'.[44]

Larmor's strategy was to treat of the molecular level which had been to a large extent by-passed in Maxwell's *Treatise* and by its Maxwellian interpreters. In particular, Maxwell and the Maxwellians had not engaged with the question of how electrical conduction occurred. Larmor, aided by a lengthy correspondence with FitzGerald, slowly came to the view that currents could be construed as the movement and interaction of microscopic particles or 'electrons'. These particles in the electromagnetic ether provided the basis for Larmor's *'electronic theory'* of matter and for the grand dynamical programme set forth in *Aether and Matter*. The mechanical had been replaced by the electromagnetic philosophy of nature.[45]

The major realignments of dynamical theory undertaken by Larmor in his bid to provide a molecular synthesis had, in his view, an important consequence for conceptions of energy:

> One effect of admitting a molecular synthesis of dynamical principles such as the one here described is to depose the conception of energy from the fundamental or absolute status that is sometimes assigned to it; if a molecular constitution of matter is fundamental, energy cannot also be so. It has appeared that we can know nothing about the aggregate or total energy of the molecules of a material system, except that its numerical value is diminished in a definite manner when the system does mechanical work or loses heat. The definite amount of energy that plays so prominent a part in mechanical and physical theory is really the mechanically available energy, which is separated out from the aggregate energy by a mathematical process of averaging, in the course of the transition from the definite molecular system to the material system considered as aggregated matter in bulk.[46]

Larmor here claimed that this energy, though 'definite', was not conserved in unchanging amount. Energy did not therefore share equal status with matter: energy 'merely possesses the statistical, yet practically exact, property, based on the partly uncoordinated character of molecular aggregation, that it cannot spontaneously increase, while it may and usually does diminish, in the course of gradual physical changes'.

Energy physics and physical chemistry: the science of energetics

Mechanical conceptualisations of energy, with work as its essential measure, dominated British and German physics in the second half of the nineteenth century. Towards the end of the century, however, different perspectives, emphasizing the independence of energy from mechanics, began to be promoted among German-speaking scientific communities as part of a widespread reaction against capitalistic materialism during the 'Great Depression' (c.1873–96). The more radical versions, especially the 'Energetics' of Ostwald, originated outside mainstream German mathematical physics and soon attracted much critical attention from a younger generation of German physicists.[47]

From the 1880s Ernst Mach (1838–1916) condemned the widespread assumption that mechanics comprised the basis of all physical phenomena and argued instead for a 'phenomenological' view in which sensory perceptions would constitute the real object of physical research. He held up the principle of conservation of energy as an ideal: although mechanical theories had aided in the formulation of the principle, once established it described only a wide range of facts concisely, directly and economically with no need for mechanical hypotheses. Mach located the starting point of his own mature epistemology in the first edition of his *History and Root of the Conservation of Work* (1872):

> In this pamphlet . . . I made the first attempt to give an adequate exposition of my epistemological standpoint – which is based on a study of the physiology of the senses – with respect to science as a whole, and to express it more clearly in so far as it concerns physics. In it both every *metaphysical* and every one-sided *mechanical* view of physics were kept away, and an arrangement, according to the principle of economy of thought, of facts – of what is ascertained by the senses – was recommended.[48]

First published in 1883, Mach's *Development of Mechanics* (*The Science of Mechanics* in its English translation) went through five German and three American editions over the next quarter century. In that and other works during the same period Mach set forth his view that the interdependence of phenomena, rather than their reduction to 'metaphysical' and 'mechanical' conceptions such as atoms and molecules, was to be the goal of natural science. Mechanics had its place in science, but it was no longer fundamental: any 'view that makes mechanics the basis of the remaining branches of physics, and explains all physical phenomena by mechanical ideas, is in our judgment a prejudice'.[49]

Although not to be identified as an 'Energeticist' himself, the severity of Mach's critique of mechanical foundations (especially atomism) added considerable authority to the new doctrines of what became known as the 'Energeticist School'. Following publication of a treatise on the energy principle, *Die Lehre von der Energie* (1887), and appointment as extraordinary (that is, junior) professor of physics at the Dresden technical institute, Georg Helm in 1890 attempted to derive the equations of motion from the conservation of energy and thereby to assimilate mechanics to energetics. In the same year, Helm forged an alliance with the ambitious physical chemist, Wilhelm Ostwald.[50]

The principle of the equivalence of heat and work, together with the associated mechanical theory of heat, had long offered the possibility of a new mechanical theory of chemistry. In his *Erhaltung der Kraft* (1847) Helmholtz had shown that 'Hess's Law', that 'no matter by which way a compound may come to be formed, the quantity of heat developed through its formation is always constant', was a direct consequence of the conservation principle. Hess's work (1839–42) had been concerned to show that heats of dilution offered a reliable measure of the problematic notion of chemical affinity. Thus while chemical affinity might or might not be due to microscopic forces of attraction and repulsion (in accordance with Laplacian doctrine), its effects could now be measured unambiguously in terms of the heat developed in chemical processes.[51]

Although Joule and others in Britain had engaged with similar questions of chemical affinity and its measurement in terms of thermal and mechanical effects, the construction of a science of 'thermochemistry' was due largely to the efforts of the Danish chemist Julius Thomsen (1826–1909) and his French rival Marcellin Bertholet (1827–1907). A friend and colleague of L.A. Colding whose measurements of the equivalence of heat and work had been at first little known outside Denmark, Thomsen set out his basic thermochemical principles in 1854. Affinity was to be understood as the 'force which unites the component parts of a chemical compound'. To overcome the affinities and thus to split up a compound required a force whose quantity 'can be measured in absolute terms; it is equal to the amount of heat evolved by the formation of the compound' from its constituents.[52]

In general Thomsen claimed that 'Every simple or complex action of a purely chemical nature is accompanied by evolution of heat'. Actions which appeared to involve an absorption of heat could not be purely chemical in nature. As professor of chemistry at the University of Copenhagen from 1866, Thomsen's thermochemical research culminated in the German-language *Thermochemische Untersuchungen* (1882–4). The programme included the extension of thermochemical principles to structural (especially organic) chemistry.[53]

Equally ambitious was Berthelot's thermochemistry which originated in the mid-1860s from his extensive research in organic chemistry and which drew on previous experimental measurements of heat in chemical reactions. Berthelot's 'principle of molecular work' (*le travail moléculaire*) dating from 1865 stated an equivalence between internal work and heat changes in chemical reactions. In the same year he introduced the terms 'exothermic' and 'endothermic' for chemical reactions which involved the evolution and absorption of heat respectively.[54]

Berthelot's 'principle of maximum work' (1864–73) stated that 'Every chemical change accomplished without the intervention of external energy [*d'une énergie étrangère*] tends to the production of that body, or system of bodies, which disengages most heat'. This principle performed a parallel causal function to that of least action in physics: the tendency of a system of bodies towards a future state governed the behaviour of the system. The principle offered a rationale for spontaneous chemical reactions which, in Berthelot's view, necessarily occur if accompanied by an evolution of heat: 'Every chemical change which can be accomplished without the aid of a preliminary action, and without the intervention of external energy . . . necessarily happens if it disengages heat'. This principle, embodied in Berthelot's two-volume treatises *Essai de mécanique chimique fondée sur la thermochimie* (1879–81) and *Thermochimie. Données et lois numériques* (1897), formed the core of Berthelot's thermochemistry, the centrepiece of French physical chemistry in the late nineteenth century.[55]

The patronage, power and prestige wielded by Berthelot in France, however, was not matched by his influence abroad. Heavily criticised on every front by Thomsen in the period 1872–86, the principles of thermochemistry as a whole came under sporadic attack from a variety of sources including Lord Rayleigh who addressed the Royal Institution in a lecture (published in *Nature*) 'On the Dissipation of Energy' (1875). Rayleigh highlighted the need for chemists to take into account the principle of dissipation of energy, a principle absent from Thomsen's and Berthelot's thermochemistry.[56]

'The chemical bearings of the theory of dissipation are very important', Rayleigh told his audience, 'but have not hitherto received much attention'. Chemical transformations which violated the theory of dissipation were impossible. On the other hand, transformations involving dissipation did not necessarily occur spontaneously: 'Otherwise, the existence of explosives like gunpowder would be impossible'. Rayleigh then examined critically and rejected the thermochemical assumption that 'the development of heat is the criterion of the possibility of a proposed transformation'.[57]

Rayleigh's 1875 lecture, widely known through its appearance in *Nature*, coincided with the publication in the United States of the early thermodynamic papers of Gibbs (ch.12). While sympathetic to the extent of seeking support for a Royal Society award for Gibbs in the early 1890s, Rayleigh told the American physicist in 1892 that his thermodynamic paper 'On the Equilibrium of Heterogeneous Substances' had been 'too condensed and too difficult for most, I might say all, readers'. Rayleigh's remarks reflected much stronger views expressed privately to him by Lord Kelvin on two occasions. 'I feel very doubtful as to the merits of Willard Gibbs' applications of the "Second law of Thermodynamics"', he wrote in September 1891. 'I find *No* light or leading [sic] for either chemistry or thermodynamics in Willard Gibbs', he asserted five months later.[58]

In contrast Ostwald translated Gibbs's memoirs into German in 1892. The chemist and his disciples constructed and promoted the new science of physical chemistry largely on the basis of Gibbs's thermodynamic approach. That approach provided physical chemists with some of their most fundamental tools, including the 'phase rule' and 'chemical potential'. And Helm soon annexed Gibbs and his writings to the cause of 'Energetics', though Gibbs himself never identified with the doctrines of that school. In the French provinces, outside the kingdom of Berthelot, the young Duhem also became enamoured of Gibbs's thermodynamics in his crusade against his countryman's thermochemistry. Others, such as Hermann Helmholtz and Max Planck, began their studies in chemical thermodynamics apparently unaware of Gibbs's papers.[59]

Unlike Gibbs, nothing hindered the promotion of Helmholtz's 1882 memoir *'Die Thermodynamik chemischer Vorgänge'* ('Thermodynamics of Chemical Processes'). Against the thermochemistry of Thomsen and Berthelot, Helmholtz distinguished between the 'bound' energy (obtainable only as heat) and 'free' energy (available for useful work and conversion to other forms) of chemical reactions. The free energy was then equal to the total energy less the bound energy (expressed, as with Clausius, in terms of the product of absolute temperature and entropy change). The free energy, rather than the heat of reaction, then became the driving agent for chemical and physical processes, determining the direction of the reaction. Consequently, in all spontaneous reactions at constant volume and temperature the free energy must decrease. Similarly, only at absolute zero would Berthelot's principle of maximum work hold good. Furthermore, the new equation (subsequently known as the 'Gibbs–Helmholtz Equation') showed how at ordinary temperatures most reactions would be exothermic, but, particularly at high temperatures, some reactions could be endothermic.[60]

By the time that Thomsen and Bertholot shared in the award of the Royal Society's Davy Medal (1883), the science of thermochemistry was becoming unfashionable among the *avant garde* of chemical science. Ostwald rapidly became a powerful and persuasive leader of that new generation. Born in Riga to German parents, he remained in Latvia for more than three decades. Graduating from the University of Dorpat in 1875, he began lecturing on the subject of chemical affinity in his capacity as *Privatdozent* in 1876. By 1881 he had been appointed professor of chemistry at the Riga Polytechnic Institute. Written during his tenure at the Polytechnic, Ostwald's textbook on general chemistry, *Lehrbuch der allgemeinen Chemie* (1885–7), soon became central to his campaign of promoting the establishment of physical chemistry as a new scientific discipline.[61]

During this period, and in collaboration with the Swedish chemist S.A. Arrhenius (1859–1927), Ostwald adopted a theory of electrolytic dissociation according to which the molecules of chemical compounds dissociated into electrically charged ions when in solution. This theoretical perspective transformed Ostwald's earlier work on chemical affinities and provided a comprehensive and unified account of many physico–chemical phenomena such as the behaviour of weak acids and bases and the properties of solutions. It also shaped a whole experimental programme of quantitative physical chemistry.[62]

In 1887 Ostwald accepted the chair of physical chemistry (the only such German chair) at the University of Leipzig, a post which he retained until early retirement to his private estate and country house aptly known as *Landhaus Energie* around 1905–6. Ostwald's Leipzig years coincided with his vigorous promotion both of the discipline of physical chemistry and of a broader programme of Energetics. With Nernst as his principal assistant, the laboratory programme (linked to the theory of electrolytic dissociation) attracted many young chemists from overseas, especially from the United States.[63]

Central to his programme of physical chemistry were transformations of energy and the doctrines of chemical thermodynamics initiated in the late 1870s and early 1880s by Gibbs and Helmholtz. Beginning at Leipzig with his inaugural lecture on 'Energy and its Transformations', Ostwald promoted his Energeticist approach in a series of lectures and publications. In 1891–92, for example, Ostwald presented his 'Studies on Energetics' to Leipzig's leading scientific society. Arguing that the advance of thermodynamics contrasted with the sterility of attempts to reduce physics and chemistry to mechanics through molecular-mechanical approaches such as the kinetic theory, he emphasized the problems of providing a satisfactory mechanical interpretation of temperature. He therefore proposed substituting energy for mass in the absolute system of measurement. More radical still was his view that mass, no longer a fundamental entity, should be subsumed under energy, now postulated as the primary 'substance' in nature.[64]

The Energeticist school had formidable critics, not least in Germany itself. At the annual meeting of the German Society of Scientists and Physicians held in Lübeck (1895), Boltzmann launched a fierce attack on the presentations of Helm ('On the Present State of Energetics') and Ostwald ('The Conquest of Scientific Materialism'). As one Boltzmann partisan observed, the attack resembled 'the fight of the bull [Boltzmann] with the lithe swordsman [Ostwald]. But this time, in spite of his swordsmanship, the toreador was defeated by the bull'.[65] In a follow–up critique published in the *Annalen der Physik*, Boltzmann condemned Ostwald's Energetics not only for perceived mathematical and physical errors but also for its false promise of easy rewards. Boltzmann, however, was not prepared to defend an approach based on mechanical-molecular hypotheses:

> Probably no-one holds energy to be a reality any more; or believes that it has been proven without question that all natural phenomena can be explained mechanically ... We are much more cautious nowadays ... The precise description of natural phenomena, as independent as possible from all hypotheses, now generally is recognized to be most important ... For a long time already, even in the theory of gases, one no longer regards molecules exclusively as aggregates of material points, but rather as unknown systems determined by generalized coordinates.[66]

Presenting the Energeticists as the dogmatists, Boltzmann labelled their approach as a mere classification system which required distinct forms of energy and its laws for each different branch of physics rather than offering the unity which a mechanical conception could provide.[67]

Of the new generation of German theoretical physicists attaining academic eminence in the late nineteenth century, Max Planck (1858–1947) had initiated himself into thermodynamic perspectives with a doctoral dissertation at Munich (1880) concerned with a reformulation of entropy. By the late 1880s he had been awarded second prize by the Göttingen Philosophical Faculty for an historical and analytical essay on the conservation of energy. While professor of theoretical physics at Kiel (1885–92) he sought to give the second law of thermodynamics the universal generality accorded to the first law. Planck's approach to thermodynamics had much in common with that of Gibbs and Ostwald in its generality and in its independence from molecular physics. But he became increasingly

hostile to the Energeticists on account of what he saw as their failure to differentiate between reversible and irreversible processes in nature.[68]

Ostwald regarded entropy, unlike energy, as a derived concept which he interpreted in terms of the dissipation of energy associated with radiation. By 1893
Planck had based his own interpretation of the second law on the broad empirical
postulate that irreversible processes do exist in nature. The second law therefore
treated of the direction that a process takes and as such could not be assimilated
to the first law. Rejecting both the anti-mechanist interpretation of Ostwald
and the mechanistic reductionism of Helmholtz and Clausius, Planck gradually
moved closer to Boltzmann's probabilistic interpretation of the second law as he
investigated the energy spectrum of black body radiation. With the benefits of
retrospective rationalisation, twentieth-century physicists would present Planck's
radical introduction of 'energy quanta' into these researches around 1900 as crucial
to the shift from 'classical' to 'modern' physics.[69]

The shifting alliances within German theoretical physics left little space for the
Energeticist programme. But Ostwald directed his efforts to a much wider stage by
offering Energetics as a programme for the world, relevant to all levels of physical
and living nature, including human society and civilization itself. In his *Naturphilosophie* (1910), for example, he asserted that the law of energy conservation
'must also be regarded as operative in all the later sciences, that is, in all the
activities of organisms, so that all the phenomena of life must also take place within
the limits of the law of conservation':

> Thus, the entire mechanism of life can be compared to a water-wheel. The
> free energy corresponds to the water, which must flow in one direction
> through the wheel in order to provide it with the necessary amount of work.
> The chemical elements of the organisms correspond to the wheel, which
> constantly turns in a circle as it transfers the energy of the falling water to
> the individual parts of the machine.[70]

Furthermore, civilization or culture, understood as that which 'serves the social
progress of mankind', consisted in 'improved methods for seizing and utilizing the
raw energies of nature for human purposes'. Conflicts between nations or between
social classes acted to destroy 'quantities of free energy which are thus withdrawn
from the total of real cultural values'. As Ostwald expressed the point in his
Energetic Imperativ: 'Waste no energy. Utilize it!'[71]

Epilogue

At Clerk Maxwell's we did our papers in the dining-room and adjourned for lunch to an upper room, probably the drawing-room, where Clerk Maxwell himself presided. The conversation turned on Darwinian evolution; I can't say how it came about, but I spoke disrespectfully of Noah's flood. Clerk Maxwell was instantly aroused to the highest pitch of anger, reproving me for want of faith in the Bible! I had no idea at the time that he had retained the rigid faith of his childhood, and was, if possible, a firmer believer than Gladstone in the accuracy of Genesis.

Karl Pearson recounts his old Tripos days at Cambridge during the mid-1870s[1]

Pearson's recollection of a verbal conflict with the Cavendish professor of experimental physics suggests that Clerk Maxwell was at heart a biblical literalist, committed to the truths of the sacred text imbibed from his earliest years in Scotland. We have seen, however, that Maxwell's theological perspectives were vastly more refined than Pearson's rather condescending remarks imply. Possessing a profound distaste for the perceived anti-Christian materialism of metropolitan naturalists, Maxwell and his fellow North British scientists of energy had not only embedded their new natural philosophy in the cultures of presbyterianism but had also been ready to deploy that natural philosophy in the service of a Christianity suitable to the wants of Victorian Britain.

From Reformation times, Scottish Calvinism had been embodied in the *Westminster Confession of Faith* (1647). That text had opened with the ringing declaration that 'Although the light of nature, and the works of creation and providence do so far manifest the goodness, wisdom, and power of God, as to leave men inexcusable; yet they are not sufficient to give that knowledge of God, and of his will, which is necessary to salvation'. *The Confession of Faith* therefore proclaimed the Holy Scripture as the means by which God revealed Himself 'for the better preserving and propagating of the truth, and for the more sure establishment and comfort of the Church against the corruption of the flesh, and the malice of Satan and of the world'. This insight into a Fallen world, the visible habitation of human beings, denied any perfectibility of man by means of his own efforts, whether in the form of individual redemption or in terms of the perfection of society. *The Confession of Faith* thus spoke instead of two convenants forged between God and man:

> I. The distance between God and the creature is so great, that although reasonable creatures do owe obedience unto Him as their Creator, yet they could never have any fruition of Him as their blessedness and reward, but by some voluntary condescension on God's part, which He hath been pleased to express by way of covenant.

> II. The first covenant made with man was a covenant of works, wherein life was promised to Adam; and in him to his posterity, upon condition of perfect and personal obedience.

> III. Man, by his fall, having made himself incapable of life by that covenant, the Lord was pleased to make a second, commonly called the covenant of grace.

Only by acceptance of the gift of grace could individual lost souls hope to be saved from being 'bound over to the wrath of God, and the curse of the law, and so made subject to death, with all miseries spiritual, temporal, and eternal'.[2]

For post-Enlightenment Scottish divines such as Thomas Chalmers, these doctrines were still the bedrock upon which presbyterianism was to be maintained. Natural theological arguments from design were necessary but insufficient pointers to the redemption of sinners. No other rational or practical effort could redeem human nature, corrupted as it was by depravity and decay. Preached in a time of social instability, Chalmers's sermons often seemed preoccupied by the extent to which visible nature, rather than the human race alone, was subject to such derangement. Even the hitherto eternal perfections of the mighty solar system seemed illusory: all nature contained 'within itself the rudiments of decay'. Moreover, just as mortal man had not within himself the means of personal salvation, neither could nature regenerate from within itself: 'unless renewed by the hand of the Almighty, the earth on which we are now treading must disappear in the mighty roll of ages and of centuries'.[3] It was a harsh and uncompromising message of a universe infected with sin and death (ch.2).

When James and William Thomson began their considerations of the seeming 'losses' of useful work or mechanical effect in waterwheels and heat engines, they were already operating within an engineering tradition, exemplified by James Watt, that focused on causes of waste rather than simply on ways of increasing total motive power. From an engineering and commercial perspective, the goal was to obtain the maximum amount of work, and hence wealth, from a given quantity of fuel by eliminating causes of waste (ch.3). From a complementary presbyterian perspective, God had freely provided those sources of mechanical effect such as water and coal. As with the spiritual gift of grace, man could choose to accept, ignore or refuse such gifts. Indeed, the spiritual and the natural gifts were closely connected: once the gift of grace had been accepted, man had a moral duty to direct, and not waste, the natural gifts. The Thomson brothers' fundamental problem of recoverability regarding the useful mechanical effect which might have been produced from the fall of water or steam had a Scottish theological counter-part: once wasted or refused those material gifts of power were 'lost' irrecoverably to man. In other words, no human or natural agencies could regenerate that missed opportunity for useful work (ch.5). But given that no destruction of the basic building blocks of the creation could occur, an enduring question for the North British scientists of energy was: what became of that mechanical effect when the fall of water or heat did not produce useful work on account of those causes of waste such as friction and conduction?

The Thomsons' encounter with Joule in 1847 brought into conflict two very different cultures. While fully agreeing that 'Nothing can be lost in the operations of nature – no energy can be destroyed', William Thomson was at first only partially convinced by Joule's arguments. He could accept that Joule had shown the conversion of work into heat and had thereby provided a plausible account of fluid friction. But he was not at all persuaded of the conversion of heat into work and so could not accept Joule's claim for mutual convertibility which Joule had recently deployed to construct a dynamically balanced system of nature (ch.4). The Thomsons' reservation derived *prima facie* from their increasing commitment to the Carnot–Clapeyron theory which postulated not the conversion of heat into work but the production, in the presence of a heat engine, of work from the fall of heat from higher to lower temperature. Although weakened by the seeming 'loss' of *vis viva* when friction or conduction occurred, the theory was soon extended by the Thomsons to the construction of an absolute scale of temperature and to the prediction of a lowering, with increased pressure, in the freezing point of ice. But the fundamental reservation with Joule's claims for mutual convertibility rest-ed with the problem of the (non)recoverability of heat in the form of useful mechanical effect when conduction and friction occurred (chs 3 and 5).

When Thomson drafted his private synthesis of Joule and Carnot early in 1851, he opted for the term 'energy' to cover such older terms as '*vis viva*' and 'mechani-cal effect' or 'work'. As before, he expressed his conviction that 'no destruction of energy can take place in the material world without an act of power possessed only by the supreme ruler'. But now he also argued that 'transformations take place which remove irrecoverably from the control of man sources of power which, if the opportunity of turning them to his own account had been made use of, might have been rendered available'. This laden sentence carried two major insights. First, these transformations of energy in the material world were directly linked to man's engineering and economic concerns with the efficient production of motive power. In particular, any inefficiency or imperfection resided not in nature but in man's imperfect knowledge and skill regarding engine design. Second, the transforma-tions themselves were occurring throughout the universe and were not simply a function of human action. Thomson therefore attempted to capture this direc-tional feature of the creation in the phrase 'Everything in the material world is progressive'. On the one hand, the phrase carried the *imprimatur* of the Cambridge geological science of Sedgwick, Whewell and Hopkins in opposition to the steady-state geology of Lyell. But, through Nichol and above all the *Vestiges*, the phrase carried radical and potentially threatening connotations of developmental theory, anathema to orthodox Scottish presbyterian culture (ch.6).

Thomson's public presentation deployed 'dissipation of mechanical energy' rather than 'progression'. With its connotations of 'waste' of heat, time, effort or capital, 'dissipation of energy' captured well the 'loss' of mechanical effect which might have been used by man had he deployed efficient engines to harness those sources of mechanical effect provided in nature. Dissipation of energy was also a doctrine to which presbyterian audiences, familiar with the refrain of Chalmers and other Scottish evangelicals concerning the tendency of the visible creation towards derangement and disorder, could readily assent. The dissipation doctrine therefore resonated with traditional, harsh Calvinist views on the fallen nature of a world now inhabited by a 'drunken and dissipated crew' and within which no striving of nature or man could achieve regeneration (ch.6).[4] But while Thomson and his allies could agree that man was an imperfect and fallible creature, prone to error and temptation, and that the visible world too manifested every evidence of irreversible dissipation, they remained uneasy about the notion of an imperfect creation. The imperfect world of harsh Calvinism that had driven James Thomson into unitarianism during the late 1840s also rendered Christianity open to scathing caricatures from critics. As W.K. Clifford wrote in 1875:

> One form of this traditional conception is set forth by the popular and received theology of Christian communities. According to this the condition of the departed depends ultimately on the will of a being who a long while ago cursed all mankind because one woman disobeyed him. The curse was no mere symbol of displeasure, but a fixed resolve to keep his victims alive for ever, writhing in horrible tortures, in a place which his divine fore-knowledge had prepared beforehand. In consideration, however, of the death of his son, effected by unknowing agents, he consented to feed with the sweets of his favour such poor wretches as should betray their brethren and speak sufficiently soft words to the destroyer of their kindred. For the rest, the old curse survives in its power; condemning them to everlasting torment for a manifestation of his glory. . . . How well and nobly soever a man shall have worked for his fellows, he must end by being either the eternal syco-phant of a celestial despot, or the eternal victim of a celestial executioner.[5]

By the 1850s, however, the old theological message was being softened, at least for the lowland middle classes, by a new generation of Scottish presbyterian and episcopalian scholars associated with the universities and the intelligentsia. For Thomson, Rankine, Maxwell and their theological colleagues the emphasis had shifted from a depraved humanity to Christ as perfect humanity (ch.2). Human beings fell short of that spiritual and moral perfection. But human beings were also imperfect in terms of their knowledge of, and power over, the material world. When in 1877 Lewis Campbell (Maxwell's friend and by now professor of Greek at St Andrews) published a series of sermons under the title *The Christian Ideal*, the incarnationalist vision was very well established but his expression of it was especially eloquent:

> The great matter is the becoming like to Christ; and whoever in any body of Christians, or belonging to no body of Christians at all, is most like to Christ in living according to justice, mercy, and truth, in loving and obeying God and doing good to men, that man is most of a Christian . . . Now ideas of justice and mercy, even of truth, have stirred the minds of men in all ages in a fitful and partial way. But never till Christ came were they embodied in a perfect life. In Him they are not ideas, but divine realities . . . There is a spirit here which has power to pass through the thickest crust of dying and decaying life, and to open a way directly to the light of heaven.[6]

Elsewhere in his sermons Campbell invoked his friend's energy culture to present a vision of an ideal Christian society which, free from self-interests and jealousies, would enable a redirection of time and thought to the task of ministering to human wants: 'what a saving was there here of human energy'. He claimed indeed that one aspect of the 'mystery of evil . . . is contained in that sad word, Waste' which was manifested in so many instances of human life. Even the vast organization of the Church herself, he suggested, spent 'energies . . . on party rivalries and outworn controversies', energies which should be 'drawn into one focus with the single aim of lifting the life of man, of continuing the work of Christ in relation to the circumstances of our age'.[7]

With respect to the material world, Thomson and Rankine had adapted Carnot's theory and set up an ideal of a perfect thermo-dynamic engine against which existing and future engines could be assessed (chs 5 and 8). All such engines were liable to some incomplete restoration if run in reverse. Friction, spillage, and conduction produced 'waste', ensuring that the engine fell short of the ideal. No human engineers could ever hope to construct such a perfect engine. But the question of nature's ultimate perfection remained. Rankine's 1852 speculation regarding nature's reconcentration of energy suggested that the universe as a whole might function as a perfectly reversible thermo-dynamic engine, thereby limiting 'dissipation' to the visible portion only and asserting that the creation did not, in Chalmers's language, contain within itself the seeds of its own destruction. Thomson, on the other hand, preferred to point to an infinite universe of energy with an 'endless progress, through endless space' in which the 'dissipation of energy' was characterized not as imperfection in nature but as an irreversible stream of energy from concentration to diffusion (ch.7). Stewart and Tait took this perspective much further, locating the visible and transitory universe within their scheme of an unseen universe in which the law of dissipation of energy might not hold as an ultimate principle (ch.12).

Whatever the ultimate condition of the universe, however, all members of the North British group agreed that the directionality of energy flow (whether expressed as 'progression' or 'dissipation' in the material world) characterized the visible creation and that this doctrine was the strongest weapon in the armoury against anti-Christian materialists and naturalists. By his direct involvement with the history and meaning of energy physics, John Tyndall had rapidly assumed the status of *bête noir* for the scientists of energy. To begin with, Tyndall's elevation of Mayer gave Tait a golden opportunity to caricature the German physician as the embodiment of speculative and amateurish metaphysics and to set him against the trustworthy and gentlemanly producer of experimental knowledge from Manchester (ch.9). But Tyndall's associations with other scientific naturalists such as Huxley and Spencer made him into far more than a wayward London professor. However much Tyndall might profess views above those of rank materialism, his opposition to dogmatic Christianity and his seeming commitment to scientific determinism throughout both inanimate and living nature made him a ready, if subtle embodiment of materialism (ch.12).

For the North British group, and especially for Thomson and Maxwell, the core doctrine of 'materialism' was reversibility. In a purely dynamical or mechanical system there was no difference between running forwards or backwards. If, then, the visible world were a purely dynamical system, we could have a cyclical world which would run in either direction. But the doctrine of irreversibility killed all such cyclical cosmologies stone dead (chs 6 and 12). The ramifications of the doctrine of irreversibility were indeed manifold. At one level Thomson and his allies deployed it to construct estimates of the past ages of earth and sun which would police geological and biological theorizing in general and undermine Darwin's doctrine of natural selection in particular (chs 7 and 9). At another level, they would use it to reinforce, as Maxwell put it, 'the doctrine of a beginning' (ch.12).

Thomson's earliest version of this doctrine, based on Fourier's mathematical theory of heat conduction applied to the body of the earth, dated from the 1840s (ch.6) but was reiterated in Stewart's *The Conservation of Energy* (1874). As Clifford pointed out in 1875, philosophers and theologians, notably Stanley Jevons in *The Principles of Science*, had read Thomson's inference as one whereby we 'have here evidence of a limit of a state of things which could not have been produced by the previous state of things according to the known laws of nature'. That reading, indeed, was not markedly different from Thomson and Maxwell's own (chs 6 and 12). It was a reading also taken up by a prolific Irish author, Joseph John Murphy, who played an active part in the reorganization of the Church of Ireland after its disestablishment by Gladstone (1868) and who was a close associate of James Thomson through their membership of the Belfast Literary Society in the 1860s. Murphy's two-volume *Habit and Intelligence* (1869) treated its readers to a whole chapter on 'The Motive Powers of the Universe' concerned with the laws of energy, especially dissipation.[8]

The doctrine of dissipation served a more explicit theological purpose in Murphy's *The Scientific Bases of Faith* (1873). 'Either the universe is from everlasting, or it had an absolute beginning in time', he argued. But while metaphysical reasoning yielded no solution, inductive science, in the guise of the nebular theory, pointed to an absolute beginning. Yet although the nebular theory 'is in the highest degree probable', a less hypothetical form of the argument centred on the process of energy dissipation which 'has been constantly going on from the first beginning of things; but it cannot have been going on through actually infinite time, because if it were so, an infinite quantity of motive power must have been expended and destroyed in every finite part of the universe; and the laws of force [energy] exclude the possibility of any such infinite supply of motive power'.[9]

Reviewing this kind of argument in the *Fortnightly Review* (1875), Clifford observed that the 'same doctrine has been used by Mr Murphy, in a very able book, to build upon it an enormous superstructure. I think the restoration of the Irish Church was one of the results of it'. But Clifford claimed that Thomson had been speaking not of 'the known laws of nature' but of 'the known laws of the conduction of heat'. Thus we should not conclude from Thomson's analysis that at the point in question 'the laws of nature began to be what they are; that is the point where the earth began to solidify; that is a process which is not a process of the deduction of heat, and so the thing cannot be given by the equation'.[10] On the other hand, Thomson's theological audiences in North and West Britain had undoubtedly read him that way and it is unlikely that he or Maxwell ever protested against such readings. What mattered for the scientists of energy, indeed, was not a specific beginning in time, but the death of cyclical, self-regenerating cosmologies running counter to basic Christian tenets.

Parallel North British concerns about the importance of free will as a directing agency in a universe of mechanical energy provided a principal context for Maxwell's statistical interpretation of the second law of thermodynamics in 1867. Framing his insight in terms of a microscopic creature possessed of free will to direct the sorting of molecules, his interpretation showed that the meaning of the second law was indeed relative to human beings who were imperfect in their ability to know molecular motions and to devise tools to control them. Available energy then became 'energy which we can direct into any desired channel' while dissipated energy was 'energy which we cannot lay hold of and direct at pleasure'. Dissipation would not therefore occur to either a creature unable to 'turn any of the energies of nature to his own account' or to one, such as Maxwell's imaginary demon, who 'could trace the motion of every molecule and seize it at the right moment'. Only to human beings, then, did energy appear to be passing 'inevitably from the available to the dissipated state' (ch.12).

In this study of the science of energy I have attempted throughout to employ the term 'energy' and its correlates as categories deployed by the historical actors themselves. Rather than present an account of 19th century energy physics in terms either of a rational reconstruction or of simultaneous discovery, I have adopted a historicist approach which places emphasis on a contextualist account of the rise of the science of energy at a particular time and place. I have shown in detail how the cultures of industrialization and presbyterianism shaped the North British formulation of the new science. In turn, the scientists of energy deployed their science in the cause of a reformed incarnationalist Christianity and in opposition to scientific naturalism as well as to older versions of moderate and evangelical presbyterianism. In wider context, their goals of international credibility were in large part attained by means of the British Association as well as by enlisting other practitioners such as Helmholtz to their cause. I have also stressed, however, that the new science entailed far more than the construction of a new way of thinking. Representing the whig, reforming and progressive values of the North British natural philosophers and engineers, the science of energy was embodied in a whole range of spaces and practices: in physical laboratories, in telegraph engineering, and in textbooks and treatises for pedagogical purposes.

North British scientists of energy made powerful and universalist claims for the new physics, arguing, for example, that, unlike force, the 'fact' that energy cannot be created or destroyed by human beings guaranteed its physical reality. Just how contingent were such claims, however, may be recognized from the contemporary critique of the sceptical Clifford in 1875:

> Every quantitative relation among phenomena can be put into a form which asserts the constancy of some quantity which can be calculated from the phenomena. 'Gravitation is inversely as the square of the distance for the same two bodies'; this may also be said in the form, 'gravitation multiplied by the square of the distance is constant for the same two bodies' ... A dream is a succession of phenomena having no external reality to correspond to them. Do we never dream of things that we cannot destroy?[11]

In the longer term, indeed, the bulwark doctrine of energy dissipation too would be taken to represent not a Christian vision of nature but a pessimistic fatalism in which, as Bertrand Russell so poignantly wrote in *A Free Man's Worship* at the end of the Great War, 'all the labours of the ages, all the devotion, all the inspirations, all the noonday brightness of human genius, are destined to extinction in the vast death of the solar system, and that the whole temple of Man's achievement must inevitably be buried beneath the debris of a universe in ruins'.[12]

Notes

Chapter 1.   Introduction: a History of Energy

1   *Phil. Mag. 18* (1884): 153–4.
2   Thomson 1881: 513; Rankine 1855b: 120–41; 1881: 209–28, on 214; Maxwell 1990–, 1: 517 [c.1857].
3   Tait 1868: 46–86; Garnett 1879: 206.
4   Latour and Woolgar 1986 [1979]: 187–233.
5   Smith 1998: 118–19.
6   Herschel 1865: 439. The history of *vis viva* is sketched in ch.3.
7   Rankine 1867: 88–92; 1881: 229–33, esp. 231.
8   J.D. Forbes to William Whewell, 2 November 1856, Add Ms a.204/116, Whewell papers, Trinity College Cambridge. I am most grateful to an anonymous referee for drawing my attention to this revealing remark. On Faraday's conception of nature see esp. Cantor 1991: 185–95. Although Faraday remained outside the science of energy, the North British group frequently drew upon his insights into electricity and magnetism in particular and retained a profound respect for him as a central icon for scientific life and practice. G.B.Airy's remarks to Faraday in 1847 perhaps encapsulate the status that Faraday had earned among his peers: 'We trouble you as a universal referee or character-counsel on all matters of science' (quoted in James 1997: 298).
9   Morrell and Thackray 1981: 58–94.
10 Thomson 1881: 513.
11 Thomson 1882–1911, 1: 174–5 [1851]. See also Brush 1970: 165–67.
12 Smith and Wise 1989: 149–202.
13 Smith and Wise 1989: 20–32; 118–20.
14 Merz 1904–12, 2: 95–6.
15 Kuhn 1959: 321. Examples of popular whig histories of physics persist. See Motz and Weaver 1989: 157–77; Lindsay 1975: 12–23.
16 Kuhn 1959: 321.
17 Kuhn 1959: 321–39.
18 Kuhn 1959: 322–3.
19 Elkana 1970b: 31–60; 1974: 175–97.

20 For example, Dahl 1963: 174–88; Heimann 1974a: 147–61; Heimann 1974b: 205–38; Heimann 1976: 277–96; Cantor 1975b: 273–90; Wise 1979a: 49–83; Cantor 1984.

21 Kuhn 1959: 345n.

22 Kuhn 1959: 323; 1962b: 760–4; Cantor 1984; Morus 1989: 2–3. More recently, Rabinbach 1990: 3–4 (and throughout) assumes the discovery of energy conservation in a traditional, unproblematic sense and thus tends to assimilate a vast range of diverse European texts to an oversimplified monolithic discourse with little or no attention paid to different constructions in different social and cultural contexts.

23 For example, Hanson 1958: 4–92; Kuhn 1962b: 760–4; Cantor 1984.

24 Kuhn 1959: 343n.

25 Maxwell to Lewis Campbell, 21 April 1862, in Campbell and Garnett 1882: 335–36; Maxwell 1990–, 1: 711–12.

26 In ordinary language, the *Oxford English Dictionary* shows the ancient pedigree of 'energy', from the Greek ενεργεια (work). In broad terms the word was often applied to human beings who displayed a high degree of mental or physical activity. But 'energy' carried a range of quite subtle overlapping meanings which included: 'vigour of expression' in speech or writing; 'activity' or actual working; 'intensity of action' or (as a personal quality) the 'capacity . . . of strenuous exertion'; natural 'power actively and efficiently displayed or exerted'; and power 'not necessarily manifested in action' but rather the 'ability or capacity to produce an effect'.

27 Larmor 1908: xxix.

Chapter 2.   From Design to Dissolution: Scotland's Presbyterian Cultures

1 Chalmers 1836–42, 7: 266–7. The words 'though of old . . . they shall be changed' are a paraphrase of Psalm 102: 25–7.

2 For accounts of Scotland in the 1830s and 1840s see Smout 1987: 7–31 (general survey); Shapin 1975: 219–43 (Edinburgh); Desmond and Moore 1991: 21–44 (Edinburgh); Smith and Wise 1989: 20–32 (Glasgow). For the wider British context see Morrell and Thackray 1981: 2–34.

3 Paley 1823: esp. 9–11 [1802] (watch analogy); Brooke 1991: 192–3 (Paley and design); Waterman 1991: 113–35 (Paley and Malthus); Hilton 1991: 170–83 *passim* (Paley and Butler). In his *Natural Theology* Paley claimed that 'there are strong intelligible reasons why there should exist in human society great disparity of *wealth* and *station*. . . . In order, for instance, to answer the various demands of civil life, there ought to be amongst the members of every civil society a diversity of education, which can only belong to an original diversity of circumstances . . . Inequalities therefore of fortune . . . may be left, as they are left, to *chance*, without any just cause for questioning the regency of a supreme Disposer of events'. See Paley 1823: 444–5 [1802]. I thank Ben Marsden for drawing my attention to this passage.

4 Smith 1976b: 302–4; Smith and Wise 1989: 90–3 (Whewell's *Bridgewater Treatise*); Brooke 1991: 198–203 (natural theology in Britain), 211, 240 (*Bridgewater Treatises*), 226–74 (rise of the historical sciences); Hilton 1991: 170–83 (Oxford and Cambridge critics of Paley); Topham 1993: 397–430 (*Bridgewater Treatises* and popular science education).

5 Smith and Wise 1989: 35. On the Scottish democratic ethos see also Saunders 1950; Davie 1961: 3–25. For general historical studies of Scottish church history see Burleigh 1960; Brown 1987; Donaldson 1987.

6 On 'Moderates' *versus* 'Evangelicals' over the election of John Leslie to the Edinburgh chair of mathematics see Clark 1960–2: 179–97; Morrell 1975: 63–82. Accusing Leslie of supporting a Humean view of causation, the Moderates, Hilton notes, came across as 'religious bigots and narrow adherents to the Westminster Confession' whereas the Evangelicals were favourable to the appointment. Committed to the design argument Moderates 'never really needed a temporal sequence. God was the cause, and his effects were conceived as operating spatially . . . [whereas evangelicals like Chalmers] inserted a temporal sequence – sin followed by punishment'. See Hilton 1991: 24–6. On the later issue of patronage see Fry 1987; Brown and Fry 1993.

7 Morrell 1969: 245–65 (Thomas Thomson and Glasgow College reform); Smith and Wise 1989: 25–32, 41–9 (James Thomson and Glasgow College reform). The Glasgow College Moderates identified strongly with Tory politics in opposition to the Whig (and Peelite Conservative) politics of the reformers, some of whom had evangelical and Free Kirk sympathies. The Rev. Thomas Brown of St John's Church, Glasgow (formerly Chalmers's Church and one regularly attended by James Thomson's family up to the Disruption) captured the deep gulf between Moderates and Evangelicals when he wrote later: '[Our ministers of the Free Church] have gone forth and scattered the seed of the Word in every corner . . . In many corners of the world the cold chilling – at best but moral – disquisitions and addresses [of the Moderates] issuing from many pulpits . . . had induced an apathetic indifference to the things of God and eternity; but the soul-melting, heart-subduing strains of the Gospel . . . aroused, and captivated, and enchained many. . . . for oh! what are all of our movements to be directed to, what are we to covet and sigh for, but that a people may be gained unto the Lord, and brands plucked from everlasting burning' (Brown 1893: 245). Not for nothing did Rev. Norman Macleod, to whom the Thomsons turned for spiritual nourishment (below), call him 'the long-winded Thomas Brown' (Macleod 1876, 1: 90).

8 Hanna 1849–52, 1: 43–4; Smith 1979: 68n.

9 Chalmers 1836–42, 2: 373, 386; Cairns 1956: 410–921.

10 Hanna 1849–52, 1: 43–4; Smith 1979: 68n.

11 Chalmers 1836–42, 1: 222–3; Smith 1979: 59–60. On Robison see Smith 1976a: 7–11.

12 Chalmers 1836–42, 1: 222–5.

13 Chalmers 1836–42, 1: 225; Smith 1979: 60. 'The first construction' theme formed a key feature of the science of energy for Thomson and Maxwell in particular (chs 6,7 and 11).

14 Chalmers 1836–42, 2: 371–2; Smith 1979: 61. Rice 1971: 23–46 discusses Chalmers's appropriation of Scottish Common Sense philosophy, hitherto the preserve of the Moderates rather than Evangelicals within the Scottish universities. See esp. Hilton 1991: 165–6, 183–87.

15 Chalmers 1836–42, 2: 389–90; Smith 1979: 62. Hilton 1991: 80 stresses that for Chalmers 'the world was merely an "arena of moral trial", an "imperfect state" capable of but a limited amount of "welfare"'. We shall see that the scientists of energy did not share Chalmers's vision of this 'utter derangement' of nature. Despite the *prima facie* congruity, the North British doctrine of energy dissipation was represented not in terms of *nature's* derangement and imperfection but in relation to 'waste' of those opportunities for useful work which the energies of nature presented to humankind. The difference is subtle but profoundly significant. The scientists of energy belonged to a post-evangelical culture whereby human beings, in their moral and material actions, had an obligation to aspire to the perfections of nature and of Christ (note 55 below and chs 5–12 *passim*). There is, nevertheless, a clear sense that man cannot recover lost

opportunities for useful work and that, concomitantly, only God can restore or regenerate the 'lost' or dissipated mechanical energies of the universe. The parallel here with
traditional presbyterianism in which only God can restore 'lost' souls is striking.

16 Chalmers 1836–42, 4: 387–8. Chalmers's 'voluntarism', with emphasis on divine will
rather than reason, had a long pedigree within Christianity. See esp. Oakley 1961: 433–
57; McGuire 1972: 523–42.

17 Chalmers 1836–42, 1: 161–87 ('On the Non-eternity of the Present Order of Things');
12: 349–72 ('Remarks on Cuvier's Theory of the Earth'); Smith 1979: 64 (on beginnings); Millhauser 1954: 65–86, esp. 66–70 (Chalmers as 'scriptural geologist'); Brooke
1991: 272 (Chalmers on *Genesis*). Hilton 1991: 23 draws attention to a distinction
provided by Yule 1976 between 'scriptural geologists' and 'geological hermeneutists'.
The former adapt the findings of geology to the Genesis story while the latter interpret Genesis to fit geology. Chalmers's formula, as an attempt to reconcile geological
and scriptural time, falls within 'geological hermeneutics'. Other examples were
the attempts to redefine the biblical 'day' in terms of much longer periods of time.
The distinction, it should be noted, was not one used by the historical actors
themselves.

18 Chalmers 1836–42, 7: 262–79 ('The Transitory Nature of Visible Things'); Smith 1979:
64–6. On Whewell and Encke's comet see Smith 1976b: 302–4.

19 Chalmers 1836–42, 7: 280–99 ('On the New Heavens and the New Earth'); Smith 1979:
66.

20 Compare Brewster 1837: 1–39, esp. 2–3, 13–15; Hilton 1991: 152–3. In this review of
William Buckland's *Bridgewater Treatise* Brewster argued that 'the Church is now feeding her flock on the green pastures of the Huttonian geology' (p.14). In appropriating
the geology of James Hutton, Brewster was implicitly rejecting Playfair's steady-state
reading of the Huttonian theory of the world (Playfair 1802) in favour of *successive*
cycles, each of which was apparently in a state of decay and each of which had been
created by God. We were now in 'the first cycle [in a probable series of cycles] of the
intellectual occupation of the globe'. But, following his friend Chalmers, Brewster held
that none of these cycles displayed permanence: 'Even the sacred volume forewarns us
of the coming day, when the elements shall melt with fervent heat, – when the earth,
and the works that are therein, shall be burnt up; – and when new heavens and a new
earth shall replace the ruins of a world' (p 3). Brewster was paraphrasing 2 Peter 3: 10–
13.

21 Hilton 1991: esp. 15–17, 31–2.

22 Chalmers 1844–5: 47 ('Political Economy of the Bible'); Hilton 1991: 85.

23 Chalmers 1844–5: esp. 33–4, 66–70, 73–114. See also Waterman 1991: 217–52. 264–73
(Chalmers and Malthus).

24 Biagioli 1993: 39.

25 Biagioli 1993: 36–54, on 51. See esp. Mauss 1990 [1950]; Douglas 1990: vii–xviii for
the argument that gifts carry obligations to reciprocate. In the case of Protestant Christianity gifts such as 'grace' appear to be freely offered by God. There can be no question
of mortal man soliciting such gifts or reciprocating with equivalent measure. Objections
to trading in spiritual matters was, after all, the *raison d'être* for the *protest* against
Catholicism in the sixteenth century. But there clearly was a two-way, unequal relationship. Acceptance of the spiritual gift brought human individuals back into solidarity
both with God and with the Christian community of true believers and carried the
obligation to act righteously thereafter.

26 *Church Hymnary* 1928: 91. The full verse reads: 'Vainly we offer each ample oblation,/

Vainly with gifts would his favour secure;/Richer by far is the heart's adoration;/Dearer to God are the prayers of the poor'.

27 Chalmers 1836–42, 11: 403–4 ('On the Superior Blessedness of the Giver to that of the Receiver').

28 Chalmers 1836–42, 11: 413; Bacon 1951 [1605]: 42.

29 Chalmers 1836–42, 11: 414–17; Hilton 1991: 102–3, 108–14 (Chalmers on charity).

30 Chalmers 1836–42, 11: 413, 419–23.

31 See the esp. discussion by Waterman 1991: 217–29.

32 Morrell 1969: 245–65; Smith and Wise 1989: 25–32; Marsden 1998: 87–117.

33 Hanna 1849–52, 4: 212–13 (Hill and Chalmers); Smith 1998: 118–46; Smith and Wise 1989: 32–49.

34 Smout 1987: 188 notes that the 1851 census indicated that at the time the Free and Established Kirks had approximately equal membership over Scotland as a whole. But in the Highlands the Free Kirk was especially strong with every minister in Aberdeen defecting from establishment at the time of the Disruption. A third force, the United Presbyterian Church, resulted in 1847 from the merger of former secessionist churches, that is, those that had left the Established Kirk many years previously. On the Free Kirk Colleges see Watt 1946; Macleod 1996: 201–3.

35 Anna Bottomley (nee Thomson) to William Thomson, c. October 1843, B201, Kelvin Collection, ULC. For natural philosophy teaching in Scottish universities see Wilson 1985: 19–26 (Edinburgh), 26–33 (Glasgow).

36 Brooke 1996: 171–86, esp. 173; Macleod 1996: 187–205; Shortland 1996: 287–300.

37 Morrell and Thackray 1981: 430–4 (Forbes and his use of the BAAS); Smith and Wise 1989: 99–103 (Forbes as a possible successor to Meikleham as Glasgow professor of natural philosophy); Smith 1998 (on university and industrial city). Several of the protagonists in the science of energy (Rankine, Gordon, Tait and Maxwell) had been pupils of Forbes at Edinburgh.

38 Thomas Chalmers to James Thomson, 14 February 1847, C80, Kelvin Collection, ULC; Smith and Wise 1989: 13–14. Chalmers and Brewster had provided the young William Thomson with introductions to the French *physicien* J.B. Biot two years earlier. William's Paris visit added experimental credibility and aided in his bid to succeed Meikleham in 1846.

39 Thompson 1910, 1: 187 (Latin inaugural dissertation).

40 *Westminster Confession* 1973: 135 [1648]. On the 'heresy' trials see Black and Chrystal 1912; Drummond and Bulloch 1978: 47; Knott 1911: 291–2 (Tait's obituary of Robertson Smith); Scott 1996: 22–4. On Miller's radical re-reading of Genesis see Brooke 1991: 272. Miller 'equated the Genesis days with successive days on which the author had received a divinely inspired vision as to how the earth would have appeared in each of its primitive epochs'. On Miller's dispute with R.S. Candlish (foremost leader of the Free Kirk after Chalmers) see Macleod 1996: 199–205. In contrast Candlish's reading of Genesis was extremely cautious: 'Let the student of science push his inquiries still further, without too hastily assuming, in the meantime, that the result to which he has been brought demands a departure from the plain sense of Scripture; and let the student of Scripture give himself to the exposition of the narrative in its moral and spiritual application, without prematurely committing himself, or it, to the particular details or principles of any scientific school'. See Candlish 1868: 19.

41 On Nichol see Schaffer 1989: 131–64. On *Vestiges* see Secord 1989: 165–94.

42 Dean Arthur Stanley quoted in Macleod 1876, 2: 305.

43 Macleod 1876, 1: 13–44, 63–85, esp. 40, 74.
44 Macleod 1876, 2: 1–56, on 2.
45 Macleod 1876, 1: 92–5, 97.
46 Macleod 1876, 1: 114–69, 282–8, on 284–5.
47 Macleod 1876, 1: 282–3.
48 Macleod 1876, 1: 208, 253–5.
49 Macleod 1876, 1: 148–9.
50 Macleod 1876, 1: 246, 279–80.
51 Smith and Wise 1989: 48.
52 Macleod 1876, 2: 29–31.
53 Caird 1858: 246–49.
54 Caird and Norman Macleod (the elder), *DNB*.
55 Macleod 1876, 2: 95–7, 136. On the rise of 'incarnationalism' see Hilton 1991: 5, 280–8,
    298–339. Reacting against those evangelical doctrines (such as Hell-fire) which
    appeared to 'cast a cloud over the perfections of God', the incarnationalist generation
    stressed Christ as 'that human character in which is incorporated the full perfection of
    every human virtue and power' (Hilton 1991: 281–82, quoting the Unitarian J.P. Hopps
    and High Churchman John Penrose respectively). At first confined to controversial
    ministers such as John McLeod Campbell and Edward Irving in the 1830s, incarnation-
    alist preaching in presbyterian North Britain eventually became the new orthodoxy
    through the work of Norman Macleod and the new Moderates.
56 Macleod 1876, 2: 139. Alexander Haldane's *Record* appeared between 1828 and 1923.
    Strongly Calvinist in content, it shared an apocalyptic other-worldliness with even more
    strident evangelical publications. See Hilton 1991: 10–11, 94–8, 211–12.
57 Macleod 1876, 2: 137–8.
58 Macleod 1876, 2: 20.

Chapter 3.   Recovering the Motive Power of Heat

1 Macleod 1876, 2: 90.
2 For example, Crouzet 1982: 278–316 (railways); Body 1971: 32–42, 65–82 (ocean
   steamships); Napier 1904: 1–165; Shields 1949; Slaven 1975: 125–33, 178–82 (Clyde
   shipbuilding and marine engineering).
3 King 1909: 1–104; Smith and Wise, 1989: 1–19, 35–6.
4 Cleland 1840: 174–75; Strang 1850: 162–69. Glasgow and the Clyde are also treated in
   Shields 1949: 12–23; Saunders 1950: 97–117; Daiches 1977: 95–179; Smith and Wise
   1989: 20–5.
5 William to Elizabeth Thomson, 24 May 1836, Add 8818/1, ULC; reproduced in King
   1909: facing p 138. See King 1909: 129 (James's early consideration of paddle blades);
   181–82 (up-river propulsion).
6 Morrell 1974: 88–92; Morrell and Thackray 1981: 202–22.
7 Murchison and Sabine 1840: xxxv. The address appears in lieu of a presidential address.
   My italics.
8 Dickinson 1935: 15–31 (early years), 131 (letter about Southern); Hills 1989: 51–52.
9 Hills 1989: 51–8; Watt 1970: 84 (Black's letter to Watt). Cardwell 1971: 40–55 seeks to
   refute the possibility that Watt's 'improvement' derived directly from Black's concept of
   latent heat but in so doing probably underestimates the *contextual* significance for Watt
   of Black's thermal researches at the University of Glasgow. As Black's letter suggests,

Watt's improvements were 'directed by' rather than derived from Black's doctrine. See also Kerker 1961: 385–6.

10 Hills 1989: 58–91 (on Watt's partnership and later innovations), 91–4; Hills and Pacey 1972: 39–43 (on indicator diagrams); Dickinson 1935 (on horse-power); Musson and Robinson 1969: 393–426 (on Boulton and Watt).

11 The Glasgow iconography of Watt is further examined in Smith 1998; 140–1 in relation to the new site for Glasgow University in the 1860s.

12 James to William Thomson, 9 July 1862, T117, Kelvin Collection, ULG. Kelland 1863: 326 confirmed James Thomson's recollection regarding tidal friction. Investigations of the tides ('tidology') had long been a favourite subject for BAAS reports and committees. Whewell was pre-eminent in this field during the 1830s. See Morrell and Thackray 1981: 513–17. In partial support of this recollection, James told his brother that he had found in an old notebook 'under the date September 1841 the following: "Some years ago it occurred to me that the tides caused on the earth by the motion of the moon round it must produce some retardation in the motion of the moon or acceleration in the motion of the earth round its axis or some other similar change". I give you the memorandum just as it stands. It is stupidly blundered in the expression . . . but assuredly I had the main idea in my mind long before'. The context for James's claim over tidal retardation was a priority dispute with John Tyndall over the priority claims of Mayer (ch.9). James's notebook appears not to have survived.

13 Russell 1840a: 186–7; 1840b: 188–90, on 189. Emmerson 1977 provides a biography.

14 Gordon 1840: 191. See Marsden 1992a: 188–235; 1992b: 323–6; 1998: 87–117 (Gordon at Glasgow University); Grattan-Guinness 1984: 21–22 (Fourneyron turbine); Scott 1970: 89–103 (Lazare Carnot).

15 Scott 1970: 155–82; Grattan-Guinness 1984: 6–33; Smith and Wise 1989: 156–57. See also Glas 1986: 249–68 (on Monge's mathematics).

16 Carnot 1803, translated and discussed in Grattan-Guinness 1984: 6–8.

17 See Kuhn 1959: 332; Cardwell 1967: 217–18; Scott 1970: 89–103; 155–9; Gillispie 1971. On the *vis viva* 'tradition' see esp. Iltis 1971; 1973a; 1973b; 1977; Hankins 1965; 1970a; 1985: 30–5. See also Laudan 1968; Heimann and McGuire 1971a; Papineau 1977.

18 Poncelet 1829, translated and discussed in Grattan-Guinness 1984: 14, 17. Coriolis is treated in Grattan-Guinness 1984: 13–15. See esp. Kuhn 1959: 332, 348–9. See also Cardwell 1967: 218–19; Scott 1970: 167–9, 172–6.

19 Grattan-Guinness 1984: 20–4.

20 Grattan-Guinness 1984: 19–20. On British interest in Morin's dynamometer see Smith and Wise 1989: 291–2. See also Constant 1983: 183–98.

21 Gordon 1840: 191–2.

22 King 1909: 190–1; Smith and Wise 1989: 51; Marsden 1992a: 63–7 (Macneill).

23 James Thomson, 'Essay on Overshot Water Wheels', Thomson Collection, QUB; Larmor 1912: xxvii–xxxiii; Smith and Wise 1989: 51–2; 292–3.

24 Morrell 1974: 88–92; Wise (with Smith) 1989–90: 221–9; *Regulations, &c of the Glasgow Philosophical Society* (Glasgow, 1812), 5–6.

25 Morrell 1974: 88–92; *Regulations* (1812), 11–12.

26 *Proc. Glasgow Phil. Soc. 1* (1841–4) provides membership lists. Desmond 1989: 14 draws attention to the vast increase in the number of medical journals issuing from steam-powered printing presses in the previous two decades.

27 *Proc. Glasgow Phil. Soc. 9* (1873–5): 1.

28 Wise (with Smith) 1989–90: 222–9.

29 Wise (with Smith) 1989–90: 227–32.

30 James to William Thomson, c.1841–2, T380, Kelvin Collection, ULC.

31 James to William Thomson, 11 October 1842, T381, Kelvin Collection, ULC. See also J.R. McClean to James Thomson, 15 November 1841 in Larmor 1912: xx–xxi.

32 James to William Thomson, 13 August 1863, T119, Kelvin Collection, ULG.

33 Whewell 1841: 145.

34 James to William Thomson, [1843], T383, Kelvin Collection, ULC; Smith and Wise 1989: 288; Fairbairn 1970: 335 [1877].

35 James to William Thomson, 1 May 1844, T398, Kelvin Collection, ULC; Larmor 1912: xxii–xxiii; Smith and Wise 1989: 288–9.

36 James to William Thomson, c.19 May 1844, T399, Kelvin Collection, ULC; Smith and Wise 1989: 290–2 (on Gordon's discussions of dynamometers and 'mechanical effect').

37 James to William Thomson, 19 June 1844, T401, Kelvin Collection, ULC.

38 James to William Thomson, 4 August 1844, T402, Kelvin Collection, ULC. The *Artisan* was a popular magazine and had been publishing articles on new forms of motive power, including the air-engine.

39 James to William Thomson, 4 August 1844, T402, Kelvin Collection, ULC.

40 James to William Thomson, 4 August 1844, T402, Kelvin Collection, ULC.

41 James to William Thomson, 23 December 1844, T404, Kelvin Collection, ULC; Larmor 1912: xxiii; Smith and Wise 1989: 292; Schaffer 1992b (Rosse's observatory).

42 James Thomson to Robert Murray, October 1845, in Larmor 1912: xxiv–xxv.

43 James Thomson to J.R. McClean, 25 December 1846 and 12 April 1847, in Larmor 1912: xxvii–xxviii.

44 William to James Thomson, 22 July 1847, T429X; James to William Thomson, 29 July 1847, T434, Kelvin Collection, ULC. See also Larmor 1912: xxviii–xxxiii.

45 Smith and Wise 1989: 105–8, 293–4.

46 Thomson 1889–94, 2: 458n [1892].

47 James to William Thomson, 22 February 1846, T415, Kelvin Collection, ULC; Smith and Wise 1989: 107–8 (Regnault and Thomson).

48 Clapeyron 1837: 348 [1834].

49 Clapeyron 1837: 349.

50 Cardwell 1971: 220–1 assumes Clapeyron's knowledge of the indicator diagram: 'The secret was . . . very closely guarded by Boulton and Watt; so well in fact that it was not until 1826 that a well-informed engineer like John Farey first saw an indicator diagram being taken in Russia . . . Some time during the ten years prior to 1834 the principle of the indicator diagram must have become known in France, or at any rate to Clapeyron . . .' On Rankine's usage see Marsden 1992a: 161–4.

51 Clapeyron 1837: 349–51 [1834]. Clapeyron did not use the later terms 'isothermal' (first and third stages) and 'adiabatic' (second and fourth stages).

52 Clapeyron 1837: 355–8. Cardwell 1971: 221–4 provides a good summary.

53 Clapeyron 1837: 359.

54 Clapeyron 1837: 359–74.

55 Clapeyron 1837: 374–5.

56 Clapeyron 1837: 347–76 (English translation). Only a partial German translation appeared in J.C. Poggendorff's *Annalen der Physik Chemie 49* (1843): 446–51. See Fox 1986: 55n.

57 Smith and Wise 1989: 296. On air engines see Daub 1974: 259–77; Redondi 1976: 243–59; 1980; Fox 1986: 35–9; Marsden 1992a: 96–140. Thomson 1847: 169 simply stated

that 'an explanation [of Carnot's theory] had been given by Professor Gordon at a previous meeting of the Society'. No record appears in the *Proceedings*.

58 *New DNB* (forthcoming). I am grateful to Ben Marsden for a pre-publication copy of his entry on Robert Stirling.

59 William Thomson to J.D. Forbes, 1 March 1847, Forbes Papers, St Andrews University Library.

60 Thomson to Forbes, 1 March 1847.

61 Thomson 1847: 169–70; 1882–1911, 5: 38–9; Smith and Wise 1989: 296–8.

62 Smith and Wise 1989: 298–9 provides full analysis of James's investigation.

63 James Thomson, 'Motive Power of Heat: Air Engine', Notebook A14[A], Thomson Collection, QUB.

64 Thomson 1882–1911, 1: 156–64 [1849].

65 Thomson 1882–1911, 1: 165–9 [1850].

66 William Thomson to Lewis Gordon, 20 December 1847, G124, Kelvin Collection, ULC.

67 Thomson 1848a: 313–17; 1882–1911, 1: 100–6.

68 Thomson 1882–1911, 1: 104. See esp. Smith and Wise 1989: 249–50, 299–301.

Chapter 4. Mr Joule of Manchester

1 Silliman 1838: 263 [1837].

2 Silliman 1838: 257–63; *DSB*.

3 Joule 1838: 122–3. Repr. Joule 1887, 1: 1–3. Joule made textual alterations to most of these early papers for the reprint. I use the original text in each case.

4 Gooding 1989: 184–5; Snelders 1990: 228–40 (on Oersted's 'discovery' of electromagnetism).

5 Sturgeon 1838–9: 429–37, on 430. See also Gee 1991: 41–72.

6 Reprinted in *Ann. Electricity 3* (1838–9): 161–3. Also quoted in Gee 1991: 67–8.

7 *Ann. Electricity 3* (1838–9): 162.

8 Gooding 1989: 195–201 (London electromagnetic 'network' of practitioners); Morus 1989: 12–46; 1992: 1–28 (Sturgeon and Faraday); 1993: 50–69 (Sturgeon and the London Electrical Society); Ginn 1991: 91–141, 270–311 (Wheatstone and Faraday).

9 Cardwell 1989: 59 states that Joule's 'was the remote, leisurely world of men like Charles Darwin and Willard Gibbs'. Rudwick 1985: 17–41 developes the term 'gentlemanly specialist' for nineteenth-century gentlemen geologists. Kargon 1977: 34–6, 49–60 treats Joule as one of a nineteenth-century breed of gentlemanly 'devotees', far more serious about their calling than their amateur and rather dilettantish predecessors.

10 Cardwell 1989: 8–9. See also Mathias 1959: esp.78–98 (on the English brewing industry and its uses of steam power).

11 Cardwell 1989: 10, 13, 30, 62.

12 Reynolds 1892: 27–9; Fox 1969: 73–4; Cardwell 1989: 5–7, 14–17; Kargon 1977: 11–14 (on Dalton and Manchester); Morrell and Thackray 1981: 398–9 (on Dalton and the BAAS); Thackray 1972. Dalton 1965 [1808] provides a good example of the Daltonian scientific style.

13 Dalton 1965: 1 [1808]. See also Fox 1971: 109–15 for fuller treatment of Dalton's views on 'caloric'.

14 See esp. Kargon 1977: 1–3; Cardwell 1989: 1–12.

15 Kargon 1977: 3–33; Cardwell 1989: 4–5.

16 Kargon 1977: 36–41; Cardwell 1989: 22, 25–8; Morus 1992: 1–2.

17 Cardwell 1989: 27–8.

18 Cardwell 1989: 28 (disputes with Harris and Faraday); Cantor 1991: 140–1, 144–5; Morus 1992: 1–28 (Sturgeon *versus* Faraday).

19 Joule 1838: 122–3. Altogether Joule contributed over a dozen letters and papers to Sturgeon's *Annals* in the period 1838–42.

20 Joule 1838: 123. On 'duty' see Hills 1989: 36–7, 107–8.

21 Sturgeon 1838–9: 436.

22 Joule 1838–9: 437–9. Repr. Joule 1887, 1: 4–6.

23 Joule 1839–40a: 131–35. Repr. Joule 1887, 1: 14.

24 Joule 1839–40b: 474–81; 1840: 187–98, 471–2. Repr. Joule 1887, 1: 19–42. Forrester 1975: 277n notes the similarity in presentation between Joule and Faraday.

25 Joule 1839–40b: 481. See also Cardwell 1989: 35. Gooding 1980b: 1–29; Cantor 1991: 185–95 examine the fundamental differences between Faraday and Joule. Forrester 1975: 273–301, on the other hand, locates Joule and Faraday in the same chemical or electrochemical community. While Joule made much use of Faraday as resource and role model, the two natural philosophers operated with very different assumptions and in very different local contexts.

26 Joule 1887, 1: 59–60 [1840]. See also Fox 1969: 83–5; Cardwell 1989: 35.

27 Cardwell 1989: 48, 69 accounts for Joule's rejection in 1840 and 1844 in terms of the Royal Society's conservatism and Joule's personality. On the Royal Society see Morrell and Thackray 1981: 47–63 (Babbage); Morus 1989: 113–67 (Grove).

28 Joule 1842a: 219, 221. Repr. Joule 1887, 1: 46–53.

29 Joule 1842a: 219–20.

30 Joule 1842a: 220. See Jacobi 1837: 503–31.

31 Joule 1842a: 220–1.

32 Joule 1842a: 221–2.

33 Joule 1842a: 222–23. See also Steffens 1979: 20–23.

34 Joule 1842a: 223–4.

35 Joule 1842a: 224.

36 Gooding 1980b: 1–29 and Cantor 1991: 161–95 explore Faraday's metaphysical and theological beliefs in relation to his natural philosophy.

37 Joule 1842b: 287; 1887, 1: 60–121. See Cardwell 1989: 39–45.

38 Kargon 1977: 5–14, 41–9; Cardwell 1989: 49–52; Forrester 1975: 273–301.

39 Joule 1846: 103. Repr. Joule 1887, 1: 119. See Cardwell 1989: 45.

40 Joule 1846: 109–10, 105.

41 Joule 1846: 104–5.

42 Joule 1846: 105n.

43 Joule 1843a: 204–8. Repr. Joule 1887, 1: 102–7. See Forrester 1975: 277–9; Morrell and Thackray 1981: 98, 396–411; Cardwell 1989: 52–3, 59.

44 Joule 1843b.

45 Joule 1843b: 263. Repr. Joule 1887, 1: 123–4.

46 Joule 1843b: 435. Repr. Joule 1887, 1: 149. For detailed discussion see Cardwell 1989: 53–9.

47 Joule 1843b: 441.

48 Wise (with Smith) 1989–90: 249–50. Compare Kuhn 1959: 321–56.

49 Joule 1843b: 441–2.

50 Joule 1887, 1: 157 [1843]; Rumford 1798: 70–1, 99–100. See also Watanabe 1959: 141–4; Olson 1970: 273–304; Fox 1971: 99–103; Cardwell 1971: 95–107; Goldfarb 1977: 25–36.

51 Rumford, *DSB*; Berman 1978: 6–18, 29–31 (reappraisal of Rumford's role in the foundation of the Royal Institution).
52 Rumford 1798: 8–102. Thomson 1830: 336–8 also provides a detailed description prior to his critical assessment.
53 Thomson 1830: 338–41. Lardner 1833: 395–7 was more favourable to Rumford's case. On 'gentlemanly witnessing', both 'actual' and 'virtual', see Shapin 1988: 373–404; Shapin and Schaffer 1985: 22–79.
54 Joule 1843b: 442.
55 Joule 1887, 1: 171 [1844].
56 Joule 1887, 1: 172.
57 Joule 1845a: 381–2. See also Cardwell 1989: 67–68.
58 Sibum 1994. Reynolds 1892: 78–86 and Cardwell 1989: 62–7 provide detailed discussion of the experiment.
59 Joule 1845a: 381. Repr. Joule 1887, 1: 187.
60 Joule 1845a: 381. For Faraday's distrust of hypotheses see Cantor 1991: 205–13.
61 Joule 1846: 110–11.
62 Joule 1845a: 381–2.
63 John to William Thomson, 2 July 1845, T520, Kelvin Collection, ULC; Morrell and Thackray 1981: 429. It was at the 1845 BAAS that William Thomson met Faraday for the first time. One consequence of their mutual encounter was that it put Faraday on the track of the 'magneto-optical effect'. See Smith and Wise 1989: 256. I thank an anonymous referee for drawing my attention to this event.
64 Joule 1845b: 31. Repr. Joule 1887, 1: 202. Sibum 1994 examines in context Joule's paddle-wheel experiment. See also Cardwell 1989: 75–6.
65 Hodgkinson 1846: 137–56; Joule 1845c: 207. Repr. Joule 1887, 1: 204 [1845].
66 Joule 1887, 1: 273 [1847]. Reynolds 1892: 2–16 reprints the lecture with the claim that it offers the 'first exposition of the law of the universal conservation of energy'. On the issue of 'dynamic equilibrium' see Wise (with Smith) 1989–90: 250. In Joule to Thomson, 26 March 1860, J154, ULG, Joule explicitly stated that he is still a Tory, opposed to the policies of Whigs and Radicals. Quoted in Cardwell 1989: 188. See also Cardwell 1989: 209.
67 Joule 1887, 1: 273. See Cardwell 1989: 69–75 (Playfair collaboration).
68 Cardwell 1989: 70, 88 (St Andrews chair); *Athenaeum* no. 1027 (3 July 1847): 711; *Literary Gazette* no. 1588 (26 June 1847). In a letter to Lyon Playfair (Cardwell 1989: 70), Joule revealed that he had received testimonials from the chemists Graham and Becquerel but not from Faraday 'on the alleged ground that one Johnson cheated him some years ago!'
69 *Athenaeum* no. 1027 (1847): 710–11; *Literary Gazette* no.1588. The official summary of Joule's presentation in fact appeared in Section B (Chemistry). There the respective mechanical equivalents appeared differently as 781.5 (water) and 782.1 (oil). Although small, the change would scarcely have aided Joule's credibility. See *BAAS Report 17* (1847): 55.
70 Mayer 1848: 385–7; Joule 1849: 132–5; Joule to Thomson, 17 March 1851, J77, Kelvin Collection, ULC. See also Cardwell 1989: 122–7.
71 Discussed in Smith 1994.
72 Lenoir 1982: esp.103–11.
73 Liebig 1842: 221; Lenoir 1982: 158–67, esp. 164. See also Holmes 1964: vii–cxvi; Lipman 1967: 167–85; Heimann 1976: 280–4.
74 Wise 1981a: 271–5. The extent of Mayer's debt to *Naturphilosophie* is discussed in

Hiebert 1959: 394; Heimann 1976: 293–4; Wise 1981a: 269–75; and in detail in Caneva 1993: 275–319.

75 On Mayer's early work see Turner 1974: 235–7; Heimann 1976: 279; Caneva 1993: 3–8.

76 Mayer 1893b: 100–1 [1841]. Lindsay 1973: 60–6 provides a translation. With this and other Mayer papers, however, he frequently transforms Mayer's original language into that of later energy physics. See, esp. for contexts, Turner 1971: 137–82; 1974: 235–6; 1982: 129–62; Heimann 1976: 284; Jungnickel and McCormmach 1986, 1: esp. 78–112; Caneva 1993: 19–22.

77 Mayer 1862: 371–7 [1842]. Translated by G.C. Foster. See Caneva 1993: 23–46. See also Holmes 1973: 344 (Liebig's *Annalen*); Turner 1974: 236; Heimann 1976: 286–90.

78 Hutchison 1977: 279–304.

Chapter 5.   Constructing a Perfect Thermo-dynamic Engine

1 J.P. Joule to G.G. Stokes, 10 July 1847, J73, Stokes Collection, ULC. See also Cardwell 1989: 86.

2 Thomson 1882–1911, 1: 119n [1849].

3 Smith 1998: 118–46. Thomson also consulted Faraday regarding the Royal Institution's apparatus in June 1847. See esp. Thomson to Faraday, 11 June 1847, in Faraday 1991–, 3: 630.

4 William Thomson to G.G. Stokes, 7 April 1847, K19, Stokes Collection, ULC; Wilson 1990, 1: 20. See esp. Smith and Wise 1989: 263–75 (discussion by Wise).

5 Wilson 1990, 1: 29–32.

6 William to Dr James Thomson (father), 1 July 1847, T367, Kelvin Collection, ULC.

7 Retrospective accounts of this first meeting between of Joule and Thomson are given in Joule 1887, 2: 215; Thomson 1882–1911, 1: 618 [1882]; Thompson 1910, 1: 264. Further analysis in Wise 1979a: 78–80; Smith and Wise 1989: 302–4. See also Cardwell 1989: 82–9.

8 J.P. Joule to William Thomson, 29 June 1847, J59, Kelvin Collection, ULC.

9 Lewis Gordon to William Thomson, 2 July 1847, G120, Kelvin Collection, ULC; William to James Thomson, 12 July 1847, T429, Kelvin Collection, ULC.

10 James to William Thomson, 24 July 1847, T433, Kelvin Collection, ULC; Larmor 1912: xxx–xxxi. See Smith and Wise 1989: 307–9.

11 J.P. Joule to G.G. Stokes, 10 and 24 July 1847, J73 and J74, Stokes Collection, ULC. See also Cardwell 1989: 87.

12 William Thomson to J.D. Forbes, 5 December 1847, F190, Kelvin Collection, ULC.

13 William Thomson to J.P. Joule, 27 October 1848, J62, Kelvin Collection, ULC.

14 J.P. Joule to William Thomson, 6 November 1848, J63, Kelvin Collection, ULC.

15 Joule 1850: 61–82.

16 Thomson 1882–1911, 1: 102n [1848].

17 Thomson 1848b: 9–10. Repr. Thomson 1882–1911, 1: 91–2. On Neumann, see Jungnickel and McCormmach 1986, 1: 84–9, 148–53.

18 Smith and Wise 1989: 309–10.

19 J.P. Joule to William Thomson, 6 October 1848, J61, Kelvin Collection, ULC.

20 William Thomson to J.P. Joule, 27 October 1848, J62, Kelvin Collection, ULC.

21 J.P. Joule to William Thomson, 6 October 1848, J61, Kelvin Collection, ULC.

22 William Thomson to J.P. Joule, 27 October 1848, J62, Kelvin Collection, ULC. Parts of the press copy are illegible.

23 Joule 1845a: 382–3. Repr. Joule 1887, 1: 189.

24 Smith 1976b: 293–319.

25 William Thomson to J.P. Joule, 27 October 1848, J62, Kelvin Collection, ULC.

26 William Thomson to J.P. Joule, 27 October 1848, J62, Kelvin Collection, ULC. See Smith and Wise 1989: 312–14.

27 J.P. Joule to William Thomson, 6 November 1848, J63, Kelvin Collection, ULC.

28 Fox 1986: 2.

29 Fox 1986: 12–17.

30 Carnot 1986: 62–3 [1824]. Carnot's manuscript notes on political economy are published (with commentary) in Fox 1978: 273–312. For a consideration of the analogy between the transfer of caloric and the transfer of wealth, with minimum waste, see Fox 1986: 161–2.

31 Carnot 1986: 61–62 [1824]. Fox 1986: 116 (n.3) points out that Carnot's figures for the increase in coal production are probably exaggerated. Fox 1986: 117 (n.4) discusses the context of steam navigation on French and North American rivers. Body 1971: 23–64 documents the growth of cross-channel and ocean steam navigation in this period.

32 See Fox 1986: 13–14 (Lazare Carnot), 6–7 (Watt), 18–20 (Clément). See also Fox 1970: 233–53 (expansive principle); Lervig 1985: 147–96 (Clément's lectures).

33 See *Institut de France. Académie des Sciences. Procés – verbaux des séances de l'Académie. Tome VIII* (1824–27), pp 101, 118; Fox 1986: 25–6, 30–2; Redondi 1980.

34 Carnot 1986: 61 [1824]. The broader context for what Donald Cardwell has termed a 'cosmology of heat' is treated in Cardwell 1971: 89–120, 186–92.

35 Carnot 1986: 61 [1824].

36 Thomson 1849: 541–74; 1882–1911, 1: 113–14.

37 Thomson 1882–1911, 1: 114.

38 Carnot 1986: 63 [1824]. See Kuhn 1961b: 567–74 (Carnot and air-engines); Fox 1986: 2–12; 118–19 (n.9) (the French engineering context); 11, 21, 39 (air-engines in France). I thank Ben Marsden for drawing my attention to the significance of air-engines as a major theme in the period 1810–60.

39 Carnot 1986: 64–5 [1824]. For discussion of Carnot's probable sources see Fox 1970: 233–53; 1971: 180–3; 1976: 162–64; 1986: 119–21 (n.13 and n.14); Pacey 1974: 135–45. On 'equilibrium systems' employed in diverse sciences (from equations in chemistry to the lever in mechanics) during the late eighteenth and early nineteenth centuries see Wise (with Smith) 1989–90: 266–94. On 'progressionist' models see Wise (with Smith) 1989–90: 391–449; 221–61.

40 Carnot 1986: 64–5 [1824].

41 Carnot 1986: 76n [1824]. For translation of Carnot's notes (with commentary) see Fox 1986: 26–7, 30–1, 181–212. Only with the benefit of hindsight, however, can we (p 30) claim that Carnot provided a statement of the conservation of energy (p 30). On the contemporary assumption that heat was not consumed in an engine see Pacey 1974: 135–45; Fox 1986: 121 (n.15).

42 Thomson 1882–1911, 1: 114–15 [1849].

43 Thomson 1882–1911, 1: 116–17.

44 Wise 1981b: 19–70. Initially Thomson had little enthusiasm for Faraday's conception of electricity but gradually came to admire the London professor's approach and to assimilate to his own. See esp. Smith and Wise 1989: 203–36.

45 William Smith, Lectures for 2 and 4 November 1849, in 'Notes of the Glasgow College natural philosophy class taken during the 1849–50 session', MS Gen. 142, Kelvin Collection, ULG. Quoted in Smith and Wise 1989: 314.

46 Thomson 1882–1911, 1: 117 [1849].

47 Thomson 1882–1911, 1: 117–18.

48 Carnot 1986: 66, 70 [1824].

49 Carnot 1986: 66–70. Fox 1986: 124 (n.22) notes that here the proof is incomplete because the cycle is not closed; the rigorous version follows on pp 73–7. See esp. Klein 1974: 213–19.

50 Thomson 1882–1911, 1: 118 [1849]. See Marsden 1992a: 117–25 (Thomson and 'perfect' thermo-dynamic engines).

51 Thomson 1882–1911, 1: 118–19 [1849].

52 See esp. Hilton 1991: 5, 298–339 for the shift from 'atonement' to 'incarnation' and the specific themes of ch.2 above. Norman Macleod is a prime example of a Scottish churchman who owed much to the evangelical Chalmers but whose presbyterianism carried a far 'softer' message of Christianity with emphasis on Christ as God incarnate. I thank Ben Marsden for discussion and suggestions on these points.

53 Thomson 1882–1911, 1: 118n.

54 William to Dr James Thomson (father), 13 May 1846, T338, Kelvin Collection, ULC; Young 1845, 1: 57–62; Murchison and Sabine 1840: xxxv; Smith 1976c: 250n.

55 Carnot 1986: 69n [1824].

56 Thomson 1882–1911, 1: 120 [1849].

57 Thomson 1882–1911, 1: 149–50; Lewis Gordon to William Thomson, 27 April 1847, G118, Kelvin Collection, ULC. See esp. Marsden 1992a: 96–109 (reappraisal of air-engine history), 105–8 (Stirling's air-engine), 109–13 (Carnot and Clapeyron on air-engines), 114–25 (Scottish interest in air-engines), 126–33 (Joule's air-engine papers), 141–87 (Napier and Rankine air-engine). See also Bryant 1973: 152–65; Daub 1974: 259–77; Cardwell and Hills 1976: 1–20.

58 William Smith, Lecture for 17 January 1850, in 'Notes of the Glasgow College natural philosophy class taken during the 1849–50 session', MS Gen. 142, Kelvin Collection, ULG.

59 William Thomson to J.D. Forbes, 10 January 1850, Forbes Papers, St Andrews University Library.

60 William Smith, Lecture for 17 January 1850, in 'Notes of the Glasgow College natural philosophy class taken during the 1849–50 session', MS Gen. 142, Kelvin Collection, ULG. See also William Thomson to Elizabeth King [nee Thomson], 31 January 1850, Add 8818/3, ULC.

61 For the context of the Glasgow class-room and laboratory see Smith 1998: 118–46. Thomson also corresponded in detail with Faraday in the period 10–14 January 1850 on these experiments concerning the lowering of the freezing point. See Faraday 1991–, 4 (forthcoming). I thank Frank James for this information.

62 Jungnickel and McCormmach 1986, 1: 163–5.

63 Clausius 1851: 3–4 [1850]. On Clausius' 1850 paper see Jungnickel and McCormmach 1986, 1: 165–7; Yagi 1981: 77–94; 1984: 177–95.

64 Clausius 1851: 4 [1850].

65 Clausius 1851: 102–19.

Chapter 6. '*Everything in the Material World is Progressive*'

1 Hertzman 1845: 163–4.

2 William Thomson, preliminary draft for the 'Dynamical Theory of Heat', PA 128, Kelvin Collection, ULC. Published in Smith 1976c: 280–8, on 281.

3 Tait 1881: xx; Marsden 1992a: 14–18; Graham 1842: 85–6.
4 Tait 1881: xx–xxi; Marsden 1992a: 19–22.
5 Marsden 1992a: 22–35, 43–4.
6 Tait 1881: xxi; Marsden 1992a: 35–40.
7 Rankine 1840: 48–65; Marsden 1992a: 44–54.
8 Marsden 1992a: 41–3; Wilson 1985: 19–26, esp. 22.
9 Tait 1881: xxi–xxii; Marsden 1992a: 55–95.
10 Tait 1881: xxii; Marsden 1992a: 74–5; Smith and Wise 1989: 134; Smith 1998: 118–46.
11 Tait 1881: xxii–xxxvi.
12 Marsden 1992a: 97; Rankine to Thomson, 19 August 1850, R18, Kelvin Collection, ULC; Rankine 1881: 16, 234. Thomson's draft report survives as PA 119, Kelvin Collection, ULC.
13 Rankine to Thomson, 19 August 1850, R18, Kelvin Collection, ULC; Rankine 1881: 234 [1853]. The earliest version of his RSE paper is Rankine 1850: 275–88.
14 Rankine 1853b: 147–90; 1881: 234–84. See also Hutchison 1981: 1–26.
15 Rankine 1881: 250 [1853]; Smith and Wise 1989: 321; Hutchison 1973: 341–64.
16 Rankine 1881: 253 [1853].
17 W.J.M. Rankine to William Thomson, 19 August 1850, R18; Thomson's report, PA 119, Kelvin Collection, ULC. Rankine 1864a: 313 offered his later opinion: 'a mechanical hypothesis does not form an *indispensable* part of thermo-dynamics, more than of any other physical science; but if a hypothetical theory of thermo-dynamics is to be used, it appears to me that its fundamental principles must be such as I have described'.
18 William Thomson, 15 August 1850, NB34, Kelvin Collection, ULC.
19 W.J.M. Rankine to William Thomson, 19 August 1850, R18, Kelvin Collection, ULC. Rankine was very anxious in this letter to demonstrate his priority over Clausius.
20 W.J.M. Rankine to William Thomson, 9 September 1850, R19, Kelvin Collection, ULC.
21 W.J.M. Rankine to William Thomson, 17 March 1851, R23, Kelvin Collection, ULC.
22 Rankine 1881: 300–6 [1853]; Clausius 1872a: 109.
23 Smith and Wise 1989: 327.
24 Rankine 1881: 260–1 [1853].
25 Thomson 1850a: 387; 1882–1911, 1: 170–1.
26 Thomson 1850a: 388–9; 1882–1911, 1: 172–3.
27 J.P. Joule to William Thomson, 6 February 1851, J76, Kelvin Collection, ULC.
28 Thomson 1853a: 264 [1851]; 1882–1911, 1: 178.
29 Thomson 1853a: 264–5 [1851]; 1882–1911, 1: 178–9.
30 Thomson, preliminary draft. Published in Smith 1976c: 286–7.
31 Smith 1976c: 286.
32 James Thomson, Notebook (1848), A14(B), James Thomson Collection, Queen's University Library, Belfast. Extract published in Smith 1976c: 278–9, on 279.
33 Smith 1976c: 285.
34 Thomson 1853a: 261 [1851]; 1882–1911, 1: 175.
35 Thomson 1851a: 48–52.
36 Thomson 1853a: 262 [1851]; 1882–1911, 1: 174–5.
37 Thomson 1853a: 264 [1851]; 1882–1911, 1: 175.
38 Thomson, preliminary draft. Published in Smith 1976c: 280–8, on 281.
39 Smith 1976c: 281.
40 Smith 1976c: 282.

41 See Smith 1976b: 312.

42 See also Thomson 1863a: 3–4; 1882–1911, 3: 297–9; Smith and Wise 1989: 566–8.

43 Smith and Wise 1989: 167–8.

44 Smith and Wise 1989: 192–3. See also Herivel 1975: 197–202.

45 Sedgwick 1831: 307; Whewell 1835: 28, 304; Smith 1985: 49–73; Rudwick 1971: 209–37; Lawrence 1978: 101–28.

46 Smith 1985: 73–83; 1989: 27–52; Smith and Wise 1989: 195–9.

47 Thomson 1845: 67–72; 1882–1911, 1: 39–47, on 39; Smith and Wise 1989: 193–4.

48 Thomson 1882–1911, 1: 39. On Thomson's 'scientific latitudinarianism' elsewhere in his work see Smith and Wise 1989: 177n, 229–30, 234, 237, 279–80, 342, 355, 357, 463, 518n.

49 Thomson 1882–1911, 1: 40.

50 Thomson 1882–1911, 1: Fuller analysis of the mathematical argument is provided in Smith and Wise 1989: 193–5.

51 For the surviving part of the inaugural dissertation see Thompson 1910, 1: 187.

52 On the election campaign see Smith and Wise 1989: 99–116. The democratic presbyterian philosophical heritage is discussed in Davie 1961.

53 For example Anna Bottomley (nee Thomson) to William Thomson, c.October 1843, B201, Kelvin Collection, ULC. On the Famine, cholera, and typhus as they affected the family see Smith and Wise 1989: 135–9.

54 William Thomson to G.G. Stokes, 14 January 1849, K29, Stokes Collection, ULC; W.W. Herringham to William Thomson, March 1847, H97, Kelvin Collection, ULC; Macleod 1876, 1: 153–4.

55 Fleming 1851: 220. I thank Anne Scott for this source.

56 Secord 1989: 168–72.

57 James Thomson to Robert Douglas, 15 August 1846, in Larmor 1912: xxxvii.

58 See Secord 1989: 170; Smith and Wise 1989: esp.20–55.

59 Hertzman 1845: 163–4.

60 Secord 1989: 166; Yeo 1989: 1–27; Morrell and Thackray 1981: esp.244–5.

61 Adam Sedgwick to Charles Lyell, 9 April 1845, in Clark and Hughes 1890, 2: 83–5.

62 Tennyson 1971: 261, 265–6 [1850].

63 Owen 1894, 1: 254; Yeo 1989: 2–12.

64 King 1850: 39–40.

65 McFarlane 1852: xviii–xix, 172–73, 179.

66 Compare the discussion in Smith and Wise 1989: 89–99. On the wider context of the relationship between Christianity and political economy in Britain see Hilton 1991; Waterman 1991.

67 Thompson 1910, 1: 42.

68 Maxwell 1990–, 2: 564–5, on 565.

69 Thomson, preliminary draft. Published in Smith 1976c: 283.

70 Thomson 1853a: 265 [1851]; 1882–1911, 1: 179.

71 Thomson 1853a: 265–6 [1851]; 1882–1911, 1: 179–80; Smith and Wise 1989: 330–1.

72 Thomson 1853a: 475 [1851]; 1882–1911, 1: 222–3.

73 Thomson 1853a: 475 [1851]; Thomson 1882–1911, 1: 223.

74 Thomson 1852a: 109; 1882–1911, 1: 505–6. See Kidwell 1981: 457–76 (on Pouillet).

75 Thomson 1852a: 110–12; 1882–1911, 1: 507–9.

76 Thomson 1852a: 112–13; 1882–1911, 1: 509–10.

77 Dr James to William Thomson, 21 February 1842, T197, Kelvin Collection, ULC.

78 Thomson 1852b: 139; 1882–1911, 1:511; William Thomson, Draft of 'On a Universal

Tendency in Nature to the Dissipation of Mechanical Energy', PA137, Kelvin Collection, ULC.

79 Thomson 1852b: 139–40; 1882–1911, 1: 512.

80 Thomson 1852b: 141–2; 1882–1911, 1: 513–14. See esp. Smith and Wise 1989: 512–21 ('Mind over matter').

81 Thomson 1852b: 142; 1882–1911, 1: 514.

82 *Phil. Mag.* 4 (1852): 8–21, 105–17, 168–76, 256–60, 304–6.

Chapter 7. '*The Epoch of Energy': the New Physics and the New Cosmology*

1 Maxwell 1876–7: 390; 1890, 2: 594–5.

2 Thomson 1882–1911, 1: 182–3n [1852].

3 Koenigsberger 1906: 109–13, 144–6.

4 Maxwell 1876–7: 389; 1890, 2: 594.

5 Maxwell to Lewis Campbell, 21 April 1862, in Campbell and Garnett 1881: 335–6; Maxwell 1990–, 1: 711–12.

6 For example Elkana 1974: 11–12, 114. More recent Helmholtz scholarship is generally exempt from this tendency. See esp. Bevilacqua 1993: 292–333.

7 Koenigsberger 1906: vii–xvii. For the history of the *Physikalisch-technische Reichsanstalt* and Helmholtz's role, see esp. Cahan 1989.

8 On Clausius's critique see Koenigsberger 1906: 117–20; Heimann 1974b: 234–5; Jungnickel and McCormmach 1986, 1: 163; Bevilacqua 1993: 317–18.

9 Koenigsberger 1906: 1–18; Turner 1972. See Jungnickel and McCormmach 1986, 1: 31 (Fechner's translation of Biot).

10 Koenigsberger 1906: 21–4, on 21–2.

11 Koenigsberger 1906: 25–8.

12 Lenoir 1982: 103–11; 197–9. Helmholtz's physiological contexts are further explored in Tuchman 1993: 17–49 (German medical community); Olesko and Holmes 1993: 50–108 (early physiological researches).

13 Translated in Lenoir 1982: 200.

14 Quoted in Koenigsberger 1906: 37.

15 Lenoir 1982: 200–2. A decrease in water extract was balanced by an increase in ethanol extract (p. 201). While doubtless applicable to the frogs' state of mind, the physiological term 'irritability' [*Irritabilität*] was also used by Mayer. See Mayer 1893a: 117 [1845]; Lindsay 1973: 135.

16 Translated and discussed in Lenoir 1982: 203.

17 Translated in Lenoir 1982: 204–12.

18 Koenigsberger 1906: 31; Cahan 1989: 37.

19 Quoted in Koenigsberger 1906: 37. See also Jungnickel and McCormmach 1986, 1: 156.

20 Translated in also Jungnickel and McCormmach 1986, 1: 157. See also Koenigsberger 1906: 37–38.

21 Jungnickel and McCormmach 1986, 1: 158.

22 Helmholtz 1853: 114 [1847] (translated by John Tyndall). I use this contemporary translation below, but cross-check where appropriate with the original, reprinted as Helmholtz 1889 [1847] (Ostwald's *Klassiker der Exacten Wissenschaften*).

23 Heimann 1974b: 205–38 explores the possible Kantian sources for Helmholtz, especially in relation to causality and the doctrine of central forces. Helmholtz himself later admitted (Helmholtz 1881: 53) that the Introduction had been influenced by

Kant's views on causality as the precondition for the lawlikeness of all natural phenomena.

24 Helmholtz 1853: 114 [1847].

25 See Wise 1981a: 271–3, 296–7.

26 Helmholtz 1853: 114–17 [1847]; Helmholtz 1889: 1–5. See Bevilacqua 1993: 304–9.

27 Helmholtz 1853: 117 [1847].

28 Helmholtz 1853: 118–19 [1847]; Bevilacqua 1993: 309.

29 Helmholtz 1853: 119–20 [1847]. Further analysis in Bevilaqua 1993: 309–19.

30 Helmholtz 1853: 121–2 [1847]. Compare Heimann 1974b: 214–16.

31 Helmholtz 1853: 124–5 [1847].

32 Helmholtz 1853: 127 [1847]. See Bevilacqua 1993: 319–21.

33 Helmholtz 1853: 130–31 [1847].

34 Helmholtz 1853: 129–31 [1847]. Helmholtz was referring to Joule 1845c: 205–7. Repr. Joule 1887, 1: 202–5. See Helmholtz 1889: 21n.

35 Cardwell 1989: 93–4 claims that Helmholtz had here 'dismissed' Joule's 1845 paper 'unfairly' and that his 1847 memoir presented a misleading picture of Joule. The attention given to Joule, however, suggests that Helmholtz was conferring upon Joule's research a good deal more credibility than many British natural philosophers of the period. See esp. Bevilaqua 1993: 325–6 (Helmholtz's faulty conversion between British and continental units).

36 Helmholtz 1853: 131 [1847]; Bevilaqua 1993: 325–6.

37 Helmholtz 1853: 131–3 [1847].

38 Helmholtz 1853: 133–4 [1847]. Helmholtz here discussed favourably Ampère's 'mechanical' hypothesis which Joule had rejected (ch.4).

39 Helmholtz 1853: 135 [1847]. See Joule 1845a: 369–83. Repr. Joule 1887, 1: 172–89.

40 Helmholtz 1853: 135–8 [1847]. See Holtzmann 1846: 189–217 [1845].

41 Helmholtz 1853: 138–60 [1847]. See esp. Jungnickel and McCormmach 1986, 1: 159–60; Bevilacqua 1993: 326–32. See also Elkana 1974: 126–7.

42 Helmholtz 1853: 160–61 [1847].

43 William Thomson, Draft of 'On a Universal Tendency in Nature to the Dissipation of Mechanical Energy', PA 137, Kelvin Collection, ULC.

44 Thomson, Draft of 'On a Universal Tendency . . .', PA 137, Kelvin Collection, ULC.. See Smith and Wise 1989: 384.

45 I thank Norton Wise for discussion of these issues.

46 Thomson, Draft of 'On a Universal Tendency..,' PA 137, Kelvin Collection, ULC.

47 Rankine 1853a: 106. In Rankine's *Miscellaneous Scientific Papers* (1881) the wording of his definitions was altered, most probably by Tait in his editorial and advisory capacity. Strikingly the last statement became: 'The law of the *Conservation of Energy* is already known – viz., that the sum of all the energies of the universe, actual and potential, is unchangeable'. The original, relatively modest claim had been that there was a known law according to which the sum of actual and potential energies was conserved. The 1881 claim, on the other hand, strongly affirmed that *all* the energies of the universe were actual and potential and that they were subject to the *Conservation of Energy*. Rankine's original statement thus served as a resource from which the 'scientists of energy' would construct between the early 1850s and early 1880s a canonical 'Law of Conservation of Energy' which governed all things in the universe.

48 Rankine 1881: 203 [1853].

49 Thomson 1853b: 102n; 1882–1911, 1: 554n; Joule to Thomson, 3 February 1853, J135, Kelvin Collection, ULC.

50 Rankine 1872: 160. See also Rankine 1864b: 404; Maxwell 1871: 92.

51 Morrell and Thackray 1981: esp. 429. The Scottish evangelical David Brewster was of course also one of the founding generation of BAAS gentlemen.

52 Airy 1851: xlix.

53 Cawood 1979: 493–518 (on Sabine and the 'Magnetic Crusade'); Smith and Wise 1989: 276–9 (on Sabine and Thomson). Sabine was also a patron of John Tyndall (ch.9) and other non-Oxbridge men.

54 Sabine 1852: xlviii.

55 Sabine 1852: xlviii.

56 Rankine 1852a: 12.

57 Rankine 1852a: 12. Rankine 1852b: 359.

58 Helmholtz 1856: 503 [1854].

59 Rankine 1852b: 359–60. My italics.

60 I am especially grateful to Andrew Reader and Ben Marsden for discussion of these issues. Further discussion in ch.8 below.

61 Thomson 1889–94, 1: 349–68, on 349–50 [1862]; Spencer 1883 [1857]: 30.

62 Smith 1989b: 33–34.

63 Hopkins 1835: 1–11; Smith 1985: 49–83; Smith and Wise 1989: 195–8.

64 Hopkins 1836: 227–36; 272–81; 357–66; Smith 1989b: 36–7; Smith and Wise 1989: 553–9.

65 For example, Hopkins 1839: 381–423; 1847: 33–92; Smith 1989b: 38–9. See also Brush 1979: 225–54.

66 See Smith and Wise 1989: 553, 557–9 on the joint experimental work of Hopkins, Joule and Fairbairn into the effects of pressure on the melting points of various substances.

67 Hopkins 1853: lii–liii.

68 Hopkins 1853: liii–liv.

69 Hopkins 1853: liv.

70 Hopkins 1853: xlv–xlvi; Wise 1982: 182–7 (Cambridge-style 'dynamical theory'); Morrell and Thackray 1981: 256–66.

71 Hopkins 1853: xlvi. See Morrell and Thackray 1981: esp. 228–32 (Dalton at the BAAS).

72 Hopkins 1853: xlvi–xlvii.

73 William Thomson, 'Notes on thermodynamics probably about 1864 in a long letter or statement addressed to Tait', NB52, Kelvin Collection, ULC.

74 Yeo 1989: 12 (quoting Murchison); Smith and Wise 1989: 518–23 (Thomson and Moigno).

75 Thomson 1854a: 61; 1882–1911, 2: 37. See James 1982: 161.

76 Thomson 1854a: 59–63; 1882–1911, 2: 37. 34–40; Smith and Wise 1989: 502–18.

77 Thomson 1882–1911, 2: 23–5; Burchfield 1975: 24–5.

78 Thomson 1854a: 59–63; 1882–1911, 2: 34–40, on p 40.

79 Thomson 1861: 27–8; 1882–1911, 2: 38–40; Smith and Wise 1989: 520–1, 526–33.

Chapter 8.   The Science of Thermodynamics

1 Thomson 1882–1911, 1: 232 [1854]. Comprised the opening to 'Thermo-electric Currents' (Part VI of 'On the Dynamical Theory of Heat').

2 Maxwell 1878: 258; Rankine 1859a: 299–448; 1908: 299–448 [1859]; 1857: 338–54.

3 This chapter is much indebted to the work of, and detailed discussions with, Ben Marsden.

4 Rankine 1871: 59 (extract from a letter written by Rev. Norman Macleod).

5 Rankine 1871: 60–3, on 60 (extracts from letters of Rev. W.G. Fraser); Gordon 1872–5: 296–306, on 296, 305 ('Obituary Notice of Professor Rankine'); Napier 1904: 3–5, 35 (the Napiers' presbyterian context).

6 Rankine 1871: 4–6 (Elder's early career); Napier 1904: 85–8, 150–1, 182–207 (the Napiers and Clyde shipbuilders); Bonsor 1975, 1: 72–84, 201–8, 218–45 (Cunard, Collins and Inman); Slaven 1975: 125–33, 178–82 (Clyde shipbuilding). See also Smith and Wise 1989: 727–32. Another Tod & MacGregor iron screw steamer, the 1950-ton *Glasgow*, was completed for the Atlantic trade from Glasgow in September 1851 following the sale of the *City of Glasgow* to Inman. A drawing of the launch appeared in the *Illustrated London News*. Once again the effect was to enhance the credibility of Glasgow's shipbuilders and engineers. See Bonsor 1975, 1: 250–4.

7 Rankine 1871: 6, 31, 37–8; Bonsor 1975, 1: 276–7 (also quoting from the *North British Daily Mail*).

8 Rankine 1871: 38; Bonsor 1983: 144–65; Lingwood 1977: 97–114.

9 Rankine 1871: 39, 65–7 (letter from William Just).

10 Gordon 1872–5: 304.

11 Rankine 1855c: 201–2; 1858: 1–11; Marsden 1992a: 257–64; 1992b. See also Channell 1982: 39–52.

12 Rankine 1871: 30, 28.

13 Rankine 1871: 25–8.

14 Rankine 1851b: 60–4; 1881: 288–99, on 293. Compare Cardwell and Hills 1976: 1–20, esp. 8–11. See also Bryant 1973: 152–65.

15 Rankine 1855a: 2, 8.

16 Rankine 1855a: 8.

17 Rankine 1855a: 19–20.

18 Rankine 1855a: 20–2.

19 Rankine 1855a: 22–3.

20 Lewis Gordon to William Thomson, 27 April 1847 and 2 and 9 October 1848, G118 and G125–6, Kelvin Collection, ULC; Marsden 1992a: 116–17. The pun also links to Robert *Stirling* Newall.

21 Thomson 1882–1911, 1: 149–50 [1849].

22 James Prescott Joule to William Thomson, 6 November 1850 and 28 April 1851, J72 and J80, Kelvin Collection, ULC; Marsden 1992a: 96.

23 Bonsor 1975, 1: 334–5; Marsden 137–40.

24 Macquorn Rankine to J.R. Napier, 7 February 1853, Napier Papers, Glasgow University Archives; Marsden 1992a: 143–4.

25 Rankine to Napier, 7 February 1853; Marsden 1992a: 145.

26 Rankine to Napier, 7 February 1853; Marsden 1992a: 146–7.

27 Unsorted Documents from the Napier Papers, Glasgow University Archives discussed in Marsden 1992a: 147–51.

28 Marsden 1992a: 150.

29 Rankine to Napier, 1 and 23 June 1853, Napier Papers, Glasgow University Archives; Marsden 1992a: 152–3.

30 *Specification* (Patent A.D. 1853 No. 1416), Napier Papers, Glasgow University Archives; Marsden 1992a: 153–56.

31 *Specification*, p 9. Quoted in Marsden 1992a: 155.

32 John to Macquorn Rankine, 10 August 1853, Acc. 8660, National Library of Scotland; Rankine to Napier, 27 January 1854, Napier Papers, Glasgow University Archives; Marsden 1992a: 157–60.

33 Rankine 1881: 339–40 [1854]; Marsden 1992a: 161–70.

34 Marsden 1992a: 161–4.

35 Rankine 1855a: 8–9; Marsden 1992a: 166–7; Daub 1974: 259–77.

36 Rankine 1881: 371–4 [1854]; Marsden 1992a: 164–7.

37 Rankine to Napier, 27 January and 17 July 1854, Napier Papers, Glasgow University Archives; Marsden 1992a: 171–3.

38 John Ericsson to J.O. Sargent, 28 April 1854, in Church 1890: 195–97. Also quoted in Bonsor 1975, 1; 334.

39 Bonsor 1975: 335.

40 Rankine 1855b.

41 Rankine 1881 [1855]: 213; Olson 1975: 271–86.

42 Macquorn Rankine, *Synopsis of Lectures on Heat-engines Delivered at Glasgow in March and April 1855, in Connection with Professor Lewis D.B. Gordon's Course of Civil Engineering and Mechanics*, Napier Papers, Glasgow University archives; Marsden 1992a: 247–54.

43 Rankine 1857: 338–54, on 353–4.

44 Maxwell 1878: 258.

45 P.G. Tait to James Clerk Maxwell, 6 December 1867, Maxwell Collection, ULC.

46 Maxwell 1878: 278.

47 Thomson 1882–1911, 1.

48 Thomson 1882–1911, 1: 181; Clausius 1851: 102 [1850]; Rankine 1881: 301 [1851]; Clausius 1872a: 109.

49 Clausius 1856a: 81–6 [1854]; 1851: 103 [1850].

50 Clausius 1856a: 86–98 [1854]. See esp. Daub 1977 (sources for Clausius' 'entropy').

51 Clausius 1862a: 85–92; See esp. Klein 1969: 135–41 (modified notation).

52 Clausius 1867: 327 [1865]. English readers would probably have been first introduced to 'entropy' in Clausius 1866: 1–17.

53 Clausius 1867: 327–9 [1865], 329n, 354–7. See also Klein 1969: 140–1.

54 Clausius 1867: 357 [1865].

55 Clausius 1867: 364–65 [1865].

Chapter 9. North British *versus* Metropolis: *Territorial Controversy in the History of Energy*

1 Tait 1873a: 86.

2 Tait 1873a: 86.

3 Maxwell to Rayleigh, 15 March 1871, in Rayleigh 1924: 49–50; Everitt 1975: 175. For an excellent contextual study of the problems of establishing physics laboratories in the period see Schaffer 1998.

4 Maxwell to Tait, 10 March 1873, Maxwell Collection, ULC.

5 Tyndall's scientific naturalism is examined in Turner 1981: 169–80. On Huxley's anti-clerical 'New Reformation', centred on 'Darwinism', see Desmond and Moore 1991: 472, 488–92; Desmond 1994: esp. 4, 252, 270, 293, 366–7. On the X-Club see esp. Macleod 1970: 305–22; Barton 1976; 1990: 53–81; Desmond and Moore 1991: 525–7; Desmond 1994: 327–30; 371–2. See also Jensen 1970: 63–72; 1981: 157–68.

6 Wilson 1987: 74–99 provides a contextual survey of Stokes's religious perspective.

7 J.P. Joule to William Thomson, 13 May 1861, J269, Kelvin Collection, ULC. On the Huxley–Wilberforce debate see Desmond and Moore 1991: 492–9. Scholars of Darwinism recognize that the significance of the debate has been overplayed and that the legend was constructed 20 or more years after the event.

8 William Thomson, Research notebook, NB45, Kelvin Collection, ULC.

9 Thomson 1861: 27–8.

10 Thomson 1889–94, 1: 361. See Burchfield 1974: 301–21 (Darwin and geological time); 1975 (broader 'age of the earth' debates in Victorian Britain).

11 Thomson 1863a: 2–3; 1882–1911, 3: 29. On Phillips see Smith and Wise 1989: 561–5. See also Thomson 1862: 610–11.

12 Tait, *DNB*.

13 Knott 1911: 3–12.

14 Knott 1911: 12–15; Larmor 1912: xl–xlviii (on James Thomson); Wilson 1991: 272–3 (on Tait and Andrews).

15 Maxwell to William Thomson, 22 February 1856, M95, ULC; Maxwell 1990–, 1: 399–401, on 399. See also Campbell and Garnett 1882: 247–57; Knott 1911: 16–17.

16 Morrell and Thackray 1981: 430–34.

17 Knott 1911: 70–1; Gooday 1989: ch.3; 1990: 28, 41.

18 Knott 1911: 205–7.

19 Tait 1860: 4–5, 22. See Marsden 1992a: 188–235; 1998 (territorial problems encountered by Lewis Gordon at the University of Glasgow).

20 Tait 1860: 5–25; Wilson 1985: 19–26 (Forbes' teaching at Edinburgh); 1991: 275–6 ('Tait's teaching in relation to 'energy').

21 Tait 1860: 25. On Grove see Maxwell 1874: 302–4; Cantor 1975b: 273–90; Morus 1989: 1–11, 46–215; 1991b: 589–621.

22 Tait to Thomson, 2 December 1862, T31, Kelvin Collection, ULG. Partly reproduced in Lloyd 1970: 216.

23 Tait 1860: 29, 32. See Wise (with Smith) 1989–90: 221–61 (Thomson and political economy); Marsden 1992a: 288–99; 1992b: 340–6 (Rankine and political economy).

24 Tait 1860: 30–1; Wilson 1991: 275.

25 Eve and Creasey 1945: 1–5; *Nature* 10(1874): 299. A useful but less well known source is the collection of essays on Tyndall by edited by Brock, McMillan and Mollan 1981.

26 Eve and Creasey 1945: 6–17.

27 Eve and Creasey 1945: 18–21.

28 Eve and Creasey 1945: 21–9; Jungnickel and McCormmach 1986, 1: 214–15, 238–40 (on Marburg).

29 Tyndall 1873: xv. See also Eve and Creasey 1945: 33.

30 Eve and Creasey 1945: 34–9.

31 Eve and Creasey 1945: 32, 35–6, 39–41. See Cantor, Gooding, James 1991: 29 (Friday Evening Discourses).

32 Eve and Creasey 1945: 43–58, 69–77. 'Metropolitan science' requires sharper characterization with respect to Tyndall. The work of Iwan Morus has shown the central position of the London 'professoriate' (located particularly at University College, King's College and the Royal Institution) of which Faraday himself was a key member. In contrast to the diverse and extensive groups of scientific (especially electrical) practitioners in the metropolis (popular electrical showmen, public lecturers, instrument makers and so on), the professoriate was numerically small, relatively well-funded, and of secure social position. Tyndall initially inherited this legacy in the 1850s after an early life of insecur-

ity. His alliances with the X-Club in the 1860s strengthen that position because, as Ruth Barton has shown, the X-Club members are highly successful in infiltrating positions of power within the prestigious Royal Society. See esp. Morus 1992; Barton 1990.

33 Eve and Creasey 1945: 82 (Edinburgh chair), 84–5, 174–8 (glacial controversies); Smith 1989b: 43–51 (Forbes–Hopkins controversy); Hevly 1996: 66–86 (glacier motion as 'heroic science').

34 Eve and Creasey 1945: 38, 54–5, 61–2 (Tyndall–Thomson relations); Smith and Wise 1989: 649–712 (Thomson and telegraphy).

35 Tyndall 1862a: 57. See also Tyndall 1863d; Eve and Creasey 1945: 67, 94, 106 (Tyndall's *Heat as a Mode of Motion*).

36 Tyndall 1862a: 58–59. The seeming discrepancy in the figures is presumably due to the impossibility of utilising the full mechanical equivalent in a pound of coal.

37 Tyndall 1862a: 61–65. For Mayer's cosmology, see Mayer 1863: 241–8, 387–409, 417–28 [1848]; James 1982: 157–61. Mayer's stable economy of nature supposed that the effects of tidal retardation would be balanced by the effects of terrestrial cooling (contraction). The rotational velocity of the earth would thus remain constant over time.

38 Tyndall 1862a: 65; Eve and Creasey 1945: 94–5.

39 Carlyle 1908: 384 [1841]. See also Eve and Creasey 1945: 17, 26–7, 36–7, 73–7, 120–7, 262–3 (Tyndall and Carlyle).

40 Joule to Thomson, 9 July 1862, J166, Kelvin Collection, ULG; partly reproduced in Lloyd 1970: 214–15.

41 Joule 1862: 121–3. Lloyd 1970: 211–25 provides a survey, with selections from the relevant correspondence, of the Joule–Mayer and Tait–Tyndall controversy. Marc Séguin (1786–1875), *DSB*, was principally involved in French civil engineering projects, especially railways. Only in the later 1840s, however, did he use his results in *De l'influence* to arrive at a 'mechanical equivalent of heat'.

42 Tyndall 1862b: 173–5.

43 *Witness* 11, 14 January 1862. Quoted in Desmond 1994: 300–1.

44 Morrell and Thackray 1981: 534–5 (Brewster and Forbes).

45 William to James Thomson 18 July 1868, T476, Kelvin Collection, ULC.

46 Knott 1911: 291 (Tait's obituary notice of Robertson Smith).

47 *Glasgow University Calendars for the Sessions 1850–51* to *1869–70*, Glasgow University Archives. Other energy topics were the production of motive power, the economy of batteries, the measurement of temperature, the dynamics of life, the mechanical energy of electricity in motion, the mechanical and physical principles of gunnery, the tides, the winds, and the history of dynamics.

48 Tait 1863a: 264.

49 Norman Macleod to unknown correspondent, 11 June 1863, in Macleod 1876, 2: 150–2, on 150.

50 Thomson and Tait 1862: 606–7.

51 Thomson and Tait 1862: 601.

52 Thomson and Tait 1862: 601–6.

53 Thomson and Tait 1862: 601.

54 Thomson and Tait 1862: 603. Thomson and Tait's 'historical' summary bore little resemblance to the sequence of Joule's researches and the presentation of his early results (ch.4). Tyndall (below) soon identified this weakness.

55 Tait 1863a: 265.

56 Thomson and Tait 1862: 604n.

57 Tyndall 1863a: 220. The 'soft tone' of *Good Words* upon which Tyndall poured scorn

was a reflection of the shift within evangelical Christianity in Britain from doctrine of atonement to incarnationalism, that is, from God as the harsh judge to Christ as perfect humanity (ch.2). For the intellectual context of incarnationalism see Hilton 1991: 298–339.

58 Tyndall 1863a: 221.

59 See esp. Shapin and Schaffer 1985: 22–79. I thank Ben Marsden for discussion.

60 Tait to Thomson, 2 March 1863, T35, Kelvin Collection, ULG. Partly reproduced in Lloyd 1970: 217. Sanitary engineering was a major feature of Victorian city life in this period.

61 Tait 1863b: 263.

62 Tyndall 1863b: 264–5.

63 Tyndall 1863b: 368–9.

64 Tyndall 1863b: 372–73. The Tait-Tyndall controversy brought a number of other claimants for 'energy conservation' out of the historical woodwork. See Colding 1864: 56–64; 1871: 1–20 [1850]. 'Discovery' of the work of the Danish engineer L.A. Colding (dating from 1840–3) gave Tait a fresh opportunity to undermine Tyndall's claims for Mayer's originality. So also did F. Mohr whose work had been published in Liebig's *Annalen* as early as 1837. See Mohr 1876: 110–14 [1837]. On Colding see also Dahl 1963: 174–88. Verdet, physics professor at the Ecole Polytechnique from 1862 until 1866, delivered a series of lectures on heat at the Sorbonne in 1864–5. His posthumous textbook, *La théorie mécanique de la chaleur* (Paris, 1868), was compiled from notes taken by two students during these lectures.

65 Tyndall 1863b: 375–7.

66 Thomson 1863: 429.

67 H.H. Lancaster (editor) to William Thomson, 3 November 1863, L13, Kelvin Collection, ULC; Smith and Wise 1989: xxi.

68 Tait 1868: 47–8; 1864b: 338. The Tait-Tyndall controversy rumbled on for another year. See Tyndall 1864: 25–51; Joule 1864: 150–2; Tait 1864c: 288–92; Tait 1876: 27–51. In the early 1870s Tyndall's German friend Clausius clashed with Tait over priority issues (ch.12).

Chapter 10.   Newton Reinvented: Thomson and Tait's Treatise on Natural Philosophy

1 Quoted in Thompson 1910, 1: 471.

2 See Morrell and Thackray 1981: 267–76 (hierarchy of BAAS sections); 439–44, 460–6 (gentlemen of geology).

3 On Thomson and Maxwell's reforming roles as examiners in Cambridge see for example Campbell and Garnett 1882: 356–8.

4 Thomson and Tait 1867: viii; Smith and Wise 1989: 348–9.

5 Maxwell 1879: 215; 1890, 2: 782. 'Household words' may have been a subtle reference to the popular journal of that name edited by Charles Dickens from 1850 until 1859 for a largely middle, rather than upper, class readership in Victorian Britain.

6 Tait to Thomson, 20 June 1864, T6X, Kelvin Collection, ULC.

7 Maxwell 1879: 213; 1890, 2: 777. Maxwell's 'text-as-space' metaphor served to emphasize the authoritative nature and grand scale of the project. See Smith and Agar 1998: 1–23.

8 Tait to Thomson, 10 December 1861, T2, Kelvin Collection, ULG.

9 Tait to Thomson, 12 December 1861, T6B; 25 December 1861, T6C, Kelvin Collec-

tion, ULC; 13 December 1861, T3, Kelvin Collection, ULG. On McCosh see Hoeveler 1981.

10 Tait to Thomson, 25 December 1861, T6C, Kelvin Collection, ULC.

11 Tait to Thomson, 28 December 1861, T6D, Kelvin Collection, ULC; Tait to Thomson, 6 January 1862, in Knott 1911: 177–8. The proposed popular version was advertised as *Elements of Natural Philosophy* (forthcoming) in the *Glasgow University Calendar* (1863–4). See Thompson 1910, 1: 465–6. 10.

12 Tait to Thomas Andrews, 20 January 1862 and 9 September 1862, in Knott 1911: 178–9; Smith and Wise 1989: 352. It is tempting to think that Tait's 'savages and gorillas' included the scientific naturalists, though he later referred to himself and Thomson as a 'couple of sold gorillas'. ,

13 Tait to Thomson, 28 December 1861, T6D, Kelvin Collection, ULC; to Thomas Andrews, 20 January 1862, in Knott 1911: 178; A. MacMillan to Tait, 19 May 1865, T60, Kelvin Collection, ULC; Tait to Thomson, January 1870, in Thompson 1910, 1: 473.

14 Thomson and Tait 1867: vi. 'Kinematics' was also used by a Cambridge network in the 1840s and especially by Robert Willis's *Principles of Mechanism* (1841) which complemented Whewell's *Mechanics of Engineering*. See Smith and Wise 1989: 199–202.

15 Maxwell 1879: 213; 1890, 2: 777; Smith and Wise 1989: 365.

16 Thomson and Tait 1867: vi.

17 Thomson and Tait 1867: vi.

18 Tait 1863a: 121–6. For condemnation of Tait see for example Cardwell 1971: 282.

19 William Smith, Lectures for 8, 18 March and 10 April 1850, in 'Notes of the Glasgow College natural philosophy class taken during the 1849–50 session', MS Gen, 142, Kelvin Collection, ULG.

20 William Smith, Lecture for 4 March 1850. See Smith and Wise 1989: 383.

21 Maxwell to William Garnett, 9 July 1877, in Campbell and Garnett 1882: 397–9, on 398. See esp. Smith and Wise 1989: 383–4 (analysis by Wise).

22 Tait 1863a: 121.

23 Tait 1863a: 122–3 (Latin original); Thomson and Tait 1867: 185 (translation).

24 Thomson and Tait 1867: 187–8.

25 Thomson and Tait 1867: 187–8.

26 *Scotsman* 8 November 1870. See Smith 1998.

27 Tait to Thomson, 7 October 1862, T26, Kelvin Collection, ULG.

28 Smith and Wise 1989: 359–60 (on Thomson's role).

29 Thomson and Tait 1867: 161. Smith and Wise 1989: 360–72 provides a full contextual account of Thomson and Tait's 'kinematics'.

30 Thomson and Tait 1867: 161–2.

31 See Maxwell 1990–, 2: 395 ('T&T' errata chap II' despatched with James Clerk Maxwell to P.G. Tait, 18 July 1868, Maxwell Collection, ULC). See also Knott 1911: 195.

32 Maxwell 1990–, 2: 395.

33 Thomson and Tait 1867: 164.

34 Thomson and Tait 1867: 164–5. See esp. Smith and Wise 1989: 385–86.

35 Thomson and Tait 1867: 166–8. See esp. Smith and Wise 1989: 356–58.

36 Thomson and Tait 1867: 176–8.

37 William Thomson, 'Expenditure of work and generation of kinetic energy thereby', 19 November [1862], PA146B, Kelvin Collection, ULC.

38 Thomson, 'Expenditure of work . . .', 19 November [1862], PA146B, Kelvin Collection, ULC. See esp. Smith and Wise 1989: 356–57.

39 Thomson and Tait 1867: 178.
40 Thomson and Tait 1867: 184; David Murray 'Lecture notes in *classe physica*, bench II, 1862–63', MS Murray 326, ULG: 177; Smith and Wise 1989: 381–2.
41 Thomson and Tait 1867: 185–6.
42 Thomson and Tait 1867: 187.
43 Thomson and Tait 1867: 188–9.
44 Thomson and Tait 1867: 189. On the history of potentials see Wise 1989: 342–56.
45 Thomson and Tait 1867: 190–91.
46 Thomson and Tait 1867: 194–5.
47 Thomson and Tait 1867: 195; Smith and Wise 1989: 396–444 (vortex theory of matter).
48 Thomson and Tait 1867: 200.
49 Hankins 1970a: 190–3; Smith and Wise 1989: 372–3.
50 Lagrange 1888–9, 1: 255 [1788]. Quoted in Smith and Wise 1989: 373. For the historical origins of 'virtual velocities' see Hiebert 1962: 7–57; Hankins 1970a: 195–6.
51. Lagrange 1888–9, 1: 21–2 [1788]. Quoted in Smith and Wise 1989: 374.
52 Smith and Wise 1989: 374. See also Fraser 1983: 197–241 (on Lagrange's mechanics).
53 Thomson and Tait 1867: 203.
54 Thomson and Tait 1867: 231–304. See esp. Wise 1988: 85–8. See Smith and Wise 1989: 390–5.
55 Wise 1988: 85–8.
56 Wise 1988: 87.
57 Wise 1988: 85–6.
58 Wise 1988: 86–7.
59 Wise 1988: 87–8.
60 Maxwell 1879: 215; 1890, 2: 782.
61 Maxwell 1879: 215–16.
62 Maxwell 1879: 216. See esp. Moyer 1977: 251–68.
63.Thompson 1910, 1: 471–3 (Helmholtz-Wertheim translation). No adequate study of *Treatise* audiences yet exists. Andrew Warwick's forthcoming study of Cambridge mathematics in the nineteenth century, however, will show the extent to which the *Treatise* was both taken up for Tripos purposes and by the developing schools of mathematical physics within the University.

Chapter 11.   Gentleman of Energy: the Natural Philosophy of James Clerk Maxwell

1 William Thomson to G.G. Stokes, 15 October 1879, K238, Stokes Collection, ULC; Wilson 1990, 2: 485.
2 Tait 1873: 478; Knott 1911: 259. Weber's conception of action-at-a-distance differed from that of Laplace. As Wise 1981a: 276 explains, 'Laplace and his associates described force as though it emanated from one particle of matter and acted on another particle at a distance – a description that implicitly tied force to a particle as its source'. Weber, on the other hand, 'insisted that force existed only as a pairwise relation between particles. The pair of particles, therefore, formed his fundamental unit of analysis'.
3 James Clerk Maxwell to R.B. Litchfield, 29 May 1857, in Campbell and Garnett 1882: 268–9, on 269. The 'memory' is probably a reference to the time of William Laud, Archbishop of Canterbury from 1633 under Charles I. In 1638 Laud decreed that the Anglican Prayer Book and form of service were to be implemented in Scotland. The presbyterian Scots resisted episcopacy with the National Covenant (1638) and two

'Bishops' wars' (1639–40). 'Covenanters' were especially strong in south-west Scotland where the persecution drove some families, including William Thomson's ancestors, to flee to Ulster.

4 Campbell and Garnett 1882: 2–23, on 7–8; Everitt 1975: 37–42. My italics.

5 Campbell and Garnett 1882: 24–87.

6 As recorded by F.J.A. Hort to Lewis Campbell, 4 February 1882, in Campbell and Garnett 1882: 417–21, on 420.

7 Ramsay, *DNB*; Needham 1993: 691.

8 Campbell and Garnett 1882: 55–6; 83–4.

9 Campbell and Garnett 1882: 105–15; Everitt 1975: 43–50.

10 Everitt 1975: 50, 88.

11 Campbell and Garnett 1882: 36, 131n; Thompson 1910, 1: 77, 82, 114, 125 (Blackburn).

12 Campbell and Garnett 1882: 131.

13 James Clerk Maxwell to Lewis Campbell, 16 September 1850, in Campbell and Garnett 1882: 144–5; Maxwell 1990–, 1: 205 (extract). See William Thomson to G.G. Stokes, 18–19 July 1850, K42B, Stokes Collection, ULC; Wilson 1990, 1: 108 and also Thompson 1910, 1: 222–3 (early references to Clerk Maxwell).

14 See Thomson 1850b: 23; 1851d 177–86; Eve and Creasey 1945: 290–7.

15 Campbell and Garnett 1882: 612 (poem entitled 'A Vision of a Wrangler, of a University, of Pedantry, and of Philosophy' dated 10 November 1852).

16 Campbell and Garnett 1882: 146–246 (at Cambridge), 247–61 (Aberdeen chair); Everitt 1975: 51–7. Like Thomson, Maxwell was coached by William Hopkins. See Maxwell to Campbell, 5 June 1852 in Maxwell 1990–, 1: 212.

17 Maxwell to Litchfield, 23 August 1853, in Campbell and Garnett 1882: 191–2 (*Vestiges* and *Theological Essays*); Campbell and Garnett 1882: 223–46 ('Apostles'); 232–3 (critique of Newman).

18 John to James Clerk Maxwell, 13 March 1853, in Campbell and Garnett 1882: 185.

19 See Schaffer 1994: 135–72.

20 Maxwell to Campbell, 22 December 1857, in Campbell and Garnett 1882: 297; to Litchfield, 29 May 1857, in Campbell and Garnett 1882: 268–9, on 269.

21 James Clerk Maxwell to Rev. C.B. Tayler, 8 July 1853, in Campbell and Garnett 1882: 188–9. On religious revival in Aberdeen see Campbell and Garnett 1882: 260–1. *The Confession of Faith* contained a chapter entitled 'Of Free Will' which explained that 'Man, by his fall into a state of sin, hath wholly lost all ability of will to any spiritual good . . .' Although a converted sinner was freed from 'natural bondage under sin', 'by reason of his remaining corruption, he doth not perfectly' will and do spiritual good. Thus the 'will of man is made perfect and immutably free to do good alone in the state of glory only' (pp 51–3).

22 Maxwell to Campbell, 22 December 1857, in Campbell and Garnett 1882: 293–4.

23 Maxwell to Litchfield, 5 March 1858, in Campbell and Garnett 1882: 305–6. Hilton 1991: esp.31–2 examines the relationship between moderate evangelical and utilitarian economics in the period up to c.1840. While sharing values of individualism and Free Trade, evangelicals held that the market 'would yield not material but moral and spiritual benefits, would engender remorse, which in turn would foster self-denial'.

24 Maxwell to Litchfield, 5 March 1858, in Campbell and Garnett 1882: 306–7.

25 Reported by Hort to Campbell, 4 February 1882, in Campbell and Garnett 1882: 421.

26 Maxwell to Litchfield, 5 March 1858, in Campbell and Garnett 1882: 306.

27 James Clerk Maxwell to Katharine Dewar, 6 May 1858, in Campbell and Garnett 1882: 309–10.

28 Maxwell to Katharine Dewar, 16 May 1858, in Campbell and Garnett 1882: 312.

29 James Clerk to Katharine Maxwell, 13 April 1860, in Campbell and Garnett 1882: 328–29.

30 Maxwell to Litchfield, 5 March 1858, in Campbell and Garnett 1882: 304.

31 Maxwell to Thomson, 20 February 1854, M87, Kelvin Collection, ULC; Maxwell 1937: 3–4; 1990–, 1: 237–8, on 237.

32 Wise 1981b: 19–70; Smith and Wise 1989: 203–36.

33 Wise 1977; 1981b: 19–70; Smith and Wise 1989: 237–81, 396–494 (chs 8,12 and13 by Wise).

34 See Smith and Wise 1989: 189–90 for Thomson's problems with the *CMJ*.

35 Maxwell to Thomson, 13 November 1854, M89, Kelvin Collection, ULC; Maxwell 1937: 7–11; 1990–, 1: 254–63, on 254. At the end of this letter Maxwell apologized for 'a long screed of electricity but I find no other man to apply to on the subject . . . '

36 Maxwell 1856a: 160–2; 1990–, 1: 367–9, on 367. On Maxwell's use of 'analogy' in relation to Scottish philosophy, see Olson 1975: 287–312; Siegel 1991: 12–28.

37 Maxwell 1990–, 1: 254–5 [1854]. See Caneva 1980: 121–38; Wise 1989: 345; Siegel 1991: 6–8 (on Ampère).

38 Maxwell 1990–, 1: 255–56 [1854]. On Faraday see Gooding 1978: 117–49 (denial of action-at-a-distance); 1980a: 91–120; 1981: 231–75; 1982: 243–59; 1989: 183–223 (magnetic fields).

39 Harman 1990: 255n provides a historical synopsis of these terms. See Cantor, Gooding, James 1991: esp. 61–2 (Faraday's new language and way of thinking). Frank James dates the letter from Thomson to Faraday as 19 June 1852. See Faraday 1991–, 4 (forthcoming).

40 James to John Clerk Maxwell, 5 May 1855 in Campbell and Garnett 1882: 211–12 and Maxwell 1990–, 1: 294; Maxwell to Thomson, 15 May 1855, M90, Kelvin Collection, ULC; Maxwell 1937: 11–16; 1990–, 1: 305–13, on 305–6; Maxwell to Thomson 13 September 1855, M91, Kelvin Collection, ULC; Maxwell 1937: 17–19; 1990–, 1: 319–24, on 320; See esp. Wise 1981a: 276–87 (Weber's velocity- and acceleration-dependent law of electrical forces and his ether theory in German cultural and philosophical context).

41 James Clerk Maxwell to Mark Pattison, 13 April 1868, in Maxwell 1990–, 2: 362–8, on 365–6.

42 Maxwell 1990–, 1: 322–3. Read on 10 December 1855 and 11 February 1856, Maxwell's paper coincided with a period of intense debate over electric telegraph theory and its relation to the practical problems of laying and working long underwater cables (especially the trans-Atlantic). See Smith and Wise 1989: 446–58, 660–1; Hunt 1991a: 1–19 (Faraday, Thomson and telegraph theory).

43 Maxwell 1890, 1: 155–6 [1856]. For discussion of this paper see Everitt 1975: 87–93; Wise 1979b: 1313–15; Harman 1982: 85, 88; 1990: 12–15; Siegel 1991: 30–3.

44 Maxwell 1890, 1: 155 [1856]. See Becher 1980: 1–48 (Whewell and Cambridge mathematics). Whewell classed electricity, magnetism and heat under 'progressive mathematical studies' in contrast to 'permanent mathematical studies' which included long-established branches of science such as mechanics and optics. See Smith and Wise 1989: 61–65.

45 Smith 1985: 49–83 (Cambridge geological style); Smith and Wise 1989: 192–202 (Cambridge 'kinematical' science).

46 Maxwell 1861–2: 162; 1890, 1: 452.

47 Maxwell 1890, 1: 157 [1856]. Wise 1989: 343–5 explains that Laplace's unnamed function V yielded the components of force at any point by simple differentiation with respect to three coordinate axes. Green (1828) and Gauss (1839–41) independently labelled V the potential function. Unlike Green, Gauss followed a Laplacian perspective in treating the potential as a formalism for describing forces. On Gauss' potential theory see Jungnickel and McCormmach 1986, 1: 66–9. On Thomson's 'discovery' of Green in 1845, see Smith and Wise 1989: 215–16; 254–5.

48 Thomson 1856: 150–8; Smith and Wise 1989: 407–8; Siegel 1991: 33–5. For full historical treatment of the 'Faraday effect' see Knudsen 1976: 235–81; James 1989: 137–61.

49 Thomson 1856: 152.

50 James Clerk Maxwell to C.J. Monro, 20 May 1857, in Maxwell 1990–, 1: 505–7, on 507. Maxwell to Thomson, 17 December 1856, M97, Kelvin Collection, ULC; Maxwell 1990–, 1: 482–6, on 486. Maxwell to Jane Cay (cousin), 28 November 1857, in Campbell and Garnett 1882: 292–3; Maxwell 1990, 1: 575. On James Thomson's vortex turbines and jet pumps in relation to William Thomson's vortex theory of matter, see Smith and Wise 1989: 412–17.

51 Maxwell to Thomson, 17 December 1856, M97, Kelvin Collection, ULC; Maxwell 1990–, 1: 482–6, on 485–6; Maxwell to C.J. Monro, 5 June 1857, in Maxwell 1990–, 1: 524–5, on 525.

52 Maxwell 1990–, 1: 517 [c.1857]. See also ch.1 above.

53 Maxwell to Faraday, 9 November 1857, in Maxwell 1990–, 1: 548–52, on 549. See Faraday 1857: 225–39; Cantor 1991: 174–95, 245–56 (Faraday's 'conservation of force').

54 Faraday to Maxwell, 13 November 1857, in Campbell and Garnett 1882: 288–90. See also Faraday 1859: 166–9; Rankine 1859b: 250–3, 347–8. As fellow scientists of energy, Rankine and Maxwell shared very similar perspectives here. Maxwell to Tait, 3 August 1868, Maxwell Collection, ULC; Maxwell 1990– ,2: 417 noted that 'Rankine in a very short statement [Rankine 1859b] in the Phil. Mag. on Conservation has expressed several things very well about energy, force and effect'.

55 Maxwell to Faraday, 9 November 1857, in Maxwell 1990–, 1: 550.

56 Maxwell 1990, 1: 552.

57 Maxwell to Thomson, 24 November 1857, M10, Kelvin Collection, ULG; Maxwell 1990–, 1: 560–3, on 562.

58 Campbell and Garnett 1882: 274–8, 314–28; Everitt: 51–7.

59 Maxwell 1990–, 1: 666 [1860]. On Rankine's energetics, see Rankine 1855b: 120–41; 1881: 209–28.

60 Maxwell to Stokes, 8 October 1859, M411, Stokes Collection, ULC; Maxwell 1990–, 1: 619–22, on 619.

61 Maxwell 1861–2: 162; 1890, 1: 452. See esp. Siegel 1991: 35–41.

62 Tait 1864b: 359.

63 Maxwell 1861–2: 347; 1890, 1: 488. Helmholtz shared the concern about Weber's theory and its seeming violation of energy conservation. See Helmholtz 1872: 530–7 and note 100 (below).

64 Maxwell 1861–2: 161–75, 345; 1890, 1: 451–66, 485. See esp. Siegel 1991: 57–65.

65 Maxwell 1861–2: 281–91, 338–48, on 283; 1890, 1: 467–88, on 468. See esp. Siegel 1991: 65–73 which rather underplays the engineering context. See Marsden 1993: 116–17 (review of Siegel).

66 Maxwell 1861–2: 285–6; 1890, 1: 471.

67 Maxwell 1861–2: 346.

68 Maxwell 1861–2: 346; Maxwell to Thomson, 15 October 1864, M17, Kelvin Collection,

ULG; Maxwell to Tait, 23 December 1867, Maxwell Collection, ULC; Maxwell 1990–, 2: 176–81, 335–9; Knott 1911: 215–16. See Siegel 1991: 73–7 (difficulties with the extended model), 39–44 ('reality' of the molecular vortices).

69 Maxwell 1861–2: 13–14; 1890, 1: 489–91; Siegel 1991: 77–83.

70 Maxwell 1861–2: 14–15; 1890, 1: 491–2. See Bromberg 1967–8: 218–34; Siegel 1986: 99–146 (Maxwell's 'displacement current').

71 Maxwell 1861–2: 15–20; 1890, 1: 492–8.

72 Maxwell 1861–2: 15, 20–22; 1890, 1: 492, 498–9; Siegel 1991: esp. 81–3.

73 Maxwell 1861–2: 22–4; 1890, 1: 499–502. See Maxwell to Faraday, 19 October 1861, in Maxwell 1990–, 1: 683–9; Maxwell to Monro, c.20 October 1860, in Maxwell 1990–, 1: 690–1; Maxwell to Thomson, 10 and 17 December 1861, M99 and M100, Kelvin Collection, ULC; Maxwell 1990–, 1: 692–8, 699–702. See Schaffer 1994: 135–72 esp.144–8. Siegel 1991: 83–143 gives a detailed analysis.

74 C.J. Monro to Maxwell, 23 October 1861, in Campbell and Garnett 1882: 329; Maxwell 1990–,1: 690n.

75 Challis 1861: 250–4. See also Challis 1862: 313–22. Challis probably had comparatively little credibility among elite mathematicians and mathematical physicists such as Stokes and Thomson. See esp. Wilson 1987: 132–45 for the disputes between Challis and Stokes in the 1840s.

76 Maxwell to H.R. Droop, 28 December 1861, in Campbell and Garnett 1882: 329–30, on 330.

77 Maxwell to Charles Hockin, 7 September 1864, in Campbell and Garnett 1882: 340. See also Hopley 1957: 265–72; 1958: 197–210; 1959: 91–108 (BAAS measurements).

78 Maxwell to Tait, 23 December 1867, Maxwell Collection, ULC; Knott 1911: 215–16.

79 Maxwell to Stokes, 15 October 1864, M423, Stokes Collection, ULC; Stokes 1907, 2: 25–26. See esp. Wise 1982: 175–205 (British 'dynamical theory'); 1987: 363–8 (Cambridge mathematical physics).

80 Stokes 1907, 2: 26. See Schaffer 1994: 148–50. On Jenkin's introduction to Thomson, see Smith and Wise 1989: 699–700.

81 Maxwell 1865a: 152.

82 Maxwell 1890, 1: 527 [1865].

83 Maxwell 1890, 1: 526–27.

84 Maxwell 1865a: 152–3.

85 Maxwell 1890, 1: 528 [1865].

86 Maxwell 1890, 1: 528 [1865]. Thomson's paper had the dramatic title 'Note on the Possible Density of the Luminiferous Medium and on the Mechanical Value of a Cubic Mile of Sunlight'. See Thomson 1855: 36–40; 1882–1911, 2: 28–33.

87 Maxwell 1890, 1: 528 [1865]. See Ruse 1976: 121–31; Smith 1985: 69–73 (*verae causae*).

88 Maxwell 1890, 1: 533. See Turner 1956: 36–47; Simpson 1970: 249–63; Chalmers 1973–4: 107–64, esp. 149–54; Olson 1975: 287–321; Wise 1982: 194–5 (Lagrangian methods in relation to Maxwell's 'dynamical theory' and its philosophical contexts).

89 Maxwell 1890, 1: 533–4; Maxwell 1865a: 155. See Hunt 1991b: 245–7 (from Maxwell's equations to 'Maxwell's Equations').

90 Maxwell 1890, 1: 564 [1865].

91 Rayleigh 1869: 1; Tait to Maxwell, 27 November 1867, Maxwell Collection, ULC. The Cambridge Mathematical Tripos examinations took place in the Senate House of the University.

92 Tait to Maxwell, 6 December 1867; Maxwell to Tait, 11 December 1867, Maxwell Collection, ULC; Maxwell 1990–, 2: 328–34.

93 Maxwell to Tait, 12 March 1868, Maxwell Collection, ULC; Maxwell 1990–, 2: 353–5.

94 Maxwell 1990–, 2: 353–5. The reference is to Jonathan Swift's *Gulliver's Travels* (Book III: 'A Voyage to Laputa', chapter 10) where the naive Gulliver visits the Kingdom of Luggnagg and imagines himself born one of the immortal Struldbruggs: 'I should then see … the *perpetual Motion*, the *universal Medicine*, and many other great Inventions brought to the utmost Perfection'. Maxwell also commented on Riemann's view in Maxwell 1873, 2: 435. On Riemann's theory see Wise 1981a: 287–91; Jungnickel and McCormmach 1986, 1: 174–81.

95 Maxwell to Tait, 12 March 1868, Maxwell Collection, ULC.

96 Maxwell to Tait 3 August 1868; 10 December 1869, Maxwell Collection, ULC; Maxwell 1873, 1: xv–xxix. The Maxwell–Thomson correspondence concerning the *Treatise* is included in Maxwell 1990–, 2: 422–515.

97 Maxwell to Tait, 19 October 1871 and 29 June 1872, Maxwell Collection, ULC; Maxwell 1990–, 2: 680, 732–3.

98 Maxwell 1873, 1: x–xiii. See Hunt 1991b: 13–14 (on the disorderly appearance of the *Treatise*).

99 Thomson 1889–94, 2: 160–1; Schaffer 1994: 154. For Thomson's critique of Maxwellian theory see Wise and Smith 1987: 323–48.

100 Weber 1872: 1 [1871]. See Helmholtz 1872: 530–7. See also Clausius 1876: 218–21. See esp. Wise 1981a: 276–87 (Weber), 287–91 (Riemann), 294–5 (Carl Neumann), 295–301 (Helmholtz); Buchwald 1985: 177–86 (Helmholtz).

101 Maxwell to Tait, 7 November 1871, Maxwell Collection, ULC. See Maxwell 1873, 2: 426–38 (critique of Continental action-at-a-distance theories).

102 Maxwell 1873, 1: 88–89, 296–98, 315–17; 1873, 2: 14–15, 42–43.

103 Maxwell 1873, 1: xi. See esp. Wise 1989: 343–47 (potential theory); Hunt 1991b: 14, 115 (Maxwell's 'potentials' which were to prove a contentious issue for his successors).

104 Maxwell 1873, 2: 180–1.

105 Maxwell 1873, 2: 184–5.

106 Maxwell 1873, 2: 185, 194; Maxwell to Tait, 29 June, 1872, Maxwell Collection, ULC; Maxwell 1990–, 2: 732–3.

107 Wise 1982: 194–5 (Maxwell's physical analogy); Hunt 1991b: 245–7 (Maxwell's thirteen general equations and their relation to the four 'Maxwell equations').

108 Maxwell 1873, 2: 437–8.

Chapter 12. Demons versus Dissipation

1 James Clerk Maxwell to Mark Pattison, 7 April 1868, in Maxwell, 1990–, 2: 358–61, on 360–1. As Rector of Lincoln College, Oxford, Pattison played a leading role in the reform of the Oxford syllabus by the removal of Butler's *Analogy* from its central position. See Hilton 1991: 337–9.

2 Maxwell 1990–, 2: 361.

3 Maxwell 1890, 2: 646. The passage highlighted two interlinked limitations of human faculties: an inability to *know* molecular motions and an inability to devise tools to *control* them. See Smith and Wise 1989: 623.

4 See Campbell and Garnett 1882: 409.

5 Clausius 1857: 108. See Jungnickel and McCormmach 1986, 1: 186–93 (Clausius's appointment at Zurich); Brush 1958a: 185–96; 1974: 35–6; Garber 1970: 299–319;

Harman 1982a: 128–33 (Clausius and the 'kinetic theory of gases'); Daub 1967: 293–303; Clark 1976: 41–105 ('atomism' and 'thermodynamics').

6 Clausius 1857: 108–9, 127. For other 'kinetic theorists' see Brush 1957a: 188–98 (John Herapath); 1957b (J.J. Waterston); Bernstein 1963: 206–16 (Maxwell and the history of 'kinetic theories'); Daub 1970c: 105–6; 1971: 512–15 (Waterston, Krönig and others); Talbot and Pacey 1966: 133–49 (Herapath and Joule); Truesdell 1975–76: 1–66.

7 Clausius 1857: 119–23. See esp. Porter 1981: 77–116 (statistical methods and gas theory); 1986: 18–88 (statistical methods in social context).

8 Clausius 1857: 121–7. See Jungnickel and McCormmach 1986, 1: 194–5.

9 Clausius 1859: 81 [1858]. Clausius's 1857 paper prompted Joule to republish his own work on heat as motion (read to the Manchester Lit. & Phil. in October 1848 and originally printed in the *Memoirs*) in the *Phil. Mag.* See Joule 1857 [1851]: 211–16.

10 Clausius 1859: 82–4 [1858].

11 Clausius 1859: 84–91 [1858], on 90. See Jungnickel and McCormmach 1986, 1: 195–6; Garber 1970: 302–3; Brush 1958a: 193–6; 1974: 36–7.

12 Maxwell 1860: 19–20. See Garber 1970: 303.

13 Maxwell to Stokes, 30 May 1859, M410, Stokes Collection, ULC; Stokes 1907, 2: 8–11; Maxwell 1990–, 1: 606–11.

14 Maxwell 1990–, 1: 610.

15 Maxwell to Stokes, 8 October 1859, M411, Stokes Collection, ULC; Stokes 1907, 2: 11–14; Maxwell 1990–, 1: 619–22, on 619.

16 Maxwell 1860: 20. This interpretation corresponds to that of Wise 1982: 196–8. See also Turner 1955: 226–38 ('physical analogy'); Dorling 1970: 229–48 ('non-speculative' foundations for kinetic theory); Garber 1970: 299–319; Brush 1958b: 243–55; 1974: 36–9; Everitt 1975: 131–63 (Maxwell and the 'kinetic theory of gases').

17 Smith and Wise 1989: 256–63 ('mechanical representation'); 402–9, esp. 408 ('dynamical illustration'). See also Knudsen 1976: 235–81.

18 Smith and Wise 1989: 200–2 (Stokes), 349, 354, 379–80 (Thomson and Tait); 396–444 (Thomson's continuum theories of matter), 533–4 (Thomson's continuum convictions in relation to his cosmology and theology).

19 Maxwell 1860: 20.

20 Maxwell 1860: 22.

21 Maxwell 1990–, 1: 610; 1860: 22–8; Clausius 1860: 434–6.

22 Maxwell 1868: 129–45 [1867], 185–217; Clausius 1862b: 417–35, 512–34. See Smith and Wise 1989: 417–25 (Thomson's vortex atom).

23 Maxwell 1868: 129–30 [1867].

24 Maxwell 1868: 131–3 [1867].

25 Maxwell 1868: 136 [1867].

26 Maxwell's position here reflected his strong emphasis on the limitations of human faculties in the physical world and the moral frailties of human beings in the spiritual life. 'Perfect' knowledge, power and love were embodied in the Incarnation of God as Christ. See also note 3 above. More specifically, however, Maxwell drew out consequences from his 'Dynamical Theory of Gases' in 1866 which appeared to violate Thomson's 'thermo-dynamic axiom' by implying that 'by means of material agency mechanical effect is derived from the gas under gravity by cooling it below the temperature of the coldest of the surrounding objects'. For Maxwell 'there remains as far as I can see a collision between Dynamics & thermodynamics'. This tension between 'dynamics' and 'thermodynamics' (centred on the character of the Second Law) almost certainly provided a context for Maxwell's 'demon' (1867) discussed below (next sec-

tion). See esp. Maxwell to Thomson, 27 February 1866, M19, Kelvin Collection, ULC and Maxwell 1990–, 2: 267–71; Thomson to Stokes, 13 October 1866, RR.6.179, Royal Society of London and Wilson 1990, 1: 327–30 for the 'paradox'.

27 Stokes 1990, 1: 197–98. Wise 1982: 185–90 discusses how Maxwell 1890, 2: 713–41 [1879] examined Boltzmann's theorem on the average distribution of energy in a system of material points. Emphasizing that the problem concerned 'the motion of a system whose motion is not completely known', Maxwell *represented* the system in terms of material particles subject to the general laws of dynamics and 'the equation of energy' (total energy equal to the sum of the potential and kinetic energy). The development of a generalized dynamical analogy, drawing upon 'energy methods' from Thomson and Tait and paralleling his own Lagrangian treatment of electromagnetism, yielded results independent of unobservable coordinates (chs 10 and 11).

28 Stokes to Thomson 23 May 1854, S372, Kelvin Collection, ULC; Wilson 1990, 1: 162.

29 Maxwell to Thomson, 15 May 1855, M90, Kelvin Collection, ULC; Maxwell 1990–, 1: 305–13, on 307.

30 Thomson to Stokes, 23 May 1857, K97, Stokes Collection, ULC; Wilson 1990, 1: 223–4.

31 Maxwell to Thomson, 24 November 1857, Kelvin Collection, ULG; Maxwell 1990–, 1: 560–3, on 560–61. Maxwell had published a short paper on the phenomenon in the *Cambridge and Dublin Mathematical Journal* three years earlier. He there explained that the direction of the rotation 'is determined, not by the irregularites of the paper [as commonly assumed], but by the initial circumstances of projection'. See Maxwell 1854: 145–8, on 145.

32 Smith and Wise 1989: 396–425 (full contextual analysis of Thomson's vortex atom theory).

33 Helmholtz 1861: 357. See esp. Smith and Wise 1989: 612–21.

34 Maxwell to Lewis Campbell, 21 April 1862, in Campbell and Garnett 1882: 314–15. See esp. Porter 1981: 77–116; 1986: 193–208.

35 Tait to Maxwell, 6 December 1867; Maxwell to Tait, 11 December 1867, Maxwell Collection, ULC; Maxwell 1990–, 2: 328–33. Partly printed in Knott 1911, 213–14.

36 Maxwell 1990–, 2: 331–2. On Maxwell's 'demon' see esp. Klein 1970: 84–97; Wise 1982: 199; Smith and Wise 1989: 621–5. See also Heimann 1970: 189–211; Daub 1970a: 213–27; Brush 1974: 40–1.

37 Maxwell to J.W. Strutt, 6 December 1870, in Maxwell 1990–, 2: 582–3. Also in Strutt 1924: 47.

38 Knott 1911: 214–15. The text is undated. Harman 1982: 140 claims that Maxwell 'objected to Thomson's use of the term 'demon' . . . there were no supernatural resonances to the argument'. See also Harman's footnote in Maxwell 1990–, 2: 332. A better interpretation of Maxwell and Thomson's deployment of these fictitious entities is to consider a hierarchy of skills (from self-acting 'valve' to intelligent beings).

39 Thomson 1882–1911, 5: 11–20 [1874], on 12n. In a private draft (1879) Thomson imagined Maxwell's 'demon' as a 'super-humanly agile and clever <laboratory> assistant'. See William Thomson, Research notebook, spring 1879, NB 70, Kelvin Collection, ULC.

40 Thomson 1882–1911, 5: 11–15; Smith and Wise 1989: 626–7.

41 Smith and Wise 1989: 625–33 explores these very subtle issues of free will and irreversibility, instability and continuum theory (discussion primarily by Wise).

42 Tait to Thomson, 24 April 1863, in Knott 1911: 183. On Stewart see *DSB*; Gooday 1989: ch.7.

43 Tyndall 1874: lxxxviii–xciv.

44 [Stewart and Tait] 1875: 28–9. On *The Unseen Universe* see Heimann 1972: 73–9; Smith 1979: 69–70n.

45 [Stewart and Tait] 1875: 26–7, 64, 93.

46 [Stewart and Tait] 1875: 157–59, 172.

47 [Stewart and Tait] 1875: 117–19, 140–53. The mathematician William Kingdon Clifford (1845–79), professor of applied mathematics at University College London, one-time second wrangler and High Churchman but subsequent admirer of Darwin and Spencer, provided a sharp critique of these views in his review of *The Unseen Universe*. See Clifford 1875b: 776–93.

48 [Stewart and Tait] 1875: 165.

49 Thompson 1910, 1: 480; Maxwell 1890: 756–62. Maxwell (p.762) concluded his review in *Nature* (1878) of Stewart and Tait's *Paradoxical Philosophy. A Sequel to* The Unseen Universe with the remarks that: 'The progress of science . . . has added nothing of importance to what has always been known about the physical consequences of death, but has rather tended to deepen the distinction between the visible part, which perishes before our eyes, and that which we are ourselves, and to show that this personality, with respect to its nature as well as its destiny, lies quite beyond the range of science'.

50 Tait 1877: v–vi; Maxwell 1878: 257–8.

51 Gibbs 1906, 2: 262–7 [1889]. Quoted in Klein 1969: 131.

52 Rankine 1859a: iii, 299–448; Maxwell 1878: 258. Hirn 1867 consisted of a two-part memoir rather than a textbook.

53 Tait 1868: iii, viii.

54 Tait 1868: iii (Tait was referring specifically to Tyndall 1863d and Stewart 1866); Tait to Maxwell, c.20 July 1868, Maxwell Collection, ULC; Maxwell 1990–, 2: 407–8.

55 Maxwell 1878: 257.

56 Tait to Maxwell, 6 December 1867, Maxwell Collection, ULC; Maxwell 1990–, 2: 328.

57 Clausius 1866: 1–17; 1867: 327–65 [1865]; 1868: 405–19 [1867].

58 Tait 1868: 100.

59 Maxwell to Tait, 13 October 1876, Maxwell Collection, ULC.

60 Maxwell 1871: 307–9 (demon); Klein 1970: 89–90.

61 Maxwell 1871: 87–94, 281 (kinetic and potential energy), 287–89 (kinetic theory of gases).

62 Clausius 1872a: 106. See also Daub 1970b: 321–54 (Tait-Clausius controversy).

63 Maxwell to Tait, 12 February 1872, Maxwell Collection, ULC.

64 Tait 1872a: 338.

65 Clausius 1872b: 443.

66 Tait 1872b: 516–18. See also Clausius 1872c: 117.

67 Maxwell to Tait, 14 February 1871, Maxwell Collection, ULC; Maxwell 1990–, 2: 609–10. See Clausius 1870: 122–27 where he assigned the name *virial*, from the Latin *vis* (force), to the mean value of a magnitude proportional to the forces acting in the system. He then showed that the mean *vis viva* of the system was equal to its virial. He also enunciated two theorems: (1) 'The sum of the *vis viva* and the ergal is constant and (2) 'The mean *vis viva* is equal to the virial'. See also Bierhalter 1982: 199–211.

68 Maxwell 1878: 278. My italics.

69 Maxwell to Tait, 1 December 1873 and 12 October 1874, Maxwell Collection, ULC; Maxwell 1990–, 2: 944–8, on 945–6. On Gibbs, see Klein 1972: 386–93.

70 Klein 1972: 386–8.

71 Klein 1972: 388–9.

72 Gibbs 1906, 1: 354 [1878]; Klein 1972: 389.

73 Klein 1972: 389–90.

74 Klein 1972: 390. Klein summarizes the formidable range of application of Gibbs's memoir as seen by later admirers: 'In this single memoir of some 300 pages he vastly extended the domain covered by thermodynamics, including chemical, elastic, surface, electromagnetic, and electrochemical phenomena in a single system . . . Gibbs's memoir showed how the general theory of thermodynamic equilibrium could be applied to phenomena as varied as the dissolving of a crystal in a liquid, the temperature dependence of the electromotive force of an electrochemical cell, and the heat absorbed when the surface of discontinuity between two fluids is increased' (Klein 1972: 389–90).

75 Maxwell 1908: 819–20 [1876]. At Larmor's behest, the address was reprinted in the *Phil. Mag.* (1908).

76 Maxwell 1908: 820–21. In his introduction to the reprint (1908), Larmor pointed out that the 'general factorization' of energy into magnitudes and intensities linked strongly to earlier papers of Maxwell and Rankine. See esp. Rankine 1859b: 250–3, 347–8; Maxwell 1890, 2: 259–60 [1871].

77 Maxwell to Tait, c. December 1873; 12 February 1872; 24 July 1873, Maxwell Collection, ULC; Maxwell 1990–, 2: 946, 710. Clausius had become professor of physics at the University of Bonn in 1869. He remained in Bonn until his death in 1888. See Jungnickel and McCormmach 1986, 1: 244–5 (Bonn appointment).

78 Maxwell to Tait, c.August 1873, Maxwell Collection, ULC; Maxwell 1990–, 2: 915–16.

79 Maxwell to Tait, 12 October 1874; 29 July 1876, Maxwell Collection, ULC. Spencer promoted the doctrine that all things tended from homogeneity to heterogeneity (uniformity to variety). Tait's public clash with Spencer is recorded in Knott 1911: 278–88. For Maxwell's handling of Spencer see esp. Maxwell to Spencer, 5 & 17 December 1873, in Maxwell 1990–, 2: 956–63.

80 Maxwell to Tait, 1 December 1873, Maxwell Collection, ULC; Maxwell 1990–, 2: 944–8. Aristophanes' *Nephelokokkygia* concerns an imaginary situation or land, especially as the product of impractical or wishful thinking. In the famous myth, Icarus' wings, made partly of wax, melted when he flew too near the sun. I thank Amalia Hatjievgeniadu and Ben Marsden for these comments.

81 Maxwell to Tait, 13 October 1876, Maxwell Collection, ULC.

82 Maxwell to Tait 15 and 28 February 1878, Maxwell Collection, ULC.

83 Maxwell to Tait, 13 October 1876, Maxwell Collection, ULC; Maxwell 1878: 278 (entropy). See also Hutchison 1981: 71–134 (Rankine and 'entropy').

84 Maxwell to Tait, 13 October 1876, Maxwell Collection, ULC.

85 Maxwell 1878: 279.

86 See esp. Maxwell to Thomson, 27 February 1866, M19, Kelvin Collection, ULC and Maxwell 1990–, 2: 267–71; Thomson to Stokes, 13 October 1866, RR.6.179, Royal Society of London and Stokes 1990, 1: 327–30 for the 'paradox' whereby Thomson's 'thermo-dynamic' axiom appeared to be violated by deductions from Maxwell's 'Dynamical Theory of Gases' (note 26 above).

87 Maxwell 1878: 279.

88 Maxwell 1878: 279–80.

89 See, for example, Szily 1872; 1876a; 1876b. On German attempts to reduce the second law to mechanical principles, see esp. Klein 1970: 87–8 (Boltzmann and Clausius); 1972: 58–82 (Boltzmann, Helmholtz and others); 1972–73: 58–82; Bierhalter 1981a: 195–205 (Boltzmann); 1981b: 207–20 (Clausius); 1981c: 71–84 (Helmholtz); 1983–84: 95–100 (Clausius and Helmholtz); 1985: 17–41; 1987: 77–99. On statistical mechanics and

irreversibility, see esp. Brush 1967–68: 145–83; 1974: 1–88; 1976: 603–30. See also Daub 1969: 318–30; Garber 1972–73: 11–39 ('probability' in physics).

90 Tait 1892: 245–6, 439, 532; Knott 1911: 273–6. Kelvin's cryptic comment translates as 'Dear Tait, I agree with what you have written, yours sincerely, Kelvin'.

Chapter 13. Energy and Electricity: 'the Apparatus of the Market Place'

1 Maxwell 1990–, 1: 520; Maxwell to C.J. Munro, 5 June 1857, in Maxwell 1990–, 1: 524–5.

2 William Thomson to James Clerk Maxwell, 24 August 1872, Maxwell Collection, ULC; Maxwell 1990–, 2: 749n; Maxwell, Draft of 'Dimensions of Physical Quantities', Maxwell Collection, ULC. Both quotes in Schaffer 1994: 135–72, on 136.

3 See, for example, Hunt 1994: 48–63.

4 Thomson 1889–94, 2: 161 [1871]; Hunt 1991a: 1–19; 1991b: 53–6; Smith and Wise 1989: 445–63; 649–83.

5 Maxwell 1873, 1: viii; 1954, 1: vii–viii [1891].

6 Maxwell 1873, 1: vi; 1954, 1: vi [1891].

7 Gooday 1989; 1990: 25–51. See also Sviedrys 1976: 405–36.

8 Lefebvre 1991: 416–18.

9 Jungnickel and McCormmach 1986, 1: 63–4.

10 Quoted in Jungnickel and McCormmach 1986, 1: 64–5.

11 Jungnickel and McCormmach 1986, 1: 65–75.

12 Morrell and Thackray 1981: 353–70, 523–31; Cawood 1979: 493–518; Smith 1989b: 29–34.

13 Morrell and Thackray 1981: 512–31; Wise and Smith 1986: 147–73; Smith 1989b: 27–34; Gooday 1989: ch.1; 1990: 25–51; Wise 1994.

14 Maxwell 1873, 1: viii; 1954, 1: vii [1891].

15 Maxwell 1873, 1: 11, 131–3 (Weber), 65–66, 70–1 (Gauss).

16 Weber 1861: 226 [1851].

17 Weber 1861: 226–27 [1851].

18 Weber 1861: 227–30 [1851].

19 Weber 1861: 230–40, 261–9 [1851]. See Wise 1981b: 276–83 (Weber's law).

20 William to Dr James Thomson (father), 1 October 1842, T220; John to William Thomson, 18 February 1843, T499, Kelvin Collection, ULC; Smith and Wise 1989: 203–12. See also Jungnickel and McCormmach 1986, 1: 66–9 (Gauss's potential theory).

21 Wise and Smith 1986: 152–5; Smith and Wise 1989: 240–3.

22 Thompson 1910, 2: 1154.

23 Wise and Smith 1986: 155–6; Smith and Wise 1989: 240, 243–5.

24 Wise and Smith 1986: 156–8.

25 William Thomson, 8 April 1845, NB33, Kelvin Collection, ULC, 177–9; Wise and Smith 1986: 158–9; Smith and Wise 1989: 248–50.

26 See Cantor 1991: 185–95 (Faraday's opposition to 'mechanical philosophy').

27 Smith and Wise 1989: 250–2.

28 Thomson 1860: 326–38; 1872: 254.

29 Wise and Smith 1986: 168–9.

30 Wise and Smith 1986: 159; Smith and Wise 1989: 248–50.

31 Thomson 1851c: 551–62; 1882–1911, 1: 490–502. See also Thomson 1851b: 429–44 ('mechanical theory of electrolysis') and Thomson 1854b: 192–7 ('mechanical values' of distributions of electricity and magnetism).

32 Thomson 1851c: 562.

33 Smith 1998: 118–46; Smith and Wise 1989: 649–50; Hunt 1994 (esp. on location of faults).

34 Wise and Smith 1986: 164–7; Hunt 1994.

35 Maxwell to Lewis Campbell, 4 September 1857, in Campbell and Garnett 1882: 278–80; Maxwell 1990–, 1: 539.

36 On the Whitehouse affair see esp. Hunt 1996: 155–69; Smith and Wise 1989: 667–78; Hunt 1994.

37 Siemens 1861: 25–38, on 25.

38 Siemens 1861: 25.

39 Matthiessen 1861: 195–6 (quoting Siemens).

40 Siemens 1861: 25–38.

41 I thank Ben Marsden for highlighting these issues.

42 By identifying two undated letters from Jenkin to Thomson (discussed below), Hunt 1994 solves the mystery of how both Thomson *and* Clark and Bright came to be credited with instigating the BAAS Committee. See also Thompson 1910, 1: 417–18; Smith and Wise 1989: 687. Tait applies Jenkin's nickname in Tait to Maxwell, 9 November 1874, Maxwell Collection, ULC.

43 Stevenson 1887, 1: xi-clix; Smith and Wise 1989: 698–705 (patents), 536–7, 580–1 (evolution), 617, 628–31 (free will).

44 Jenkin to William Thomson, [1861], J37, Kelvin Collection, ULC (identified by Bruce Hunt).

45 Jenkin to William Thomson, [1861], J37, Kelvin Collection, ULC. See BAAS 1862: 125; 1863: 111.

46 Jenkin to Thomson, [1861], J37, Kelvin Collection, ULC. In a second undated letter (J36), Jenkin told Thomson that he thought 'I have done a good deal for the absolute units which (by force of telling others about them) I am beginning really to understand'. He also discussed apparatus for experimental determination of absolute resistance.

47 BAAS 1862: 125–6, 135. Tunbridge 1992: 24–33 surveys the work of the BAAS Committee (with special focus on Thomson).

48 BAAS 1862: 129.

49 BAAS 1862: 129–30. See Hunt 1994 (subsequent alteration to the 1862 standard).

50 BAAS 1862: 131–5, 135–50.

51 BAAS 1862: 133, 156–9.

52 BAAS 1863: 111; Thompson 1910, 1: 419 (Thomson's authorship). Jenkin and Maxwell drew up their comprehensive 'Elementary Relations Between Electrical Measurements' for the BAAS in the same year. See Maxwell and Jenkin 1865: 436–60, 507–25 [1863].

53 BAAS 1863: 111–12.

54 Siemens 1866: 326. Quoted in Schaffer 1994: 141–2.

55 Siemens 1966: 167. Quoted in Schaffer 1994: 142.

56 BAAS 1863: 112–13; Smith and Wise 1989: 687–9 (absolute system).

57 BAAS 1863: 112–13; Smith and Wise 1989: 179, 463–71 ('saving of brains').

58 BAAS 1863: 113–15; Smith and Wise 1989: 689–93.

59 See Schaffer 1994: 140 for an account of the King's College technology.

60 BAAS 1863: 115–22.

61 BAAS 1865: 308–13.

62 BAAS 1867: 476–79; 1869: 434–8; 1870: 14–15; Smith and Wise 1989: 694.

63 Schaffer 1994: 142–3.

64 Schaffer 1994: 145, 148.

65 Schaffer 1994: 145–8.
66 Schaffer 1994: 150–61; Maxwell 1873, 2: 372–3; 1891, 2: 418; James Clerk Maxwell to Fleeming Jenkin, 27 August 1863, in Campbell and Garnett 1882: 336–37; J.P. Joule to James Clerk Maxwell, [1876], Maxwell Collection, ULC. Both quotes in Schaffer 1994: 156.
67 Schaffer 1994: 161–3. On Ayrton see esp. Gooday 1994: 239–82 (energy metering); 1991b: 73–111 (teaching space).
68 Schaffer 1994: 163–4; 1992a: 40.
69 Rayleigh 1924: 123–4.
70 Thompson 1910, 2: 773–5. See esp. Schaffer 1992a: 23–49 (Cavendish Laboratory measurements of the ohm).
71 Siemens 1882: 4–8, on 8. See also Siemens 1879: 352–6.
72 Larmor 1908: xxix.

Chapter 14.   Sequel: Transforming Energy in the Late Nineteenth Century

1 Lodge 1885a: 482.
2 Heaviside to FitzGerald, [March 1895], FitzGerald Collection, Royal Dublin Society. Quoted and discussed in Hunt 1991b: 2–3, 228. The first section of the present chapter draws extensively on Hunt's excellent study. Andrew Warwick's forthcoming study of nineteenth century Cambridge mathematics and mathematical physics, showing the major reinterpretations of Maxwell's *Treatise* by Cambridge wranglers, will provide an important part of the sequel to the North British story presented in previous chapters.
3 Kelvin to FitzGerald, 9 April 1896, in Thompson 1910, 2: 1064. See Smith and Wise 1989: 488–94.
4 Kelvin to Larmor, 9 October 1906, L37, Kelvin Collection, ULC.
5 Hunt 1991b: 122, 202–3, 216–17.
6 Hunt 1991b: 6.
7 Hunt 1991b: 8–11.
8 Hunt 1991b: 15–23.
9 Hunt 1991b: 25–6. On Foster, see Gooday 1989: ch.4; 1990: 28.
10 Duhem 1954: 70–1 [1904]; Hunt 1991b: 87–8, 91n.
11 Lodge 1879: 277.
12 Lodge 1879: 279n.
13 Lodge 1879: 283n.
14 Lodge 1879: 283–4n.
15 Lodge 1879: 284n. Lodge's discussion relates closely to Maxwell's statistical interpretation of the second law of thermodynamics (ch.12). Jenkin, in an essay posthumously published as 'Is One Man's Gain Another Man's Loss?' (1887) asked 'If at every sale every one grows richer, where does poverty hail from?' His answer asserted that poverty was not the child of commerce but rather 'is the unwelcome offspring of that Physical Necessity which Sir William Thomson calls the Dissipation of Energy', paraphrased by saying 'that everything wears out'. Thus 'Everything we have is for ever losing value by decay'. See Jenkin 1887, 2: 141; Schaffer 1994: 136.
16 Hunt 1991b: 28–40.
17 FitzGerald to Rowland, 13 July 1882, Rowland Collection, Johns Hopkins University. Quoted and discussed in Hunt 1991b: 41. See Buchwald 1985b: 73–7 (Rowland and Maxwell's electromagnetic theory). See also Miller 1972: 5–27. In 1868 Maxwell told

Tait that 'There are two kinds of Dissipation of E. one of which is possible in a strictly conservative universe provided it is ∞, namely the propagation of undulations to ∞ from a vibrating body ... The other dissipation is conversion to heat. Either kind causes a steady periodic driving power to produce a motion converging to a steady periodicity'. See Maxwell to Tait, 3 August 1868, Maxwell Collection, ULC; Maxwell 1990–, 2: 416–18, on 417–18.

18 Hunt 1991b: 41–7.

19 Hunt 1991b: 49–53.

20 Hunt 1991b: 56–61.

21 Hunt 1991b: 61–8; Smith and Wise 1989: 446–58.

22 Hunt 1991b: 109–10.

23 Lodge 1885a: 482; Hunt 1991b: 111–12.

24 Hunt 1991b: 112; Buchwald 1985b: 41–53; Lodge 1885a: 482. At the conclusion of his *Treatise*, Maxwell raised the issue of continuity of energy transfer (ch.11). See Maxwell 1873, 2: 437–8.

25 Poynting 1920: 175 [1884]. Quoted and discussed in Hunt 1991b: 112–14.

26 Lodge 1885a: 482–3. Hunt 1991b: 114n points out that initially Lodge did not appreciate Poynting's insight. It is likely that FitzGerald, an early enthusiast, quickly persuaded Lodge of its value.

27 Lodge 1885a: 483.

28 Lodge 1885a: 483–84. Lodge had earlier participated in a controversy over action at a distance in the *Phil. Mag.* See esp. Lodge 1881a: 36–7, 220; Preston 1881a: 38–9, 218–20; Allen 1881: 74–5; Browne 1881: 129–31, 379–81; Preston 1881b: 391–3; Lodge 1881b: 529–34. See also Browne 1883a: 35–42; 1883b: 387–93; Tait 1883: 439–47.

29 Lodge 1885a: 484.

30 Lodge 1885a: 484n.

31 Thomson 1987: 206 [1884].

32 See esp. Hunt 1991b: 72–8. See also Moyer 1977: 251–68.

33 FitzGerald to Lodge, 1 January 1885, Lodge Collection, University College London. Quoted and discussed in Hunt 1991b: 78–87, on 79. See FitzGerald 1885: 438–43 (mechanical models of aether).

34 Smith and Wise 1989: 396–444 (Thomson's vortex theory of matter), 458–94 (Thomson's elastic-solid ether); Hunt 1987: 299–321; 1991b: 78–87, 96–104, 114–19 (FitzGerald's ether models).

35 Poynting 1920: 264 [1893]. Quoted and discussed in Hunt 1991b: 94–5.

36 Hunt 1991b: 119–22.

37 Hunt 1991b: 122–8, 245–7 (full derivation).

38 Hunt 1991b: 162–74.

39 Hunt 1991b: 153–8. For Hertz's own contexts see esp. Buchwald 1994. On Helmholtz's electrodynamics see Buchwald 1993: 334–73; Kaiser 1993: 374–402. See also D'Agostino 1975: 261–323 (on Hertz); Darrigol 1993: 189–280 (on Maxwellian theory in Germany).

40 FitzGerald 1902: 231 [1888]. Quoted and discussed in Hunt 1991b: 158–62.

41 Hunt 1991b: 160–1.

42 Hunt 1991b: 197–9.

43 *Nature* 48(1893): 538–9. Quoted in Hunt 1991b: 199–200.

44 Larmor 1900: 276; Hunt 1991b: 227; Buchwald 1985b: 133–40. See also Topper 1970–1: 393–410; 1980: 31–57 (on J.J. Thomson).

45 Hunt 1991b: 209–39; Buchwald 1985b: 141–73; Warwick 1991: 29–91 (on Larmor). See

also MacCormmach 1969: 41–87; 1970: 458–97 (on Lorentz and the 'electron'); Doran 1975: 133–260.

46 Larmor 1900: 286.

47 See esp. Jungnickel and McCormmach 1986, 2: 211–27.

48 Mach 1911: 9 [1909]; Hiebert 1973: 595–607; Jungnickel and McCormmach 1986, 2: 216–17.

49 Mach 1911: 9n [1909]; 1893: 495–6 [1889]; Jungnickel and McCormmach 1986, 2: 216.

50 Jungnickel and McCormmach 1986, 2: 218–19.

51 Schelar 1966: 101; Kragh 1984: 255.

52 Thomsen 1853–4: 34. Translated in Schelar 1966: 110; Kragh 1984: 256–7.

53 Kragh 1984: 255–72.

54 Crosland 1970: 69.

55 Berthelot 1875: 52, 73, 212. Translated in Schelar 1966: 113–14.

56 Dolby 1984: 381–91; Kragh 1984: 259–64. See also Schelar 1966: 112–16.

57 Rayleigh 1875: 454–5.

58 Rayleigh to Gibbs, 5 June 1892, quoted in Klein 1972: 390; Kelvin to Rayleigh, 13 September 1891 and 9 February 1892, quoted in Rayleigh 1924: 172.

59 Klein 1972: 390; Dolby 1976: 303; 1984: 385 (Ostwald and Gibbs); Dolby 1984: 387–9 (Duhem).

60 Turner 1972: 244–5; Kragh 1993: 403–31. See also Schelar 1966: 119–20.

61 Hiebert 1978: 455–6.

62 Hiebert 1978: 458–9.

63 Hiebert 1978: 456, 459–60.

64 Hiebert 1978: 462–4; 1971: 75; Jungnickel and McCormmach 1986, 2: 219. See also Leegwater 1986: 314–37 (Ostwald's chemical energetics); Holt 1970: 386–9; 1977: 146–50 (Ostwald's energetic perspectives).

65 Hiebert 1971: 67–70, on 69 (quoting Arnold Sommerfeld).

66 Boltzmann 1896: 64–65. Translated in Hiebert 1971: 70.

67 Jungnickel and McCormmach 1986, 2: 221.

68 Hiebert 1971: 71–5; Jungnickel and McCormmach 1986, 2: 223. See also Kangro 1975: 7–10.

69 Hiebert 1971: 75–9. See Klein 1962: 459–79 (Planck and early quantum theory); Garber 1976a: 89–126; Kuhn 1984: 231–52; Jungnickel and McCormmach 1986, 2: 228–31 (Planck's electromagnetism and thermodynamic researches), 256–65 (black body radiation researches). Planck's thermodynamics found its most celebrated expression in his treatise on the subject, first published in 1897 and translated into English in 1903. See Planck 1903 [1897].

70 Ostwald 1911: 136, 171.

71 Ostwald 1911: 184–5; 1912: 13.

Epilogue

1 Pearson 1936: 32. I thank Roger Mallion and Ben Marsden for this reference. Karl Pearson (1857–1936), third wrangler in 1879, venerated Darwin and became one of the most celebrated statisticians of the period (especially for the application of statistics to biological science).

2 *Westminster Confession* 1973: 19, 41–3 [1648].

3 Chalmers 1836–42, 7: 266–7.

4 McFarlane 1852: xviii–xix.
5 Clifford 1875b: 780. On James Thomson's unitarianism see Smith and Wise 1989: 138.
6 Campbell 1877: 15, 213–14.
7 Campbell 1877: 83–4, 88, 105.
8 Murphy 1869, 1: 55–67; Larmor 1912: xli, liv. Murphy also drew on the publications of Grove and Tyndall but the style was North British.
9 Murphy 1873: 49–51.
10 Clifford 1875a: 478–82.
11 Clifford 1875a: 787–8.
12 Quoted in Burtt 1949: 23 [1924].

Bibliography

Airy, G.B. 1851. '[Presidential] address'. *BAAS Rep. 21*: xxxix–liii.

Alexander, H.G. ed. 1956. *The Leibniz-Clarke Correspondence*. Manchester: Manchester University Press.

Allen, A.J.C. 1881. 'Note on Mr Browne's Paper "On Action at a Distance"'. *Phil. Mag. 11*: 74–75.

Arnold, D.H. 1983a. 'The *mécanique physique* of Siméon Denis Poisson: the evolution and isolation in France of his approach to physical theory (1800–1840). II. The Laplacian Programme'. *Arch. Hist. Ex. Sci. 28*: 267–87.

——. 1983b. 'IV. Disquiet with respect to Fourier's treatment of heat'. *Arch. Hist. Ex. Sci. 28*: 299–320.

——. 1983c. 'VII. *Mécanique physique*'. *Arch. Hist. Ex. Sci. 29*: 37–52.

——. 1983d. 'VIII. Applications of the *mécanique physique*'. *Arch. Hist. Ex. Sci. 29*: 53–72.

——. 1983e. 'IX. Poisson's closing synthesis: *Traité de physique mathématique*'. *Arch. Hist. Ex. Sci. 29*: 73–94.

Babbage, Charles. 1830. *Reflections on the Decline of Science in England and on Some of its Causes*. London: Fellowes. Repr. New York: Kelley, 1970.

Bacon, Francis. 1951. *The Advancement of Learning* and *New Atlantis*. Oxford: Oxford University Press. First published 1605 and 1627.

Barton, Ruth. 1976. 'The X Club: science, religion, and social change in Victorian England'. PhD diss., University of Pennsylvania.

——. 1990. '"An Influential Set of Chaps": the X-Club and Royal Society politics 1864–85'. *BJHS 23*: 53–81.

Becher, H.W. 1980. 'William Whewell and Cambridge mathematics'. *HSPS 11*(i): 1–48.

Berg, Maxine. 1980. *The Machinery Question and the Making of Political Economy 1815–1848*. Cambridge: Cambridge University Press.

Berman, Morris. 1978. *Social Change and Scientific Organization: The Royal Institution*. London: Heinemann.

Bernstein, H.T. 1963. 'J. Clerk Maxwell on the history of the kinetic theory of gases, 1871'. *Isis 54*: 206–16.

Berthelot, P.E.M. 1875. 'Principes généraux de la thermochimie. Cinquiéme mémoire. Sur le troisième principe de la thermochemie ou principe de travail maximum'. *Annales de chimie et de physique 4*: 52–73.

Bevilacqua, Fabio. 1993. 'Helmholtz's *Ueber die Erhaltung der Kraft*'. In David Cahan, ed., *Hermann von Helmholtz and the Foundations of Nineteenth-century Science*', pp 291–333. Berkeley, Los Angeles and London: University of California Press.

Biagioli, Mario. 1993. *Galileo, Courtier*. Chicago and London: University of Chicago Press.

Bierhalter, Günther. 1981a. '*Boltzmanns mechanische Grundlegung des zweiten Hauptsatzes der Wärmelehre aus dem Jahre 1866*'. *Arch. Hist Ex. Sci.* 24: 195–205.

——. 1981b. '*Clausius' mechanische Grundlegung des zweiten Hauptsatzes der Wärmelehre aus dem Jahre 1871*'. *Arch. Hist. Ex. Sci.* 24: 207–20.

——. 1981c. '*Zu Hermann von Helmholtzens mechanischer Grundlegung der Wärmelehre aus dem Jahre 1884*'. *Arch. Hist. Ex. Sci.* 25: 71–84.

——. 1982. '*Das Virialtheorem in seiner Beziehung zu den mechanischen Grundlegungen des zweiten Hauptsatzes der Wärmelehre*'. *Arch. Hist. Ex. Sci.* 27: 199–211.

——. 1983–4. '*Die v. Helmholtzschen Monozykel-Analogien zur Thermodynamik und das Clausissche Disgregationkonzept*'. *Arch. Hist. Ex. Sci.* 29: 95–100.

——. 1985. '*Die mechanischen Entropie und Disgregationskonzepte aus dem Jahrhundert: ihre Grundlegen, ihr Versagen und ihr Entstehungshintergrund*'. *Arch. Hist. Ex. Sci.* 32: 17–41.

——. 1992. '*Von L. Boltzmann bis J.J. Thomson: die Versuche einer mechanischen Grundlegen der Thermodynamik (1866–1890)*'. *Arch. Hist. Ex. Sci.* 44: 25–75.

Black, Joseph. 1803. *Lectures on the Elements of Chemistry Delivered in the University of Edinburgh*. Ed. John Robison. Edinburgh: Creech.

Black, J.S. and G.W. Chrystal. 1912. *William Robertson Smith*. London: Black.

Blackwell, R.J. 1966. 'Descartes' laws of motion'. *Isis 57*: 220–34.

Body, Geoffrey. 1971. *British Paddle Steamers*. Newton Abbot: David & Charles.

Boltzmann, Ludwig. 1896. 'Ein Wort der Mathematik an die Energetik'. *Ann. Phys. Chem. 57*: 39–71.

Bonsor, N.R.P. 1975. *North Atlantic Seaway*. Vol. 1. Newton Abbot: David & Charles.

——. 1983. *South Atlantic Seaway*. Jersey: Brookside Publications.

Brewster, David. 1837. 'Dr Buckland's *Bridgewater Treatise* – Geology and Mineralogy'. *Edinburgh Review 65*: 1–39.

British Association. 1862. 'Provisional report of the committee appointed by the British Association on standards of electrical resistance'. *BAAS Rep. 32*: 125–63.

——. 1863. 'Report of the committee appointed by the British Association on standards of electrical resistance'. *BAAS Rep. 33*: 111–76.

——. 1865. 'Report of the Committee on Standards of Electrical Resistance'. *BAAS Report 35*: 308–13.

——. 1867. 'Report of the Committee on Standards of Electrical Resistance'. *BAAS Rep. 37*: 474–522.

——. 1869. 'Report of the Committee on Standards of Electrical Resistance'. *BAAS Rep. 39*: 434–8.

——. 1870. 'Report of the Committee on Standards of Electrical Resistance'. *BAAS Rep. 40*: 14–15.

——. 1891–94. 'Report of a committee, consisting of Messrs. J. Larmor and G.H. Bryan, on the present state of our knowledge of thermodynamics, specially with regard to the second law'. *BAAS Rep. 61*: 85–122; *64*: 64–106.

Brock, W.H., N.D. McMillan and R.C. Mollan (eds). 1981. *John Tyndall. Essays on a Natural Philosopher*. Dublin: Royal Dublin Society.

Bromberg, Joan. 1967–8. 'Maxwell's displacement current and his theory of light'. *Arch. Hist. Ex. Sci. 4*: 218–34.

Brooke, J.H. 1979. 'Nebular contraction and the expansion of naturalism'. *BJHS 12*: 200–11.

———. 1991. *Science and Religion. Some Historical Perspectives*. Cambridge: Cambridge University Press.

———. 1996. 'Like minds: the God of Hugh Miller'. In Michael Shortland, ed., *Hugh Miller and the Controversies of Victorian Science*, pp 171–86. Oxford: Oxford University Press.

Brown, C.G. 1987. *A Social History of Religion in Scotland since 1730*. London: Methuen.

Brown, S.C. 1967. *Benjamin Thompson – Count Rumford. Count Rumford on the Nature of Heat*. London: Pergamon Press.

Brown, S.J. 1993. 'The Ten Years' Conflict and the Disruption of 1843'. In S.J. Brown and Michael Fry, eds., *Scotland in the Age of Disruption*, pp 1–27. Edinburgh: Edinburgh University Press.

Brown, Thomas. 1893. *Annals of the Disruption; with Extracts from the Narratives of Ministers who left the Scottish Establishment in 1843*. Edinburgh: Macniven and Wallace.

Browne, W.R. 1881. 'On action at a distance'. *Phil. Mag. 11*: 129–31, 379–81.

———. 1883a. 'On central forces and the conservation of energy'. *Phil. Mag. 15*: 35–42.

———. 1883b. 'On the reality of force'. *Phil. Mag. 16*: 387–93.

Brücke, Ernst. 1858. 'On gravitation and the conservation of force'. *Phil. Mag. 15*: 81–90.

Brush, S.G. 1957a. 'The development of the kinetic theory of gases. I. Herapath'. *Ann. Sci. 13*: 188–98.

———. 1957b. 'The development of the kinetic theory of gases. II. Waterston'. *Ann. Sci. 13*: 273–82.

———. 1958a. 'The development of the kinetic theory of gases. III. Clausius'. *Ann. Sci. 14*: 185–96.

———. 1958b. 'The development of the kinetic theory of gases. IV. Maxwell'. *Ann. Sci. 14*: 243–55.

———. 1967–8. 'Foundations of statistical mechanics 1845–1915'. *Arch. Hist. Ex. Sci. 4*: 145–83.

———. 1970. 'The wave theory of heat: a forgotten stage in the transition from the caloric theory to thermodynamics'. *BJHS 18*: 145–67.

———. 1973. 'The development of the kinetic theory of gases. VII. Heat conduction and the Stefan–Boltzmann Law'. *Arch. Hist. Ex. Sci. 11*: 38–96.

———. 1974. 'The development of the kinetic theory of gases. VIII. Randomness and irreversibility'. *Arch. Hist. Ex. Sci. 12*: 1–88.

———. 1976. 'Irreversibility and indeterminism: Fourier to Heisenberg'. *Journal of the History of Ideas 37*: 603–30.

———. 1978. *The Temperature of History. Phases of Science and Culture in the Nineteenth Century*. New York: Burt Franklin.

———. 1979. 'Nineteenth-century debates about the inside of the earth: solid, liquid or gas?' *Ann. Sci. 36*: 225–54.

———. 1987. 'The nebular hypothesis and the evolutionary worldview'. *Hist. Sci. 25*: 245–78.

Bryant, Lynwood. 1973. 'The role of thermodynamics in the evolution of heat engines'. *Tech. Cult. 14*: 152–65.

———. 1976. 'The development of the diesel engine'. *Tech. Cult. 17*: 432–46.

Buchwald, J.Z. 1980. 'The Hall effect and Maxwellian electrodynamics in the 1880s. Part I:

The discovery of a new electric field'. *Centaurus 23*: 51–99; 'Part II: The unification of theory, 1881–1893'. *Centaurus 23*: 118–62.

——. 1981. 'The quantitative ether in the first half of the nineteenth century'. In G.N. Cantor and M.J.S. Hodge, eds, *Conceptions of Ether. Studies in the History of Ether Theories 1740–1900*, pp 215–37. Cambridge: Cambridge University Press.

——. 1985a. 'Oliver Heaviside, Maxwell's apostle and Maxwellian apostate'. *Centaurus 28*: 288–330.

——. 1985b. *From Maxwell to Microphysics. Aspects of Electromagnetic Theory in the Last Quarter of the Nineteenth Century*. Chicago: University of Chicago Press.

——. 1989. *The Rise of the Wave Theory of Light. Optical Theory and Experiment in the Early Nineteenth Century*. Chicago: University of Chicago Press.

——. 1993. 'Electrodynamics in context. Object states, laboratory practice, and anti-romanticism'. In David Cahan, ed., *Hermann von Helmholtz and the Foundations of Nineteenth-century Science*, pp 334–73. Berkeley, Los Angeles and London: University of California Press.

——. 1994. *The Creation of Scientific Effects: Heinrich Hertz and Electric Waves*. Chicago: University of Chicago Press.

Burchfield, J.D. 1974. 'Darwin and the dilemma of geological time'. *Isis 65*: 301–21.

——. 1975. *Lord Kelvin and the Age of the Earth*. London: Macmillan.

Burleigh, J.H.S. 1960. *A Church History of Scotland*. Oxford: Oxford University Press.

Cahan, David. 1982. 'Werner Siemens and the origin of the Physikalisch–Technische Reichs-anstalt, 1872–1887'. *HSPS 12*(ii): 253–83.

——. 1985. 'The institutional revolution in German physics, 1865–1914'. *HSPS 15*(ii): 1–65.

——. 1989. *An Institute for an Empire. The Physikalisch-Technische Reichsanstalt 1871–1918*. Cambridge: Cambridge University Press.

Caird, John. 1858. *Sermons*. Edinburgh: Blackwood.

Cairns, David. 1956. 'Thomas Chalmers's *Astronomical Discourses*: a study in natural theology'. *Scottish Journal of Theology 9*: 410–21.

Campbell, Lewis. 1877. *Some Aspects of the Christian Ideal*. London: Macmillan.

Campbell, Lewis and William Garnett. 1882. *The Life of James Clerk Maxwell. With a Selection from his Correspondence and Occasional Writings and a Sketch of his Contributions to Science*. London: Macmillan.

Candlish, R.S. 1868. *The Book of Genesis Expounded in a Series of Discourses*. 2 vols. Edinburgh: Black.

Caneva, K.L. 1978. 'From galvanism to electrodynamics: the transformation of German physics and its social context'. *HSPS 9*: 63–159.

——. 1980. 'Ampère, the etherians, and the Oersted connexion'. *BJHS 13*: 121–38.

——. 1993. *Robert Mayer and the Conservation of Energy*. Princeton: Princeton University Press.

Cantor, Geoffrey. 1970. 'The changing role of Young's ether'. *BJHS 5*: 44–62.

——. 1971. 'Henry Brougham and the Scottish methodological tradition'. *SHPS 2*: 69–89.

——. 1975a. 'The reception of the wave theory of light in Britain: a case study illustrating the role of methodology in scientific debate'. *HSPS 6*: 109–32.

——. 1975b. 'William Robert Grove, the correlation of forces, and the conservation of energy'. *Centaurus 19*: 273–90.

——. 1983. *Optics after Newton. Theories of Light in Britain and Ireland, 1704–1840*. Manchester: Manchester University Press.

——. 1984. 'Locating the first law of thermodynamics'. Unpublished paper precirculated to conference on 'New Perspectives in Nineteenth-century Science', University of Kent at Canterbury.

——. 1991. *Michael Faraday: Sandemanian and Scientist. A Study of Science and Religion in the Nineteenth Century*. London: Macmillan.

Cantor, Geoffrey, David Gooding and F.A.J.L. James. 1991. *Faraday*. London: Macmillan.

Cardwell, D.S.L. 1965. 'Power technologies and the advance of science, 1700–1825'. *Tech. Cult. 6*: 188–207.

——. 1967. 'Some factors in the early development of the concepts of power, work and energy'. *BJHS 3*: 209–24.

——. 1971. *From Watt to Clausius. The Rise of Thermodynamics in the Early Industrial Age*. London: Heinemann.

——. 1977. 'Theories of heat and the rise of physics'. *Hist. Sci. 15*: 138–45.

——. 1989. *James Joule. A Biography*. Manchester: Manchester University Press.

Cardwell. D.S.L and R.L. Hills. 1976. 'Thermodynamics and practical engineering in the nineteenth century'. *Hist. Tech. 1*: 1–20.

Carlyle, Thomas. 1908. *Sartor Resartus and On Heroes and Hero Worship*. London: Dent.

Carnot, L.N.M. 1803. *Principes fondamentaux de l'équilibre et du mouvement*. Paris: Deterville.

Carnot, Sadi. 1960 [1824]. 'Reflections on the motive power of fire'. Trans. R.H. Thurston. In E. Mendoza, ed., *Reflections on the Motive Power of Fire by Sadi Carnot and Other Papers on the Second Law of Thermodynamics by E. Clapeyron and R. Clausius*, pp 1–59. New York: Dover.

——. 1986 [1824]. *Reflexions on the Motive Power of Fire. A Critical Edition with the Surviving Scientific Manuscripts*. Trans. and ed. Robert Fox. Manchester: Manchester University Press.

Cawood, John. 1979. 'The Magnetic Crusade: Science and politics in early Victorian Britain'. *Isis 70*: 493–518.

Challis, James. 1861. 'On theories of magnetism and other forces, in reply to remarks by Professor Maxwell'. *Phil. Mag. 21*: 250–4.

——. 1862. 'On the principles of theoretical physics'. *Phil. Mag. 23*: 313–22.

Chalmers, A.F. 1973. 'The limitations of Maxwell's electromagnetic theory'. *Isis 64*: 469–83.

——. 1973–4. 'Maxwell's methodology and his application of it to electromagnetism'. *SHPS 4*: 107–64.

——. 1986. 'The heuristic role of Maxwell's mechanical model of electromagnetic phenomena'. *SHPS 17*: 415–27.

Chalmers, Thomas. 1836–42. *The Works of Thomas Chalmers*. 25 vols. Glasgow: Collins.

——. 1844–45. 'Political economy of the Bible'. *North British Review 2*: 1–52.

Channell, D.F. 1982. 'The harmony of theory and practice: the engineering science of W.J.M. Rankine'. *Tech. Cult. 23*: 39–52.

Church Hymnary. 1928. *The Church Hymnary. Revised Edition*. Oxford: Oxford University Press.

Church, W.C. 1890. *The Life of John Ericsson*. 2 vols. London: Sampson Low.

Clapeyron, Emile. 1837 [1834]. 'Memoir on the Motive Power of Heat'. In Richard Taylor, ed., *Scientific Memoirs, Selected from the Transactions of Foreign Academies of Science and Learned Societies, and from Foreign Journals*, vol. 1, pp. 347–76. London: Taylor and Francis.

Clark, I.D.L. 1960–62. 'The Leslie controversy, 1805'. *Records of the Scottish Church Historical Society 14*: 179–97.

Clark, J.W. and T. McK. Hughes. *The Life and Letters of the Reverend Adam Sedgwick*. 2 vols. Cambridge: Cambridge University Press.

Clark, Peter. 1976. 'Atomism versus thermodynamics'. In Colin Howson, ed., *Method and Appraisal in the Physical Sciences*, pp. 41–105. Cambridge: Cambridge University Press.

Clausius, Rudolf. 1851 [1850]. 'On the moving force of heat, and the laws regarding the nature of heat itself which are deducible therefrom'. *Phil. Mag. 2*: 1–21, 102–19.

——. 1856a [1854]. 'On a modified form of the second fundamental theorem in the mechanical theory of heat'. *Phil. Mag. 12*: 81–98.

——. 1856b [1854]. 'On the application of the mechanical theory of heat to the steam-engine'. *Phil. Mag. 12*: 241–65, 338–54, 426–43.

——. 1857. 'On the nature of the motion which we call heat'. *Phil. Mag. 14*: 108–27.

——. 1859 [1858]. 'On the mean length of the paths described by the separate molecules of gaseous bodies on the occurrence of molecular motion: together with some other remarks upon the mechanical theory of heat'. *Phil. Mag. 17*: 81–91.

——. 1860. 'On the dynamical theory of gases'. *Phil. Mag. 19*: 434–6.

——. 1862a. 'On the application of the theorem of the equivalence of transformations to the internal work of a mass of matter'. *Phil. Mag. 24*: 81–97, 201–13.

——. 1862b. 'On the conduction of heat by gases'. *Phil. Mag. 23*: 417–35, 512–34.

——. 1866. 'On the determination of the energy and entropy of a body'. *Phil. Mag. 32*: 1–17.

——. 1867. *The Mechanical Theory of Heat, with its Applications to the Steam-engine and to the Physical Properties of Bodies*. Ed. T.A. Hirst. London: John Van Voorst.

——. 1868 [1867]. 'On the second fundamental theorem of the mechanical theory of heat; a lecture delivered before the Forty-first Meeting of the German Scientific Association, at Frankfort on the Maine, September 23, 1867'. *Phil. Mag. 35*: 405–19.

——. 1870. 'On a mechanical theorem applicable to heat'. *Phil. Mag. 40*: 122–7.

——. 1872a. 'A contribution to the history of the mechanical theory of heat'. *Phil. Mag. 43*: 106–15.

——. 1872b. 'On the objections raised by Mr. Tait against my treatment of the mechanical theory of heat'. *Phil. Mag. 43*: 443–6.

——. 1872c. 'A necessary correction of one of Mr Tait's remarks'. *Phil. Mag. 44*: 117.

——. 1872d. 'On the connexion of the second proposition of the mechanical theory of heat with Hamilton's principle'. *Phil. Mag. 44*: 365–9.

——. 1874. 'On different forms of the virial'. *Phil. Mag. 48*: 1–11.

——. 1875. 'On the theorem of the mean ergal, and its application to the molecular motions of gases'. *Phil. Mag. 50*: 26–46, 101–17, 191–200.

——. 1876. 'On the bearing of the fundamental law of electrodynamics toward the principle of the conservation of energy, and on a further simplification of the former'. *Phil. Mag. 1*: 218–21.

Cleland, James. 1840. 'On the population, trade and commerce of the city of Glasgow'. *BAAS Rep. 10*: 174–5.

Clifford, W.K. 1875a. 'The first and last catastrophe'. *Fortnightly Review 17*: 465–84.

——. 1975b. 'The unseen universe'. *Fortnightly Review 17*: 776–93.

Colding, L.A. 1864. 'On the history of the principle of the conservation of energy'. *Phil. Mag. 27*: 56–64.

——. 1871 [1850]. 'On the universal powers of nature and their mutual dependence'. *Phil. Mag. 42*: 1–20.

Constant, E.W. 1983. 'Scientific theory and technological testability: science, dynamometers, and water turbines in the 19th Century'. *Tech. Cult. 24*: 183–98.

Cooper, M.L. and V.M.D. Hall. 1982. 'William Robert Grove and the London Institution, 1841–1845'. *Ann. Sci. 39*: 229–54.

Coriolis, G.G. de. 1829. *Du calcul de l'effet des machines, ou considérations sur l'emploi des moteurs et sur leur évaluation, pour servir d'introduction a l'étude spéciale des machines.* Paris: Carilian-Goeury.

Cropper, W.H. 1988. 'James Joule's work in electrochemistry and the emergence of the first law of thermodynamics'. *HSPBS 19*(i): 1–15.

Crosland, Maurice. 1967. *The Society of Arcueil. A View of French Science at the Time of Napoleon I.* London: Heinemann.

———. 1970. 'Berthelot, Pierre Eugéne Marcellin'. *DSB 2*: 63–72.

———. 1992. *Science Under Control. The French Academy of Sciences 1795–1914.* Cambridge: Cambridge University Press.

Crosland, Maurice and Crosbie Smith. 1978. 'The transmission of physics from France to Britain: 1800–1840'. *HSPS 9*: 1–61.

Crouzet, François. 1982. *The Victorian Economy.* Trans. Anthony Forster. London: Methuen.

D'Agostino, Salvo. 1975. 'Hertz's researches on electromagnetic waves'. *HSPS 6*: 261–323.

Dahl, P.F. 1963. 'Colding and the conservation of energy'. *Centaurus 8*: 174–88.

Daiches, David. 1977. *Glasgow.* London: Andre Deutsch.

Dalton, John. 1965 [1808]. *A New System of Chemical Philosophy.* London: Peter Owen.

Darrigol, Olivier. 1993. 'The electrodynamic revolution in Germany as documented by early German expositions of Maxwell's theory'. *Arch. Hist. Ex. Sci. 45*: 189–280.

Daub, E.E. 1967. 'Atomism and thermodynamics'. *Isis 58*: 293–303.

———. 1969. 'Probability and thermodynamics: the reduction of the second law'. *Isis 60*: 318–30.

———. 1970a. 'Maxwell's demon'. *SHPS 1*: 213–27.

———. 1970b. 'Entropy and dissipation'. *HSPS 2*: 321–54.

———. 1970c. 'Waterston, Rankine, and Clausius on the kinetic theory of gases'. *Isis 61*: 105–6.

———. 1971. 'Waterston's influence on Krönig's kinetic theory of gases'. *Isis 62*: 512–15.

———. 1974. 'The regenerator principle in the Stirling and Ericsson hot air engine'. *BJHS 7*: 259–77.

———. 1977. 'Sources for Clausius' entropy concept: Reech and Rankine'. In E.G. Forbes, ed., *Proceedings of the XVth International Congress of the History of Science. Edinburgh 10–19 August 1977*, pp 342–58. Edinburgh: Edinburgh University Press.

Davie, G.E. 1961. *The Democratic Intellect. Scotland and her Universities in the Late Nineteenth Century.* Edinburgh: Edinburgh University Press.

Desmond, Adrian. 1989. 'Lamarckism and democracy: corporations, corruption and comparative anatomy in the 1830s'. In J.R. Moore, ed., *History, Humanity and Evolution*, pp 99–130. Cambridge: Cambridge University Press.

———. 1994. *Huxley: the Devil's Disciple.* London: Michael Joseph.

———. 1997. *Huxley: Evolution's High Priest.* London: Michael Joseph.

Desmond, Adrian and James Moore. 1991. *Darwin.* London: Michael Joseph.

Dickinson, H.W. 1935. *James Watt. Craftsman and Engineer.* Cambridge: Cambridge University Press.

Dolby, R.G.A. 1976. 'Debates over the theory of solution: a study of dissent in physical chemistry in the English-speaking world in the late nineteenth and early twentieth centuries'. *HSPS 7*: 297–404.

——. 1984. 'Thermochemistry versus thermodynamics: the nineteenth century controversy'. *Hist. Sci. 22*: 375–400.

Donaldson, G. 1987. *Scottish Church History*. Edinburgh: Scottish Academic Press.

Doran, B.G. 1975. 'Origins and consolidation of field theory in nineteenth-century Britain: from the mechanical to the electromagnetic view of nature'. *HSPS 6*: 133–260.

Dorling, Jon. 1970. 'Maxwell's attempts to arrive at non-speculative foundations for the kinetic theory'. *SHPS 1*: 229–48.

Douglas, Mary. 1990. 'Foreward: no free gifts'. In Marcel Mauss, *The Gift*, pp vii–xviii. London: Routledge.

Drummond, A.L. and James Bulloch. 1978. *The Church in Victorian Scotland 1873–1874*. Edinburgh: The Saint Andrew Press.

Duhem, Pierre. 1954 [1904]. *The Aim and Structure of Physical Theory*. Trans. P.P. Wiener. Princeton: Princeton University Press.

Elkana, Yehuda. 1970a. 'Helmholtz' "Kraft": an illustration of concepts in flux'. *HSPS 2*: 263–98.

——. 1970b. 'The conservation of energy: a case of simultaneous discovery?' *Arch. int. d'hist. des sci. 90–1*: 31–60. Repr. Elkana 1974: 175–97.

——. 1974. *The Discovery of the Conservation of Energy*. London: Hutchinson.

Emmerson, G.S. 1977. *John Scott Russell. A Great Victorian Engineer and Naval Architect*. London: Murray.

Eve, A.S. and Creasey, A.S. 1945. *Life and Work of John Tyndall*. London: Macmillan.

Everitt, C.W.F. 1975. *James Clerk Maxwell. Physicist and Natural Philosopher*. New York: Scribner's Sons.

Ewart, Peter. 1813. 'On the measure of moving force'. *Mem. Man. Lit. & Phil. 2*: 105–258.

Fairbairn, William. 1970 [1877]. *The Life of Sir William Fairbairn, Bart*. Ed. William Pole. Repr. ed. A.E. Musson. Newton Abbot: David & Charles.

Faraday, Michael. 1857. 'On the conservation of force'. *Phil. Mag. 13*: 225–39.

——. 1859. 'On the conservation of force'. *Phil. Mag. 17*: 166–9.

——. 1991-. *The Correspondence of Michael Faraday. Volume 1, 1811–1831; Volume 2, 1832–1840; Volume 3, 1841–December 1848*. Ed. Frank A.J.L. James. Stevenage: Institution of Electrical Engineers.

Feffer, Stuart M. 1989. 'Arthur Schuster, J.J. Thomson, and the discovery of the electron'. *HSPBS 20*(i): 33–61.

FitzGerald, G.F. 1885. 'On the structure of mechanical models illustrating some properties of the aether'. *Phil. Mag. 19*: 438–43.

——. 1902. *The Scientific Writings of the Late George Francis FitzGerald*. Ed. Joseph Larmor. Dublin: Hodges and Figgis.

Fleming, John. 1851. 'Natural Science'. In *Inauguration of New College, Edinburgh . . . 1851*, pp 215–32.

Forrester, John. 1975. 'Chemistry and the conservation of energy: the work of James Prescott Joule'. *SHPS 6*: 273–313.

Fourier, Joseph. 1878 [1822]. *The Analytical Theory of Heat*. Trans. Alexander Freeman. Cambridge: Cambridge University Press. First published in Paris as *Théorie analytique de la chaleur*.

Fox, Robert. 1968. 'The background to the discovery of Dulong and Petit's Law'. *BJHS 4*: 1–22.

——. 1969. 'James Prescott Joule (1818–1889)'. In John North, ed., *Mid-nineteenth-century Scientists*, pp 72–103. Oxford: Pergamon Press.

——. 1970. 'Watt's expansive principle in the work of Sadi Carnot and Nicolas Clément'. *Notes and Records RSL. 24*: 233–53.

——. 1971. *The Caloric Theory of Gases. From Lavoisier to Regnault*. Oxford: Clarendon Press.

——. 1974. 'The rise and fall of Laplacian physics'. *HSPS 4*: 89–136.

——. 1976. 'The challenge of a new technology: Theorists and the high-pressure steam engine before 1824'. In *Table ronde du Centre National de la Recherche Scientifique. Sadi Carnot et l'essor de la thermodynamique. Paris, Ecole Polytechnique, 11–13 juin 1974*. Paris: Editions du Centre de la Recherche Scientifique.

— (ed.). 1978. *Réflexions sur la puissance du feu . . . .* Paris: Libraire Philosophique J. Vrin. Contains a complete transcription of Sadi Carnot's manuscripts.

——. 1986. 'Introduction'. In Robert Fox, ed., *Sadi Carnot. Reflexions on the Motive Power of Fire. A Critical Edition with the Surviving Scientific Manuscripts*, pp 1–57. Manchester: Manchester University Press.

Fox, R.W. 1840. 'Report on some observations on subterranean temperature'. *BAAS Rep. 10*: 309–19.

Fraser, Craig. 1983. 'J.-L. Lagrange: early contributions to the principles and methods of mechanics'. *Arch. Hist. Ex. Sci. 28*: 197–241.

——. 1985. 'D'Alembert's principle: the original formulation and application in Jean d'Alembert's *Traité de dynamique* (1743)'. *Centaurus 28*: 31–61, 145–59.

Friedman, R.M. 1977. 'The creation of a new science: Joseph Fourier's analytical theory of heat'. *HSPS 8*: 73–99.

Fry, Michael. 1987. *Patronage and Principle. A Political History of Modern Scotland*. Aberdeen: Aberdeen University Press.

Gabbey, Alan. 1971. 'Force and inertia in seventeenth-century dynamics'. *SHPS 2*: 1–67.

——. 1989. 'Newton and Natural Philosophy'. In R.C. Olby, G.N. Cantor, J.R.R. Christie and M.J.S. Hodge, eds., *Companion to the History of Modern Science*, pp 243–63. London: Routledge.

Garber, Elizabeth. 1970. 'Clausius and Maxwell's kinetic theory of gases'. *HSPS 2*: 299–319.

——. 1972–3. 'Aspects of the introduction of probability into physics'. *Centaurus 17*: 11–39.

——. 1976a. 'Some reactions to Planck's Law, 1900–1914'. *SHPS 7*: 89–126.

——. 1976b. 'Thermodynamics and meteorology (1850–1900)'. *Ann. Sci. 33*: 51–65.

Garnett, William. 1879. 'Energy'. *Enc. Brit.* [9th ed.] *8*: 205–11.

Gavroglu, Kostas. 1990. 'The reaction of the British physicists and chemists to Van der Waals' early work and to the law of corresponding states'. *HSPBS 20*(ii): 199–237.

Gee, Brian. 1991. 'Electromagnetic engines. Pre-technology and development immediately following Faraday's discovery of electromagnetic rotations'. *Hist. Tech. 13*: 41–72.

Gibbs, J.W. 1906. *The Scientific Papers of J. Willard Gibbs*. 2 vols. London: Longmans, Green. Repr. New York: Dover, 1961.

Gillispie, C.C. 1971. *Lazare Carnot Savant*. Princeton: Princeton University Press.

Ginn, W.T. 1991. 'Philosophers and artisans: The relationship between men of science and instrument makers in London 1820–1860'. Ph.D. diss., University of Kent at Canterbury.

Glas, Eduard. 1986. 'On the dynamics of mathematical change in the case of Monge and the French Revolution'. *SHPS 17*: 249–68.

Goldfarb, S.J. 1977. 'Rumford's theory of heat: A reassessment'. *BJHS 10*: 25–36.

Gooday, Graeme. 1989. 'Precision measurement and the genesis of physics teaching laboratories in Victorian Britain'. PhD. diss., University of Kent at Canterbury.

———. 1990. 'Precision measurement and the genesis of physics teaching laboratories in Victorian Britain'. *BJHS 23*: 25–51.

———. 1991a. '"Nature" in the laboratory: Domestication and discipline with the microscope in Victorian life science'. *BJHS 24*: 307–41.

———. 1991b. 'Teaching telegraphy and electrotechnics in the physics teaching laboratory: William Ayrton and the creation of academic space for electrical engineering 1873–84'. *Hist. Tech. 13*: 73–111.

———. 1994. 'The morals of energy metering: Constructing and deconstructing the precision of the Victorian electrical engineer's ammeter and voltmeter'. In M. Norton Wise, ed., *The Values of Precision*, pp 239–82. Princeton: Princeton University Press.

Gooding, David. 1978. 'Conceptual and experimental bases of Faraday's denial of electrostatic action at a distance'. *SHPS 9*: 117–49.

———. 1980a. 'Faraday, Thomson, and the concept of the magnetic field'. *BJHS 13*: 91–120.

———. 1980b. 'Metaphysics versus measurement: the conversion and conservation of force in Faraday's physics'. *Ann. Sci. 37*: 1–29.

———. 1981. 'Final steps to the field theory: Faraday's study of magnetic phenomena, 1845–1850'. *HSPS 11*(ii): 231–75.

———. 1982. 'A convergence of opinion on the divergence of lines: Faraday and Thomson's discussion of diamagnetism'. *Notes and Records RSL. 36*: 243–59.

———. 1989. '"Magnetic curves" and the magnetic field: experimentation and representation in the history of a theory'. In David Gooding, Trevor Pinch and Simon Schaffer, eds, *The Uses of Experiment. Studies in the Natural Sciences*, pp 183–223. Cambridge: Cambridge University Press.

Gordon, Lewis. 1840. 'On the turbine water-wheel'. *BAAS Rep. 10*: 191–2.

———. 1872–75. 'Obituary notice of Professor Rankine'. *Proc. RSE 8*: 296–306.

Graham, Thomas. 1842. *Elements of Chemistry, Including the Applications of the Science in the Arts*. London: Bailliere.

Grattan-Guinness, Ivor. 1984. 'Work for the workers: advances in engineering mechanics and instruction in France, 1800–1830'. *Ann. Sci. 41*: 1–33.

Grattan-Guinness, Ivor (in collaboration with J.R. Ravetz). 1972. *Joseph Fourier 1768–1830. A Survey of his Life and Work, Based on a Critical Edition of his Monograph on the Propagation of Heat, Presented to the Institut de France in 1807*. Cambridge, MA: MIT Press.

Graukroger, Stephen. 1982. 'The metaphysics of impenetrability: Euler's conception of force'. *BJHS 15*: 132–54.

Hachette, J.N.P. 1811. *Traité elémentaire des machines*. Paris: Klostermann.

Hankins, T.L. 1965. 'Eighteenth-century attempts to resolve the *vis viva* controversy'. *Isis 56*: 281–97.

———. 1967. 'The reception of Newton's second law of motion in the eighteenth century'. *Arch. int. d'hist. des sciences 78–79*: 43–65.

———. 1970a. *Jean d'Alembert. Science and the Enlightenment*. Oxford: Clarendon Press.

———. 1970b. 'The concept of hard bodies in the history of physics'. *Hist. Sci. 9*: 119–28.

———. 1980. *Sir William Rowan Hamilton*. Baltimore: Johns Hopkins University Press.

——. 1985. *Science and the Enlightenment*. Cambridge: Cambridge University Press.

Hanna, William. 1849–52. *Memoirs of the Life and Writings of Thomas Chalmers*. 4 vols. Edinburgh.

Hanson, N.R. 1958. *Patterns of Discovery. An Inquiry into the Conceptual Foundations of Science*. Cambridge: Cambridge University Press.

Heimann [Harman], P.M. 1969–70. 'Maxwell and the modes of consistent representation'. *Arch. Hist. Ex. Sci. 6*: 171–213.

——. 1970. 'Molecular forces, statistical representation and Maxwell's demon'. *SHPS. 1*: 189–211.

——. 1971. 'Maxwell, Hertz, and the nature of electricity'. *Isis 62*: 149–57.

——. 1972. 'The *Unseen Universe*: Physics and the philosophy of nature in Victorian Britain'. *BJHS 6*: 73–9.

——. 1974a. 'Conversion of forces and the conservation of energy'. *Centaurus 18*: 147–61.

——. 1974b. 'Helmholtz and Kant: the metaphysical foundations of *Ueber die Erhaltung der Kraft*'. *SHPS 5*: 205–38.

——. 1976. 'Mayer's concept of "Force": the "Axis" of a new science of physics'. *HSPS 7*: 277–96.

——. 1982a. *Energy, Force, and Matter. The Conceptual Development of Nineteenth-century Physics*. Cambridge: Cambridge University Press.

——. 1982b. *Metaphysics and Natural Philosophy. The Problem of Substance in Classical Physics*. Brighton: Harvester Press.

——. 1987. 'Mathematics and reality in Maxwell's dynamical physics'. In Robert Kargon and Peter Achinstein, eds., *Kelvin's Baltimore Lectures and Modern Theoretical Physics*, pp 267–96. Cambridge, MA: MIT Press.

Heimann, P.M. and J.E. McGuire. 1971a. 'Cavendish and the *vis viva* controversy: a Leibnizian postscript'. *Isis 62*: 225–7.

——. 1971b. 'Newtonian forces and Lockean powers: concepts of matter in eighteenth-century thought'. *HSPS 3*: 233–306.

Helmholtz, Hermann. 1853 [1847]. 'On the conservation of force; a physical memoir'. In John Tyndall and William Francis, eds, *Scientific Memoirs. Natural Philosophy*, pp 114–62. London: Taylor and Francis.

——. 1856 [1854]. 'On the interaction of natural forces'. *Phil. Mag. 11*: 489–518.

——. 1861. 'On the application of the law of the conservation of force to organic nature'. *Proc. RI 3*: 347–57.

——. 1872. 'On the theory of electrodynamics'. *Phil. Mag. 44*: 530–37.

——. 1873. *Popular Lectures on Scientific Subjects*. Trans. E. Atkinson. London: Longmans, Green.

——. 1881. *Popular Lectures on Scientific Subjects*. Second series. Trans. E. Atkinson. London: Longmans, Green.

——. 1889. *Über die Erhaltung der Kraft* (Ostwald's *Klassiker der exakten Wissenschaften Nr. 1*). Leipzig: Engelmann.

——. 1908–12. *Popular Lectures on Scientific Subjects*. Trans. E. Atkinson. New edn. 2 vols. London: Longmans, Green.

Hempstead, C.A. 1991. 'An appraisal of Fleeming Jenkin (1833–1885), Electrical Engineer'. *Hist. Tech. 13*: 119–44.

Herivel, John. 1972–3. 'The influence of Fourier on British mathematics'. *Centaurus 17*: 40–57.

——. 1975. *Joseph Fourier. The Man and the Physicist*. Oxford: Clarendon Press.

Herschel, J.F.W. 1865. 'On the origin of force'. *Fortnightly Review 1*: 435–42.

Hertzman, Rudolph. 1845. 'Thoughts of a silent man', *The Broadway Journal 1*: 163–64.

Hevly, Bruce. 1996. 'The heroic science of glacier motion'. *Osiris 11*: 66–86.

Hiebert, E.N. 1959. 'Commentary on the papers of Thomas S. Kuhn and I. Bernhard Cohen'. In M. Clagett, ed., *Critical Problems in the History of Science*, pp 391–400. Madison: Wisconsin University Press.

——. 1962. *Historical Roots of the Principle of Conservation of Energy*. Madison: University of Wisconsin Department of History.

——. 1971. 'The energetics controversy and the new thermodynamics'. In D.H.D. Roller, ed., *Perspectives in the History of Science and Technology*, 67–86. Norman: University of Oklahoma Press.

——. 1973. 'Ernst Mach'. *DSB 8*: 595–607.

——. 1978. 'Ostwald, Friedrich Wilhelm'. *DSB 15*: 455–69.

Hills, R.L. 1989. *Power from Steam. A History of the Stationary Steam Engine*. Cambridge: Cambridge University Press.

Hills, R.L. and A.J. Pacey. 1972. 'The measurement of power in early steam-driven textile mills'. *Tech. Cult. 13*: 39–43.

Hilton, Boyd. 1991. *The Age of Atonement. The Influence of Evangelicalism on Social and Economic Thought 1785–1865*. Oxford: Clarendon Press. First published 1988.

Hirn, G.-A. 1867. *Mémoire sur la thermodynamique*. Paris: Gauthiers-Villars.

Hodgkinson, Eaton. 1840. 'On the temperature of the Earth in the deep mines in the neighbourhood of Manchester'. *BAAS Rep. 10*: 17.

——. 1846. 'Some account of the late Mr Ewart's paper on the measure of moving force; and of the recent applications of the principle of living forces to estimate the effects of machines and movers'. *Mem. Man. Lit. & Phil. 7*: 137–56.

Hoeveler, J.D. 1981. *James McCosh and the Scottish Intellectual Tradition. From Glasgow to Princeton*. Princeton: Princeton University Press.

Holmes, F.L. 1964. 'Introduction'. In F.L. Holmes, ed., *Animal Chemistry, or Organic Chemistry in its Application to Physiology and Pathology*, pp vii–cxvi. New York: Johnson Repr. Corp.

——. 1973. 'Justus von Liebig'. *DSB 8*: 329–50.

——. 1989. 'The complementarity of teaching and research in Liebig's laboratory'. *Osiris 5*: 121–64.

Holt, N.R. 1970. 'A Note on Wilhelm Ostwald's "Energism"'. *Isis 61*: 386–9.

——. 1977. 'Wilhelm Ostwald's "The Bridge"'. *BJHS 10*: 146–50.

Holtzmann, K.H.A. 1846 [1845]. 'On the heat and elasticity of gases and vapours'. In Richard Taylor, ed., *Scientific Memoirs, Selected from the Transactions of Foreign Academies of Science and Learned Societies, and from Foreign Journals*, vol. 4 pp 189–217. London: Taylor and Francis.

Home, R.W. 1968. 'The third law of motion in Newton's mechanics'. *BJHS 6*: 39–51.

——. 1983. 'Poisson's memoirs on electricity: academic politics and a new style in physics'. *BJHS 16*: 239–59.

Hopkins, William. 1835. 'Researches in physical geology'. *Transactions of the Cambridge Philosophical Society 6*: 1–84.

——. 1836. 'An abstract of a memoir on physical geology, with a further exposition of certain points connected with the subject'. *Phil. Mag. 8*: 227–36, 272–81, 357–66.

——. 1839. 'Researches in physical geology – first series'. *Phil. Trans.* 381–423.

——. 1847. 'Report on the geological theories of elevation and earthquakes'. *BAAS Rep. 17*: 33–92.

——. 1853. '[Presidential] address'. *BAAS Rep. 23*: xli–lvii.

Hopley, I.P. 1957. 'Maxwell's work on electrical resistance. I. The determination of the absolute unit of resistance'. *Ann. Sci. 13*: 265–72.

——. 1958. 'Maxwell's work on electrical resistance. II. Proposals for the re-determination of the B.A. unit of 1863'. *Ann. Sci. 14*: 197–210.

——. 1959. 'Maxwell's determination of the number of electrostatic units in one electro-magnetic unit of electricity'. *Ann. Sci. 15*: 91–108.

Hoyer, Ulrich. 1975. 'How did Carnot calculate the mechanical equivalent of heat?' *Centaurus 19*: 207–19.

Hunt, B.J. 1986. 'Experimenting on the ether: Oliver J. Lodge and the great whirling machine'. *HSPBS 16*(i): 111–34.

——. 1987. '"How my model was right": G.F. FitzGerald and the reform of Maxwell's theory'. In Robert Kargon and Peter Achinstein, eds, *Kelvin's Baltimore Lectures and Modern Theoretical Physics*, pp 299–321. Cambridge, MA: MIT Press.

——. 1988. 'The origins of the FitzGerald contraction'. *BJHS 21*: 67–76.

——. 1991a. 'Michael Faraday, cable telegraphy and the rise of field theory'. *Hist. Tech. 13*: 1–19.

——. 1991b. *The Maxwellians*. Ithaca and London: Cornell University Press.

——. 1994. 'The ohm is where the art is: British telegraph engineers and the development of electrical standards'. *Osiris 9*: 48–63.

——. 1996. 'Scientists, engineers and Wildman Whitehouse: measurement and credibility in early cable telegraphy'. *BJHS 29*: 155–69.

Hutchison, Keith. 1973. 'Der Ursprung der Entropiefunktion bei Rankine and Clausius'. *Ann. Sci. 30*: 341–64.

——. 1977. 'Mayer's hypothesis: a study of the early years of thermodynamics'. *Centaurus 20*: 279–304.

——. 1981. 'W.J.M. Rankine and the rise of thermodynamics'. *BJHS 14*: 1–26.

Iltis, Carolyn. 1970. 'D'Alembert and the *vis viva* controversy'. *SHPS 1*: 135–44.

——. 1971. 'Leibniz and the *vis viva* controversy'. *Isis 62*: 21–35.

——. 1973a. 'The decline of Cartesianism in mechanics: the Leibnizian–Cartesian debates'. *Isis 64*: 356–73.

——. 1973b. 'The Leibnizian-Newtonian debates: natural philosophy and social psychology'. *BJHS 6*: 343–77.

——. 1977. 'Madame du Châtelet's metaphysics and mechanics'. *SHPS 8*: 29–48.

Jacobi, M.H. 1837. 'On the application of electro-magnetism to the movement of machines'. In Richard Taylor, ed., *Scientific Memoirs, Selected from the Transactions of Foreign Academies of Science and Learned Societies, and from Foreign Journals*, vol. 1, pp 503–31. London: Taylor and Francis.

James, F.A.J.L. 1982. 'Thermodynamics and sources of solar heat, 1846–1862'. *BJHS 15*: 157–81.

——. 1983. 'The conservation of energy, theories of absorption and resonating molecules, 1851–1854: G.G. Stokes, A.J. Angstrom and W. Thomson'. *Notes and Records RSL. 38*: 79–107.

——. 1984. 'The physical interpretation of the wave theory of light'. *BJHS 17*: 47–60.

——. 1985. '"The optical mode of investigation": light and matter in Faraday's natural philosophy'. In David Gooding and Frank James (eds), *Faraday Rediscovered. Essays on the Life and Work of Michael Faraday, 1791–1867*, pp 137–61. Basingstoke: Macmillan.

——. 1997. 'Faraday in the pits, Faraday at sea: the role of the Royal Institution in changing the practice of science and technology in nineteenth-century Britain'. *Proc. RI 68*: 277–301.

Jenkin, Fleeming. 1866. 'Reply to Dr Werner Siemens's paper "On the question of the unit of electrical resistance". *Phil. Mag. 32*: 161–77.

——. 1887. *Papers Literary, Scientific, &c. by the Late Fleeming Jenkin.* 2 vols. Ed. Sidney Colvin and J.A. Ewing. London: Longmans, Green.

Jensen, J.V. 1970. 'The X Club: fraternity of Victorian scientists'. *BJHS 5*: 63–72.

——. 1981. 'Tyndall's role in the "X Club"'. In Brock, W.H., N.D. McMillan and R.C. Mollan (eds). *John Tyndall. Essays on a Natural Philosopher*, pp 157–68. Dublin: Royal Dublin Society.

Jones, Gordon. 1969. 'Joule's early researches'. *Centaurus. 13*: 198–219.

Joule, J.P. 1838. 'Description of an electro-magnetic Engine'. *Ann. Electricity 2*: 122–3.

——. 1838–39. 'Description of an electro-magnetic engine, with experiments'. *Ann. Electricity. 3*: 437–9.

——. 1839–40a. 'Investigations in magnetism and electro-magnetism'. *Ann. Electricity 4*: 131–5.

——. 1839–40b. 'On electro-magnetic forces'. *Ann. Electricity 4*: 474–81.

——. 1840. 'On electro-magnetic forces'. *Ann. Electricity 5*: 187–98, 470–2.

——. 1842a. 'On a new class of magnetic forces'. *Ann. Electricity 8*: 219–24.

——. 1842b. 'On the heat evolved by metallic conductors of electricity, and in the cells of a battery during electrolysis'. *Ann. Electricity 8*: 287–301.

——. 1842c. 'On the electric origin of the heat of combustion'. *Ann. Electricity 8*: 302–15.

——. 1843a. 'On the electric origin of chemical heat'. *Phil. Mag. 22*: 204–8.

——. 1843b. 'On the calorific effects of magneto-electricity, and on the mechanical value of heat'. *Phil. Mag. 23*: 263–76, 347–55, 435–43.

——. 1845a. 'On the changes of temperature produced by the rarefaction and condensation of air'. *Phil. Mag. 26*: 369–83.

——. 1845b. 'On the mechanical equivalent of heat'. *BAAS Rep. 15*: 31.

——. 1845c. 'On the existence of an equivalent relation between heat and the ordinary forms of mechanical power'. *Phil. Mag. 27*: 205–7.

——. 1846. 'On the heat evolved during the electrolysis of water'. *Mem. Man. Lit. & Phil. 7*: 87–112.

——. 1849. '*Sur l'équivalent mécanique du calorique*'. *Comptes rendus 28*: 132–35.

——. 1850. 'On the mechanical equivalent of heat'. *Phil. Trans*: 61–82.

——. 1853. 'On the œconomical production of mechanical effect from chemical forces'. *Phil. Mag. 5*: 1–5.

——. 1857 [1851]. 'Some remarks on heat, and the constitution of elastic fluids'. *Phil. Mag. 14*: 211–16.

——. 1862. 'Note on the history of the dynamical theory of heat'. *Phil. Mag. 24*: 121–3.

——. 1863. 'On the dynamical theory of heat'. *Phil. Mag. 26*: 145–7.

——. 1864. 'Note on the history of the dynamical theory of heat'. *Phil. Mag. 28*: 150–2.

——. 1887. *The Scientific Papers of James Prescott Joule.* 2 vols. London: The Physical Society. Repr. London: Dawsons, 1963.

Joyce, Patrick. 1987. 'The historical meanings of work: an introduction'. In Patrick Joyce, ed., *The Historical Meanings of Work*, pp 1–30. Cambridge: Cambridge University Press.

Jungnickel, Christa and Russell McCormmach. 1986. *Intellectual Mastery of Nature. Theoretical Physics from Ohm to Einstein.* 2 vols. Chicago: University of Chicago Press.

Kaiser, Walter. 1993. 'Helmholtz's instrumental role in the formation of classical electrodynamics'. In David Cahan, ed., *Hermann von Helmholtz and the Foundations of Nineteenth-century Science*, pp 374–402. Berkeley, Los Angeles and London: University of California Press.

Kangro, Hans. 1975. 'Planck, Max Karl Ernst Ludwig'. *DSB 11*: 7–17.

Kargon, R.H. 1977. *Science in Victorian Manchester. Enterprise and Expertise*. Manchester: Manchester University Press.

Kelland, Philip. 1840. 'On the conduction of heat'. *BAAS Rep. 10*: 15–16.

——. 1841. 'On the present state of our theoretical and experimental knowledge of the laws of conduction of heat'. *BAAS Rep. 11*: 1–25.

——. 1863. 'On the conservation of energy'. *Phil. Mag. 26*: 326.

Kent, Andrew. 1964. 'Thomas Thomson (1773–1852). Historian of chemistry'. *BJHS 2*: 59–63.

Kerker, Milton. 1960. 'Sadi Carnot and the steam engine engineers'. *Isis 51*: 257–70.

——. 1961. 'Science and the steam engine'. *Tech. Cult. 2*: 381–90.

Kidwell, P.A. 1981. 'Prelude to solar energy: Pouillet, Herschel, Forbes and the solar constant'. *Ann. Sci. 38*: 457–76.

King, David. 1850. *The Principles of Geology Explained, and Viewed in their Relations to Revealed and Natural Religion*. London: Johnstone and Hunter.

King, Elizabeth. 1909. *Lord Kelvin's Early Home*. London: Macmillan.

Klein, M.J. 1962. 'Max Planck and the beginnings of the quantum theory'. *Arch. Hist. Ex. Sci. 1*: 459–79.

——. 1969. 'Gibbs on Clausius'. *HSPS 1*: 127–49.

——. 1970. 'Maxwell, his demon, and the second law of thermodynamics'. *American Scientist 58*: 84–97.

——. 1972. 'Gibbs, Josiah Willard'. *DSB 5*: 386–93.

——. 1972–73. 'Mechanical explanation at the end of the nineteenth century'. *Centaurus 17*: 58–82.

——. 1974. 'Closing the Carnot cycle'. In *Table ronde du Centre National de la Recherche Scientifique. Sadi Carnot et l'essor de la thermodynamique. Paris, Ecole Polytechnique, 11–13 juin 1974*, pp 213–19. Paris: Editions du Centre de la Recherche Scientifique.

Knott, C.G. 1911. *Life and Scientific Work of Peter Guthrie Tait*. Cambridge: Cambridge University Press.

Knudsen, Ole. 1971–2. 'From Lord Kelvin's Notebook: Ether speculations'. *Centaurus 16*: 41–53.

——. 1976. 'The Faraday effect and physical theory, 1845–1873'. *Arch. Hist. Ex. Sci. 15*: 235–81.

Koenigsberger, Leo. 1906. *Hermann von Helmholtz*. Trans. F.A. Welby. Oxford: Clarendon. Repr. New York: Dover, 1965.

Kragh, Helge. 1984. 'Julius Thomsen and classical thermochemistry'. *BJHS 17*: 255–72.

Krips, H. 1986. 'Atomism, Poincaré and Planck'. *SHPS 17*: 43–63.

Kuhn, T.S. 1959. 'Energy conservation as an example of simultaneous discovery'. In M. Clagett, ed., *Critical Problems in the History of Science*, pp 321–56. Madison: Wisconsin University Press. Repr. T.S. Kuhn, ed., *The Essential Tension*, pp. 66–104. Chicago: Chicago University Press.

——. 1961a. 'The function of measurement in modern physical science'. *Isis 52*: 161–93.

——. 1961b. 'Sadi Carnot and the Cagnard Engine'. *Isis 52*: 567–74.

——. 1962a. *The Structure of Scientific Revolutions*. Chicago: Chicago University Press.

——. 1962b. 'The historical structure of scientific discovery'. *Science 136*: 760–4.

——. 1984. 'Revisiting Planck'. *HSPS 14*(ii): 231–52.

Lagrange, J.-L. 1888–9 [1788]. *Mécanique analytique*. In *Oeuvres de Lagrange*, vols.11–12. 14 vols. Paris: Gauthier-Villars.

Laidler, K.J. 1985. 'Chemical kinetics and the origins of physical chemistry'. *Arch. Hist. Ex. Sci. 32*: 43–75.

Landes, D.S. 1969. *The Unbound Prometheus. Technological Change and Industrial Development in Western Europe from 1750 to the Present*. Cambridge: Cambridge University Press.

Laplace, P.S. de. 1798–1827. *Traité de mécanique céleste*. 5 vols. Paris: Duprat.

——. 1878–1912. *Oeuvres completes de Laplace*. 14 vols. Paris: Gauthiers-Villars.

Lardner, Dionysius. 1833. *A Treatise on Heat*. London: Longman.

Larmor, Joseph. 1900. *Aether and Matter. A Development of the Dynamical Relations of the Aether to Material Systems on the Basis of the Atomic Constitution of Matter Including a Discussion of the Earth's Motion on Optical Phenomena*. Cambridge: Cambridge University Press.

——. 1908. 'Lord Kelvin'. *Proc. Roy. Soc. 81*: iii–lxxvi.

——. 1912. 'Biographical sketch [of James Thomson]'. In Sir Joseph Larmor and James Thomson, eds., *Collected Papers in Science and Engineering*, pp xiii–xci. Cambridge: Cambridge University Press.

Latour, Bruno and Steve Woolgar. 1986. *Laboratory Life. The Construction of Scientific Facts*. Princeton: Princeton University Press. First published 1979.

Laudan, L.L. 1968. 'The *vis viva* controversy, a post-mortem'. *Isis 59*: 131–43.

Lavoisier, A.-L. 1965 [1790]. *Elements of Chemistry. In a New Systematic Order, Containing All the Modern Discoveries*. Trans. Robert Kerr. New York: Dover.

Lawrence, Phillip. 1978. 'Charles Lyell versus the theory of central heat'. *Journal of the History of Biology 11*: 101– 28.

Leegwater, Arie. 1986. 'The development of Wilhelm Ostwald's chemical energetics'. *Centaurus 29*: 314–7.

Lefebvre, Henri. 1991. *The Production of Space*. Trans. Donald Nicholson-Smith. Oxford: Blackwell. First published 1974.

Lenoir, Timothy. 1982. *The Strategy of Life. Teleology and Mechanics in Nineteenth Century German Biology*. Dordrecht: D. Reidel.

——. 1986. 'Models and instruments in the development of electrophysiology, 1845–1912'. *HSPBS 17*(i): 1–54.

Lervig, Philip. 1972. 'On the structure of Carnot's theory of heat'. *Arch. Hist. Ex. Sci. 9*: 222–39.

——. 1982–83. 'What is heat? C. Truesdell's view of thermodynamics. A Critical Discussion'. *Centaurus 26*: 85–122.

——. 1985. 'Sadi Carnot and the steam engine: Nicolas Clément's lectures on industrial chemistry, 1823–28'. *BJHS 18*: 147–96.

Liebig, Justus von. 1842. *Animal Chemistry, or Organic Chemistry in Its Application to Physiology and Pathology*. Cambridge: John Owen. Repr. New York: Johnson Repr. Corp., 1964.

Lindsay, R.B. 1973. *Julius Robert Mayer. Prophet of Energy*. Oxford: Pergamon Press.

——. 1975. *Energy: Historical Development of the Concept*. Stroudsburg, Pennsylvania: Dowden, Hutchinson and Ross.

Lingwood, J.E. 1977. 'The steam conquistadores. A history of the Pacific Steam Navigation Company'. *Sea Breezes 51*: 97–115.

Lipman, T.O. 1967. 'Vitalism and reductionism in Liebig's physiological thought'. *Isis 58*: 167–85.

Lloyd, J.T. 1970. 'Background to the Joule–Mayer Controversy'. *Notes and Records RSL. 25*: 211–25.

Lodge, O.J. 1879. 'An attempt at a systematic classification of the various forms of energy'. *Phil. Mag. 8*: 277–86.

———. 1881a. 'On action at a distance'. *Phil. Mag. 11*: 36–7, 220.

———. 1881b. 'On action at a distance, and the conservation of energy'. *Phil. Mag. 11*: 529–34.

———. 1885a. 'On the identity of energy: in connection with Mr Poynting's paper on the transfer of energy in an electromagnetic field; and on the two fundamental forms of energy'. *Phil. Mag. 19*: 482–94.

———. 1885b. *Elementary Mechanics Including Hydrostatics and Pneumatics*. Revised ed. London and Edinburgh: W.&R. Chambers.

Lushington, Edmund. 1870. 'General introductory address'. In *Introductory Addresses Delivered at the Opening of the University of Glasgow Session 1870–71*. Glasgow: Edinburgh and London: Blackwood.

McCormmach, Russell. 1969. 'Einstein, Lorentz, and the electron theory'. *HSPS 2*: 41–87.

———. 1970. 'H.A. Lorentz and the electromagnetic view of nature'. *Isis 61*: 458–97.

McFarlane, P. 1852. *Exposure of the Principles of Modern Geology, in a Review of King's "Geology and Religion". Wherein is Shown the Utter Incompatibility of these principles with Both Science and Sacred History*. Edinburgh: Grant.

McGuire, J.E. 1968. 'The origin of Newton's doctrine of essential qualities'. *Centaurus 12*: 233–60.

———. 1972. 'Boyle's conception of nature'. *JHI 33*: 523–42.

Mach, Ernst. 1893 [1883]. *The Science of Mechanics: a Critical and Historical Account of its Development*. Trans. T.J. McCormack. La Salle, Illinois: Open Court. First German ed. 1883.

———. 1911 [1909]. *History and Root of the Principle of the Conservation of Energy*. Trans. from 2nd ed. P.E.B. Jourdain. Chicago: Open Court. First ed. Prague, 1872.

MacLeod, Christine. 1988. *Inventing the Industrial Revolution. The English Patent System, 1660–1800*. Cambridge: Cambridge University Press.

Macleod, Donald. 1876. *Memoir of Norman Macleod, D.D.* 2 vols. London: Daldy, Isbister.

Macleod, Donald. 1996. 'Hugh Miller, the Disruption and the Free Church of Scotland'. In Michael Shortland, ed., *Hugh Miller and the Controversies of Victorian Science*, pp 187–205. Oxford: Oxford University Press.

MacLeod, R.M. 1970. 'The X-Club: a social network of science in late-Victorian England'. *Notes and records RSL 24*: 305–22.

Marsden, Ben. 1992a. 'Engineering science in Glasgow. W.J.M. Rankine and the motive power of air'. PhD diss., University of Kent at Canterbury.

———. 1992b. 'Engineering science in Glasgow: economy, efficiency and measurement as prime movers in the differentiation of an academic discipline'. *BJHS 25*: 319–46.

———. 1993. Review of Siegel 1991. *BJHS 26*: 116–17.

———. 1998. '"A most important trespass": Lewis Gordon and the Glasgow Chair of Civil Engineering and Mechanics 1840–1855'. In Jon Agar and Crosbie Smith, *Making Space for Science. Territorial Themes in the Shaping of Knowledge*, pp 87–117. Basingstoke: Macmillan.

Mathias, Peter. 1959. *The Brewing Industry in England 1700–1830*. Cambridge: Cambridge University Press.

——. 1969. *The First Industrial Nation. An Economic History of Britain 1700–1914*. London: Methuen.

Matthiessen, Augustus. 1861. 'Some remarks on Dr Siemens's paper "On standards of electrical resistance, and on the influence of temperature on the resistance of metals"'. *Phil. Mag. 22*: 195–202.

Mauss, Marcel. 1990. *The Gift. The Form and Reason for Exchange in Archaic Societies*. Trans. W.D. Halls. London: Routledge. First published 1950.

Maxwell, J.C. 1854. 'On a particular case of the descent of a heavy body in a resisting medium'. *CDMJ 9*: 145–8.

——. 1856a. 'On Faraday's lines of force' (abstract). *Proc. Camb. Phil. Soc. 1*: 160–6.

——. 1856b. 'On Faraday's lines of force'. *Trans. Camb. Phil. Soc. 10*: 27–83.

——. 1860. 'Illustrations of the dynamical theory of gases'. *Phil. Mag. 19*: 19–32; *20*: 21–37.

——. 1861–62. 'On physical lines of force'. *Phil. Mag. 21*: 161–75, 281–91, 338–48; *23*: 12–24, 85–95.

——. 1865a. 'A dynamical theory of the electromagnetic field'. *Phil. Mag. 29*: 152–7.

——. 1865b. 'A dynamical theory of the electromagnetic field'. *Phil. Trans. 155*: 459–512.

——. 1868 [1867]. 'On the dynamical theory of gases'. *Phil. Mag. 35*: 129–45, 185–217.

——. 1871. *Theory of Heat*. London: Longmans, Green.

——. 1873. *A Treatise on Electricity and Magnetism*. 2 vols. Oxford: Clarendon.

——. 1874. 'Grove's "Correlation of Physical Forces"'. *Nature 10*: 302–4.

——. 1875. *Theory of Heat*. 2nd edn. London: Longmans, Green.

——. 1876–77. 'Hermann Ludwig Ferdinand Helmholtz'. *Nature 15*: 389–91.

——. 1877. *Matter and Motion*. London: Society for Promoting Christian Knowledge. Repr. London: Sheldon Press, 1925.

——. 1878. 'Tait's "Thermodynamics"'. *Nature 17*: 257–9, 278–80.

——. 1879. 'Thomson and Tait's Natural Philosophy'. *Nature 20*: 213–16.

——. 1881. *An Elementary Treatise on Electricity*. Ed. William Garnett. Oxford: Clarendon.

——. 1891. *A Treatise on Electricity and Magnetism*. 2 vols. 3rd edn. Oxford: Clarendon.

——. 1890. *The Scientific Papers of James Clerk Maxwell*. 2 vols. Ed. W.D. Niven. Cambridge: Cambridge University Press. Repr. New York: Dover, 1965.

——. 1908. 'On the equilibrium of heterogeneous substances (1876)'. *Phil. Mag. 16*: 818–24.

——. 1937. *Origins of Clerk Maxwell's Electric Ideas as Described in Familiar Letters to William Thomson*. Ed. Sir Joseph Larmor. Cambridge: Cambridge University Press.

——. 1954 [1891]. *A Treatise on Electricity and Magnetism*. 2 vols. 3rd edn. New York: Dover.

——. 1990-. *The Scientific Letters and Papers of James Clerk Maxwell*. 3 vols. Ed. P.M. Harman. Cambridge: Cambridge University Press.

Maxwell, J.C., and Fleeming Jenkin. 1865 [1863]. 'On the elementary relations between electrical measurements'. *Phil. Mag. 29*: 436–60, 507–25.

Mayer, J.R. 1848. '*Sur la transformation de la force vive en chaleur, et réciproquement*'. *Comptes rendu 27*: 385–87.

——. 1862 [1842]. 'Remarks on the forces of inorganic Nature'. *Phil. Mag. 24*: 371–7.

——. 1863 [1848]. 'On celestial dynamics'. *Phil. Mag. 25*: 241–8, 387–409, 417–28.

——. 1893a. *Die Mechanik der Wärme in gesammelten Schriften.* Ed. J.J. Weyrauch. Stuttgart: Cotta.

——. 1893b. *Kleinere Schriften und Briefe von Robert Mayer, nebst Mittheilungen aus seinem Leben.* Ed. J.J. Weyrauch. Stuttgart: Cotta.

Mendoza, Eric. 1975. 'A critical examination of Herapath's dynamical theory of gases'. *BJHS 8*: 155–65.

Mendoza, Eric and D.S.L. Cardwell. 1981. 'Note on a suggestion concerning the work of J.P. Joule'. *BJHS 14*: 177–80.

Merz, J.T. 1904–12. *A History of European Thought in the Nineteenth Century.* 4 vols. Edinburgh and London: Blackwood. Repr. New York: Dover, 1965.

Meyerson, Emile. 1930 [1908]. *Identity and Reality.* Trans. Kate Loewenberg. London: Allen and Unwin.

Middleton, W.E.K. 1966. *A History of the Thermometer and Its Use in Meteorology.* Baltimore, MD: Johns Hopkins Press.

Miller, J.D. 1972. 'Rowland and the nature of electric currents'. *Isis 63*: 5–27.

Millhauser, Milton. 1954. 'The scriptural geologists. An episode in the history of opinion'. *Osiris 11*: 65–86.

Mirowski, Philip. 1989. *More Heat than Light. Economics as Social Physics, Physics as Nature's Economics.* Cambridge: Cambridge University Press.

——. 1992. 'Looking for those natural numbers: dimensionless constants and the idea of natural measurement'. *Science in Context 5*: 165–88.

Mohr, F. 1876 [1837]. 'Views of the nature of heat'. *Phil. Mag. 2*: 110–14.

Moore, J.R. 1979. *The Post-Darwinian Controversies. A Study of the Protestant Struggle to Come to Terms With Darwin in Great Britain and America 1870–1900.* Cambridge: Cambridge University Press.

Morrell, Jack. 1969. 'Thomas Thomson: professor of chemistry and university reformer'. *BJHS 4*: 245–65.

——. 1972a. 'Science and Scottish university reform: Edinburgh in 1826'. *BJHS 6*: 39–56.

——. 1972b. 'The chemist breeders: the research schools of Liebig and Thomas Thomson'. *Ambix 19*: 1–46.

——. 1974. 'Reflections on the history of Scottish Science'. *Hist. Sci. 12*: 81–94.

——. 1975. 'The Leslie affair: careers, kirk, and politics in Edinburgh in 1805'. *Scottish Historical Review 54*: 63–82.

Morrell, Jack and Arnold Thackray. 1981. *Gentlemen of Science. Early Years of the British Association for the Advancement of Science.* Oxford: Clarendon.

Morris, R.J. 1972. 'Lavoisier and the caloric theory'. *BJHS 6*: 1–38.

Morus, I.R. 1989. 'The politics of power. Reform and regulation in the work of William Robert Grove'. PhD diss., University of Cambridge.

——. 1991a. 'Telegraphy and the technology of display. The electricians and Samuel Morse'. *Hist. Tech. 13*: 20–40.

——. 1991b. 'Correlation and control: William Robert Grove and the construction of a new philosophy of scientific reform'. *SHPS 22*: 589–621.

——. 1992. 'Different experimental lives: Michael Faraday and William Sturgeon'. *Hist. Sci. 30*: 1–28.

——. 1993. 'Currents from the underworld: electricity and the technology of display in early Victorian England'. *Isis 84*: 50–69.

Motz, Lloyd and Weaver, J.H. 1989. *The Story of Physics.* New York and London: Plenum Press.

Moyer, D.F. 1977. 'Energy, dynamics, hidden machinery: Rankine, Thomson and Tait, Maxwell'. *SHPS 8*: 251–68.

———. 1978. 'Continuum mechanics and field theory: Thomson and Maxwell'. *SHPS 9*: 35–50.

Muirhead, J.P. 1858. *The Life of James Watt, with Selections from his Correspondence.* London: Murray.

Murchison, Roderick, and Edward Sabine. 1840. 'Address'. *BAAS Rep. 10*: xxxv–xlviii.

Murphy, J.J. 1869. *Habit and Intelligence, in their Connexion with the Laws of Matter and Force: a Series of Scientific Essays.* 2 vols. London: Macmillan.

———. 1873. *The Scientific Bases of Faith.* London: Macmillan.

Musson, A.E. and Eric Robinson. 1969. *Science and Technology in the Industrial Revolution.* Manchester: Manchester University Press.

Myers, Greg. 1985–86. 'Nineteenth-century popularizations of thermodynamics and the rhetoric of social prophecy'. *Victorian Studies 29*: 35–66.

Napier, James. *Life of Robert Napier of West Shandon.* Edinburgh and London: Blackwood.

Navier, C.L.M.H. 1819. *Bélidor's Architecture hydraulique.* Paris: Didot.

Needham, N.R. 1993. 'Ramsay, Edward Bannerman (1793–1872)'. In N.M. de S. Cameron, ed., *Dictionary of Scottish Church History and Theology*, p 691. Edinburgh: T.&T. Clark.

Newton, Isaac. 1934 [1687]. *Principia.* 2 vols. Trans. Andrew Motte and Florian Cajori. Berkeley, Los Angeles and London: University of California Press.

———. 1952 [1730]. *Opticks. Or a Treatise of the Reflections, Refractions, Inflections & Colours of Light.* Based on the 4th edn. of 1730. New York: Dover.

Nye, M.J. 1976. 'The nineteenth-century atomic debates and the dilemma of an "Indifferent Hypothesis"'. *SHPS 7*: 245–68.

———. 1981. 'Berthelot's anti-atomism: a matter of taste?' *Ann. Sci. 38*: 585–90.

Oakley, Francis. 1961. 'Christian theology and Newtonian science'. *Church History 30*: 433–57.

Olesko, K.M. and F.L. Holmes. 1993. 'Experiment, quantification, and discovery: Helmholtz's early physiological researches'. In David Cahan, ed., *Hermann von Helmholtz and the Foundations of Nineteenth-century Science*', pp 50–108. Berkeley, Los Angeles and London: University of California Press.

Olson, R.G. 1970. 'Count Rumford, Sir John Leslie, and the study of the nature and propagation of heat at the beginning of the nineteenth century'. *Ann. Sci. 26*: 273–304.

———. 1975. *Scottish Philosophy and British Physics 1750–1880. A Study in the Foundations of the Victorian Scientific Style.* Princeton: Princeton University Press.

Ostwald, Wilhelm. 1911. *Natural Philosophy.* Trans. Thomas Seltzer. London: Williams and Norgate.

———. 1912. *Der Energetische Imperativ.* Leipzig: Akademische Verlagsgesellschaft.

Owen, Richard. 1894. *The Life of Richard Owen by his Grandson.* 2 vols. London: Murray.

Pacey, A.J. 1974. 'Some early heat engine concepts and conservation of heat'. *BJHS 7*: 135–45.

Paley, William. 1823. *The Works of William Paley, D.D.* 5 vols. London: Thomas Tegg.

Papineau, David. 1977. 'The *vis viva* controversy: Do meanings matter?' *SHPS 8*: 111–42.

Pearson, Karl. 1936. 'Old tripos days at Cambridge'. *Mathematical Gazette 20*: 27–36.

Planck, Max. 1903 [1897]. *Treatise on Thermodynamics*. Trans. Alexander Ogg. London: Longmans, Green.

——. 1914 [1913]. 'New paths of physical knowledge: being the address delivered on commencing the rectorate of the Friedrich-Wilhelm University, Berlin, on October 15th 1913'. *Phil. Mag. 28*: 60–71.

——. 1925. *A Survey of Physics. A Collection of Lectures and Essays*. Trans. R. Jones and D.H. Williams. London: Methuen.

Playfair, John. 1802. *Illustrations of the Huttonian Theory of the Earth*. Edinburgh: Creech.

[Playfair, John]. 1808. 'Wollaston's Bakerian Lecture'. *Edinburgh Review 12*: 120–30.

Poncelet, J.V. 1829. *Cours de mécanique industrielle, fait aux artistes et ouvriers messins, pendant les hivers de 1827 à 1828, et de 1828 à 1829*. Metz: Tavernier.

Porter, T.M. 1981. 'A statistical survey of gases: Maxwell's social physics'. *HSPS 12*(i): 77–116.

——. 1986. *The Rise of Statistical Thinking, 1820–1900*. Princeton: Princeton University Press.

Poynting, J.H. 1920. *Collected Scientific Papers*. Cambridge: Cambridge University Press.

Preston, S.T. 1881a. 'On action at a distance'. *Phil. Mag. 11*: 38–9, 218–20.

——. 1881b. 'On the importance of experiments in relation to the mechanical theory of gravitation'. *Phil. Mag. 11*: 391–3.

Rabinbach, Anson. 1990. *The Human Motor. Energy, Fatigue, and the Origins of Modernity*. Berkeley and Los Angeles: University of California Press.

Rankine, W.J.M. 1840. 'On the laws of the conduction of heat and on their application to some geothermal problems'. In *The Edinburgh Academic Journal for MDCCCXL, Consisting of Contributions in Literature and Science by Alumni of the University of Edinburgh*, pp 48–65. Edinburgh: Black.

——. 1850. 'Abstract of a paper on the hypothesis of molecular vortices, and its application to the mechanical theory of heat'. *Proc. RSE. 2*: 275–88.

——. 1851a. 'On the centrifugal theory of elasticity, as applied to gases and vapours'. *Phil. Mag. 2*: 509–42.

——. 1851b. 'On the economy of single-acting expansive steam engines, and expansive machines generally; being supplements to a paper on the mechanical action of heat'. *Proc. RSE 3*: 60–4.

——. 1852a. 'On the reconcentration of the mechanical energy of the universe'. *BAAS Rep. 22*: 12.

——. 1852b. 'On the reconcentration of the mechanical energy of the universe'. *Phil. Mag. 4*: 358–60.

——. 1853a. 'On the general law of the transformation of energy'. *Phil. Mag. 5*: 106–17.

——. 1853b. 'On the mechanical action of heat, especially in gases and vapours'. *Trans. RSE. 20*: 147–90; *Phil. Mag. 7*: 1–21, 111–22, 172–85.

——. 1853c. On the mechanical effect of heat and of chemical forces'. *Phil. Mag. 5*: 6–9.

——. 1854. 'On the geometrical representation of the expansive action of heat, and the theory of thermo-dynamic engines'. *Phil. Trans*: 115–76.

——. 1855a. 'On the means of realizing the advantages of the air-engine'. *Edinburgh New Phil. J. 1855*: 1–32.

——. 1855b. 'Outlines of the science of energetics'. *Edinburgh New Phil. J. 2*: 120–41.

——. 1857. 'Heat, theory of the mechanical action of, or thermo-dynamics'. In J.P. Nichol, ed., *Cyclopaedia of the Physical Sciences*, pp 338–54.

——. 1859a. *A Manual of the Steam Engine and Other Prime Movers*. London and Glasgow: Griffin.

——. 1859b. 'On the conservation of energy'. *Phil. Mag. 17*: 250–3; 'Note to a letter "On the Conservation of Energy"'. *Phil. Mag. 17*: 347–8.

——. 1864a. 'On the hypothesis of molecular vortices'. *Phil. Mag. 27*: 313.

——. 1864b. 'On the history of energetics'. *Phil. Mag. 28*: 404.

——. 1867. 'On the phrase "Potential Energy", and on the definitions of physical quantities'. *Phil. Mag. 33*: 88–92.

——. 1871. *A Memoir of John Elder Engineer and Shipbuilder*. Edinburgh and London: Blackwood.

——. 1872. 'On actual energy'. *Phil. Mag. 43*: 160.

——. 1881. *Miscellaneous Scientific Papers*. Ed. W.J. Millar. London: Griffin.

——. 1908. *A Manual of the Steam Engine and Other Prime Movers*. 17th edn. Revised by W.J. Millar. London: Griffin.

Rayleigh, Lord [J.W. Strutt]. 1869. 'On some electromagnetic phenomena considered in connexion with the dynamical theory'. *Phil. Mag. 38*: 1–15.

——. 1875. 'On the dissipation of energy'. *Nature 11*: 454–5.

——. 1877. *The Theory of Sound*. 2 vols. London: MacMillan.

Rayleigh, Lord [R.J. Strutt]. 1924. *John William Strutt. Third Baron Rayleigh*. London: Edward Arnold.

Redondi, Pietro. 1976. '*Sadi Carnot et la recherche technologique en France de 1825 à 1850*'. *Revue d'histoire des sciences 29*: 243–59.

——. 1980. *L'accueil des idées de Sadi Carnot et la technologie Française de 1820 à 1860. De la légende a l'histoire*. Paris: Vrin.

Reynolds, Osborne. 1892. *Memoir of James Prescott Joule*. Manchester: Manchester Lit. & Phil.

Rice, D.F. 1971. 'Natural theology and the Scottish philosophy in the thought of Thomas Chalmers'. *Scottish Journal of Theology 24*: 23–46.

Robinson, Eric and A.E. Musson. 1969. *James Watt and the Steam Revolution. A Documentary History*. London: Adams and Dart.

Rudwick, M.J.S. 1971. 'Uniformity and progression. Reflections on the structure of geological theory in the age of Lyell'. In D.H.D. Roller, ed., *Perspectives in the History of Science and Technology*, pp 209–37. Norman, Oklahoma.

——. 1985. *The Great Devonian Controversy. The Shaping of Scientific Knowledge Among Gentlemanly Specialists*. Chicago: Chicago University Press.

Rumford, Count of [Benjamin Thompson]. 1798. 'An inquiry concerning the source of the heat which is excited by friction'. *Phil. Trans. 88*: 80–102.

Ruse, Michael. 1976. 'Charles Lyell and the philosophers of science'. *BJHS. 9*: 121–31.

Russell, C.A. 1981. *Time, Chance and Thermodynamics*. Milton Keynes: Open University Press.

Russell, J.S. 1840a. 'On the temperature of most effective condensation in steam vessels'. *BAAS Rep. 10*: 186–7.

——. 1840b. 'Additional notice concerning the most œconomical and effective proportion of engine power to the tonnage of the hull in steam vessels, and more especially in those designed for long voyages'. *BAAS Rep. 10*: 188–90.

Sabine, Edward. 1852. '[Presidential] address'. *BAAS Rep. 22*: xli–lxiii.

Saunders, L.J. 1950. *Scottish Democracy: 1800–1840. The Social and Intellectual Background*. London: Oliver and Boyd.

Schaffer, Simon. 1980. 'Natural philosophy'. In G.S. Rousseau and Roy Porter, eds., *The Ferment of Knowledge: Studies in the Historiography of Eighteenth–century Science*, pp 55–91. Cambridge: Cambridge University Press.

——. 1983a. 'Natural philosophy and public spectacle in the eighteenth century'. *Hist. Sci.* *21*: 1–43.

——. 1983b. 'History of physical science'. In Pietro Corsi and Paul Weindling, eds., *Information Sources in the History of Science and Medicine*, pp 285–314. London: Butterworth Scientific.

——. 1989. 'The nebular hypothesis and the science of progress'. In J.R. Moore, ed., *History, Humanity and Evolution*, pp 131–64. Cambridge: Cambridge University Press.

——. 1992a. 'Late Victorian metrology and its instrumentation: a manufactory of ohms'. In Robert Bud and S.E. Cozzens, eds., *Invisible Connections. Instruments, Institutions, and Science*, pp 23–56. Bellingham, Washington: Spie Optical Engineering Press.

——. 1992b. 'The leviathan of Parsonstown: Reflections on astronomical authority'. Unpublished paper presented at conference on 'Astronomy in Nineteenth Century Britain', National Maritime Museum, Greenwich, 14 November 1992.

——. 1994. 'Accurate measurement is an English science'. In M. Norton Wise, ed., *The Values of Precision*, pp. 135–72. Princeton: Princeton University Press.

——. 1998. 'Physics laboratories and the Victorian country house'. In Jon Agar and Crosbie Smith, *Making Space for Science. Territorial Themes in the Shaping of Knowledge*, pp 149–80. Basingstoke: Macmillan.

Schelar, V.M. 1966. 'Thermochemistry and the third law of thermodynamics'. *Chymia 11*: 99–124.

Schofield, R.E. 1970. *Mechanism and Materialism. British Natural Philosophy in an Age of Reason*. Princeton: Princeton University Press.

Scott, Anne. 1996. 'The ascent of man: Henry Drummond and free kirk audiences in late-Victorian Scotland'. MA diss., University of Kent at Canterbury.

Scott, W.L. 1959. 'The significance of "hard bodies" in the history of scientific thought'. *Isis 50*: 199–210.

——. 1970. *The Conflict between Atomism and Conservation Theory 1644 to 1860*. London and New York: MacDonald/Elsevier.

Secord, J.A. 1989. 'Behind the veil: Robert Chambers and *Vestiges*'. In J.R. Moore, ed., *History, Humanity and Evolution*, pp 165–94. Cambridge: Cambridge University Press.

Sedgwick, Adam. 1831. 'Address of the President'. *Proceedings of the Geological Society of London 1*: 281–316.

Shapin, Steven. 1975. 'Phrenological knowledge and the social structure of early nineteenth-century Edinburgh', *Ann. Sci. 32*: 219–43.

——. 1981. 'Of gods and kings: natural philosophy and politics in the Leibniz-Clarke disputes'. *Isis 72*: 187–215.

——. 1988. 'The house of experiment in seventeeth-century England'. *Isis 79*: 373–404.

——. 1994. *A Social History of Truth. Civility and Science in Seventeenth-century England*. Chicago and London: University of Chicago Press.

Shapin, Steven, and Simon Schaffer. 1985. *Leviathan and the Air Pump. Hobbes, Boyle, and the Experimental Life*. Princeton: Princeton University Press.

Shields, John. 1949. *Clyde Built*. Glasgow: Maclellan.

Shortland, Michael. 1996. 'Hugh Miller's contribution to the *Witness*: 1840–56'. In Michael

Shortland, ed., *Hugh Miller and the Controversies of Victorian Science*, pp 287–302. Oxford: Oxford University Press.

Sibum, H.O. 1994. 'Reworking the mechanical value of heat: instruments of precision and gestures of accuracy in early Victorian England'. *SHPS 26*: 73–106.

Siegel, Daniel. 1975. 'Completeness as a goal in Maxwell's electromagnetic theory'. *Isis 66*: 361–68.

———. 1981. 'Thomson, Maxwell, and the universal ether in Victorian physics'. In G.N. Cantor and M.J.S. Hodge, eds, *Conceptions of Ether. Studies in the History of Ether Theories 1740–1900*, pp 240–68. Cambridge: Cambridge University Press.

———. 1986. 'The origin of the displacement current'. *HSPBS 17*(i): 99–146.

———. 1991. *Innovation in Maxwell's Electromagnetic Theory. Molecular Vortices, Displacement Current, and Light*. Cambridge: Cambridge University Press.

Siemens, C.W. 1879. 'On the transmission and distribution of energy by the electric current'. *Phil. Mag. 7*: 352–6.

———. 1882. '[Presidential] Address'. *BAAS Rep. 52*: 1–33.

Siemens, Werner. 1861. 'Proposal for a new reproducible standard measure of resistance to galvanic currents'. *Phil. Mag. 21*: 25–38.

———. 1866. 'On the question of the unit of electrical resistance'. *Phil. Mag. 31*: 325–36.

Silliman, Benjamin. 1838 [1837]. 'Notice of the electro-magnetic machine of Mr. Thomas Davenport'. *Ann. Electricity 2*: 257–64.

Silliman, R.H. 1963. 'William Thomson: smoke rings and nineteenth-century atomism'. *Isis 54*: 461–74.

Simpson, T.K. 1966. 'Maxwell and the direct experimental test of his electromagnetic theory'. *Isis 57*: 411–32.

———. 1970. 'Some observations on Maxwell's *Treatise on Electricity and Magnetism*. On the role of the "Dynamical Theory of the Electromagnetic Field" in Part IV of the *Treatise*'. *SHPS 1*: 249–63.

Slaven, Anthony. 1975. *The Development of the West of Scotland: 1750–1960*. London: Routledge and Kegan Paul.

Smith, Adam. 1976 [1776]. *An Inquiry into the Nature and Causes of the Wealth of Nations*. 2 vols. Eds R.H. Campbell, A.S. Skinner and W.B. Todd. Oxford: Clarendon.

Smith, Crosbie. 1976a. ' "Mechanical Philosophy" and the emergence of physics in Britain: 1800–1850'. *Ann. Sci. 33*: 3–29.

———. 1976b. 'Natural philosophy and thermodynamics: William Thomson and "The Dynamical Theory of Heat" '. *BJHS 9*: 293–319.

———. 1976c. 'William Thomson and the creation of thermodynamics: 1840–1855'. *Arch. Hist. Ex. Sci. 16*: 231–88.

———. 1978. 'A new chart for British natural philosophy: the development of energy physics in the nineteenth century'. *Hist. Sci. 16*: 231–79.

———. 1979. 'From design to dissolution: Thomas Chalmers' debt to John Robison'. *BJHS 12*: 59–70.

———. 1985. 'Geologists and mathematicians: the rise of physical geology'. In P.M. Harman, ed., *Wranglers and Physicists. Studies on Cambridge Physics in the Nineteenth Century*, pp 49–83. Manchester: Manchester University Press.

———. 1989a. 'Energy'. In R.C. Olby, G.N. Cantor, J.R.R. Christie and M.J.S. Hodge, eds., *Companion to the History of Modern Science*, pp 326–41. London: Routledge.

———. 1989b. 'William Hopkins and the shaping of dynamical geology: 1830–1860'. *BJHS. 22*: 27–52.

——. 1993. 'Imperial measurement: the academic and industrial contexts of William Thomson's physical laboratory'. Unpublished paper presented at workshop on 'Practical Electricians, High Science, and the Third Way', Dibner Institute for the History of Science and Technology, MIT, 16–17 April 1993.

——. 1994. 'Rejection and recovery: the strange case of Julius Robert Mayer'. Unpublished paper presented at conference on 'The Outsider in Science', Birkbeck College, 14 May 1994.

——. 1998. '"No where but in a great town": William Thomson's spiral of class-room credibility'. In Jon Agar and Crosbie Smith, *Making Space for Science. Territorial Themes in the Shaping of Knowledge*, pp 118–46. Basingstoke: Macmillan.

Smith, Crosbie and M. Norton Wise. 1989. *Energy and Empire. A Biographical Study of Lord Kelvin*. Cambridge: Cambridge University Press.

Smith, Crosbie and Jon Agar (eds). 1998. *Making Space for Science. Territorial Themes in the Shaping of Knowledge*. Basingstoke: Macmillan.

Smout, T.C. *A Century of the Scottish People 1830–1950*. London: Fontana Press.

Snelders, H.A.M. 1990. 'Oersted's discovery of electromagnetism'. In Andrew Cunningham and Nicolas Jardine, eds., *Romanticism and the Sciences*, pp 228–40. Cambridge: Cambridge University Press.

Spencer, Herbert. 1887. 'Progress: its law and cause'. In Herbert Spencer (ed.), *Essays: Scientific, Political, and Speculative Vol. I*. London: Williams and Norgate.

Spurgin, C.B. 1987. 'Gay-Lussac's gas-expansivity experiments and the traditional misteaching of "Charles's law"'. *Ann. Sci. 44*: 489–505.

Stallo, J.B. 1960 [1881]. *The Concepts and Theories of Modern Physics*. Ed. P.W. Bridgman. Cambridge, MA: Harvard University Press.

Steffens, H.J. 1979. *James Prescott Joule and the Concept of Energy*. New York and Folkestone: Neale Watson/Dawson.

Stevenson, R.L. 1887. 'Memoir [of Fleeming Jenkin]'. In Sidney Colvin and J.A. Ewing, eds. *Papers Literary, Scientific, &c. by the Late Fleeming Jenkin*, pp xi–cliv. 2 vols. London: Longmans, Green.

Stewart, Balfour. 1866. *An Elementary Treatise on Heat*. Oxford: Clarendon.

——. 1874. *The Conservation of Energy. Being an Elementary Treatise on Energy and its Laws*. London: King.

[Stewart, Balfour, and P.G. Tait]. 1875. *The Unseen Universe or Physical Speculations on a Future State*. 3rd. edn. London: MacMillan.

Stokes, G.G. 1907. *Memoir and Scientific Correspondence of the Late Sir George Gabriel Stokes*. 2 vols. Ed. Joseph Larmor. Cambridge: Cambridge University Press.

Strang, John. 1850. 'On the progress of Glasgow, in population, wealth, manufactures, &c'. *BAAS Rep. 20*: 162–9.

Strutt, J.W. and R.J. See Rayleigh, Lord.

Sturgeon, William. 1838–39. 'Historical sketch of the rise and progress of electro-magnetic engines for propelling machinery'. *Ann. Electricity 3*: 429–37.

Süsskind, Charles. 1965. 'Hertz and the technological significance of electromagnetic waves'. *Isis 56*: 342–5.

Sviedrys, Romualdas. 1976. 'The rise of physics laboratories in Britain'. *HSPS 7*: 405–36.

Szily, C. 1872. 'On Hamilton's principle and the second proposition of the mechanical theory of heat'. *Phil. Mag. 43*: 339–43.

——. 1876a. 'The second proposition of the mechanical theory of heat deduced from the first'. *Phil. Mag. 1*: 22–31.

——. 1876b. 'On the dynamical signification of the quantities occurring in the mechanical theory of heat'. *Phil. Mag. 2*: 254–69.

Tait, P.G. 1860. 'The position and prospects of physical science. A public inaugural lecture delivered on November 7, 1860'. Edinburgh: Edmonston and Douglas.
——. 1863a. 'On the conservation of energy'. *Proc. RSE. 5*: 121–6.
——. 1863b. 'Reply to Prof. Tyndall's remarks on a paper on "Energy" in *Good Words*'. *Phil. Mag. 25*: 263–6.
——. 1863c. 'On the conservation of energy'. *Phil. Mag. 25*: 429–31; *26*: 144–5.
——. 1864a. 'The dynamical theory of heat'. *North British Review 40*: 40–69.
——. 1864b. 'Energy'. *North British Review 40*: 337–68.
——. 1864c. 'On the history of thermo-dynamics'. *Phil. Mag. 28*: 288–92.
——. 1865. 'Note on the history of energy'. *Phil. Mag. 29*: 55–7.
——. 1868. *Sketch of Thermodynamics*. Edinburgh: Edmonston and Douglas.
——. 1871. 'Address [as President of Section A]'. *BAAS Rep. 41*: 1–8.
——. 1872a. "Reply to Professor Clausius'. *Phil. Mag. 43*: 338.
——. 1872b. 'On the history of the second law of thermodynamics'. *Phil. Mag. 43*: 516–18.
——. 1873a. 'Thermo-electricity'. *Nature 8*: 86–88; 122–3.
——. 1873b. 'Clerk-Maxwell's Electricity and Magnetism'. *Nature 7*: 478–80.
——. 1876. *Lectures on Some Recent Advances in Physical Sciences*. London: MacMillan.
——. 1877. *Sketch of Thermodynamics*. 2nd edn. Edinburgh: Douglas.
——. 1879. 'On the dissipation of energy'. *Phil. Mag. 7*: 344–8.
——. 1881. 'Memoir [of Rankine]'. In W.J. Millar. ed., *Miscellaneous Scientific Papers by W.J. Macquorn Rankine*, pp xix–xxxvi. London: Griffin.
——. 1883. 'On the laws of motion'. *Phil. Mag. 16*: 439–47.
——. 1892. 'Poincaré's thermodynamics'. *Nature 45*: 245–6, 439, 532.
Talbot, G.R. and A.J. Pacey. 1966. 'Some early kinetic theories of gases: Herapath and his predecessors'. *BJHS 3*: 133–49.
Tennyson, Alfred. 1971. '*In memoriam*'. In T.H. Warren, ed., *Tennyson. Poems and Plays*, pp 230–66. Oxford: Oxford University Press.
Thackray, Arnold. 1970. *Atoms and Powers. An Essay on Newtonian Matter-theory and the Development of Chemistry*. Cambridge, MA: Harvard University Press.
——. 1972. *John Dalton: Critical Assessments of His Life and Science*. Cambridge, MA: Harvard University Press.
Thompson, S.P. 1910. *The Life of William Thomson. Baron Kelvin of Largs*. 2 vols. London: MacMillan.
Thomsen, Julius. 1853–4. 'Die Grundzüge eines thermochemisches Systems', *Ann. Phys. Chim. 88*: 349–62; *90*: 261–88; *91*: 83–104; *92*: 34–57.
[Thomson, James]. 1825. 'State of science in Scotland'. *Belfast Magazine 1*: 269–78.
Thomson, Thomas. 1830. *An Outline of the Sciences of Heat and Electricity*. Edinburgh and London: Blackwood. 2nd. edn., 1840.
Thomson, William [Lord Kelvin]. 1845. 'Note on some points in the theory of heat'. *CMJ 4*: 67–72.
——. 1847. 'Notice of Stirling's Air-engine'. *Glas. Phil. Soc. Proc. 2*: 169–70.
——. 1848a. 'On an absolute thermometric scale, founded on Carnot's theory of the motive power of heat, and calculated from the results of Regnault's experiments on the pressure and latent heat of steam'. *Phil. Mag. 33*: 313–17.
——. 1848b. 'On the theory of electromagnetic induction'. *BAAS Rep. 18*: 9–10.

——. 1849. 'An account of Carnot's theory of the motive power of heat, with numerical results deduced from Regnault's experiments on steam'. *Trans. RSE 16*: 541–74.

——. 1850a. 'On a remarkable property of steam connected with the theory of the steam-engine'. *Phil. Mag. 37*: 386–9.

——. 1850b. 'On the theory of magnetic induction in crystalline substances'. *BAAS Rep. 20*: 23.

——. 1851a. 'On the dynamical theory of heat; with numerical results deduced from Mr Joule's "Equivalent of a Thermal Unit" and Mr Regnault's "Observations on Steam"'. *Proc. RSE 3*: 48–52.

——. 1851b. 'On the mechanical theory of electrolysis'. *Phil. Mag. 2*: 429–44.

——. 1851c. 'Applications of the principle of mechanical effect to the measurement of electro-motive forces, and of galvanic resistance, in absolute units'. *Phil. Mag. 2*: 551–62.

——. 1851d. 'On the theory of magnetic induction in crystalline substances'. *Phil. Mag. 1*: 177–86.

——. 1852a. 'On the mechanical action of radiant heat or light. On the power of animated creatures over matter. On the sources available to man for the production of mechanical effect'. *Proc. RSE 3*: 108–13; *Phil. Mag. 4*: 256–60.

——. 1852b. 'On a universal tendency in nature to the dissipation of mechanical energy'. *Proc. RSE 3*: 139–42; *Phil. Mag. 4*: 304–6.

——. 1853a [1851]. 'On the dynamical theory of heat; with numerical results deduced from Mr Joule's "Equivalent of a Thermal Unit" and M. Regnault's "Observations on Steam"'. Parts I-V. *Trans. RSE 20*: 261–98, 475–82; *Phil. Mag. 4*: 8–21, 105–17, 168–76; *9*: 523–31.

——. 1853b. 'On the restoration of mechanical energy from an unequally heated space'. *Phil. Mag. 5*: 102–5.

——. 1854a. 'On the mechanical antecedents of motion, heat, and light'. *BAAS Rep. 24*: 59–63; *MPP 2*: 34–40.

——. 1854b. 'On the mechanical values of distributions of electricity, magnetism and galvanism'. *Phil. Mag. 7*: 192–7.

——. 1855. 'Note on the possible density of the luminiferous medium and on the mechanical value of a cubic mile of sunlight'. *Phil. Mag. 9*: 36–40.

——. 1856. 'Dynamical illustrations of the magnetic and the helicoidal rotatory effects of transparent bodies on polarized light'. *Proc. Roy. Soc. 8*: 150–8.

——. 1860. 'Measurement of the electromotive force required to produce a spark in air between parallel metal plates at different distances'. *Phil. Mag. 20*: 316–26.

——. 1861. 'Physical considerations regarding the possible age of the Sun's heat'. *BAAS Rep. 30*: 27–8.

——. 1862. 'On the secular cooling of the Earth'. *Proc. RSE 4*: 610–11.

——. 1863a. 'On the secular cooling of the Earth'. *Phil. Mag. 25*: 1–14.

——. 1863b. 'Note on Professor Tyndall's "Remarks on the Dynamical Theory of Heat'. *Phil. Mag. 25*: 429.

——. 1872. *Reprint of Papers on Electrostatics and Magnetism*. London: Macmillan.

——. 1879. 'On thermodynamic motivity'. *Phil. Mag. 7*: 348–52.

——. 1881. 'On the sources of energy available to man for the production of mechanical effect'. *BAAS Rep. 51*: 513–18; *PL 2*: 433–50.

——. 1882–1911. *Mathematical and Physical Papers*. 6 vols. Cambridge: Cambridge University Press.

——. 1889–94. *Popular Lectures and Addresses*. 3 vols. London: Macmillan.

——. 1987 [1884]. 'Notes of lectures on molecular dynamics and the wave theory of light'. In Robert Kargon and Peter Achinstein, eds. *Kelvin's Baltimore Lectures and Modern Theoretical Physics. Historical and Philosophical Perspectives*, pp 7–263. Cambridge, MA: MIT Press.

Thomson, William and P.G. Tait. 1862. 'Energy'. *Good Words. 3*: 601–7.

——. 1867. *Treatise on Natural Philosophy*. Oxford: Clarendon.

Topham, Jon. 1993. 'Science and popular education in the 1830s: the role of the *Bridgewater Treatises*'. *BJHS 25*: 397–430.

Topper, D.R. 1970–71. 'Commitment to mechanism: J.J. Thomson, the early years'. *Arch. Hist. Ex. Sci. 7*: 393–410.

——. 1980. '"To Reason by Means of Images": J.J. Thomson and the mechanical picture of Nature'. *Ann. Sci. 37*: 31–57.

Trenn, T.J. 1979. 'The central role of energy in Soddy's holistic and critical approach to nuclear science, economics, and social responsibility'. *BJHS 12*: 261–76.

Truesdell, C. 1960–62. 'A program toward rediscovering the rational mechanics of the age of reason'. *Arch. Hist. Ex. Sci. 1*: 3–36.

——. 1975–76. 'Early kinetic theories of gases'. *Arch. Hist. Ex. Sci. 15*: 1–66.

Tuchman, Arleen. 1993. 'Helmholtz and the German medical community'. In David Cahan, ed., *Hermann von Helmholtz and the Foundations of Nineteenth-century Science*', pp 17–49. Berkeley, Los Angeles and London: University of California Press.

Tunbridge, Paul. 1992. *Lord Kelvin. His Influence on Electrical Measurements and Units*. London: Peregrinus.

Turner, F.M. 1981. 'John Tyndall and Victorian scientific naturalism'. In Brock, W.H., N.D. McMillan and R.C. Mollan (eds), *John Tyndall. Essays on a Natural Philosopher*, pp 169–80. Dublin: Royal Dublin Society.

Turner, Joseph. 1955. 'Maxwell on the method of physical analogy'. *BJPS 6*: 226–38.

——. 1956. 'Maxwell on the logic of dynamical explanation'. *Philosophy of Science 23*: 36–47.

Turner, R. S. 1971. 'The growth of professorial research in Prussia, 1818 to 1848 – Causes and context'. *HSPS 3*: 137–82.

——. 1972. 'Helmholtz, Hermann von'. *DSB 6*: 241–53.

——. 1974. 'Mayer, Julius Robert'. *DSB 11*: 235–44.

——. 1982. 'Justus Liebig versus Prussian chemistry: reflections on early institute-building in Germany'. *HSPS 13*(i): 129–62.

Tyndall, John. 1862a. 'On force'. *Phil. Mag. 24*: 57–66.

——. 1862b. 'On Mayer, and the mechanical theory of heat'. *Phil. Mag. 24*: 173–5.

——. 1863a. 'Remarks on an article entitled "Energy" in *Good Words*'. *Phil. Mag. 25*: 220–4.

——. 1863b. 'Remarks on the dynamical theory of heat'. *Phil. Mag. 25*: 368–87.

——. 1863c. 'Remarks on Professor Tait's last letter to Sir David Brewster'. *Phil. Mag. 26*: 65–7.

——. 1863d. *Heat. A Mode of Motion*. London: Longmans, Green. 6th edn., 1880.

——. 1864. 'Notes on scientific history'. *Phil. Mag. 28*: 25–51.

——. 1873. 'Introduction'. In Hermann Helmholtz, *Popular Lectures on Scientific Subjects*, pp xv–xvi. London: Longmans, Green.

——. 1874. '[Presidential] Address'. *BAAS Rep. 44*: lxvii–xcvii.

Von Tunzelmann, G.N. 1978. *Steam Power and British Industrialization to 1860*. Oxford: Clarendon.

Warwick, Andrew. 1991. 'On the role of the FitzGerald–Lorentz contraction hypothesis in the development of Joseph Larmor's electronic theory of matter'. *Arch. Hist. Ex. Sci.* 43: 29–91.

Watanabe, Masao. 1959. 'Count Rumford's first exposition of the dynamic aspect of heat'. *Isis 50*: 141–4.

Waterman, A.M.C. 1991. *Revolution, Economics & Religion. Christian Political Economy 1798–1833*. Cambridge: Cambridge University Press.

Watt, James. 1970. *Partners in Science. Letters of James Watt and Joseph Black*. Ed. Eric Robinson and Douglas McKie. London: Constable.

Weber, Wilhelm. 1861 [1851]. 'On the measurement of electric resistance according to an absolute standard'. *Phil. Mag. 22*: 226–40, 261–9.

——. 1872. 'Electrodynamic measurements. Sixth memoir, relating specially to the principle of the conservation of energy'. *Phil. Mag. 43*: 1–20, 119–49.

Westfall, R.S. 1971. *Force in Newton's Physics. The Science of Dynamics in the Seventeeth Century*. London and New York: MacDonald/American Elsevier.

——. 1980. *Never at Rest. A Biography of Isaac Newton*. Cambridge: Cambridge University Press.

Westminster Confession. 1973. *The Westminster Confession of Faith [with] The Larger and Shorter Catechisms*. Glasgow: Free Presbyterian Publications. First published 1648.

Wheeler, L.P. 1951. *Josiah Willard Gibbs. The History of a Great Mind*. New Haven: Yale University Press.

Whewell, William. 1835. 'Report on the recent progress and present condition of the mathematical theories of electricity, magnetism and heat'. *BAAS Rep. 5*: 1–34.

——. 1837. *History of the Inductive Sciences from the Earliest to the Present Times*. 3 vols. London: J.W. Parker.

——. 1841. *The Mechanics of Engineering, Intended for Use in Universities, and in Colleges of Engineers*. London: J.W. Parker.

Wilson, D.B. 1982. 'Experimentalists among the mathematicians: physics in the Cambridge natural sciences tripos, 1851–1900'. *HSPS 12*(ii): 325–71.

——. 1985. 'The educational matrix: physics education at early-Victorian Cambridge, Edinburgh and Glasgow Universities'. In P.M. Harman, ed., *Wranglers and Physicists. Studies on Cambidge Physics in the Nineteenth Century*, pp 12–48. Manchester: Manchester University Press.

——. 1987. *Kelvin and Stokes. A Comparative Study in Victorian Physics*. Bristol: Adam Hilger.

——. 1990. *The Correspondence Between Sir George Gabriel Stokes and Sir William Thomson, Baron Kelvin of Largs*. 2 vols. Ed. D.B. Wilson. Cambridge: Cambridge University Press.

——. 1991. 'P.G. Tait and Edinburgh Natural Philosophy, 1860–1901'. *Ann. Sci. 48*: 267–87.

Wise, M. Norton. 1979a. 'William Thomson's mathematical route to energy conservation: a case study of the role of mathematics in concept formation'. *HSPS 10*: 49–83.

——. 1979b. 'The mutual embrace of electricity and magnetism'. *Science 203*: 1310–18.

——. 1981a. 'German concepts of force, energy, and the electromagnetic ether: 1845–1880'. In G.N. Cantor and M.J.S. Hodge, eds., *Conceptions of Ether. Studies in the History of Ether Theories 1740–1900*, pp 269–307. Cambridge: Cambridge University Press.

——. 1981b. 'The flow analogy to electricity and magnetism – Part I'. *Arch. Hist. Ex. Sci. 25*: 19–70.

——. 1982. 'The Maxwell literature and British dynamical theory'. *HSPS 13*(i): 175–205.

——. 1987. 'What did 19th century British physics owe to Cambridge?' *HSPBS 17*(ii): 363–8.

——. 1988. 'Mediating machines'. *Science in Context 2*: 77–113.

——. 1989. 'Electromagnetic theory in the nineteenth Century'. In R.C. Olby, G.N. Cantor, J.R.R. Christie and M.J.S. Hodge, eds., *Companion to the History of Modern Science*, pp 342–56. London: Routledge.

——. 1992. 'Exchange value: Fleeming Jenkin measures energy and utility'. Unpublished paper.

Wise, M. Norton (ed.). 1994. *The Values of Precision*. Princeton: Princeton University Press.

Wise, M. Norton (with the collaboration of Crosbie Smith). 1989–90. 'Work and waste: political economy and natural philosophy in nineteenth century Britain (I)-(III)'. *Hist. Sci. 27*: 263–301; 391–449; *28*: 221–61.

Wise, M. Norton and Crosbie Smith. 1986. 'Measurement, work and industry in Lord Kelvin's Britain'. *HSPBS 17*(i): 147–73.

——. 1987. 'The practical imperative: Kelvin challenges the Maxwellians'. In Robert Kargon and Peter Achinstein, eds., *Kelvin's Baltimore Lectures and Modern Theoretical Physics*, pp 323–48. Cambridge, MA: MIT Press.

Woodruff, A.E. 1962. 'Action at a distance in nineteenth-century electrodynamics'. *Isis 53*: 439–59.

——. 1968. 'The contributions of Hermann von Helmholtz to electrodynamics'. *Isis 59*: 300–11.

Yagi, Eri. 1981. 'Analytical approach to Clausius's first memoir on mechanical theory of heat (1850)'. *Historia Scientiarum* no. *20*: 77–94.

——. 1984. 'Clausius's mathematical method and the mechanical theory of heat'. *HSPS 15*(i): 177–95.

Yeo, Richard. 1989. 'Science and intellectual authority in mid-nineteenth-century Britain: Robert Chambers and *Vestiges of the Natural History of Creation*'. In Patrick Brantlinger, ed., *Energy & Entropy. Science and Culture in Victorian Britain. Essays from Victorian Studies*, pp 1–27. Bloomington and Indianapolis: Indiana University Press.

Youmans, E.L., ed. 1886. *The Correlation and Conservation of Forces*. New York: Appleton.

Young, Thomas. 1807. *A Course of Lectures on Natural Philosophy and the Mechanical Arts*. 2 vols. London: Johnson.

——. 1845. *A Course of Lectures on Natural Philosophy and the Mechanical Arts*. Ed. Philip Kelland. 2 vols. 2nd edn. London: Taylor and Walton.

Yule, J.D. 1976. 'The impact of science on British religious thought in the second quarter of the nineteenth century'. PhD diss., University of Cambridge.

Index